P9-DNP-699

Cartography
Thematic Map Design

Second Edition

Borden D. Dent

Georgia State University

Wm. C. Brown Publishers

Basal text: 10/12 Garamond light
Display type: Garamond light, book, bold
Typesetting system: Penta/Autologic
Composition: Wm. C. Brown Publishers
Printing: Wm. C. Brown Publishers
Binding: Worzalla Publishing Company
Art sources: Vantage Art, Incorporated (technical art)
Richard Pusey (maps)
Four-color separation: Mas Graphics

Book Team

Editor *Jeffrey L. Hahn*
Developmental Editor *Lynne M. Meyers*
Production Coordinator *Peggy Selle*

Wm. C. Brown Publishers

President *G. Franklin Lewis*
Vice President, Publisher *George Wm. Bergquist*
Vice President, Publisher *Thomas E. Doran*
Vice President, Operations and Production *Beverly Kolz*
National Sales Manager *Virginia S. Moffat*
Advertising Manager *Ann M. Knepper*
Marketing Manager *David F. Horwitz*
Executive Editor *Edward G. Jaffe*
Production Editorial Manager *Colleen A. Yonda*
Production Editorial Manager *Julie A. Kennedy*
Publishing Services Manager *Karen J. Slaght*
Manager of Visuals and Design *Faye M. Schilling*

Cover design by John R. Rokusek

Library of Congress Catalog Card Number: 89–50282

ISBN 0–697–07991–0

Printed in the United States of America by Wm. C. Brown Publishers, 2460 Kerper Boulevard, Dubuque, IA 52001

10 9 8 7 6 5 4 3 2 1

For Jeanne, Andrew, and Jeffrey

Contents

Appendices

Preface

This book, as in the first edition, provides teachers and students with an up-to-date text on modern thematic map design. The emphasis remains with the principles and concepts of small-scale quantitative thematic mapping, devoted to the ideas of good design and successful graphic communication. A central focus is on teaching quantitative map making for upper division college cartography and geography students and to provide material so that others can understand what professional cartographers do. A section has been added that defines geographic cartography and this supports the general philosophy of this book. Throughout the text the idea that the student is designer is stressed. This text provides students with techniques of good design for making a variety of quantitative map products, using accepted practices of the modern cartographer.

The pedagogical elements of this book, including its style, level, and treatment, make it an ideal text for the one- or two-quarter or semester course in introductory thematic cartography. Several features incorporated in the first edition are retained in this new edition. For example, an examination of geographical concepts, important to the student cartographer and geographer, are largely overlooked in other cartography texts but are included here in Chapter 4 (The Nature of Geographic Phenomena and the Selection of Thematic Map Symbols). And the discussion of Census geography, brought up to date, is retained in this edition. This was a very popular feature in the earlier version. As before, ideas are conveyed in a straight forward way. Chapter previews highlight each chapter's main ideas, and end-of-chapter glossaries provide learning tools to assist the student in mastery of the subject. End-of-chapter readings offer diverse suggestions for in-depth further study. The text is fully and accurately referenced.

New features in *Cartography: Thematic Map Design* include updated and expanded references, new boxed articles highlighting microcomputer uses in the classroom, new discussion on modern laser printing, new treatment on choropleth map classing techniques, and a complete new chapter on flow mapping. Several new illustrations are provided to complement those successful ones used in the first edition. A new chapter organization is used that groups the chapters into parts that provide for a better organizational structure to the book. The book and text have been redesigned to offer a fresh and modern look that reflects concern for readability and aesthetic appeal. The organization of the text on the printed page in a clear and meaningful way suggests easily the hierarchy of ideas and text structure.

As expressed before in the earlier edition, the instructor may wish to organize the chapter presentations in a different order. One of the main features of this text over other cartography texts is the way in which the material is organized into stand-alone chapters. For example, each of the thematic map types is assigned its own chapter. The instructor can easily assign only portions of the text and skip others, and yet maintain an orderly flow of material. Students too appreciate this organization.

This text is organized into three parts. PART ONE contains the chapters that give background in the fundamentals of map communication and design, mapping and map projections, geography, and data preprocessing. Chapter 1 presents the fundamentals of design. Chapter 2 provides information about map projections useful for the thematic cartographer and Chapter 3 articulates approaches to the employment of map projections. A thematic cartographer requires background into the nature of geography, the subject of Chapter 4. A real task in thematic map design is to make data mappable, and various techniques for this are presented in Chapter 5 on preprocessing geographical data.

PART TWO groups chapters that treat six different thematic mapping techniques. Each chapter is devoted to one technique. Chapter 6 looks at the choropleth mapping technique and Chapter 7 examines the common dot map. Proportional symbol mapping is discussed in Chapter 8. The isarithmic map is presented in Chapter 9 and mapping by value-by-area methods is retained in Chapter 10. The flow map, new to this edition, is placed in Chapter 11. The map forms represented in these chapters display the more common quantitative thematic maps found today. The decision to add the chapter on flow mapping is based on my belief that this form of mapping offers the student unique problems that are useful pedagogical experiences in cartographic design.

Designing Thematic Maps is a group of five chapters making up PART THREE that treat several important design aspects of thematic mapping. Chapter 12 introduces graphic tools of map production and reproduction. Chapter 13 and 14 discuss the elements of map composition and total map organization, including important material on the figure-ground organization. Using typographics in map design is the subject of Chapter 15, and finally color in map design and production are included in Chapter 16. Appendices for geographical tables, census geography definitions, and map sources round out the text.

When the first edition of *Cartography: Thematic Map Design* appeared in late 1984, the role of the microcomputer in cartography was not yet determined, although one could make a reasonable guess. Today, its potential is easily recognized, from data manipulation, design, to production. In fact, it has revolutionized small-scale cartography. Several of the illustrations in this edition (for example, Figures 1.11, 1.16, 5.19, and 8.15) were designed and produced using relatively low-cost desktop graphic editors and laser printers. They were expressly included to illustrate the capabilities of the microcomputer for the small-scale thematic cartographer.

A large cartographic literature was reviewed in the preparation of the manuscript for this edition. I have detected from this literature three trends in the last five years: the choropleth map (and its symbolization) is a focus of academic interest in quantitative mapping, there is a growing interest among professional cartographers in the history of thematic mapping, and there is a strong growth in the number of articles and books that deal with automated mapping (line generalization, printer and CRT symbolization, program reviews, and the like). Subjects such as isarithmic mapping, proportional symbol mapping, cartogram, and others, have had less appeal for cartographic researchers. Whenever possible and when appropriate, this new literature has been incorporated into the text of this new edition.

The temptation throughout the preparation of this second edition was to expand it to include the diverse and rich map reading and cognitive literature. To be accurate, the thematic map designer requires knowledge about the map reading processes and cognitive abilities of the map reader. As a practical solution this book only provides introductory material on map making and does not specifically address more complex questions concerning the cognitive aspects of map reading. I feel this is beyond the purpose of this introductory text and might be best dealt with by others in more advanced treatments. However, I do not deny their importance in cartographic design. At the same time, no collection of theory about our cognitive map reading processes appear to account for all our abilities. Not knowing these has not prevented us from making maps.

I omitted materials in the first edition that were related to computer mapping and its products. This edition also omits computer mapping, as this subject is better left to more specialized texts. However, this edition does include new material (mostly in the form of boxed articles) on mapping programs for microcomputers particularly useful in cartographic instruction. The philosophical approach of the first edition, that is that the principles of thematic map design are fundamental whether one is engaged in manual or automated mapping, remains the same for this text. Other specialized subjects in thematic mapping, such as terrain mapping, are again not included. These subjects, although quite fascinating, deserve book-length treatment and are best dealt with elsewhere.

The production of a book of this sort is not the effort of just one individual. There are a number of persons to whom I would like to express appreciation. Dean Clyde W. Faulkner of the College of Arts and Sciences at Georgia State University, and Dr. Truman A. Hartshorn, chairman of the Department of Geography, were both very instrumental in providing institutional support for the effort. The many librarians at Georgia State University assisted greatly in obtaining rare materials. I want to especially thank my cartographic colleagues Richard Groop, Dennis Fitzsimons, and Richard Lindenberg for their reading of the early versions of the manuscript and for their very helpful ideas on how to make this a better product. My associates in the department offered encouragement along the way, as did my students. Ed Jaffe, Jeffrey Hahn, Lynne Meyers, and Peggy Selle, all editors at Wm. C. Brown, were a great professional team that offered suggestions and support along the way. Their belief in this book was encouraging enough. The work of other cartographers, largely upon which this book rests, gave me the inspiration to keep this project alive. At varying times Cynthia Fox, Keisha Greene, Chuck Runge, and Janette Heck all participated in the typing and assembling of the manuscript. Jeff McMichael and Elizabeth Cheney both contributed in the dark room production of many of the illustrations found in the text. I would also like to thank my many former teachers at the Baltimore Polytechnic Institute, who impressed upon me and my generation that working up to your own potential, regardless of how meager or abundant that might be, is a duty we all have. To them I extend an expression of sincere gratitude.

Borden D. Dent
Atlanta, Georgia
May 1989

Part

I

Thematic Mapping Essentials

The first chapter in this part contains an introduction to the world of the thematic map and presents a backdrop to this increasingly used form of map. Various map examples highlight the rich variety. A discussion is included about map design and the role the cartographer plays in the design activity. After the introductory chapter, this part groups chapters that provide material necessary to begin the thematic mapping task. Thematic maps contain two chief components, the base map and the thematic overlay. The next two chapters provide background and techniques to organize and develop the base map portion of the complete thematic map, and include a discussion of the variety of map projections useful for the thematic designer. A knowledge of map projections and their characteristics is required for successful map design.

Before the thematic cartographer can begin mapping, a fundamental awareness of the nature of geographical inquiry and the ways geographical data present themselves must be grasped. Ability to recognize data forms assists the map designer in the important activity of symbol design logic. Another important part of the cartographer's training is a knowledge of census or enumeration data, notably that provided by the United States Census Bureau. A brief background is provided here. For most cartographic design tasks the cartographer needs to develop and transform original data into forms that can be mapped. To do this, a knowledge of at least descriptive statistics is essential. The last chapter in this part presents the fundamental ideas for beginning statistics and concludes this part on thematic mapping essentials.

Chapter

1

Introduction to Thematic Mapping

Chapter Preview

Maps are graphic representations of the cultural and physical environment. Two subclasses of maps exist: general-purpose (reference) maps and thematic maps. This text concerns the design of the thematic map, which shows the spatial distribution of some geographical phenomenon. Map scale, or the amount of reduction of the real world, is critical to the cartographer, since it determines selection and generalization. Map making must always be viewed in the context of communication. This perspective on the role of the map makes clear the two important transformations in the cartographic process. The first involves the abstraction of nonmapped data into mapped form, and the second deals with how the map user deduces spatial information from the map's symbols. Cartographic abstraction includes selection, classification, simplification, and symbolization. Thematic map design is the aggregate of all the mental processes that lead to solutions in the abstraction phase of cartographic communication. Creativity and ideation and visualization methods assist the designer in producing a better map. Cartographic design, in summary, is a process of selection, dealing with the combination of such elements as scale, symbolization, color, and typography.

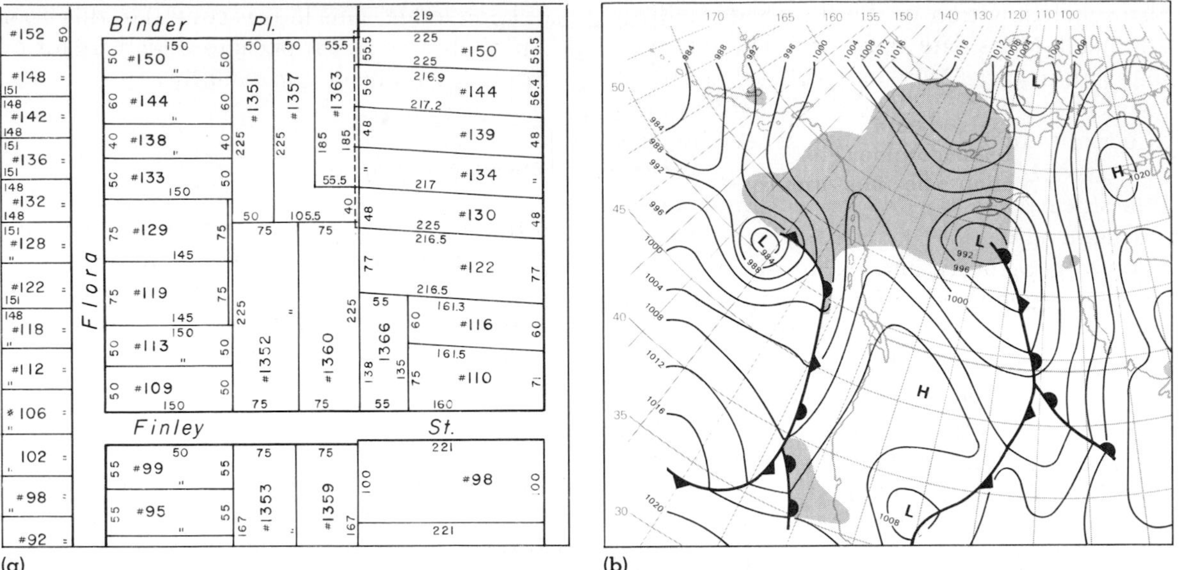

Figure 1.1. Maps satisfy a variety of needs.
A record of city land lots is kept on cadastral maps (a), and the National Weather Service produces a daily weather map, as in (b). These are but two of the hundreds of uses of maps.

The purpose of this book is to introduce several principles of thematic map design. The thematic map, only one of many map forms, will be more precisely defined later in this chapter. For now, some general comments about all maps will help set the stage. Maps seem to be everywhere we look: in daily newspapers and weekly newsmagazines, in books, on television, on trains, in kiosks, and even on table place mats. There is also great variety in types of maps. Some are greatly detailed and look like engineering drawings; others appear as freehand sketches or simple way-finding diagrams. Some show the whole world, others an area no larger than your back yard.

As our society has become more complex, the needs and uses for maps of all kinds have increased. Local governments and planning agencies use them for plotting environmental and resource data. Soil, geology, and water resource professionals use them daily in their work and planning. Public utility and engineering firms consult technical maps in order to complete their tasks. Land use maps are utilized by planners; detailed cadastral maps are indispensable to city tax recorders. (See Figure 1.1.) Astronauts have used maps to help them land on the moon. The list of uses and users is virtually endless. In response to the rapidly increasing uses for maps, the United States government established the National Mapping Program in 1973 to coordinate the efforts of the many federal agencies involved in map production.[1] As a result, these agencies have become more responsive to the needs of map users, both public and private, across the country.

In recent years, we have also seen a proliferation of state atlases. These have particular relevance to our discussion because they consist mainly of thematic maps, collected to assist educators and other professionals in displaying historical and geographical information about the state.

Map making is an interesting subject to study and an activity that has enjoyed a long history, closely tied to the history of civilization itself. Maps date as far back as the fifth or sixth century B.C.[2] Map making has come to be respected as a disciplined field of study in its own right. In this country, map making has been closely associated with geography curricula since the early decades of this century. Although still linked in many ways to the study of geography, and rightfully so, *cartography* is recognized more and more as a distinct field. With the growth of map making has come a partitioning of the field into several component parts, each with its own scope, educational requirements, technology, and philosophical underpinnings. The practice of mapping today is far different from that used by our ancestors, although there are some common bonds.

In this chapter, we will look at the foundations of the thematic map. Thematic maps came late in the development of cartography; they were not widely introduced until the early nineteenth century. Today, such maps are produced quickly and cheaply, primarily because the difficult process of base mapping has already been done by others, but also because of the benefits of computer technology. The last 30 years have been referred to as the "era of thematic mapping," and this trend is expected to continue in the future. Thematic maps make it easier for professional geographers, planners, and other scientists and academicians to view the spatial distribution of phenomena.

The Realm of Maps

What is astounding in written language is that new, often provocative, concepts and meanings can be generated by the combination of just a few simple words ("the universe began with a big bang"). The same can be said for a map with just a few graphic elements, and this is the essence of the cartographic instruction presented here. Professional cartographers think of maps as vehicles for the transmission of knowledge. This idea is central to this book and will serve as our point of departure for introducing thematic cartography.

The Map Defined

Although there are many kinds of maps, one description can be adopted that defines all maps: "A map is a graphic representation of the milieu."[3] In this context, *milieu* is used broadly to include all aspects of the cultural and physical environment. It is important to note that this definition includes *mental abstractions* that are not physically present on the geographical landscape. It is possible, for example, to map people's attitudes, although these do not occupy physical space.

It is also assumed here that a map is a tangible product. There are such things as **mental maps,** generally described as mental images that have spatial attributes. Mental maps are developed in our minds over time by the accumulation of many sensory inputs, including tangible maps. The definition used throughout this book, however, assumes that the map is a physical object that can be touched.

What is Cartography?

The newcomer will probably be confused by the array of terms, names, and descriptions associated with map making. It is unlikely that complete agreement on terminology will ever be reached by all those involved. For present purposes, we will adopt the following definitions. First, **map making,** or mapping, refers to the production of a tangible map and is defined as "the aggregate of those individual and largely technical processes of data collection, cartographic design and construction (drafting, scribing, display), reproduction, et cetera, normally associated with the actual production of maps."[4] Mapping, then, is the process of "designing, compiling, and producing maps."[5] The map maker may also be called a *cartographer.*[6]

Maps provide us with a structure for storing geographic knowledge and experience. Without them, we would find it difficult, if not impossible, to orient ourselves in larger environments. We would be dependent upon the close, familiar world of personal experience and would be hesitant—since many of us lack the explorer's intrepid sense of adventure—to strike out into unknown, uncharted terrain. Moreover, maps give us a means not only for storing information, but for analyzing it, comparing it, generalizing or abstracting from it. From thousands of separate experiences of places, we create larger spatial clusters that become neighborhoods, districts, routes, regions, countries, all in relation to one another.

Source: Michael Southworth and Susan Southworth, *Maps: A Visual Survey and Design Guide.* A New York Graphic Society Book. Boston: Little, Brown and Company, 1982, p. 11.

A second problem is the proper definition of **cartography.** As the discipline has matured and become broader in scope, many professional cartographers have come to make a distinction between map making and cartography. In general, cartography is viewed as broader than map making, for it requires the study of the philosophical and theoretical bases of the rules for map making, including the study of map communication.[7] It is often thought to be the study of the artistic and scientific foundations of map making. The International Cartographic Association defines cartography as follows:

The art, science, and technology of making maps, together with their study as scientific documents and works of art. In this context may be regarded as including all types of maps, plans, charts, and sections, three-dimensional models and globes representing the Earth or any celestial body at any scale.[8]

This definition is broad enough in scope to be acceptable to most practitioners. It certainly will serve us adequately in this introduction.

Geographic Cartography

Geographic cartography, although certainly a part of all cartography, should be defined a bit farther. *Geographic cartography* is distinct from other branches of cartography in that it alone is the tool and product of the geographer. The geographic cartographer understands the spatial perspective of the physical environment and has the skills to abstract and symbolize this environment. The cartographer specializing in this branch of cartography is skillful in map projection selection, the mapping and understanding of areal relationships, and has a thorough knowledge of the importance of scale to the final presentation of spatial data.[9] Furthermore, geographic cartography, involving an intimacy with the abstraction of geographical reality and its symbolization to the printed page, is capable of "unraveling" or revising the process, that is, geographic cartographers are very capable at map reading.[10] For the most part, geographic cartographers are involved in producing thematic maps, whether quantitative or qualitative, and are not usually associated with the production of highly detailed, large-scale reference (topographic) maps, photogrammetric products, surveying methods, or remotely-sensed images.

On the other hand, geographic cartographers are associated with the production of thematic maps, are versed in the reading of photomaps and other remotely sensed images, and use these latter products in the preparation of their special purpose (including atlas) maps. Geographic cartography is a branch of the broader science, and geographic cartographers understand spatial methodology. Geographers, too, embrace the study of maps, as Borchert has so clearly stated:

In short, maps and other graphics comprise one of three major modes of communication, together with words and numbers. Because of the distinctive subject matter of geography, the language of maps is the distinctive language of geography. Hence sophistication in map reading

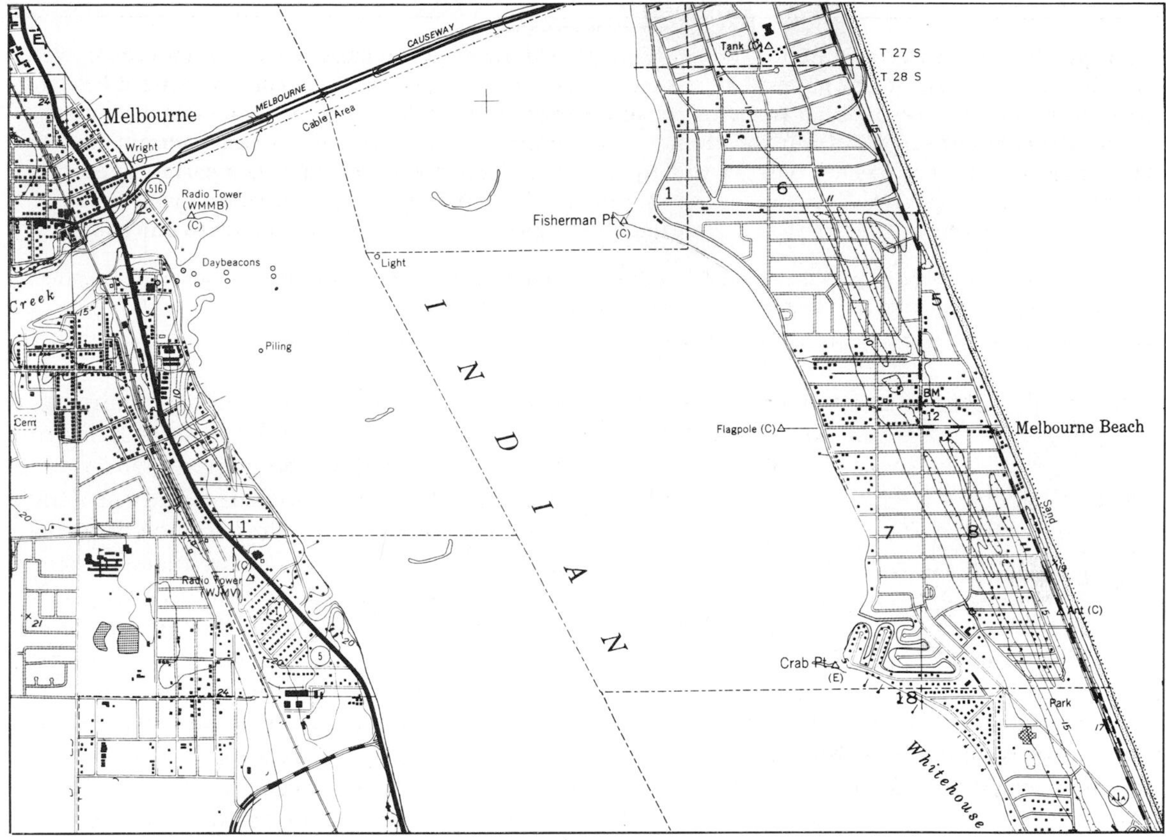

Figure 1.2. A general-purpose or reference map.
This is a portion of the Melbourne East, Florida, 7½° USGS topographic map sheet and represents a typical reference map at large scale, showing both physical and cultural features. This reproduction shows a 50-percent reduction of the original.

and composition, and ability to translate between the languages of maps, words, and numbers are fundamental to the study and practice of geography.[11]

Kinds of Maps

A review of the variety of maps will lead to a better understanding of where quantitative thematic mapping fits into the larger realm of maps and mapping. Thematic mapping has been recently estimated to be only 10 percent of the field.[12] In general, all maps may be classified as either general-purpose or thematic types.

General-purpose Maps

Another name commonly applied to the general-purpose map is *reference map.*[13] Such maps customarily display objects (both natural and man-made) from the geographical environment. The emphasis is on location, and the purpose is to show a variety of features of the world or a portion of it.[14] Examples of such maps are topographic maps and atlas maps. The kinds of features found on these maps include coastlines, lakes, ponds, rivers, canals, political boundaries, roads, houses, and similar objects. (See Figure 1.2.) In this country, topographic maps are produced by the United States Geological Survey (USGS).

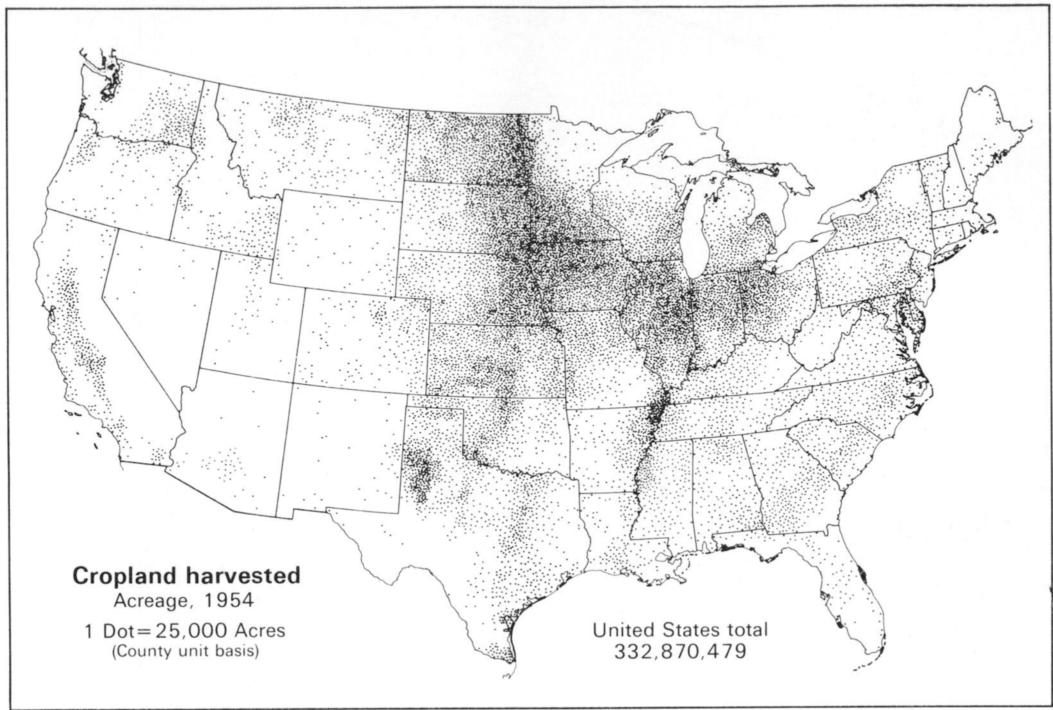

Figure 1.3. A typical thematic map.

Source: United States Department of the Interior, Geological Survey. *National Atlas of the United States,* 1956.

Historically, the general-purpose or reference map was the prevalent map form until the middle of the eighteenth century. Geographers, explorers, and cartographers were preoccupied with "filling in" the world map. Since knowledge about the world was still accumulating, emphasis was placed on this form. It was not until later, when scientists began to seize the opportunity to express the spatial attributes of social and scientific data, that thematic maps began to appear. Such subjects as climate, vegetation, geology, and trade, to mention a few, were mapped.

Thematic Maps

The other major class of map is the thematic map, also called a *special-purpose, single-topic,* or *statistical* map. The International Cartographic Association defines the thematic map this way: "A map designed to demonstrate particular features or concepts. In conventional use this term excludes topographic maps."[15]

The purpose of all thematic maps is to illustrate the "structural characteristics of some particular geographical distribution."[16] This involves the mapping of physical and cultural phenomena or abstract ideas about them. Structural features include distance and directional relationships, patterns of location, or spatial attributes of magnitude change. (See Figure 1.3.)

A thematic map, as its name implies, presents a *graphic theme* about a subject. It must be remembered that a single theme is chosen for such a map; this is what distinguishes it from a reference map. A reference map is to a thematic map what a dictionary is to an essay.

Thematic maps may be subdivided into two groups, **qualitative** and **quantitative.** The

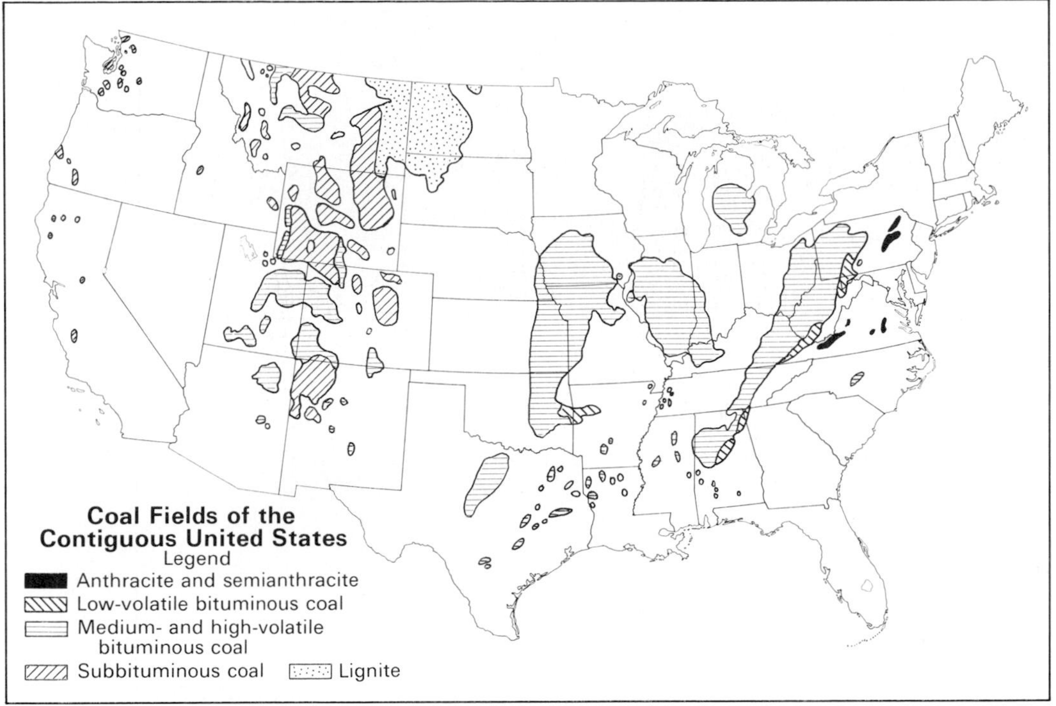

Figure 1.4. A qualitative thematic map.
The principal purpose of this kind of thematic map is to show the distribution of different nominal classes of geographical data.
Source: Department of the Navy, *Energy Fact Book* (Washington, D.C.: Navy Energy Office, 1979), p. 111.

principal purpose of a qualitative thematic map is to show the spatial distribution or location of kind or nominal data. For example, the mapping of the distribution of the principal coal fields in the United States is a qualitative thematic map. On this form of map, the reader cannot determine quantity, except as shown by relative areal extent. (See Figure 1.4.) It is not possible to know which location produces the most coal, or in what quantities.

Quantitative thematic maps, on the other hand, display the spatial aspects of numerical data. In most instances, a single variable, such as corn, people, or income is chosen, and the map focuses on the variation of the feature from place to place. These maps may illustrate numerical data on the ordinal (less than/greater than) scale or the interval/ratio (how much different) scale. (See Figure 1.5.) These measurement scales will be treated in depth in a later chapter.

In a recent book on map appreciation, the author identifies at least ten different types or groups of maps, several of which are thematic in nature: photomaps, maps of the landscape and atmosphere, population maps, political maps, maps of the municipality, journalistic, advertising and persuasive maps, and computer maps.[17] The list is not meant to be exhaustive, but representative. Journalistic maps have been investigated recently, for example, and are defined as ". . . maps which are published in the mass news media . . . either accompanying and relating to news stories or other features, or as primary news vehicle."[18] The types of maps are quite varied!

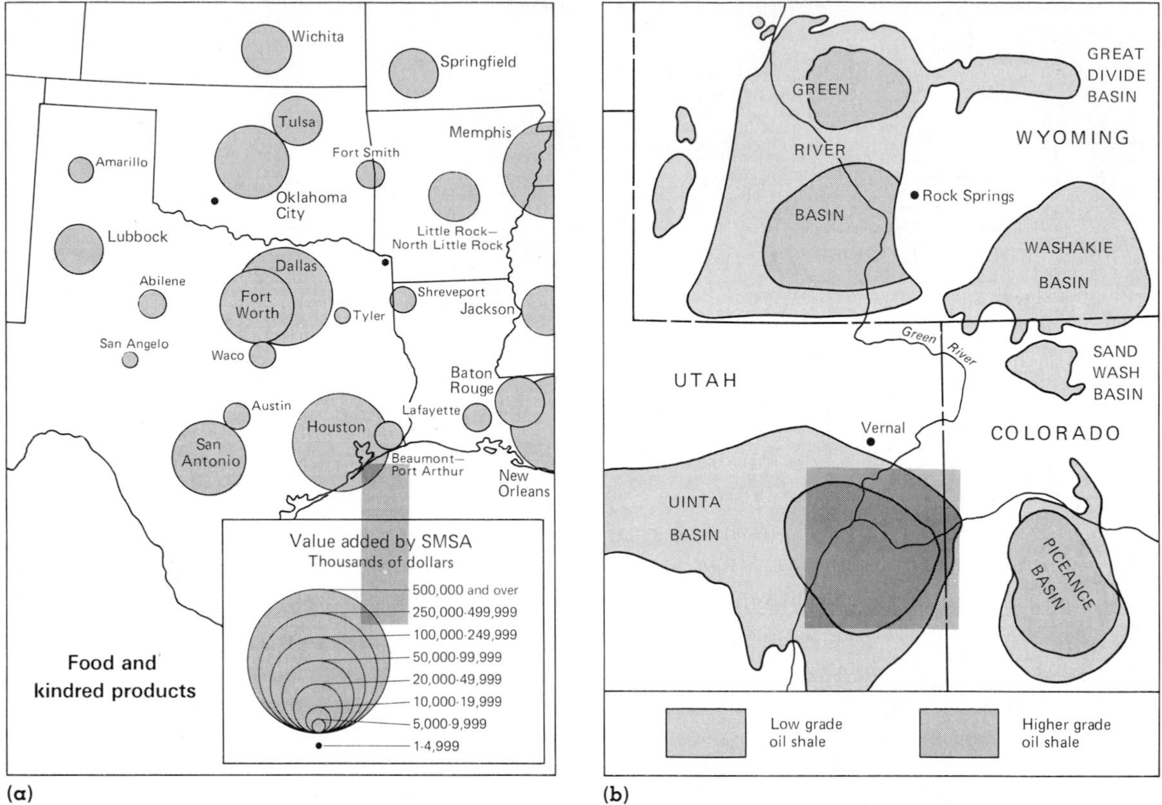

Figure 1.5. Quantitative thematic maps.
Interval/ratio (how much greater than) data, as in (a), or ordinal (less than/greater than) data, as in (b), may be represented on quantitative thematic maps.

Sources: map (a) redrawn with modifications from United States Geological Survey, *The National Atlas of the United States of America* (Washington, D.C.: USGPO, 1970), p. 197; map (b) redrawn with modifications from Department of the Navy, *Energy Fact Book* (Washington, D.C.: Navy Energy Office, 1979), p. 175.

As further examples of the different forms that thematic maps may take, the reader is directed to Figures 1.6, 1.7, and 1.8. In the first, a cartogrammic, or very abstract, view of the United States is illustrated. In this map, the country is shown as it might look if the largest states were those most often used in country music lyrics. It is quantitative (more–less), although numerically quite simple; it is not planimetric (e.g., correct in terms of exact horizontal location). Nonetheless, it is an intriguing map and portrays a concept in an interesting and forceful way.

The map illustrated in Figure 1.7 departs from the normal view of the United States in yet another way. Distance on this map is scaled to airline fares, not units of length as is customarily used. Shapes and contiguity are sacrificed, and in some places terrible distortions occur (San Francisco in Nevada!). However, it points

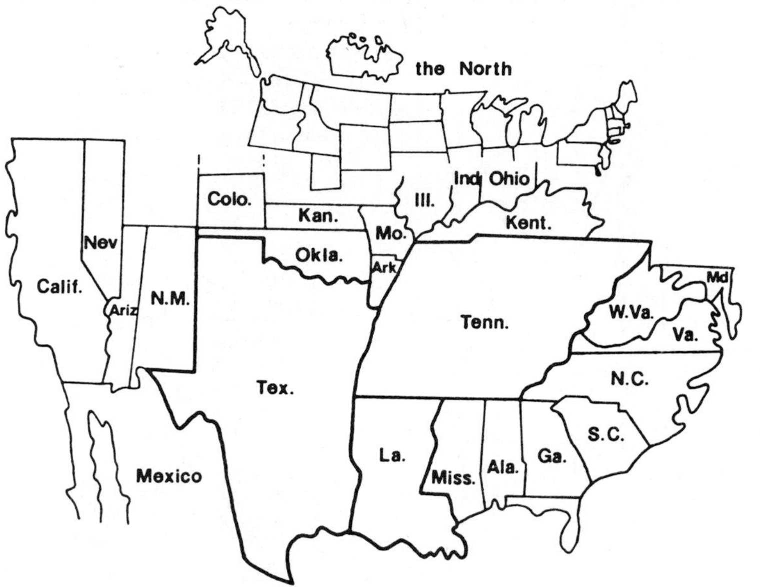

Figure 1.6. Places mentioned in country music lyrics.
A cartogram form of map, very abstract, yet one that conveys its message clearly and forcefully.
Reprinted with permission of Harpers Magazine.

out that geographical space is structured in ways other than units of length. Both of these maps point out that map content is a very important aspect of thematic map design.

Thematic maps, of course, may also be non-quantitative, as illustrated in Figure 1.8. This map depicts a portion of the Geographical Area Designator Map used by the U.S. Weather Service in their aviation weather forecasts. This map does not show any quantities at all, but purely qualitative information, and is not precise but rather generalized in its record.

Components of the thematic map. Every thematic map is composed of two important elements: a **geographic** or base map and a **thematic overlay.** (See Figure 1.9.) The user of a thematic map must integrate these two, visually and intellectually, during map reading. The purpose of the geographic base map is to provide locational information to which the thematic overlay can be related. It must be well-designed *and include only the amount of information thought necessary to convey the map's message.* Simplicity and clarity are important design features of the thematic overlay. Design strategies for each of these components are dealt with in later chapters.

Map Scale

When cartographers decide on graphical representation of the environment or a portion of it, an early choice to be made is that of **map scale.** Scale is the amount of reduction that takes place in going from real-world dimensions to the new mapped area on the map plane. Technically, map scale is defined as a

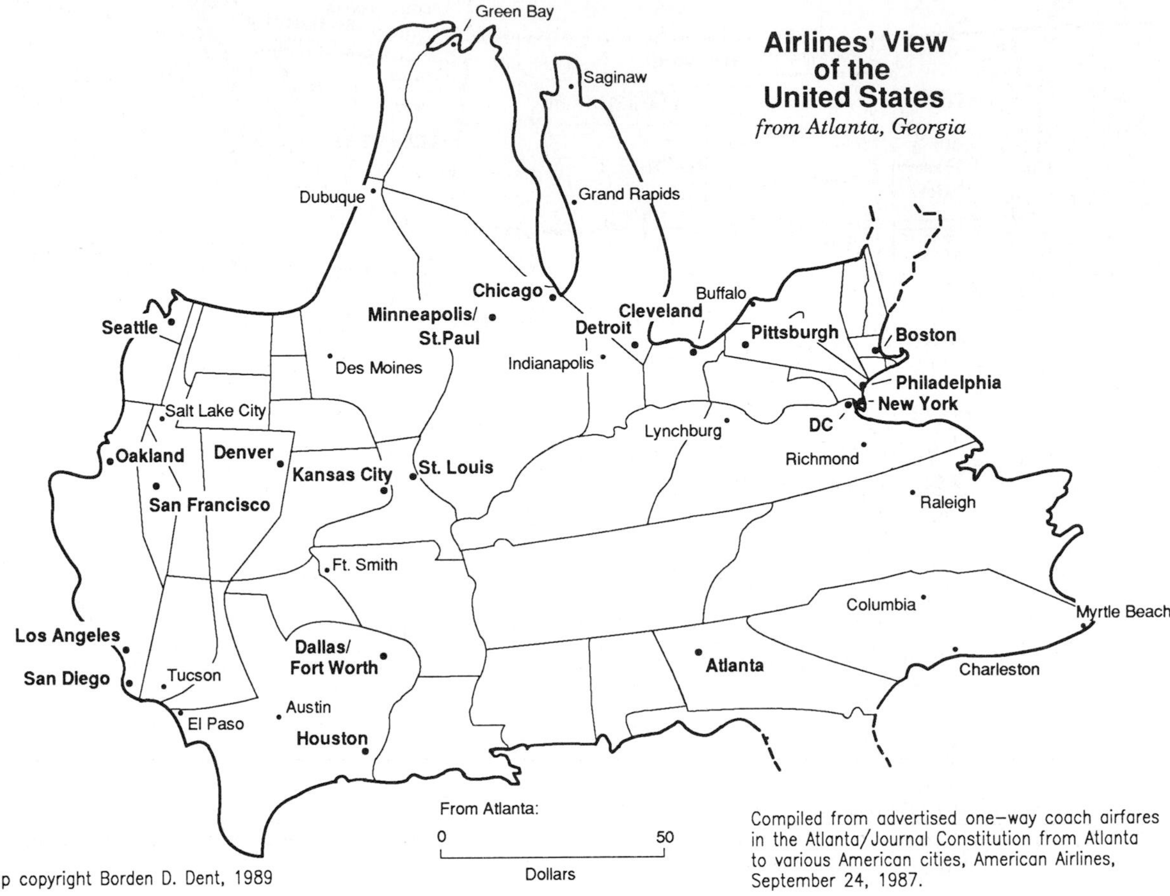

Figure 1.7. Airline's view of the United States.
Maps can be scaled to units other than distance. In this case, airline fares are used instead of miles or other linear units.
Map copyright by the author.

ratio of map distance to earth distance, with each distance expressed in the same units of measurement and customarily reduced so that unity appears in the numerator (e.g., 1:25,000). A more detailed discussion of scale appears in the next chapter. For now, our discussion is focused on simply the idea of scale and its relationship to such design considerations as generalization and symbolization.

Scale selection has important consequences for the map's appearance and its potential as a communication device. Scale operates along a continuum from large-scale to small-scale. (See Figure 1.10.) Large-scale maps show small portions of the earth's surface; *detailed information* may therefore be shown. Small-scale maps show large areas, so only *limited detail* can be carried on the map. Which final scale is selected for a given map design problem will depend on the map's purpose and physical size. The amount of geographical detail necessary to satisfy the purpose of the map will also act

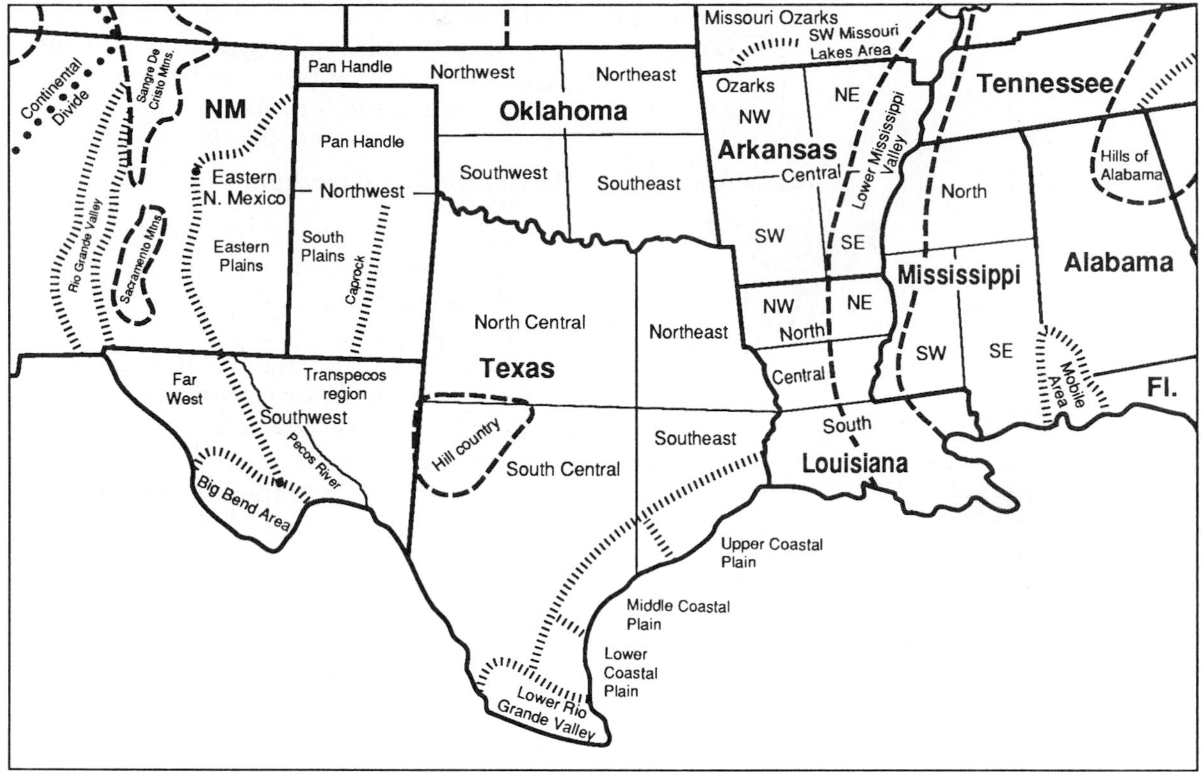

Figure 1.8. A non-quantitative thematic map.
The purpose here is to show location and approximate boundaries of places identified by the U.S. Weather Service in their aviation weather forecasts.
Map source: National Weather Service.

as a constraint in scale selection. Generally, the scale used will be a compromise between these three controlling factors.

Another important consequence of scale selection is its impact on *symbolization.* In reducing from large scale to small scale, map objects must increasingly be represented with symbols that are no longer true to scale and thus are more *generalized.* At large scales, the outline and area of a city may be shown in proportion to accurate scale—that is, occupy areas on the map proportional to mapped space. At smaller scales, whole cities may be represented by a single dot having no size relation to the map's scale.

Scale varies over the map depending on the projection used, as will be explained in Chapter 2. Scale, symbolization, and map projection are thus interdependent, and the selection of each will have considerable effect on the final map. The selection of scale is perhaps the most important decision a cartographer makes about any map.

In general, there is an inverse relationship between reference maps and thematic maps regarding scale. (See Figure 1.11.) The proportion of reference maps to thematic maps is greater at larger scales, and the reverse is true at small scales. Thematic cartographers generally work at small map scales, which requires

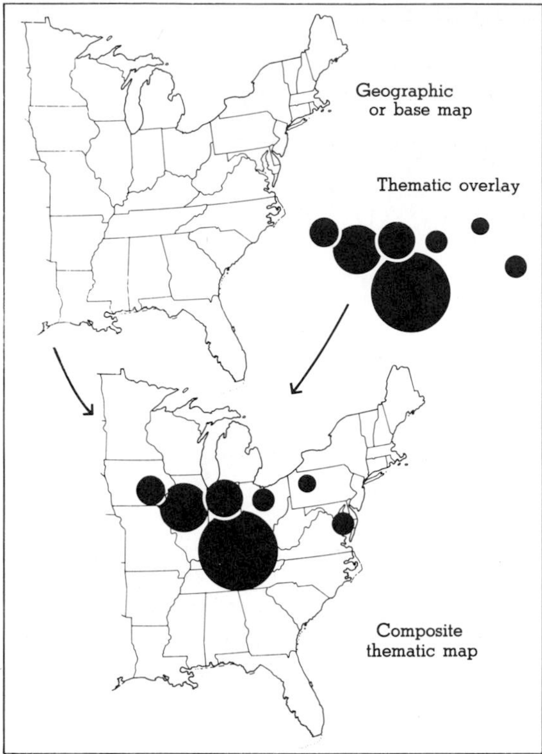

Figure 1.9. The major components of the thematic map.
The composite or completed thematic map is made up of two important components: the geographic or base map and the thematic overlay.

them to be especially attentive to the operations of cartographic generalization (discussed below). This is especially true for geographic cartographers.

Map Communication

It is customary today for cartographers to refer to mapping as a geographic information communication process. There is no question that we get information from maps, that the map is an excellent storage device, and that our behavior can be affected by reading maps. With reference to this communication structure, several important approaches to map design can be learned.

An Acceptable Paradigm

The increasing sophistication of geographic and cartographic research during the past several decades has led many scholars to approach their theoretical work from the standpoint of model building. **Models** can be theories, laws, hypotheses, or structured ideas concerning the real world; in traditional geography, such models typically include the spatial dimension.[19] **Paradigms** are large-scale models, broader in scope, and really refer to *patterns of searching* in scientific inquiry. They differ from models in that they are less structured and usually deal only with *how* the real world is being searched rather than investigating the world itself.[20] What does all this have to do with cartographic design? It is simply a way of explaining that the *cartographic paradigm* is one which includes inquiry into how map readers gain information from maps and how cartographers obtain information to put there.

Cartographic design is done best when practiced within this paradigm. The kinds of questions asked by the designer and the steps he or she will take will be controlled by the large-scale model. In practice, maps are always made with a purpose; the map reader's abilities and needs and the limitations of the graphic media all influence design decisions.[21] This is what makes cartographic design so interesting.

A model for clarity. A very generalized graphic depiction of thematic cartographic communication would include four major components: data field, map author, map, and map reader.[22] (See Figure 1.12.) Proceeding

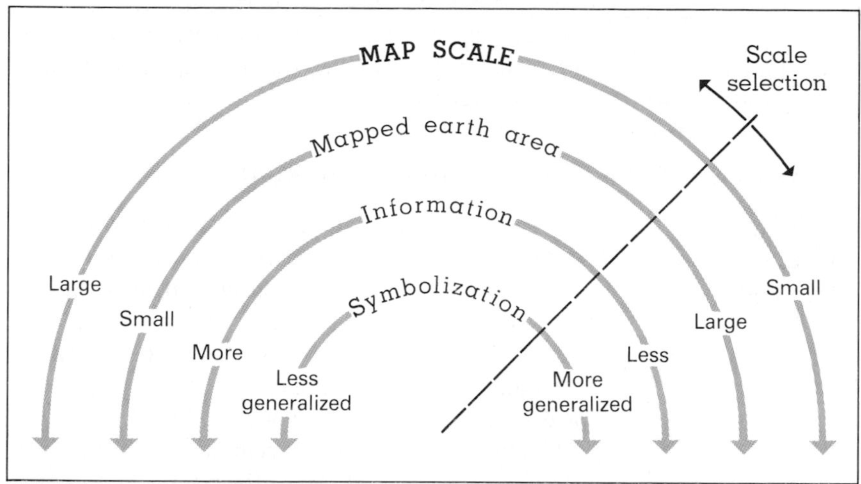

Figure 1.10. Map scale and its effect on mapped earth area, map information, and symbolization.
The selection of a map scale has definite consequences for the design. For example, small-scale maps contain large earth areas and less specific detail and must use symbols that are more generalized. Selection of map scale is a very important design consideration because it will affect other map elements.

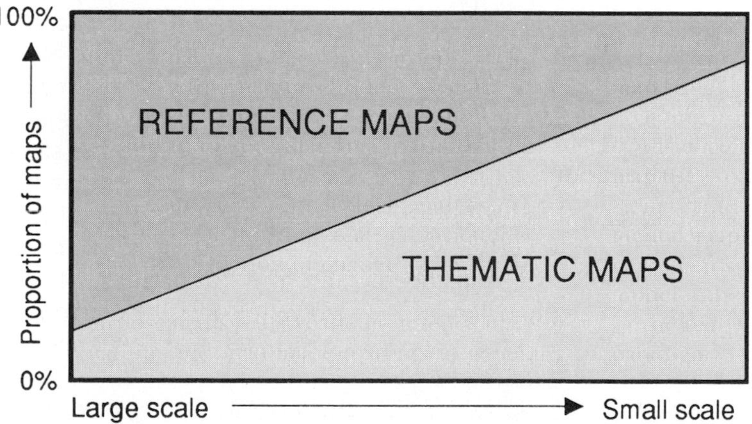

Figure 1.11. Reference maps and thematic maps operate at different scales.
The proportion of reference maps to thematic maps is greater at larger scales, and the reverse is true at smaller scales.

Redrawn from Phillip Muehrcke, "Maps in Geography," *Cartographica* 18 (1981):1–41, Figure 3.

Making a road relatively wider than it is on the Earth makes it visible; distorting distances on a projection enables the map user to see the whole Earth at once; separating features by greater than Earth distances allows representation of relative positions. Distortion is necessary in order that the map reader be permitted to comprehend the meaning of the map.

Source: Mark S. Monmonier, *Maps, Distortion, and Meaning.* Resource Paper No. 75-4 (Washington, D.C.: Association of American Geographers, 1977), 7.

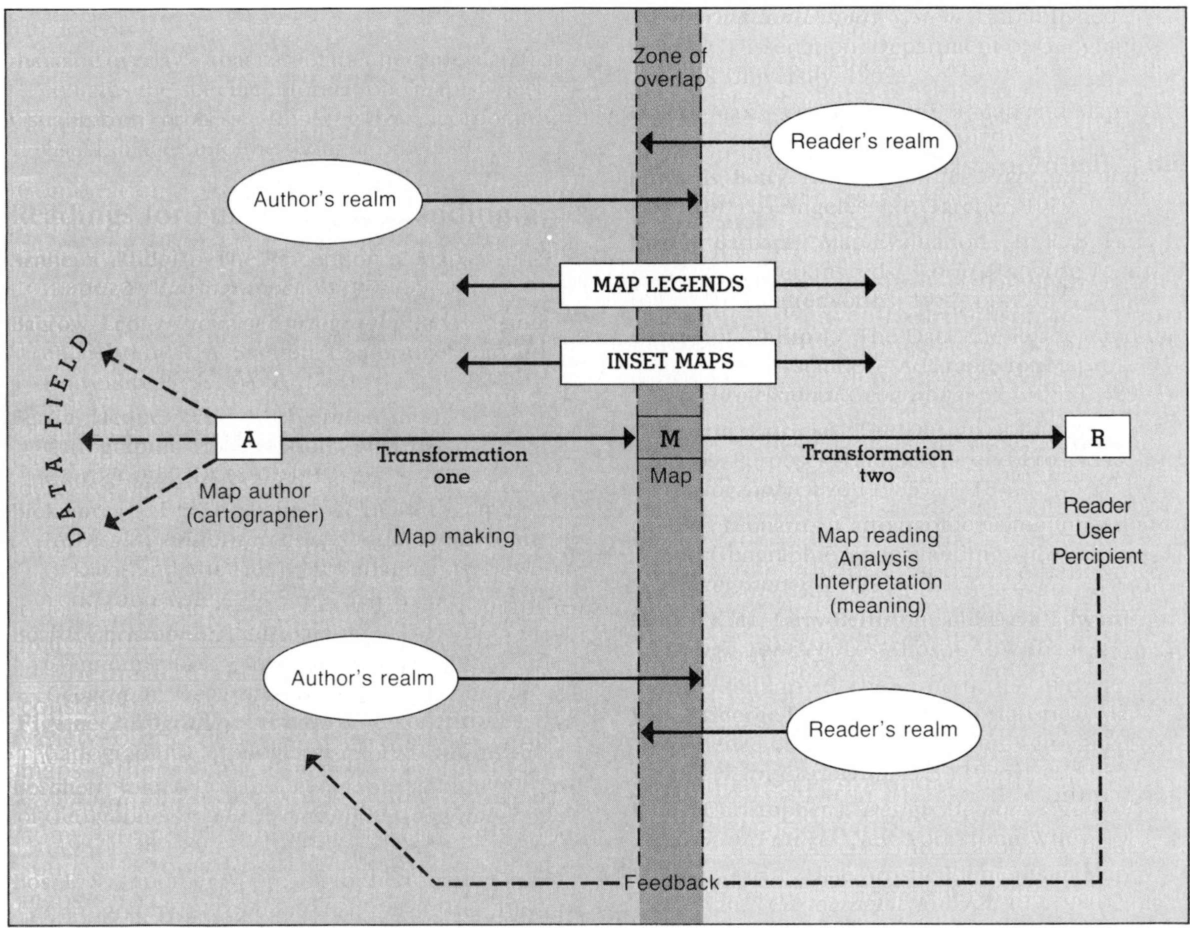

Figure 1.12. A graphic depiction of the cartographic communication process.
For communication to take place, the author's and reader's realms of general knowledge must overlap. Such devices as map legends that explain unfamiliar symbols or inset maps that serve as locational cues are often instrumental in bridging the fields of knowledge.

Table 1.1 Sources for the Initiation of a Map Message

1. A decision by intending map users or their representatives.
2. A decision by map makers in anticipation of a need.
3. A decision by a map author as a consequence of his own need for explanation.
4. A decision by a scientific body to contribute to the total information in a specialist field.
5. A decision by an individual to express himself through making a map.

Source of table: Adapted with changes for J. S. Keates, *Understanding Maps.* (New York: John Wiley, 1982), 103–104.

from the unmapped data field to the map reader involves two important transformations. The first involves the alteration of the unmapped data into a set of graphic marks (symbols) that are placed on the map. The second transformaion involves the reader's registering these marks and deducing the spatial information message from them. Map reading error develops when there is a discrepancy between the message intended by the map author and the one gained by the map reader. Taken together, these operations comprise the **cartographic process.**

A **map author** is someone who wishes to convey a spatial message. He or she may or may not be a map maker (or cartographer in the broadest use of the term).[23] When a map author wishes to structure a spatial message most effectively, he or she may employ a trained cartographer to design the map. A cartographer who originates a message is his or her own designer.

The data field comprises both numerical and non-numerical phenomena, from which the map author chooses to structure the map's message. Non-numerical objects might include base-map elements, such as coastlines, political boundaries, and other nominal geographical information deemed relevant to the structured communication at hand. Numerical data objects are the quantitative elements that form the core of the communication. These may be enumeration (census) data, results of special surveys, field data, and the like. Taken together, these objects are transformed into the graphic marks that make up the map. The transformation process is one of abstraction and requires *cartographic generalization,* which involves *selection, classification, simplification,* and *symbolization.*[24] These topics will be more fully discussed later in this chapter.

Several constraints are imposed on the map author or cartographer in developing the message. These include the purpose of the map, map format, scale, intended map audience, data quality, graphic and printing limitations, and economic considerations.[25] These may be considered to be *design constraints;* they are the main subject matter of this book.

Initiation of a map can come from a variety of sources. (See Table 1.1.) Maps are made for aims other than pure communication, as the last item in Table 1.1 suggests. Some are constructed as decorative pieces of art, purely for visual appreciation. In such cases, map content plays a lesser role in the map's design.

In the cartographic process, the abilities and needs of the map reader are of paramount importance to the map author and designer. In fact, the most important aspect of cartographic communication and its function in map design is that the map maker and map percipient are not independent of each other.[26] A **map percipient** is one who gains a spatial knowledge by looking at a map;[27] most thematic maps are designed for percipients.

Table 1.2 Reading Tasks in the Second Transformation of the Cartographic Process

1. Pre-Map Reading Tasks
   obtaining, unfolding, etc.
   orienting
2. Detection, Discrimination, and Recognition Tasks
   search
   locate
   identify
   delimit
   verify
3. Estimation Tasks
   count
   compare or contrast
   measurement
   a. direct estimation
   b. indirect estimation
4. Attitudes on Map Style
   pleasantness
   preference

Source of table: Joel L. Morrison, "Towards a Functional Definition of the Science of Cartography with Emphasis on Map Reading," *American Cartographer* 5 (1978): p. 106.

The second transformation in the cartographic process is complex and involves the human nervous system. Professor Phillip Muehrcke states that **map use** comprises reading, analysis, and interpretation.[28] In **map reading,** the viewer looks at a map and determines what is displayed and how the map maker did it. On closer inspection, the map users begins to see different patterns; this begins thoughtful *analysis.* Finally, a desire to explain these patterns leads to map *interpretation.* In this last case, causal explanation (probably not displayed on the map) are sought by the map user.

Many complex reading tasks have been identified. (See Table 1.2.) It is no wonder that map communication is so difficult! Maps become especially difficult to read when the map author or cartographer incorporates several tasks in one design. Simplicity in design is a goal and can be achieved in part by reducing the number of map-reading tasks on a single map. Reader training has been shown to improve the efficiency of map communication.[29]

This general model of the cartographic communication process includes a fundamental element not to be overlooked. The map author's knowledge must overlap that of the reader, and this mutual understanding must be embodied in the map. We are not dealing with the new spatial structures that the map author wishes to communicate to the reader, but rather with those elements necessary to "unlock" the new ones. For example, if a map contains words in a foreign language, communication will be impeded. Likewise, symbolization (especially unconventional forms) must be fully explained to the map user. This is the function of the map legend: to bridge any gaps that may exist between realms. Difficult formats that yield awkward shapes will also produce problems, as will unfamiliar or undetailed base maps. (See Figure 1.13.) In cases such as these, communication can be made more efficient by providing **inset maps** to bridge the gap in geographical knowledge.

Feedback is helpful to the map author in that it allows him or her to alter the design by incorporating positive changes suggested by the map user. Unfortunately, most cartographic designers are separated from users by time and space, so it is difficult to make use of feedback. The wise designer will make use of it when possible to improve designs.

The Importance of Meaning

Recent discussion among professional cartographers has emphasized the importance of meaning in cartographic communication. The *needs* of the map user are very important in the

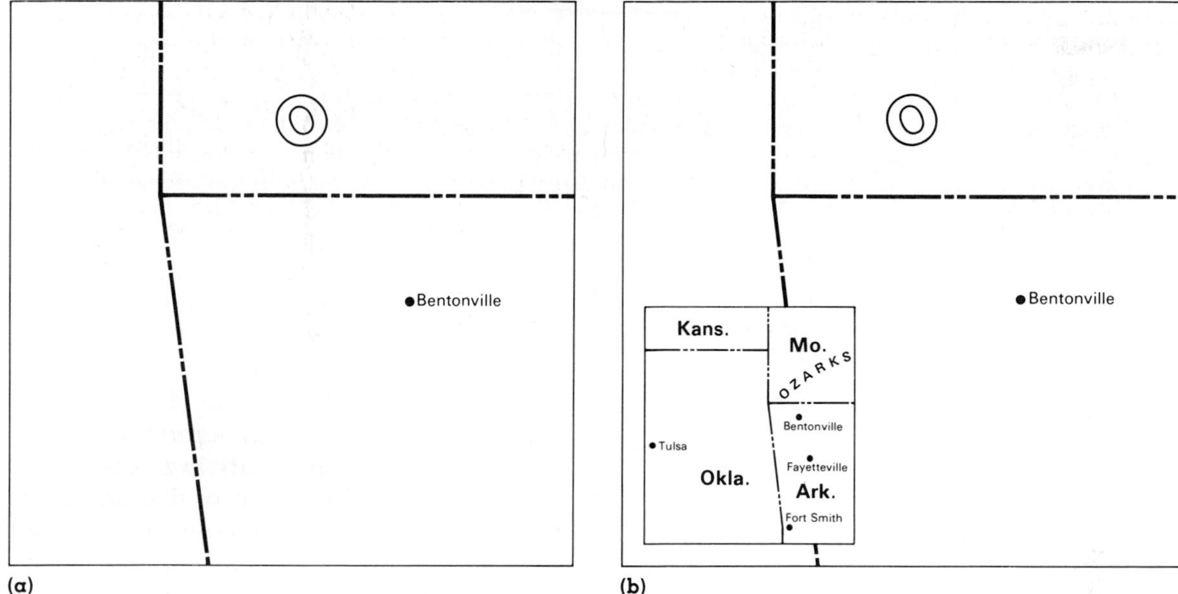

Figure 1.13. Confusion results from inadequate map detail.
Awkward formats can lead to poor base maps, as in (a). The map reader's needs can be more fully met with the provision of an inset map to provide a greater number of locational cues, as in (b).

transfer of knowledge between map author and user. It is essential to include map objects of specific relevance to the context of the map's message. For example, John Dornbach incorporated the users' requirement in his map design strategy during the development of new aeronautical charts.[30] The pilots of modern jet aircraft need maps with specific information that can be easily perceived. Not just any map will do; maps can fail if the designer does not look to the needs of the user.

It becomes difficult for designers when the needs of the user are brought into the design process. Design strategy then requires the cartographer to *project ahead* to determine what sorts of map objects (and what organization of them) the user will need from the map. In a given design task, *only information that is potentially meaningful to the context should be included on the map.*[31] The designer must therefore be knowledgeable about the subject with which he or she is dealing; if not, experts should be consulted.

Geographic context is of crucial importance in designing a thematic map. Every thematic map has a purpose and must present clearly a specific spatial concept or relationship among geographical phenomena. To this end, the phenomena must be placed in an appropriate geographical context. Which mapped space, how much space, and what organizational framework should be used? These questions are difficult to answer, especially since readers are providing their own meanings, unlocked by the visual stimuli of the map. A map becomes meaningful only in relation to the previous

The possibility of placing data in inappropriate or misleading contexts is a real danger. Just as the meaning of statistics can be radically altered by judicious manipulation of their context, so too can the meaning of map data be changed according to the contextual information included. It is, therefore, essential that a cartograpaher have a broad background in a wide range of subjects to enable him to select appropriate contextual information against which the data mapped can be more readily comprehended. This ability implies that he has a knowledge of factors behind distributional patterns and some understanding of their causal interconnections.

Source: Leonard Guelke, "Cartographic Communication and Geographic Understanding," *Canadian Cartographer* 13 (1976): 107–122.

knowledge the user brings to it.[32] The probable level of knowledge of users should be taken into account before the final design of a thematic map is specified.

In sum, success in map communication depends on how well the cartographic designer has been able to interpret the requirements of the user.[33] Cartography has been called "the science of communicating information between individuals by the use of maps."[34] This statement implies that both map authors and map users are part of the process; the cartographer must pay particular attention to the needs of the user in designing each map.

Cartograpahic Abstraction and Generalization

Cartographic abstraction is that part of the mapping activity wherein the map author or cartographer *transforms* unmapped data into map form and *selects* and *organizes* the information necessary to develop the user's understanding of the concepts. "When we accept the idea that not all the available information needs to be presented, that instead information must be selected for particular purposes, then the mapping task becomes the identification of relevant elements."[35] Of course, the selection is guided by the purpose of the map.

Selection, classification, simplification, and symbolization are each part of cartographic abstraction and are **generalizing** operations. Each results in a reduction of the amount of specific detail carried on the map, yet the end result presents the map reader with enough information to grasp the conceptual meaning of the map. Generalization takes place in the context of designing a map to meet user needs. The generalization processes lead to simple visual images, which are more apt to remain in the map user's memory.[36] Unless simplicity is achieved, the map will likely be cluttered with unnecesary detail. Appropriate generalization will result in a spatial message that is efficiently structured for the reader. On the other hand, excessive generalization may cause map images to contain so little useful information that there is no transfer of knowledge. A balance must be struck by the cartographer.

Selection

The **selection** process in the generalization operation in cartographic design begins the map making activity. Selection involves early decisions regarding the geographic space to be mapped, map scale, map projection and aspect, which data variables are appropriate for the map's purpose, and any data gathering or sampling methods which must be employed. Se-

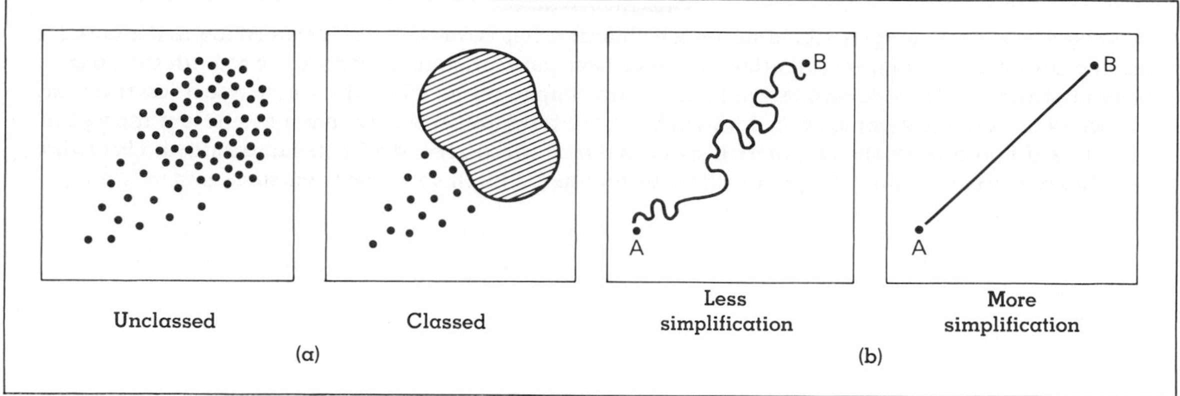

Figure 1.14. Classification and simplification in cartographic communication. More effective cartographic communication may result if data are first classified and simplified.

lection is critical and may involve working very closely with the map author or map client.[37] The selection activity, finally, requires the cartographer to be at least somewhat familiar with the map's content—what is being mapped—and this cannot be overly emphasized.

Classification

Classification is a process in which objects are placed in groups having identical or similar features. The individuality and detail of each element is lost. Information is conveyed through identification of the boundaries of the group. Classification *reduces the complexity* of the map image, helps to *organize* the mapped information, and thus enhances communication.

In thematic mapping, classification can be carried out with qualitative or quantitative information. Qualitative data might include the identification of geographical regions: e.g., the wheat belt, corn belt, or bible belt. It may be far simpler to communicate the concept of region by one large area than to show individual elements. (See Figure 1.14a.) Quantitative classification is normally in numerical data applications. Generally, an entire data array is divided into numerical classes, and each value is placed in its proper class. Only class boundaries are shown on the map. This process reduces overall information but usually results in a map that is more meaningful.

Simplification

Selection and classification are examples of **simplification,** but simplification may take other forms as well. An example might be the *smoothing* of natural or man-made lines on the map to eliminate unnecessary detail. In the selection process, the cartographer may choose to include in the map's base material a road classed as all-weather. Simplification would be the process in which its path is *straightened* between two points so that it no longer retains exact planimetric location (even though this might be possible at the given scale). It can be straightened because the purpose of this map is simply to show *connectivity* between two points, not to illustrate the road's precise locational features. (See Figure 1.14b.)

It is evident that cartography is not merely a technical art. It is for the greater part an applied art, an art governed and determined by scientific laws. But how can cartography avoid the rigid rules of mathematical precision? The decisive turning-point, according to my opinion, lies in the transition from the topographic to the general map. As long as the scale allows the objects in nature to be represented in their true proportion on the map, technical skill alone is necessary. Where this possibility ends the art of the cartographer begins. With generalization art enters into the making of maps.

Source: Max Eckert, "On the Nature of Maps and Map Logic," *Cartographica* 19 (1977): 1–7.

Symbolization

Perhaps most complex of the mapping abstractions is **symbolization.** Developing a map requires symbolization, since it is not possible to create a reduced image of the real world without devising a set of marks (symbols) that stand for real-world things. In thematic mapping, if an element is mapped, it is usually said to be *symbolized.*

There are two major classes of symbols used for thematic maps: **replicative** and **abstract.** Replicative symbols are those that are designed to look like their real-world counterparts; they are only used to stand for tangible objects. Coastlines, trees, railroads, houses, and cars are examples. Base map symbols are replicative in nature, whereas thematic-overlay symbols may be either replicative or abstract. Abstract symbols generally take the form of geometric shapes, such as circles, squares, and triangles. They are traditionally used to represent amounts that vary from place to place; they can represent anything and require sophistication of the map user. A detailed legend is required.

By its very nature, the symbolization process is a generalizing activity shaped by the influence of scale. At smaller scales it is virtually impossible to represent geographical features at true-to-scale likeness. Distortions are necessary. For example, rivers on base maps are widened, and cities that have irregular boundaries in reality are represented by squares or dots.

The cartographer's choices of symbols are not so automatic that every possible mapping problem will yield similar symbol solutions. The symbolization process in thematic mapping is more complex than that, due to a variety of factors. (See Figure 1.15.) This topic will be discussed in fuller detail in Chapter 4. It will suffice to say here that the selection of map symbols should be based on conventions and standards, appropriateness, readers' abili-

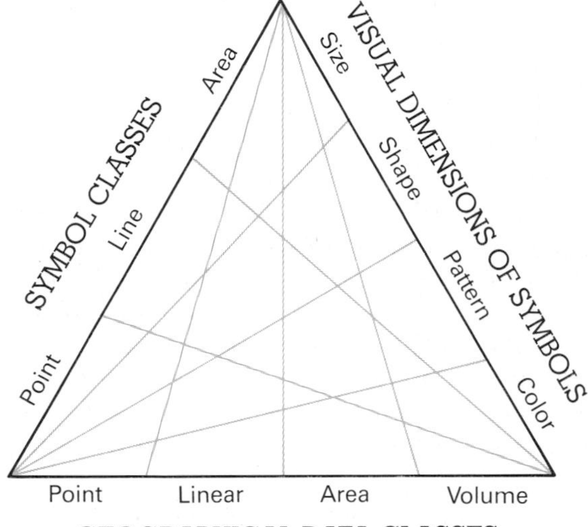

Figure 1.15. Map symbolization variables. The selection of thematic symbolization is compounded because of the complex nature of the world of symbols. Geographical data are found in one of four classes but must be symbolized in only one of three symbol classes. Furthermore, each symbol may or may not be varied, depending on four visual dimensions of symbols. The internal lines of the triangle are included only to illustrate the complexity and nature of the map symbolization process.

ties to use the symbols, ease of construction, and similar considerations. Although various attempts have been made, very little standardization exists in the realm of thematic map symbolization.

The Art in Cartography

Professional cartographers have long discussed whether cartography should be regarded as an art or as a science. Most would probably agree that it is not art in the traditional view of that activity, that is, an expressive medium. Most would certainly agree that it is a science in the sense that the methods used to study its nature are scientific. The idea that there is an art to cartography suggests that its practice as a profession involves more than the mere learning of a set of well-established rules and conventions. *The art in cartography is the cartographer's ability to synthesize the various ingredients involved in the abstraction process into an organized whole that facilitates the communication of ideas.*

One prominent British cartographer very much believes that art is a part of cartography. In his design process, there is a critical point which is ". . . a series of design decisions embodied in the symbol specifications, the point at which design moves from any general ideas to the particular."[38] What is important here is that at that point he is not primarily interested in communication (the science of) but *representation,* and a desire for the map to be not only informationally effective, but to be also aesthetically pleasing.

Each mapping problem is unique; its solutions cannot be predetermined by rigid formulae. How well the cartographic designer can orchestrate all the variables in the abstraction processes is the measure of how much of an artist he or she is. The reader must be considered; the final map solution must reduce complexity, and the map must heighten the reader's interest.[39] These considerations require an artist's skill, which can only be acquired through learning and experience.

The Quantitative Mapping Process

Quantitative mapping, as already pointed out, functions to show how much of something is present in the mapped area. The overriding abstraction in quantitative thematic mapping is in the transformation of tabular data (an aspatial format) into the spatial format of the map.[40] The qualities that the map format provide (distance, direction, shape, and location) are not easily obtainable from the aspatial tabular listing. In fact, this is the quantitative map's raison d'être. If the transformation does not add any spatial understanding, the map should not be considered an alternative form for the reader; the table will suffice.

If exact amounts are required by the reader, a quantitative thematic map is not the answer. The results of the transformations in mapping are *generalized* pictures of the original data. The map is therefore an inefficient and inaccurate form of the original data. "A statistical map is a symbolized generalization of the information contained in a table."[41] Yet the map is the only graphic means we have of showing the spatial attributes of quantitative geographical phenomena. The special process of abstracting, generalizing, and mapping data (even at the expense of losing detail) is therefore justified when the spatial dimension is to be communicated.

Thematic Map Design

Now that the cartographic communication process has been described, it is possible to develop the concepts of thematic map **design.** Cartographic design has been described as the "most fundamental, challenging, and creative aspect of the cartographic process."[42] But what is design? Is it a thing or an activity? We can say, "That map has a good design," or, "Don't bother me while I'm designing." Is there a way to relate the noun to the verb? Indeed, there is.

The cartographer is essentially an engineer attempting to construct a visual device which will effectively communicate geographical information to a percipient. . . . As an engineer, the map designer cannot solve his design problems unless he has carefully settled on the purpose of the map he is going to design.

Source: Arthur H. Robinson, "Map Design," *Proceedings* (International Symposium on Computer-Assisted Cartography, 1975): 9–14.

What is Map Design?

What is design? There are almost as many definitions as there are designers. Design is the optimum use of tools for the creation of better solutions to the problems that confront us. The end result of design "is the initiation of change in man-made things."[43] Things are not only products, such as automobiles or toasters—they can be laws, institutions, processes, opinions, and the like—but they do include physical objects, including maps. Furthermore, design is a dynamic *activity.* "To design is to conceive, to innovate, to create."[44] It is process, a sequential (and repetitive) ordering of events. To include the fashioning of everything from a child's soapbox racer to an artificial human heart, we can accept this definition: "The planning and patterning of any act towards a desired foreseeable end constitutes the design process."[45]

Design operates in at least three broad categories: product design (things), environmental design (places), and communications design (messages).[46] Cartographic design fits this classification most aptly in the category of communications, although it performs its functions with things (maps).

Map design is the aggregate of all the thought processes that cartographers go through during the abstraction phase of the cartographic process. It "involves all major decision-making having to do with specification of scale, projection, symbology, typography, color, and so on."[47] Designers speak of the *principle of synthesis:* "All features of a product must combine to satisfy all the characteristics we expect it to possess with an acceptable relative importance for as long as we wish, bearing in mind the resources available to make and use it."[48] The cartographic designer seeks an organized whole for the graphic elements on the map, to achieve an efficient and accurate transfer of knowledge between map author and map user. Map design is the *functional relationship* between these two.

To reach the communication goals set forth for a given map design problem, the cartographer looks carefully at the intended audience and then defines the map's purpose with precision.[49] This involves an investigation into the functional context of the map, its intended uses, and its communication objectives. Every manipulation of the marks on the map is planned so that the end result will yield a structured visual whole that serves the map's purpose.

Thus, cartographic design is a complex activity involving both intellectual and visual aspects. It is intellectual in the sense that the cartographer relies on the foundations of sciences such as communication, geography, and psychology during the creation of the map. Map design is visual in the sense that the cartographer is striving to reach goals of communication through a visual medium. When we speak of a map as having *a* design, we are actually looking at the product as the result of design *activity.*

The Design Process

Most designers would agree that all design takes place in sequential *steps,* ordered in a way that eventually yields the planned result. Identifying these steps or stages in design is helpful in learning how design takes place.

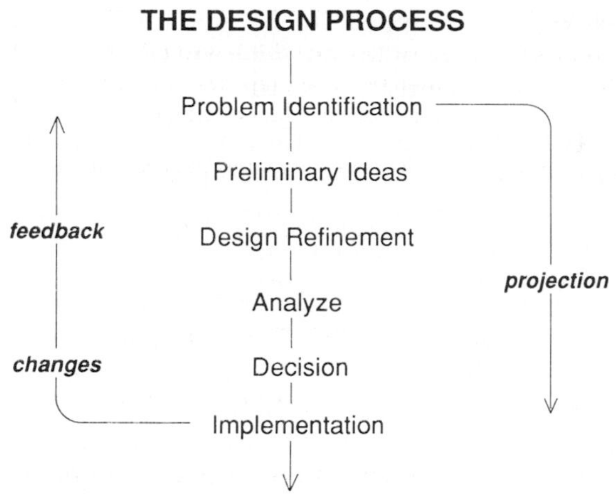

THE DESIGN PROCESS

Problem Identification
Preliminary Ideas
Design Refinement
Analyze
Decision
Implementation
feedback
changes
projection

Figure 1.16. The design process.
See text for explanation.

The **map design process,** like any act of designing, includes six essential stages: problem identification, preliminary ideas, design refinement, analysis, decision, and implementation. (See Figure 1.16.)[50]

Needs and design criteria are established in the first stage. Limitations are usually set in this stage. In the case of mapping, this stage includes the identification of map purpose and map reader and such factors as cost and technical considerations. The most creative step in the design process occurs in the second stage, where preliminary ideas are formulated. Brainstorming or having synectic (problem-solving based on creative thinking) sessions are helpful here. Many solutions to design problems are found unconsciously, and then "pop" into the mind's eye. Sorting through one's visual memory often takes place in this stage, and is especially helpful in cartographic design.

In the third stage, called design refinements, all preliminary ideas are evaluated—and may be accepted or rejected. Those ideas that are retrieved are refined and sharpened, and decisions are made that will affect the whole process. For cartographers, this stage usually involves setting down in writing details of the mapping project. For example, critical data needs are reviewed and finalized.

Models are often created in the analysis stage. For product designers, the models serve to make the drawings and sketches come alive and assist in the visualization process. Today, computer modelling in graphic form is possible, and real models are often not made. But, this remains as a prototype stage. Market research is often conducted on the prototypes and changes can be made. Map designers use this stage to develop detailed drawings and to work out problem areas (such as tricky printing problems or unique symbol schemes). Cartographers may wish to test their preliminary designs on sample readers.

The decision stage is as the name suggests. Changes, if any, are made on the prototype based on research and fact-finding from the previous stage. Ideas are rejected or accepted, and the final stage, implementation, begins. For cartographers this stage signals the beginning of final map production.

Feedback in the design process is a continuous process. Each design teaches us something about future problems and processes. Feedback is a critical element that helps designers to become efficient and to recognize that each design process may be unique, and that not every design problem will utilize the design stages in exactly the same manner.

Projection in the process assists the designer in anticipating solutions and problem areas. Projection also involves visualization, especially in cartographic design. Projection allows the designer to "see" the end product, and thereby helps decision-making along the way. Being able to project is essential, and comes with experience, regardless of the design area.

The designer may cycle through all stages in the design process as many times as are required to reach an acceptable solution to the problem. Repetition is to be expected. Successful design requires that the designer be able to see or project, in his or her "mind's eye," the final map product. This is easy enough to say but difficult to do or teach—experience may be the only teacher. Nonetheless, the ability to conceive the final solution before it is physically mapped will reduce the number of wrong choices in design.

Design Evaluation

It is indeed difficult to evaluate map design, usually because the one doing the evaluating was not intimately involved in the design process. The outsider is never sure what compromises the cartographer had to make to balance the decisions in the design process. We do not know the relationship between map author and designer, and what sacrifices and learning had to take place to get anything in the map. One cartographer does suggest, however, that a map's design should only be judged with regard to the map's purpose and intended audience.[51] This seems fair enough.

There may be a few guidelines, however, that may be followed. Southworth and Southworth, for example, list these design characteristics of successful maps.[52]

1. A map should be suited to the needs of its users.
2. A map should be easy to use.
3. Maps should be accurate, presenting information without error, distortions, or misrepresentation.
4. The fit between the map and the environment represented should be good (the map user should be able to connect map to environment and environment to map).
5. The language of the map should relate to the elements or qualities represented.
6. A map should be clear, legible, and attractive.
7. Many maps would ideally permit interaction with the user, allowing change, updating, or personalization.

Creativity and Ideation

Creativity is the ability to see relationships among elements (regardless of the design arena). Although there is no recipe for creativity, there appear to be certain activities shared by people considered to be great thinkers, scientists, or artists.[53]

1. Challenging assumptions—daring to question what most people take as truth.
2. Recognizing patterns—perceiving significant similarities or differences in ideas, events, or physical phenomena.
3. Seeing in new ways—looking at the commonplace with new perceptions, transforming the familiar into the strange, and the strange into the familiar.
4. Making connections—bringing together seemingly unrelated ideas, objects, or events in ways that lead to new concepts.
5. Taking risks—daring to try new ways, with no control over the outcome.
6. Using chance—taking advantage of the unexpected.
7. Constructing networks—forming associations for the exchange of ideas, perceptions, questions, and encouragement.

Cartographic designers can learn from this list. Conscious effort to participate in new ways of thinking during the transformation stage should become an integral part of the design procedure. The **visualization process,** or thinking by incorporating visual images into thought, occurs to the fullest when seeing, imagining, and graphic ideation come into active interplay.[54]

For the creative person, *seeing* is integral to all thought processes, and willingness to restructure visual images into new configurations is essential. Looking at a map inside out or upside down can yield solutions never thought possible. **Imagining,** or creating visual images in the mind's eye, is also helpful in developing design experience.[55] These imagination images may or may not be composed of past perceptions. Imagining is the act of "seeing" a map before it is physically produced. The ability to project in the design process calls for the strengthening of imagining talents.

Graphic ideation, or sketching, is often practiced by creative persons. The goal of this activity is to bring vague images into clear focus.[56] Designers can improve their creative potential by practicing seeing, imagining, and graphic ideation.

One real contribution of rapidly generated temporary maps (displayed on cathode ray tubes) is their potential as "what if" images. Hundreds of maps with alternative designs can be created in a short time, so that the designer can explore alternatives. Scale, projection, color, typography, classification, and other elements of design can be varied and checked almost instantly. Although this may be considered a feature in favor of the new technology, it should not replace the designer's own imagining abilities—that would reduce overall creativity in the long run.

Map Aesthetics

Although maps may not be objects of aesthetic concern, there is a trend toward giving greater consideration to the quality of appearance of thematic maps. Writing several decades ago, the great map critic John K. Wright addressed this issue:

> The quality of a map is also in part an aesthetic matter. Maps should have harmony within themselves. An ugly map, with crude colors, careless line work, and disagreeable, poorly arranged lettering may be intrinsically as accurate as a beautiful map, but it is less likely to inspire confidence.[57]

More recently, John S. Keates, a British cartographer, has remarked, "The 'art' of cartography . . . is not simply an anachronism surviving from some pre-scientific era; it is an integral part of the cartographic process."[58]

Three elements have been identified as forming the basis for the evaluation of map aesthetics: harmony, composition, and clarity.[59] *Harmony* is viewed as the relationship between different map elements (that is, how do the elements look *together?*). *Composition* deals with the arrangement of the elements and the emphasis placed on them. In other words, how does the structural balance of emphasis appear? Finally, *clarity* deals with the ease of recognition of the map's elements by the map user. "A map which lacks one or more of these three main elements, lacks beauty."[60]

Cartographic designers have a certain degree of freedom in the design process. Of course map function is the overriding concern, along with the needs of the user, but beyond that the designer is working in a subjective realm. How well he or she performs in the creative, aesthetic realm will more than likely depend on intuitive judgments, conditioned by fundamental training and experience.

The subjective elements of design have been listed as follows:[61]

- generalization—beauty of simplified shapes
- symbolization—beauty of graphic representation
- color—beauty of color accent and balance
- layout—beauty of composition
- typography—beauty of typographic appearance

This brief excursion into map aesthetics is not intended to be exhaustive. It has been included only to suggest to the reader that map design is subjective as well as objective. Beyond the rigid, scientific world of numerical cartography lies the intuitive, artistic, and aesthetic world of maps where designers can exercise their expressive talents.

The best designs . . . are *intriguing and curiosity-provoking,* drawing the viewer into the wonder of the data, sometimes by narrative power, sometimes by immense detail, and sometimes by elegant presentation of simple but interesting data. But no information, no sense of discovery, no wonder, no substance is generated by chartjunk.

Source: Edward R. Tufte, *The Visual Display of Quantitative Information.* (Cheshire, Conn.: Graphics Press, 1983), p. 121.

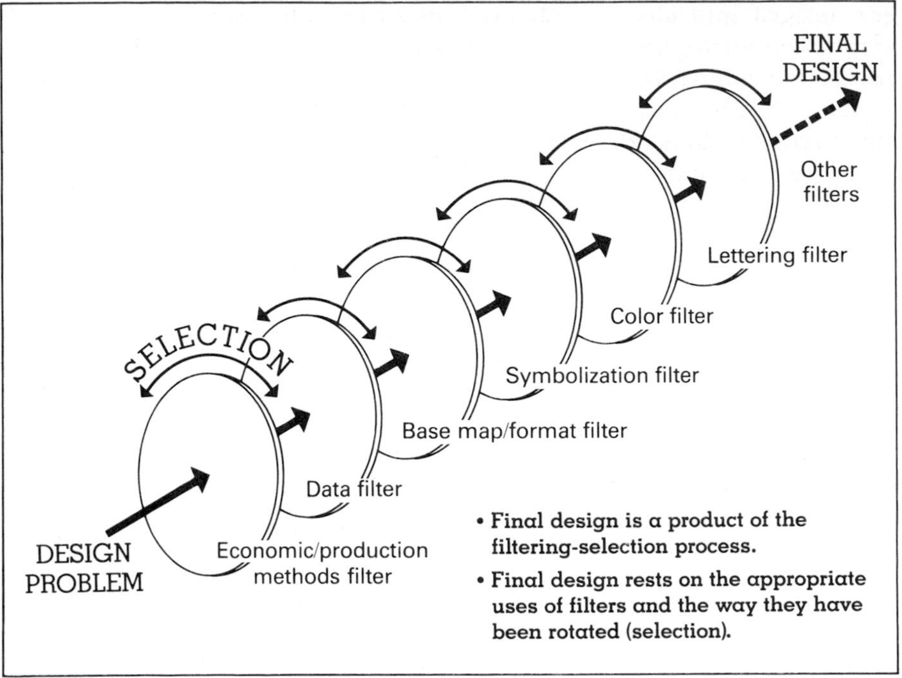

Figure 1.17. Map design as a filtering-selection process.
In this view of design, a series of filters must be rotated (selection), allowing design activity to continue until an appropriate final solution is reached.

Map Elements for Design Consideration

Cartographic designers arrive at solutions by arranging, organizing, and selecting certain map elements. These include scale, projection, symbolization, pattern, color, and typography, among others. From this perspective, map design is like a series of filters in which selections are made at each filter. (See Figure 1.17.) This analogy directs attention to the fact that map design is a complex affair involving many decisions, each of which affects all the others. Good design is simply the best solution among many, given a set of constraints imposed by the problem. The best design will likely be a simple one that works well with the least amount of trouble. The optimum solution may not be achievable, and what is good design today may be ineffective in the future. We are constantly learning more about the map user, and this will modify our future decisions as designers.

Important design principles include simplicity, appropriateness in a functional context, pleasing appearance, and considerations

Just as some realistically painted cows are full of life while others are deadly mechanical records, so some faithful maps are alive while others leave us untouched.

Source: Rudolf Arnheim, "The Perception of Maps," *American Cartographer* 3 (1976): 5–10.

of economy. The designer's tools of creativity, visualization, ideation, and problem solving are used to sift through the map elements in order to bring these principles into a proper balance. The subjects treated throughout the remainder of this book are intended to provide a framework in which to achieve these goals.

Notes

1. Morris M. Thompson, *Maps for America* (United States Department of the Interior, Geological Survey—Washington, D.C.: USGPO), p. 13.
2. Leo Bagrow, *History of Cartography,* revised ed. (Cambridge: Harvard University Press, 1966), p. 25.
3. Arthur H. Robinson and Barbara Bartz Petchenik, *The Nature of Maps: Essays Toward Understanding Maps and Meaning* (Chicago: University of Chicago Press, 1976), pp. 16–17.
4. Phillip C. Muehrcke, *Thematic Cartography* (Commission on College Geography, Resource Paper No. 19—Washington, D.C.: Association of American Geographers, 1972), p. 1.
5. Mark S. Monmonier, *Maps, Distortion, and Meaning* (Resource Paper No. 75-4—Washington, D.C.: Association of American Geographers, 1977), p. 9.
6. Robinson and Petchenik, *The Nature of Maps,* p. 19.
7. Muehrcke, *Thematic Cartography,* p. 1; and Monmonier, *Maps, Distortion, and Meaning,* p. 9.
8. E. Meynen, ed., *Multilingual Dictionary of Technical Terms in Cartography* (International Cartographic Association, Commission II—Wiesbaden: Franz Steiner Verlag, 1973), p. 1.
9. Arthur. H. Robinson, "Geographical Cartography," in Preston E. James and Clarence F. Jones, eds. *American Geography, Inventory and Prospect* (Syracuse: Syracuse University Press, 1954), pp. 553–77.
10. Phillip C. Muehrcke, "Maps in Geography," *Cartographica* 18 (1981): 1–37.
11. John R. Borchert, "Maps, Geography, and Geographers," *The Professional Geographer* 39 (1987): 387–89.
12. D. P. Bickmore, "The Relevance of Cartography," in *Display and Analysis of Spatial Data,* John C. Davis and Michael J. McCullagh, eds. (New York: John Wiley, 1975), p. 330.
13. Arthur H. Robinson, "Map Design," *Proceedings* (International Symposium on Computer-Assisted Cartography), 1975, pp. 9–14.
14. Robinson and Petchenik, *The Nature of Maps,* pp. 116–17.
15. Meynen, *Multilingual Dictionary,* p. 291.
16. Robinson, "Map Design," pp. 9–14.
17. Mark S. Monmonier and George A. Schnell, *Map Appreciation* (Englewood Cliffs, NJ: Prentice-Hall, 1988).
18. Patricia Gilmartin, "The Design of Journalistic Maps: Purposes, Parameters, and Prospects," *Cartographica* 22 (1985): 1–18; see also Jeffrey S. Murray, "The Map is the Message," *The Geographical Magazine* 59 (1987): 237–41.
19. Richard J. Chorley and Peter Haggett, eds., *Models in Geography* (London: Methuen, 1967), p. 21.
20. *Ibid.,* p. 26.
21. These ideas were first expressed by Arthur H. Robinson in *The Look of Maps* (Madison, WI: University of Wisconsin Press, 1966), p. 13.
22. The generalized model as described in this text is somewhat different than that used by Muehrcke, although the idea of transformation is incorporated in it. See Muehrcke, *Thematic Cartography,* pp. 3–4; see also Philip J. Gersmehl and Sona K. Andrews, "Teaching the Language of Maps," *Journal of Geography* 86 (1986): 267–70.
23. Monmonier, *Maps, Distortion, and Meaning,* pp. 9–10.
24. Phillip C. Muehrcke, *Map Use, Reading, Analysis, and Interpretation* (Madison, WI: JP Publications, 1978), p. 19; see also Joel L. Morrison, "Map Generalization: Theory, Practice, and Economics," *Proceedings* (International Symposium on Computer-Assisted Cartography, 1975), pp. 99–112.
25. *Ibid.,* pp. 99–112.
26. Arthur H. Robinson and Barbara Bartz Petchenik, "The Map as a Communication System," *Cartographic Journal* 12 (1975): 7–15.

27. Robinson and Petchenik, *The Nature of Maps,* p. 20.
28. Muehrcke, *Map Use, Reading, Analysis, and Interpretation,* p. 8.
29. Judy M. Olson, "Experience and the Improvement of Cartographic Communication," *Cartographic Journal* 12 (1975): 94–108.
30. John E. Dornbach, *An Analysis of the Map as an Information Display System* (unpublished Ph.D. dissertation, Department of Geography, Clark University, 1967).
31. Leonard Guelke, "Cartographic Communication and Geographic Understanding," *Canadian Cartographer* 13 (1976): 107–22.
32. Barbara Bartz Petchenik, "Cognition in Cartography," *Proceedings* (International Symposium on Computer-Assisted Cartography, 1975), pp. 183–93.
33. Guelke, "Cartographic Communication," pp. 107–22.
34. Joel L. Morrison, "Towards a Functional Definition of the Science of Cartography with Emphasis on Map Reading," *American Cartographer* 5 (1978): 97–110.
35. Gershon Weltman, ed., *Maps: A Guide to Innovative Design* (Woodland Hills, CA: Perceptronics, 1979), p. 25.
36. Rudolf Arnheim, "The Perception of Maps," *American Cartographer* 3 (1976): 5–10.
37. Rosemary E. Ommer and Clifford H. Wood, "Data, Concept and the Translation to Graphics," *Cartographica* 22 (1985): 44–62.
38. J. S. Keates, "Cartographic Art," *Cartographica* 21 (1984): 37–43.
39. Monmonier, *Maps, Distortion, and Meaning,* p. 14; see also George F. McCleary, "In Pursuit of the Map User," *Proceedings* (International Symposium on Computer-Assisted Cartography, 1975), pp. 238–49.
40. George F. Jenks, "Contemporary Statistical Maps—Evidence of Spatial and Graphic Ignorance," *American Cartographer* 3 (1976): 11–19.
41. *Ibid.*
42. Alan DeLucia, "Design: The Fundamental Cartographic Process," *Proceedings* (Association of American Geographers) 6 (1974): 83–86.
43. Christopher J. Jones, *Design Methods: Seeds of Human Futures* (New York: John Wiley, 1981), p. 6.
44. Warren J. Luzadder, *Innovative Design, with an Introduction to Graphic Design* (Englewood Cliffs, NJ: Prentice-Hall, 1975), p. 13.
45. Victor Papenek, *Design for the Real World* (New York: Pantheon Books, 1971), p. 3.
46. Norman Potter, *What is a Designer: Education and Practice* (London: Studio-Vista, 1969), p. 9–10.
47. Robinson and Petchenik, *The Nature of Maps,* p. 19.
48. W. H. Mayall, *Principles of Design* (New York: Van Nostrand Reinhold, 1979), p. 90.
49. Robinson, "Map Design," pp. 9–14.
50. Kurt Hanks, Larry Bellistan, and Dave Edwards, *Design Yourself* (Los Altos, CA: William Kaufmann, 1978), pp. 60–61.
51. Phillip C. Muehrcke, "An Integrated Approach to Map Design and Production," *American Cartographer* 9 (1982): 109–22.
52. Michael Southworth and Susan Southworth, *Maps: A Visual Survey and Design Guide,* A New York Graphic Society Book, (Boston: Little, Brown and Company, 1982), pp. 16–17.
53. The Burdick Group, *Creativity: The Human Resource* (exhibit) (San Francisco: Standard Oil Company of California, 1982).
54. Mike Samuels and Nancy Samuels, *Seeing with the Mind's Eye* (New York: Random House, 1975), p. xi; and Phillip C. Muehrcke, "Maps in Geography," in *Maps in Modern Geography: Geographical Perspectives on the New Geography,* Leonard Guelke, ed. (Cartographica, Monograph 27, 1981), p. 11.
55. Samuels and Samuels, *Seeing with the Mind's Eye,* p. 43.

56. Muehrcke, "Maps in Geography," p. 13.
57. John K. Wright, "Map Makers Are Human," *Geographical Review* 32 (1944): 527–44.
58. Keates, *Understanding Maps,* p. 127.
59. Aart J. Karssen, "The Artistic Elements in Design," *Cartographic Journal* 17 (1980): 124–27.
60. *Ibid.*
61. *Ibid.*

Glossary

abstract symbols may represent anything; usually take the form of geometrical shapes such as circles, squares, or triangles

cartographic abstraction transformational process in which map author or cartographer selects and organizes material to be mapped; identification of relevant elements

cartographic generalization transformation process of abstraction involving selection, classification, simplification, and symbolization

cartographic process the transformation of unmapped data into map form involving symbolization and the map reading activities whereby map users gain information

cartography somewhat broader in scope than map making; the study of the foundations of map making

classification placing objects in groups having identical or similar features; reduces complexity and helps organize materials for communication

creativity ability to see relationships among elements

design a dynamic activity that deals with the optimum use of tools for the creation of better solutions to life; initiation of change in man-made things

general-purpose maps also called reference maps; show natural and man-made features of general interest for widespread public use

geographic map base map component of the thematic map; used to provide locational information for the map user

graphic ideation bringing images into clear focus by sketching

imagining creating visual images in the mind's eye

inset maps small map provided along with the main map, helping to bridge the gap between map author and map user

map a graphic representation of the milieu

map analysis map use activity in which the map user begins to see patterns

map author anyone wishing to convey a spatial message; may or may not be a cartographer

map design aggregate of all thought processes that the map author or cartographer goes through during the abstraction phase of the cartographic process; activity seeking graphic solutions

map design process characterized by six stages: problem identification, preliminary ideas, design refinement, analysis, decision, and implementation.

map interpretation map use activity in which the map user attempts to explain mapped patterns

map making also called mapping; all processes associated with the production of maps; the designing, compiling, and producing of maps

map meaning supplied to the map by the user; instrumental in the communication process; not evident from the physical marks on the map

map percipient anyone who gains spatial knowledge by looking at a map

map reading map use activity in which the user simply determines what is displayed and how the map maker did it

map scale amount of reduction that takes place in going from real world to map plane; ratio of map distance to earth distance

map use comprises map reading, map analysis, and map interpretation

mental maps mental images having spatial attributes

model theories, laws, or hypotheses concerning the real world

paradigm large-scale model; specifies a searching pattern in scientific inquiry

qualitative map thematic map whose main purpose is to show locational features of nominal data

quantitative map thematic map whose main purpose is to show spatial aspects of numerical data—spatial variation of amount

replicative symbols designed to appear like their real-world counterparts

selection a generalization process in which selections of items to be mapped are made, all within the context of map purpose and map scale

simplification reducing amount of information by selection, classification, or smoothing

symbolization devising a set of graphic marks that stand for real-world things

thematic map special-purpose, single-topic map; designed to demonstrate particular features or concepts

thematic overlay that part of the thematic map that contains the specific information (map subject)

visualization process thinking by incorporating visual images into the thought processes

Readings for Further Understanding

Arnheim, Rudolf. "The Perception of Maps." *American Cartographer* 3 (1976): 5–10.

Bagrow, Leo. *History of Cartography.* Revised and enlarged by R. A. Skelton. Cambridge: Harvard University Press, 1966.

Bertin, Jacque. "Visual Perception and Cartographic Transcription." *World Cartography* XV (1979): 17–27.

Bickmore, D. P. "The Relevance of Cartography." In *Display and Analysis of Spatial Data.* John C. Davis and Michael J. McCullagh, eds. New York: John Wiley, 1975, pp. 328–67.

Board, Christopher. "Cartographic Communication." In *Maps in Modern Geography: Geographical Perspectives on the New Cartography.* Leonard Guelke, ed. Cartographica, Monograph 27, 1981, pp. 42–78.

Borchert, John R. "Maps, Geography, and Geographers." *The Professional Geographer* 39 (1987): 387–89.

Bos, E. S. "Another Approach to the Identity of Cartography." *ITC Journal* 2 (1982): 104–108.

The Burdick Group, *Creativity: The Human Resource.* (Exhibit.) San Francisco: Standard Oil Company of California, 1982.

Chorley, Richard J., and Peter Haggett, eds. *Models in Geography.* London: Methuen, 1967.

DeLucia, Alan. "Design: The Fundamental Cartographic Process." *Proceedings* (Association of American Geographers) 6 (1974): 83–86.

Dent, Borden D. "Visual Organization and Thematic Map Communication." *Annals* (Association of American Geographers) 61 (1972): 79–92.

———. "Postulates on the Nature of Map Reading." Paper presented at the Georgia Academy of Science, Annual Meetings, 1976.

Dornbach, John E. *An Analysis of the Map as an Information Display System.* Unpublished Ph.D. Dissertation, Department of Geography, Clark University, 1967.

Eckert, Max. "On the Nature of Map and Map Logic." *Cartographica* 19 (1977): 1–7.

Edwards, Betty. *Drawing on the Right Side of the Brain.* Los Angeles: J. P. Tarcher, 1979.

Farrell, Barbara. "Map Evaluation." In R. B. Parry and C. R. Perkins, eds. *World Mapping Today.* London: Butterworths, 1987, pp. 27–34.

Gersmehl, Philip J. "The Data, the Reader, and the Innocent Bystander—A Parable for Map Users." *The Professional Geographer* 37 (1985): 329–34.

Gilmartin, Patricia. "The Design of Journalistic Maps: Purposes, Parameters, and Prospects." *Cartographica* 22 (1985): 1–18.

Guelke, Leonard. "Cartographic Communication and Geographic Understanding." *Canadian Cartographer* 13 (1976): 107–122.

Hanks, Kurt; Larry Belliston, and Dave Edwards. *Design Yourself.* Los Altos, CA: William Kaufmann, 1978.

Jenks, George F. "Contemporary Statistical Maps—Evidence of Spatial and Graphic Ignorance." *American Cartographer* 3 (1976): 11–19.

Jones, Christopher J. *Design Methods: Seeds of Human Futures.* New York: John Wiley, 1981.

Karssen, Aart J. "The Artistic Elements in Map Design." *Cartographic Journal* 17 (1980): 124–27.

Keates, J. S. *Understanding Maps.* New York: John Wiley, 1982.

———. "Cartographic Art," *Cartographica* 21 (1984): 37–43.

Kolacny, A. "Cartographic Information—A Fundamental Concept and Term in Modern

Cartography." *Cartographic Journal* 6 (1969): 47–49.

Luzadder, Warren J. *Innovative Design, with an Introduction to Design Graphics.* Englewood Cliffs, NJ: Prentice-Hall, 1975.

MacEachren, Alan M. "Map Use and Map Making Education: Attention to Sources of Geographic Information." *The Cartographic Journal* 23 (1986): 115–22.

Mayall, W. H. *Principles in Design.* New York: Van Nostrand Reinhold, 1979.

McCleary, George F. "In Pursuit of the Map User." *Proceedings* (International Symposium on Computer-Assisted Cartography, 1975): 238–49.

Meine, Karl-Heintz. "Thematic Mapping: Present and Future Capabilities." *World Cartography* XV (1979): 1–16.

Meynen, E., ed. *Multilingual Dictionary of Technical Terms in Cartography.* International Cartographic Association, Commission II. Wiesbaden: Franz Steiner Verlag, 1973.

Monmonier, Mark S. *Maps, Distortion, and Meaning.* Resource Paper No. 75-4. Washington, D.C.: Association of American Geographers, 1977.

Monmonier, Mark S., and George A. Schnell. *Map Appreciation.* Englewood Cliffs, NJ: Prentice-Hall, 1988.

Morrison, Joel L. "Map Generalization: Theory, Practice and Economics." *Proceedings* (International Symposium on Computer-Assisted Cartography, 1975): 99–112.

———. "Towards a Functional Definition of the Science of Cartography with Emphasis on Map Reading." *American Cartographer* 5 (1978): 97–110.

Muehrcke, Phillip C. *Thematic Cartography.* Commission on College Geography, Resource Paper No. 19. Washington D.C.: Association of American Geographers, 1972.

———. *Map Use, Reading, Analysis, and Interpretation.* Madison, WI: JP Publications, 1978.

———. "Maps in Geography." In *Maps in Modern Geography: Geographical Perspectives on the New Cartography.* Leonard Guelke, ed. Cartographica, Monograph 27, 1981, pp. 1–41.

———. "Whatever Happened to Geographic Cartography?" *The Professional Geographer* 33 (1981): 397–405.

———. "An Integrated Approach to Map Design and Production." *American Cartographer* 9 (1982): 109–22.

Olson, Judy M. "Experience and the Improvement of Cartographic Communication." *Cartographic Journal* 12 (1975): 94–108.

Ommer, Rosemary E., and Clifford H. Wood. "Data, Concept and the Translation to Graphics." *Cartographica* 22 (1985): 44–62.

Papenek, Victor. *Design for the Real World.* New York: Pantheon Books, 1971.

Petchenik, Barbara Bartz. "Cognition in Cartography." *Proceedings* (International Symposium on Computer-Assisted Cartography, 1975): 183–93.

———. "From Place to Space: The Psychological Achievement of Thematic Mapping." *American Cartographer* 6 (1979): 5–12.

Potter, Norman. *What is a Designer: Education and Practice.* London: Studio-Vista, 1969.

Robinson, Arthur H. "Geographical Cartography." In Preston E. James and Clarence F. Jones, eds. *American Geography, Inventory and Prospect.* Syracuse: Syracuse University Press, 1954.

———. *The Look of Maps.* Madison, WI: University of Wisconsin Press, 1966.

———. "An International Standard Symbolism for Thematic Maps: Approaches and Problems." *International Yearbook of Cartography* 13 (1973): 19–26.

———. "Map Design." *Proceedings* (International Symposium on Computer-Assisted Cartography, 1975): 9–14.

———. *The Nature of Maps: Essays Toward Understanding Maps and Mapping.* Chicago: University of Chicago Press, 1976.

Robinson, Arthur H., and Barbara Bartz Petchenik. "The Map as a Communication System." *Cartographic Journal* 12 (1975): 7–15.

Salichtchev, K. A. "Some Reflections on the Subject and Method of Cartography after the Sixth International Cartographic Conference." *Cartographia* 19 (1977): 111–16.

Samuels, Mike, and Nancy Samuels. *Seeing with the Mind's Eye*. New York: Random House, 1975.

Southworth, Michael, and Susan Southworth. *Maps: A Visual Survey and Design Guide*. A New York Graphic Society Book. Boston: Little, Brown and Company, 1982.

Tufte, Edward R. *The Visual Display of Quantitative Information*. Cheshire, CT: Graphics Press, 1983.

Weltman, Gershon, ed. *Maps: A Guide to Innovative Design*. Woodland Hills, CA: Perceptronics, 1979.

Wolter, John A. "Cartography—An Emerging Discipline." *Canadian Cartographer* 12 (1975): 210–16.

Wood, Michael. "Human Factors in Cartographic Communication." *Cartographic Journal* 9 (1972): 123–32.

Wright, John K. "Map Makers Are Human." *Geographical Review* 32 (1944): 527–44.

Chapter

2

The Round Earth to Flat Map: Map Projections for Designers

Chapter Preview

The earth is an irregularly shaped body, roughly approximating a sphere but more precisely defined by a reference ellipsoid. Although the earth can be accurately described this way, it is assumed to be perfectly spherical in most applications of thematic cartography. The slight imperfections of the earth's sphere simply cannot be accommodated on page-size maps because of the great reduction. The cartographer must have a fundamental knowledge of the earth's geographic grid, as it appears on a perfect sphere, for the preparation and selection of map projections. Map projections provide geometric control over the locational accuracy and appearance of the final thematic map; the cartographer must thoroughly understand the distortion brought on by the projection process. Certain map projections are better than others for representing different parts of the earth's land areas. The Albers conical equal-area projection, for example, is uniquely suited for portraying the United States.

Thematic maps are composed of two main structural elements: the base map and the thematic overlay. The appearance of the base map and its appropriateness to the whole map depend upon the projection on which it is compiled. In choosing an appropriate projection, cartographers must know at least the fundamental concepts of plane and spherical geometry.

Map projection is one element in the total design effort. This chapter will introduce the main aspects of the earth's size and shape as a foundation for the description of its spherical or geographic coordinate system. These topics will lead to an explanation of map projections.

The Size and Shape of the Earth

It is not known exactly when the earth was first thought to be round or spherical in form, but Pythagoras (sixth century B.C.) and Aristotle (384–322 B.C.) are known to have decided that the earth was round. Aristotle based his conclusions partly on the idea, then widely held among Greek philosophers, that the sphere was a perfect shape and that the earth must therefore be spherical. Celestial observations, notably lunar eclipses, also helped him to this important conclusion. The idea of a spherical earth soon became adopted by most philosophers and mathematicians.

Greek scholars turned their attention to measurement of the earth. In fact, the earth's size was measured quite accurately by a Greek scholar living in Alexandria. **Eratosthenes** (276–194 B.C.) calculated the equatorial circumference to be 40,233 km (25,000 mi), remarkably close to today's measurement of 40,072 km (24,900 mi).[1] Another Greek mathematician, Poseidonius (c. 130–51 B.C.), measured the earth's equatorial circumference to be 38,622 km (24,000 mi).

Eratosthenes' ingenious method of measuring the earth employed simple geometrical calculations. (See Figure 2.1.) In fact, the method is still used today. Eratosthenes noticed on the day of the summer solstice that the noon sun shone directly down a well at Syene, near present-day Aswan in southern Egypt. However, the sun was not directly overhead at Alexandria but rather cast a shadow that was 7°12′ off the vertical. Applying geometrical principles, he knew that the deviation of the sun's rays from the vertical would subtend an angle of 7°12′ at the center of the earth. This angle is 1/50 the whole circumference (360°). The only remaining measurement needed to complete the calculations was the distance between Alexandria and Syene; this was estimated at 804.65 km (500 mi). Multiplying this figure by 50, he calculated the total circumference to be 40,233 km (25,000 mi).

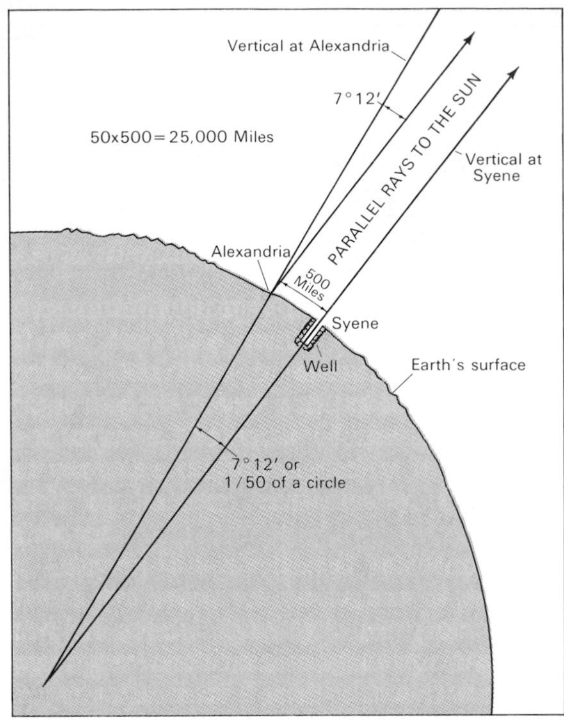

Figure 2.1. Eratosthenes' method of measuring the size of the earth.
See the text for a complete explanation.

Three assumptions on Eratosthenes' part led to error in the results, although these partially compensated for each other. Alexandria and

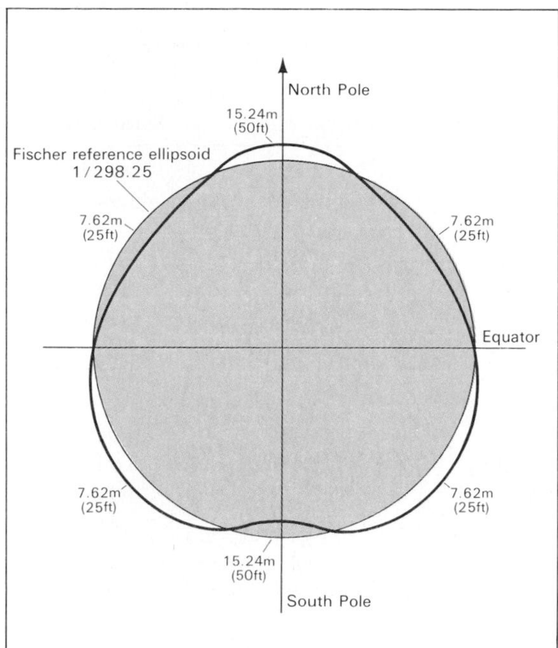

Figure 2.2. The shape of the earth.
Recent satellite measurements show that there are minor bulges and depressions away from the Fischer reference ellipsoid. We cannot detect these on the surface of the earth. The Fischer ellipsoid is the International Astronomical Union figure for the earth (1968) and has a flattening of 1:298.25.

Syene are not on the same meridian. Syene's latitude is 24°5′30″—it is not at the Tropic of Cancer, where the sun's rays are perfectly vertical at the summer solstice. Finally, the actual distance between Alexandria and Syene is 729 km (453 mi). Regardless of these sources of error, Eratosthenes made remarkably accurate calculations.

Not until the end of the seventeenth century was the notion of an imperfectly shaped earth introduced. By that time, accurate measurement of gravitational pull was possible. Most notably, Newton in England and Huygens in Holland put forward the theory that the earth was flattened at the poles and extended (bulged) at the equator. This idea was later tested by field observation in Peru and Lapland by the prestigious French Academy of Sciences. The earth was indeed flattened at the poles! It is interesting to note that the first indication of this flattening came from seamen who noticed that their chronometers were not keeping consistent time as they sailed great latitudinal distances. The unequal pull of gravity, caused by the imperfectly shaped earth, created different gravitational effects on the pendulums of their clocks.

Today we describe the shape of the earth as a **geoid** (earth-shaped). Satellite measurements of the earth since 1958 suggest that, in addition to being flattened at the poles and extended at the equator, the earth also contains great areas of depressions and bulges.[2] (See Figure 2.2.) Of course, these irregularities are not noticeable to us on the earth's surface. They do, however, have significance for precise survey work and geodetic measurement. **Geodesy** is the science of earth measurement.

For the purposes of geodesy, and for some branches of cartography, the earth's flattened shape can be best described by reference to an **ellipsoid.** This geometrical solid is generated by rotating an ellipse around its minor axis and choosing the lengths of the major and minor axes that best fit those of the real earth. (See Figure 2.3.) Various reference ellipsoids have been adopted by different countries throughout the world for their official mapping programs, based on the local precision of the ellipsoid in describing their part of the earth's surface. (See Table 2.1.)

Until recently, the United States used the **Clarke ellipsoid of 1866** as its reference ellipsoid in establishing a *datum* (starting point) for horizontal control in large-scale mapping. Because of recent advances in satellite geodesy with higher degrees of accuracy, and because many surveying errors have built–up since the last datum was adopted (North American Datum—NAD27), a new datum was selected in 1983. The NAD83 uses the **Geodetic Reference System (GRS80)** reference ellipsoid. This ellipsoid is earth-centered (center

Table 2.1 Reference Ellipsoids

Ellipsoid	Equatorial Radius (a) Nautical Miles	Equatorial Radius (a) Statute Miles	Polar Radius (b) Nautical Miles	Polar Radius (b) Statute Miles
Airy (1830)	3,443.609	3,962.56	3,432.104	3,949.32
Austrian Nat'l-South Am. (1969)	3,443.931	3.962.93	3,432.384	3,949.64
Bessel (1841)	3,443.520	3,962.46	3,432.010	3,949.21
Clarke (1866)	3,443.957	3.962.96	3,432.281	3.949.53
Clarke (1880)	3,443.980	3,962.99	3,432.245	3,949.48
Everest (1830)	3,443.454	3,962.38	3,432.006	3,949.21
Geodetic Reference System (1980)	3,443.939	3,962.94	3,432.392	3,949.65
International (1924)	3,444.054	3,963.07	3,432.459	3,949.73
Krassowsky (1940)	3,443.977	3,962.98	3,432.430	3,949.70
World Geodetic System (1972)	3,443.917	3,962.92	3,432.370	3,949.62

Note: The nautical mile as adopted by the International Hydrographic Bureau is 6,076.12 feet, and is approximately 1.1507 statute miles. The statute mile is 5,280 feet.

Source of Table: Defense Mapping Agency, Hydrographic Center, *American Practical Navigator,* Volume 1, Washington, D.C.: Defense Mapping Agency, 1977, pp. 117–120; John P. Snyder, *Map Projections—A Working Manual.* United States Geological Survey Professional Paper 1395. Washington, D.C.: USGPO, 1987, p. 12.

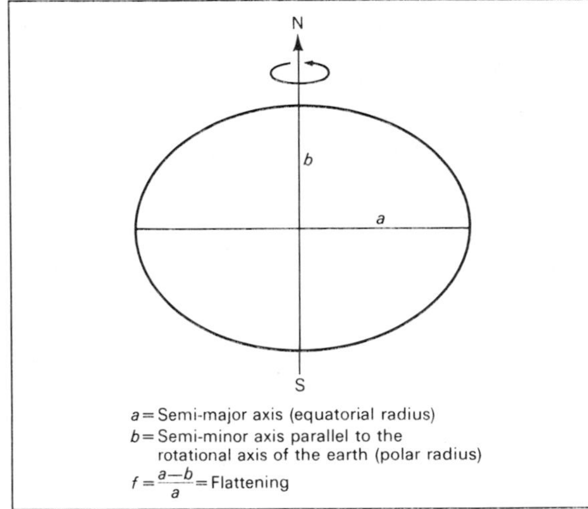

Figure 2.3. The ellipsoid.
Reference ellipsoids are chosen to represent the size and shape of the earth.

of mass), and does not have a single origin (a place where the ellipsoid theoretically touches the geoid), as does the Clarke ellipsoid of 1866.[3]

As a result of adopting the new datum, almost all places in North America will have a slightly different latitude and longitude. These shifts (some may be as much as 300 meters) will be noted at the margins of newly printed large-scale topographic maps. Table 2.2 shows the dimensions of the earth as described by the GRS80 reference ellipsoid.

Although any discussion of the true size and shape of the earth is interesting and useful to the thematic cartographer, we need not dwell on it at length, since we can assume the earth to be a perfect sphere for purposes of practically all cartography. The development of the base map begins from a small model of the real

Mean Radius (2a+b)/3 Statute Miles	Ellipticity (flattening) f=1−b/a	Where Used
3,958.15	1/299.32	Great Britain
3,958.50	1/298.25	Australia, South America
3,958,04	1/299.15	China, Korea, Japan
3,958.48	1/294.98	North America, Central America, Greenland
3,958.49	1/293.46	Much of Africa
3,957.99	1/300.80	India, Southeast Asia, Indonesia
3,958.51	1/298.25	Newly adopted for North America
3,958.63	1/297.00	Europe, Individual States in South America
3,958.56	1/298.30	U.S.S.R.
3,958.49	1/298.26	NASA; U.S. Department of Defense; oil companies

Table 2.2 Earth Measurements (after GRS80 ellipsoid)

	Statute Mi (U.S.)	Kilometers
Equatorial diameter	7,926.59	12,756.27
Polar diameter	7,900.01	12,713.50
Equatorial circumference	24,902.13	40,075.00
Polar circumference	24,818.64	39,940.64
Area*	197,000,000 sq mi	509,000,000 sq km

*Obtained by using the main radius (3,956.65 mi) of the ellipsoid.

earth. If we assume this model to be, for example, an ellipsoid whose semi-major axis is 30.48 cm (12 in), then its semi-minor axis will be 30.37 cm (11.5 in)—a difference of only .10 cm (.04 in). For most projects in thematic cartography, this introduces precision not ordinarily obtainable, nor even necessarily desirable.

Coordinate Geometry for the Cartographer

Location was the key idea behind the historical development of the earth's coordinate geometry. Pre-Christian astronomers were naturally concerned with this question as they delved

Until 1929, the ***nautical mile*** was equivalent to the length of one minute of arc along a great circle on a sphere having a surface area equal to that of the Clarke reference ellipsoid of 1866. This equaled 6082.2 feet. Actually, on the Clarke spheroid of 1866, one minute of latitude varies from 6,046 feet at the equator to 6,108 feet at the poles. This nautical standard was changed by international agreement in 1929 to 1852 meters (6,076.12 feet) and became known as the ***international nautical mile.*** For most purposes, however, one minute of latitude or along any great circle arc is customarily referred to as a nautical mile. The ***land*** or ***statute mile,*** in the United States, is 5,280 feet. This mile is also used for navigation on inland waters and lakes. One nautical mile is equal to 1.1507 statute miles. Since there are 1852 meters in a nautical mile, there are 1610.4 meters in a statute mile. Therefore, one statute mile equals 1.6 kilometers.

into questions related to the earth's size and shape. During the fifteenth, sixteenth, and seventeenth centuries, when exploration flourished, exactness in ocean navigation and location became critical. Death often awaited mariners who did not know their way along treacherous coasts. Naval military operations in the seventeenth and eighteenth centuries also required precise determination of location on the globe, as is the case today. Considerable sums of money are now spent on orbiting satellites that can beam locational information to earth. The United States, for example, has launched several satellites of a new Global Positioning System (GPS) which will serve the locational requirements of the military. Precise location within 10 meters and timing within 100-billionths of a second are possible. It is even anticipated that hand-held receiving units will be operational in the future. The determination of precise location on the earth has occupied the minds of men for a long time.

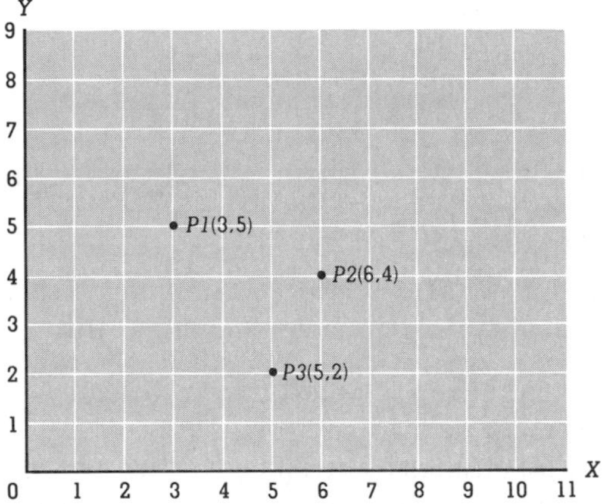

Figure 2.4. The cartesian coordinate system.
Points (P1, P2, P3, and so on) can have their exact locations defined by reference to the grid. Absolute and relative locations are therefore easy to determine.

Plane Coordinate Geometry

Perhaps the best way to introduce the earth's spherical coordinate system is to examine plane coordinate geometry. **Descartes,** a French mathematician of the seventeenth century, devised a system for geometric interpretation of algebraic relationships. This eventually led to the branch of mathematics called analytic geometry. It is from his contributions that we have **cartesian coordinate geometry.** This system of intersecting perpendicular lines on a plane contains two principal axes, called the *x*- and *y*-axis. (See Figure 2.4.) The vertical axis is usually referred to as the *y*-axis and the horizontal the *x*-axis.

The plane of cartesian space is marked at intervals by equally spaced lines. The position of any point (*Pxy*) can be specified by simply indicating the values of *x* and *y* and plotting its location with respect to the values of the cartesian plane. In this manner, each point can have its own unique, unambiguous location. Relative location can easily be shown by plotting several points in the space. Cartography uses cartesian geometry in a variety of ways. In ad-

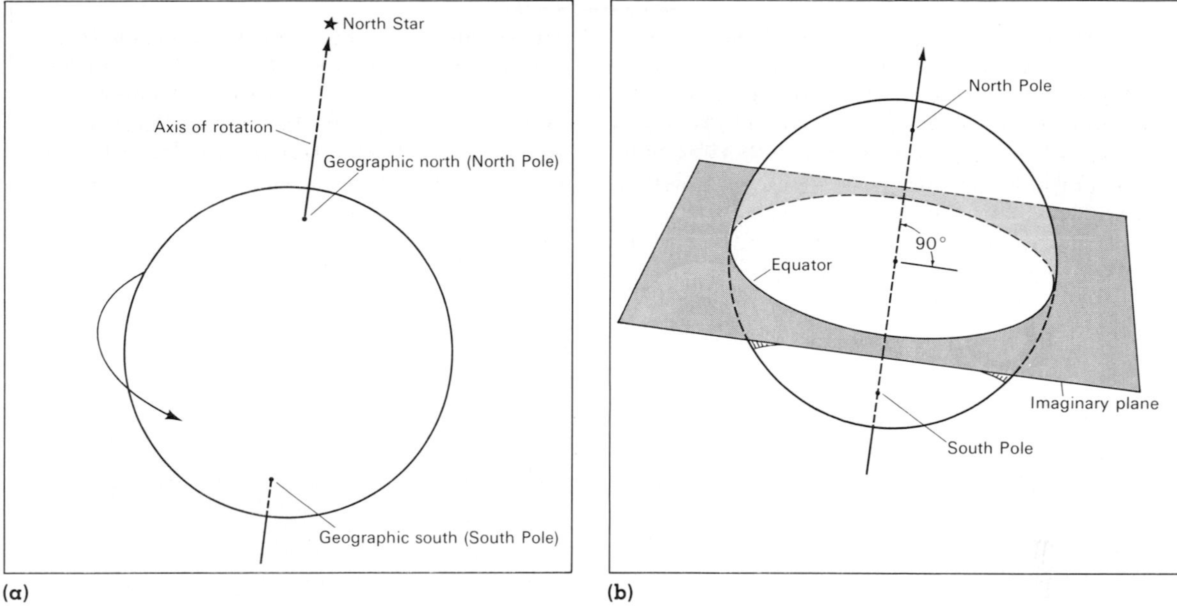

Figure 2.5. The determination of earth's poles (a) and the equator (b).

dition to being a good method of introducing the earth's spherical coordinate system (because of similarities), cartesian geometry is also of use in the development of computer maps, a process called **digitizing.** Digitizing involves specifying the locations of geographical points on a map in plane cartesian space.[4] The cartesian system is also used in hand-plotting map projections. In fact, the drafting work station that commonly employs the T-square and triangle has the cartesian system as a structural foundation. Finally, the thematic cartographer often deals with statistical concepts that are best portrayed in cartesian space.

Earth Coordinate Geometry

Concepts similar to those used in plane or cartesian coordinate geometry are incorporated in the earth's coordinate system. The earth's geometry is somewhat more complex because of its spherical shape. Nonetheless, it can be easily learned. To specify location on the earth (or any spherical body), angular measurement must be used in addition to the elements of the ordinary plane system. Angular measurement is based on a **sexagesimal** scale: division of a circle into 360 degrees, each degree into 60 minutes, and each minute into 60 seconds.

Our planet rotates about an imaginary axis, called the **axis of rotation.** (See Figure 2.5a.) If extended, one of the axes would point to a fixed star, the North Star. The place on earth where this axis of rotation emerges is referred to as **geographic north** (the North Pole). The opposite, or **antipodal point,** is called **geographic south** or the South Pole. These points 2are very important because the entire coordinate geometry of the earth is keyed to them.

If we were to pass through the earth an imaginary plane that bisected the axis of rotation and was perpendicular to it, the intersection of the plane with the surface of the earth would form a complete circle (assuming that the earth is perfectly spherical). This imaginary circle is referred to as the earth's **equator.** (See Figure 2.5b.) The North and South Poles and the equator are the most important elements of the earth's coordinate system.

Converting conventional angular measure of the geographic coordinate system into decimal degrees is a fairly simple task. Decimal degrees are often useful for navigation and other map or location work. Since each degree contains 60 minutes and each minute contains 60 seconds, there are 3600 seconds in a degree. If 52°17′32″ is to be converted, first multiply 17 × 60 (= 1020), then add 32 to 1020 (= 1052). A ratio is then computed, 1052/3600 (= .2922). The answer is 52.2922°. For practice, convert 17°52′08″ into decimal degrees. Convert 18.52° into conventional notation.

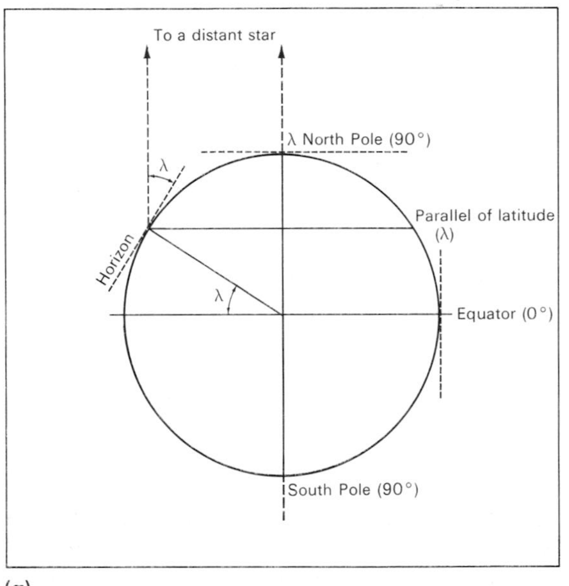

(a)

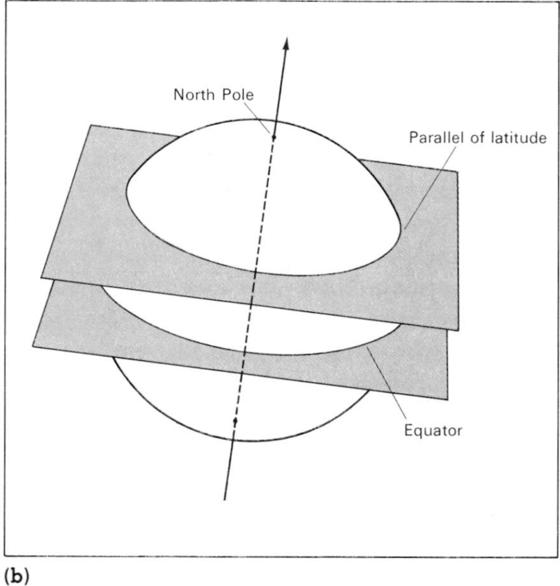

(b)

Figure 2.6. Latitude determination and developing parallels.
Latitude is a function of the angle between the horizon and the North Star (a). Parallels are generated by passing planes through the earth parallel to the equator (b).

Latitude Determination

Latitude is simply the location on the earth's surface between the equator and either the North or the South Pole. Latitude determination is easily accomplished: It is a function of the angle between the horizon and the North Star (or some other fixed star). (See Figure 2.6a.) As one travels closer to the pole, this angle increases. It can be demonstrated that the angle subtended at the center of the earth between a radius to any point on the earth's surface and the equator is identical in magnitude to the angle made between the horizon and the North Star at that location. If an imaginary plane is passed through this point parallel to the equatorial plane, it will intersect the earth's surface forming a **small circle,** or a **parallel of latitude.** (See Figure 2.6b.) There are, of course, an infinite number of these parallels, and every place on the earth can have a parallel. Conceptually, the parallels of latitude are *x*-axes similar to those of plane cartesian space.

Latitude designation is in angular degrees, from 0° at the equator to 90° at the poles. It is customary to label with a capital N or S the position north or south of the equator. Thus common latitude designations would be 82°N, 16°S, 47°N, and so on. Angular degrees are subdivided into minutes and seconds; exact latitudinal positions on the earth require minute and second determination.

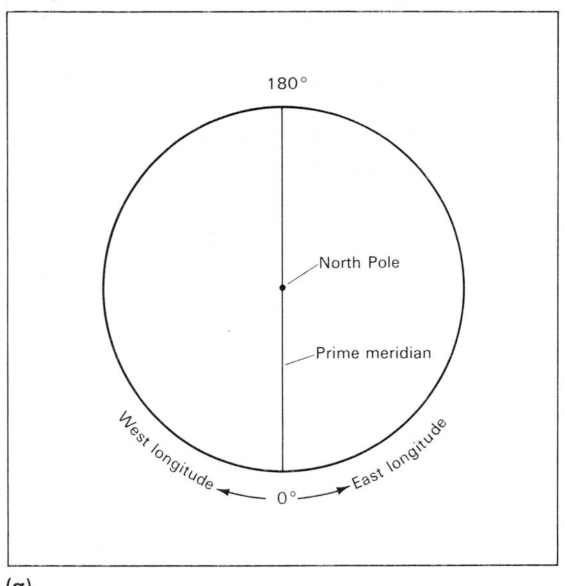

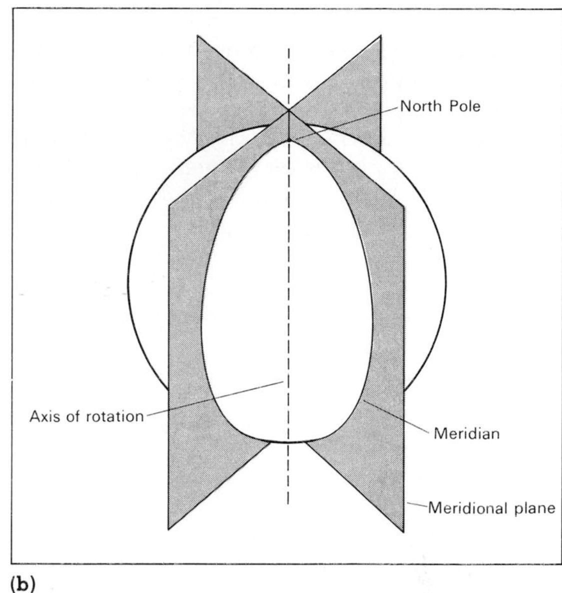

Figure 2.7. Longitude determination and developing meridians.
Longitude position is designated as 0°–180° east or west of the prime meridian (a). Meridians are generated by passing planes through the earth so that they intersect the axis of rotation in a line (b).

Longitude Determination

Longitude on the earth's surface has always been more difficult to determine than latitude. It baffled early astronomers and seamen, not so much for its concept, but for the instrumentation required to record it. Because the earth rotates on its axis, there is no fixed point at which to begin counting position. Navigators, cartographers, and others from the fourteenth to the seventeenth centuries knew that in practice they would need a fixed reference point. They also knew that the earth rotated on its axis approximately every 24 hours. Any point on the earth would thus move through 360 angular degrees in a day's time, or 15 degrees in each hour. If a navigator could keep a record of the time at some agreed-upon fixed point and determine the difference in time between the local time and the point of reference, this could be converted into angular degrees and hence position.

The concept was simple enough, but the technology of measuring time was slow in coming. In 1714, the British Parliament and its newly formed Board of Longitude announced a competition to build and test such an accurate timepiece. After several unsuccessful attempts, John Harrison built his famous **marine chronometer** in 1761. It was accurate to within 1.25 nautical miles. The puzzle of longitude was solved.[5]

At first, each country specified some place within its boundaries as the fixed reference point for reckoning longitude. By international agreement, most countries now recognize the meridian passing through the British Royal Observatory at Greenwich as the fixed or reference line. Some countries still retain local reference lines.

If an imaginary plane is passed through the earth so that it intersects the axis of rotation in a line, it will intersect the surface of the earth as a complete circle. One half of this circle, from pole to pole, is called a **meridian.**(See Figure 2.7.) The meridian passing through Greenwich is referred to as the **prime meridian** and has the angular designation 0°. Al-

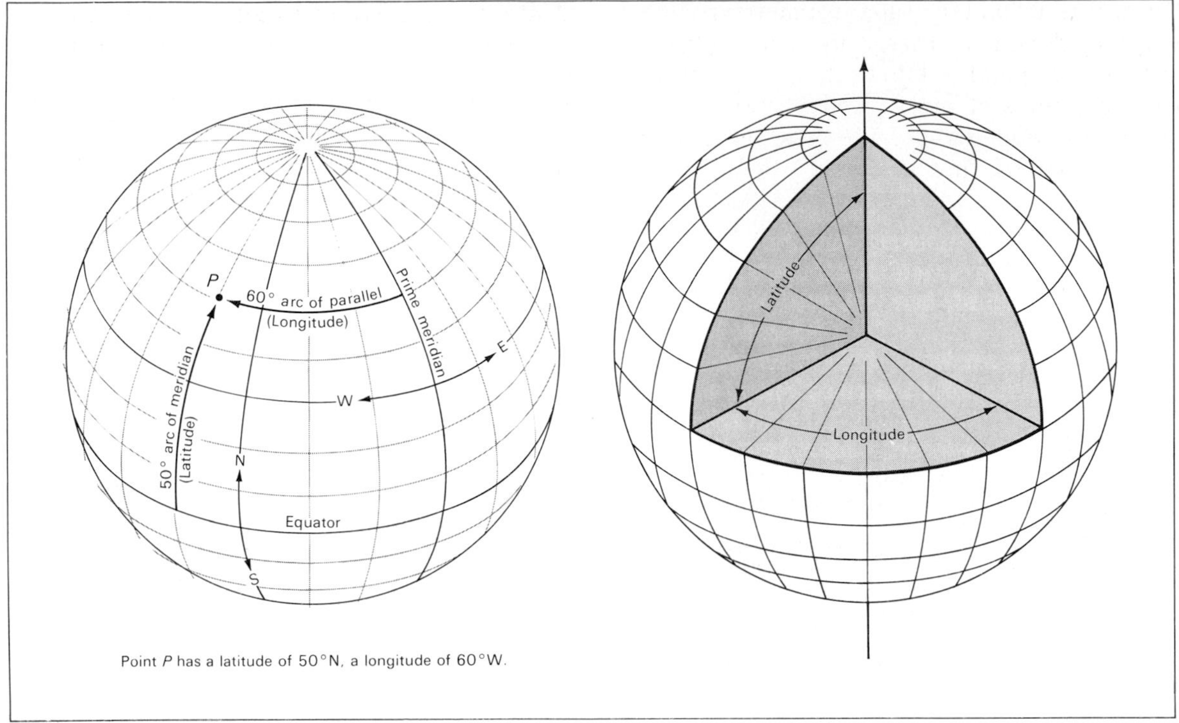

Figure 2.8. The complete geographic grid.
The similarity of the geographic grid to the cartesian grid is evident.

though there are 360 angular degrees in a circle, longitude position is designated as 0° −180° east or west of the prime meridian. Separating east and west longitude on the side of the earth opposite the prime meridian is the meridian of the International Dateline, along which the date changes. Meridians can be thought of as corresponding to the *y*-axis of the plane coordinate system.

Until the advent of radio after World War I and radar after World War II, navigation and reckoning one's position on the earth were accomplished by "shooting the stars" (celestial navigation), by marine chronometer, and by compass. These still form the basis for navigation, but radio, radar, and satellite telemetering are more often used today. The precision made possible by satellite-beamed signals is remarkable.

With the system of parallels and meridians just described, also called the **geographic grid,** the earth's spherical coordinate geometry is complete. (See Figure 2.8.) Its similarities to the plane cartesian system should be apparent. The earth coordinate system must be understood if the thematic cartographer is to select map projections wisely. From a description of the elements of spherical coordinate geometry, we now turn to the geometrical relationships among these elements. Knowledge of how these elements are arranged with respect to each other is essential.

Principal Geometric and Mathematical Relationships of the Earth's Coordinate Grid

The thematic cartographer is interested in the geographic grid and its portrayal as a map projection because it must be understood in order to render the best possible thematic map for the purpose at hand. The appropriateness of the final map depends in large measure on how well the cartographer knows the relationships of the elements of the grid. The present discussion is intended to provide the student with a basic understanding.

Linear

Lengths of lines of the spherical grid have fixed relations to each other. The most important length is the magnitude of the radius (r); from this and easily learned formulas, most other line lengths can be calculated. For example, the diameter is $2r$. For the perfect sphere, the polar radius is identical to the equatorial radius. There is only one circumference, and it is equal to $2\pi r$ ($\pi = 3.1415$). On the real earth, of course, the equatorial circumference does not equal the meridional circumferences because of flattening at the poles.

Meridonial lengths are the easiest to handle. On the perfect sphere, a meridian is one-half the circumference of the globe. Normally, we only want to deal with the length of the degree along a meridian. On the perfect sphere, this length is simply the circumference divided by 360; every degree is equal to every other degree. On the real earth, however, polar flattening causes the radius of the arc of the meridian to change, fitting it to an ellipse. Consequently, the length of the degree along the meridian is not constant. (Appendix A describes these lengths based on the Clarke ellipsoid of 1866.) The meridian is equal to one-half the circumference—or half the length of the equator. Knowing this is critical to the evaluation of projection properties.

Parallels of latitude also have fixed linear relationships. No parallel in one hemisphere is equal in length to any other. Parallels decrease in length at high latitudes. This relationship has a very definite mathematical expression, namely,

Length of parallel at latitude λ = (cosine of λ) × (length of the equator)

The length of the degree of the parallel is determined by dividing its whole extent by 360.

Table 2.3 Length of Parallels on a Perfect Sphere ($r = 1.0$) Circumference (Equator) $= \pi d = 6.28318$

Latitude (°)	Length of Parallel	Percent of Equatorial Length
0 (Equator)	6.2831	100.00
5	6.2592	99.61
10	6.1877	98.48
15	6.0690	96.59
20	5.9042	93.96
25	5.6945	90.63
30	5.4414	86.60
35	5.1468	81.91
40	4.8132	76.60
45	4.4428	70.71
50	4.0387	64.27
55	3.6039	57.35
60	3.1416	50.00
65	2.6554	42.26
70	2.1490	34.20
75	1.6262	25.88
80	1.0911	17.36
85	.5476	8.71
90	.0000	.00

The cosine of 60° is .5. Thus the length of the degree along the 60° parallel is but one-half that at the equator. (See Table 2.3. This information can be used intelligently in assessing map projections and consequent distortions. The lengths of the degree along the parallels on the real earth are listed in Appendix A.)

Angular

Inspection of a globe printed with a geographic grid will reveal important angular characteristics of the grid's elements. Most notably, it can be easily seen that meridians converge poleward and diverge equatorward. Parallels, by definition, are parallel. What is more subtle is that meridians and parallels intersect at right angles. This is an important characteristic of the grid which can help in the evaluation of projection properties.

A special line on the globe is a **loxodrome,** which has a constant compass bearing. The equator, all meridians, and all parallels are loxodromes. Other loxodromes are special, and because they maintain constant compass bearings, they intersect all meridians at equal angles. Because meridians converge, special loxodrome tends to spiral toward the pole, theoretically never reaching it. Throughout history, the loxodrome has always been important in sailing; mariners often wish to maintain the same heading throughout much of a journey. Unfortunately, loxodromes do not follow the course of the shortest distance between points on the earth, which is a **great circle arc.** Navigators would usually approximate a great circle arc by subdividing it into loxodrome segments in order to reduce the number of heading changes during travel. Great circle arcs are followed very closely today, especially in airplane navigation.

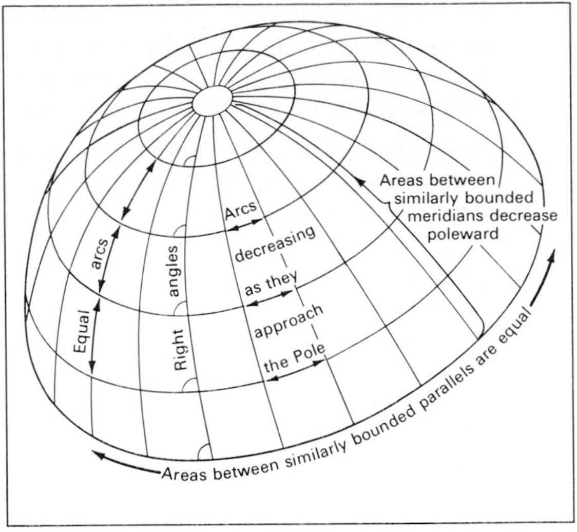

Figure 2.9. The principal geometrical relationships of the earth's coordinate system.

Area

The areas of quadrilaterals found between bounding meridians and parallels are important in understanding the areal aspects of the earth's spherical coordinate system. (See Figure 2.9.) Between two bounding meridians, for identical latitudinal extents, the quadrilateral areas decrease poleward. Any misrepresentation of this feature during projection will have a profound effect on the appearance of the final land/water areas of the map. The relationship of these areas on a perfect sphere, based on change of latitude, is represented in Table 2.4. The decrease in area relative to the lowest quadrilateral is more dramatic at the higher latitudes because of rapid meridional convergence. At lower latitudes, meridians converge slowly; as a result, the change in area is less marked.

Table 2.4 Areal Relationships of Quadrilaterals Using the Geographical Grid on a Perfect Sphere (r = 1.0) 5° Longitudinal Extent

Latitude (°)	Area	Percent of Lowest Quadrilateral
0–5	.007605	100.00
5–10	.007547	99.23
10–15	.007432	97.72
15–20	.007260	95.46
20–25	.007033	92.47
25–30	.006752	88.78
30–35	.006420	84.41
35–40	.006039	79.41
40–45	.005612	73.79
45–50	.005143	67.62
50–55	.004634	60.93
55–60	.004090	53.78
60–65	.003515	46.21
65–70	.002913	38.30
70–75	.002289	30.09
75–80	.001647	21.66
80–85	.000993	13.06
85–90	.000332	4.36

Points

Perfect spheres are considered to be *allside surfaces,* on which there are no differences from point to point. Every point is like every

Familiarity with the spherical grid and the characteristics of the arrangement of meridians and parallels is important in estimating graticule distortion on the flat map. The student is encouraged to inspect a globe and its grid. These important generalizations should be noted:

1. **Scale is everywhere the same on the globe; all great circles will have equal lengths; all meridians are of equal length and equal to the equator; the poles are points.**
2. **Meridians are evenly spaced on parallels; meridians converge toward the poles and diverge toward the equator.**
3. **Parallels are parallel and are equally spaced on the meridians.**
4. **Meridians and parallels intersect at right angles.**
5. **Quadrilaterals that are formed between any two parallels and that have equal longitudinal extent have equal areas.**
6. **The areas of quadrilaterals between any two meridians and between similarly spaced parallels decrease poleward and increase equatorward.**

other point, with the surface falling away from each point in a similar manner everywhere. In dealing with spheres and with the projection of the spherical grid onto a plane surface, we may think of points as having dimensional qualities. That is, they are commensurate figures, though infinitesimally small.

Circles on the Grid

Two special circles appear on the spherical grid. A *great circle* is formed by passing a plane through the center of the sphere. This plane forms a perfect circle where it intersects with the sphere's surface. There are certain qualities about the circle that are worth knowing: (a) The plane forming the great circle bisects the spherical surface. (b) Great circles always bisect other great circles. (c) An arc segment of a great circle is the shortest distance between two points on the spherical surface. Some of the elements of the geographic grid are great circles: All meridians are great circles, as is the equator.

Circles on the grid that are not great circles are called **small circles.** Parallels of latitude are small circles. On the earth, to travel along a meridian (N–S) is to go by the shortest distance. Traveling along a parallel (E–W) is *not* the shortest distance! Following the path of the equator, however, is an efficient way of travel.

An Introduction to Map Projections for the Designer

Map projections are one of the more interesting aspects of thematic mapping. They have occupied the thinking of mathematicians, cartographers, and geographers for many centuries, and new ones appear frequently. Projections are important for the designer, for they are the locational framework of the thematic map. Much as steel girders make up the skeleton of a modern skyscraper, the map projection provides the framework on which the remainder of the map rests. Projections can serve to focus the map reader's attention, to amplify, and to provide selective detail for the map's message. Every thematic map has a projection, and understanding them should be part of the education of a map designer.

The Map Projection Process

In thematic mapping, as in all cartography, we consider the production of a map a process of representing the earth (or a part of it) as a *reduced model of reality.* This involves at least four alterations. (See Figure 2.10.) The earth's real shape (geoid) is represented by an ellipsoid of reference through the adoption of the magnitudes of the major and minor semi-axes

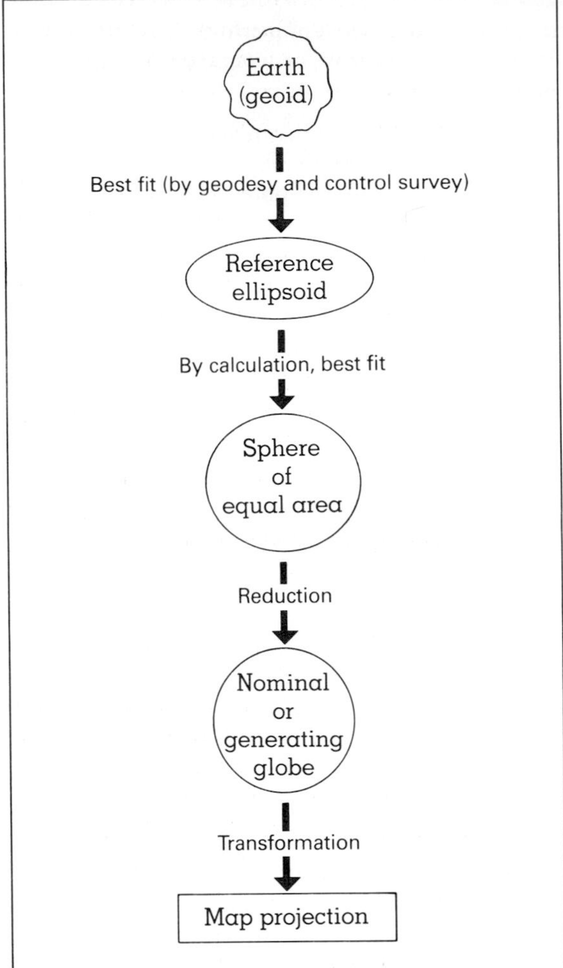

Figure 2.10. The map projection process. The final development of a map projection is the result of several best-fit and reduction alterations of the actual size and shape of the earth.

that best fit this real shape. For most of cartography, a sphere having the same surface area as the ellipsoid is chosen as a standard and its radius calculated. Cartographers reduce this spherical model to a **reference globe** (also called a *nominal* or *generating globe*), from which a map projection is generated.

Map projection is the transformation of the spherical surface to a plane surface; it occurs at the last step in the series of alterations from the earth to the flat map. *Any map projection is the systematic arrangement of the earth's (or generating globe's) meridians and parallels onto a plane surface.* Together, these spherical elements, the meridians and parallels, are called the projection **graticule.**

Nominal Scale

In Chapter 1, we introduced the concept of map scale. These additional comments will also prove useful to an understanding of scale. The generating globe chosen has a **nominal** or **defined scale.** In cartography, scale is represented by a fraction:

$$\text{map scale} = \frac{\text{map distance}}{\text{earth distance}}$$

It is generally expressed as a representative fraction (RF), and will always contain unity in the numerator. Fractions such as 1:25,000, 1:50,000, or 1:50,000,000 are examples of RF scales. They are to be read, "One unit on the map *represents* so many of the *same* units on the earth." The number in the fraction may be in any units, but both numerator and denominator will be in the same units. The cartographer should never say, "One unit on the map *equals* so many of the same units on the earth." This is incorrect and logically inconsistent.

Generating globe sizes will be chosen with a nominal scale in mind, usually one that describes the radius of the globe. A generating globe that has a 15-inch radius will have a nominal scale of 1:16,740,000:

$$\frac{\text{map distance}}{\text{earth distance}} = \frac{\text{globe radius}}{\text{earth radius}}$$
$$= \frac{15 \text{ inches}}{3963 \text{ miles} \times 63{,}360 \text{ in}}$$
$$= \frac{1}{16{,}740{,}000}$$
$$= 1{:}16{,}740{,}000$$

Every line and every point on the generating globe will have this same scale. Thus, all meridians and parallels, the equator, or other great circle arcs will be the same scale. This is a fundamental characteristic of the spherical or all-side geometrical figure.

There are three customary ways of expressing scale on a map: representative fraction, graphic, and verbal scale. *Representative fraction* (RF) is a ratio expressing the relationship of the number of units on the map to the number of the same units on the real earth. An example would be 1:500,000. The RF usually refers to the scale of a standard line and in fact changes over the map, depending on the selected projection.

On many maps, a *graphic* bar is included. This bar is usually divided into equally spaced segments and labeled with familiar linear units. This scale is read the same as an RF scale. The distance between any two divisions can be measured with a drafting scale (inches or centimeters); this distance represents the earth distance as labeled on the bar. Usually, the labeled units must be converted to those on the map and the ratio reduced to unity in the numerator. This form of scale is very useful when the map is to be reduced during reproduction since it changes in correct proportion to the amount of reduction.

Another common scale is the simple *verbal scale.* This is a simple expression on the face of the map stating the linear relationship. For example, "Five miles to the inch" is an example of a verbal scale. This scale form is easily converted to an RF scale between map and earth distances.

Regardless of the scale form chosen, *all maps usually should contain a scale.*

Two sample scale problems for practice:

1. Determine the fractional scale which is expressed verbally as:

   (A) 1 inch to 4 miles.

   Remember, the basic scale ratio is: map distance to earth distance (always expressed in the same units). The problem here is we have inches and miles. *Solution:* Convert 4 miles into inches (4 × 63,360 = 253,440), and the representative fraction becomes 1:253,440.

   (B) 4 inches to the mile.

   This is a bit more complicated. *Solution:* First, convert all units into the same measure. To change miles to inches, 1 mile × 63,360 = 63,360 inches. Now, we have 4 inches to 63,360 inches. This is not correct, though, because we need the numerator expressed as unity. Verbally, we say that if there are 4 inches to 63,360 inches, how many would there be for one? Let us simply cast two ratios (quotients) equal, like this:

   $$\underset{\textbf{ratio 1}}{\frac{4}{63{,}360}} = \underset{\textbf{ratio 2}}{\frac{1}{\text{unknown}}}$$

   We can do this because the quotient yielded in ratio 2 must be the same as in ratio 1 (we are not changing the *relationship,* only the numbers with which we are working). Then (by applying the cross multiplication rule of proportions), we get

   $$\textbf{unknown} = \frac{(1) \cdot (63360)}{4}$$

   $$\textbf{unknown} = \textbf{15,340}$$

   The answer, then, is 1:15,840.

2. Typical map scale problem:

   (A) The distance between two known points on a map is 5 miles. What is the scale of a map on which the points are 3.168 inches apart?

   Solution: At first glance, this may dismay you. Remember the first rule: cast the scale as a ratio of map distance to earth distance:

   $$\frac{\textbf{map distance}}{\textbf{earth distance}}$$

The earth distance given is 5 miles but change this to inches right away, 5 × 63360 = 316,800 inches. The map distance given is 3.168 inches, so our scale is

$$\frac{3.168}{316{,}800};$$

but this is not the correct form (numerator should be expressed as unity). So, then, do what we did in (1B) above, i.e.:

$$\begin{array}{ccc} \textbf{ratio 1} & = & \textbf{ratio 2} \\ \dfrac{3.168}{316{,}800} & & \dfrac{1}{\textbf{unknown}} \end{array}$$

then:

$$(3.168)\cdot(\text{unknown}) = (1)\cdot(316{,}800)$$

$$\text{unknown} = \frac{(1)\cdot(316{,}800)}{3.168}$$

$$\text{unknown} = 100{,}000$$

answer is: 1:100,000

Note: See the *simplicity* of all this? Just remember to cast the scale as the ratio of map distance to earth distance, note your given quantities, change the units, and solve.

Surface Transformation and Map Distortion

In the transformation process from the spherical surface to a plane, there occurs some distortion that cannot be completely eliminated. Although designers strive to develop the perfect map, free of error, *all* maps contain errors because of the transformation process. It is impossible to render the spherical surface of the generating globe as a flat map without distortion error caused by *tearing, shearing,* or *compression* of the surface. (See Figure 2.11.) The designer's task is to select the most appropriate projection so that there is a measure of control over the unwanted error.

These distortions and their consequences for the appearance of the map vary with scale. One can think of the globe as being made up of very small quadrilaterals. If each quadrilateral were extremely small, it would not differ significantly from a plane surface. For mapping small earth areas (large-scale mapping), distortion is not a major design problem. It can be ignored because it cannot be dealt with in the preparation of the final map.

As the mapped area increases to subcontinental or continental proportions, distortion becomes a significant design problem for the cartographer. In designing maps to portray the whole earth, we can no longer ignore this surface distortion. At such scales, the map designer must contend with alterations of area, shape, distance, and direction. No projection of the globe's graticule can maintain all of these properties simultaneously.

Equal-area Mapping

Map projections on which area relationships of all parts of the globe are maintained are called **equal-area** (or equivalent) projections. Linear or distance distortion often occurs in such projections. It is impossible for one projection to maintain both equivalency and conformality (shape preservation). On equal-area projections, therefore, shape is often quite skewed.

In cartography, it has become conventional to refer to scale as large, intermediate, or small. These terms are relative; rigorous numerical boundaries do not exist. Small scales are those that map large earth areas; a typical scale might be 1:30,000,000. An example would be a world map on an 8½ × 11-inch piece of paper. A large scale is 1:24,000 (or larger) and shows only small earth areas. The United States Geological Survey's topographic map series at 1:24,000 is an example. Such maps can show numerous physical and cultural features. Intermediate scales would be RFs between these two extremes.

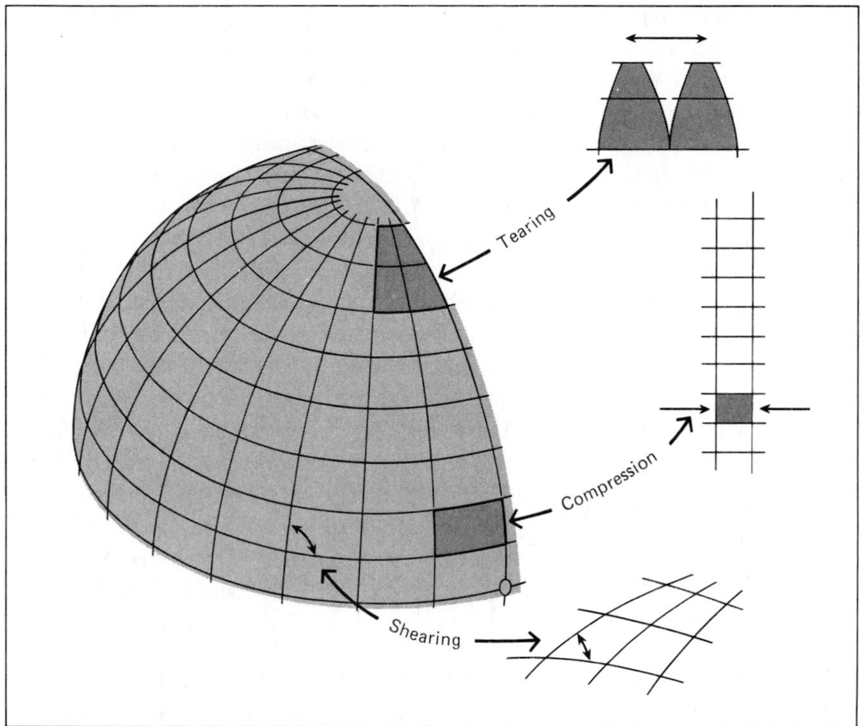

Figure 2.11. Distortion caused by the projection process.
Transformation from sphere to plane may cause distortion brought about by tearing, shearing, or compression.

Also, the intersections of meridians and parallels are not at right angles. Nonetheless, areas of similarly bounded quadrilaterals maintain correct area properties. The cartographer must select the most appropriate features for the design purpose.

Equivalent projections are very important for general quantitative thematic map work. It is usually desirable to retain area properties, because area is often part of the data being mapped. Population density is an example. Unless the map calls for unique projection attributes, equivalent projections have the greater all-around utility, regardless of the extent of the earth area mapped. Marschner wrote, several decades ago:

> It is hardly necessary to recapitulate here the arguments used in favor of equal-area representation. There is no doubt but that the preservation of a true areal expression of maps, used as a basis for land economic investigation and research, is more important than a theoretical retention of angular values. Of the three geometrical elements that can be considered in mapping, the linear, angular and areal, the last is the one around which land economic questions usually revolve. They do so for obvious reasons. Man does not inhabit a line, but occupies the area; he does not cultivate an angle of land, but cultivates and utilizes the land area. One of the principal functions of lines and angles is to define the boundary of the area. . . . They provide only the framework for controlling the relative position of features in the area, and with it a means for controlling the areal expression of the map itself.[6]

Conformal Mapping

Conformal (or orthomorphic) mapping of the sphere means that angles are preserved around points and that the shapes of small areas are preserved. The quality of orthomorphism applies to small areas (theoretically only to points). On conformal projections, meridians intersect parallels at right angles, and the scale is the same in all directions about a point. Scale may change from point to point, however. It is misleading to think that shapes over large areas can be held true. Although shapes for small areas are maintained, the shape of larger regions, such as continents, may be severely distorted. Areas are also distorted on orthographic projections.

The shape quality of mapped areas is an elusive element. If we view a continent on a globe so that our eyes are perpendicular to the globe at a point near the center of the continent, we see a shape of that continent. However, the shape of the continent is distorted because the globe's surface is falling away from the center point of our vision. We can view but one point orthographically at a time. If we select another point, the view changes, and so does our perception of the continent's shape. It becomes difficult for us to compare the shapes we see on a map to those on the globe, and it is safe to say that shapes of large areas on conformed projections are to be believed with caution.[7]

For large-scale mapping of small earth areas, distortion is not significant. Indeed, the choice between an equivalent or an orthomorphic projection becomes virtually a moot exercise. At intermediate to small scales, the selection of the projection is more critical in the design process. Even at these scales, however, it is seldom necessary to specify a conformal projection, except in rare circumstances. Mapping phenomena with circular radiational patterns may warrant such a choice. Radio broadcast areas, seismic wave patterns, or average wind directions would be examples.

Equidistance Mapping

The property of equidistance on projections refers to the preservation of great circle distances. There are certain limitations: Distance can be held true from one to all other points, or from a few points to others, but not from all points to all other points. The distance property is never global. Scale will be uniform along the lines whose distances are true. Projections that contain these properties are called **equidistant** projections.

Azimuthal Mapping

On **azimuthal projections,** true directions are shown from one central point to all other points. Directions or azimuths from points other than the central point to other points are not accurate. The quality of azimuthality is not an exclusive projection quality. It can occur with equivalency, conformality, and equidistance.

Azimuth is a directional relationship between two points on the surface of the earth and has special meaning in navigation. If we pass a great circle arc through two points, *A* and *B*, the azimuth of *B* relative to *A* is the angle formed by the intersection of the meridian and the great circle arc passing through *A*. The great circle arc will intersect other meridians along the path from *A* to *B* at different angles. Although it is the shortest distance between *A* and *B*, it is not a line of constant compass bearing (loxodrome).

Over the years, cartographers have stressed the idea that **minimum error projections** are best suited for general geographic cartography. These projections contain all of the four previously mentioned errors, to a greater or lesser degree, but as a whole provide us with a good picture of the globe or parts of it. These projections are chosen by the designer on much the same basis as a projection having a uniquely preserved property: by selection of those total qualities that best suit the mapping task.

Determining Deformation and Its Distribution Over the Projection

There are two chief methods available for determining projection distortion and its distribution over the map. One is to depict a

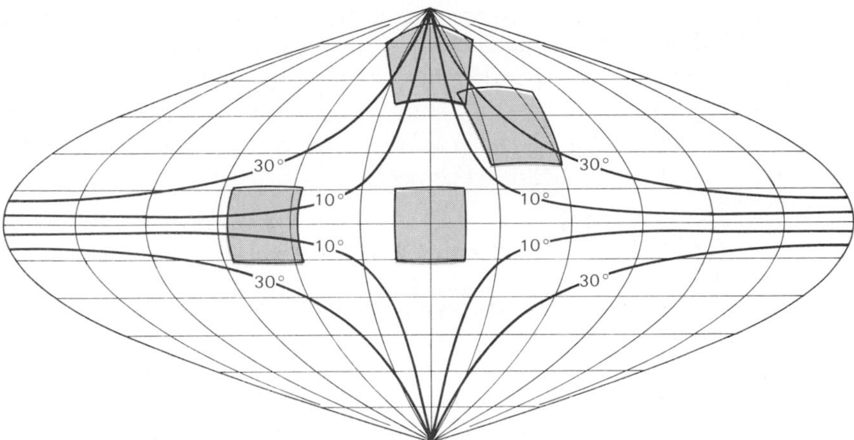

Figure 2.12. Shape distortion on an equal-area projection.
On the globe, the squares in the figure are all the same size. On this sinusoidal projection, the angular relations (and hence shape) are not preserved. Notice that distortion occurs most severely at the edges of the projection, as shown by the plot of the 2w values. (See text for explanation.)

Source: Squares have been drawn from information obtainable from Wellman Chamberlin, *The Round Earth on Flat Paper* (Washington, D.C.: National Geographic Society, 1947), p. 97.

geometrical figure (square, triangle, or circle) or familiar object (such as a person's head) and plot it at several locations on the projection graticule. (See Figure 2.12.) Distortion on the projection is readily apparent. This method is very effective and quite sufficient for most general cartographic analyses; its weakness is the lack of a general quantitative index of distortion.

Another method, conceptually and mathematically more complex, uses **Tissot's indicatrix.** Tissot, a French mathematician working in the latter part of the nineteenth century, developed a way to show distortion at points on the projection graticule.[8] Following his work, others have computed these indices and mapped them on several projections to show the patterns of distortion. The weakness of Tissot's method is that it is somewhat more complex mathematically than plotting simple geometrical shapes. Its strength lies in its quantitative ability to describe distortion.

The construct of Tissot's indicatrix consists of a very small circle, whose scale is unity (1.0), on the globe's surface. This small circle and two perpendicular radii appear on the plane map surface during transformation as a circle of the same size, a circle of different size, or an ellipse. For equal-area mapping, the circle is transformed into an ellipse with the ratio of the new semi-major and semi-minor axes such that its area remains as unity. Angular properties are not preserved. (See Figure 2.13.) On conformal projections, the small circle is transformed on the projection as a circle, although its size (area) varies over the map. Because the small circle is accurately projected as a circle, angular relationships are maintained. On some projections, the small circle becomes an ellipse that preserves neither equivalency nor conformality. Such projections are not classified as either equivalent or conformal.

For purposes of computation and explanation, the indicatrix has these qualities:

- Maximum angular distortion = 2ω
- Scale along the ellipse major axis = a
- Scale along the ellipse minor axis = b
- Maximum areal distortion = S

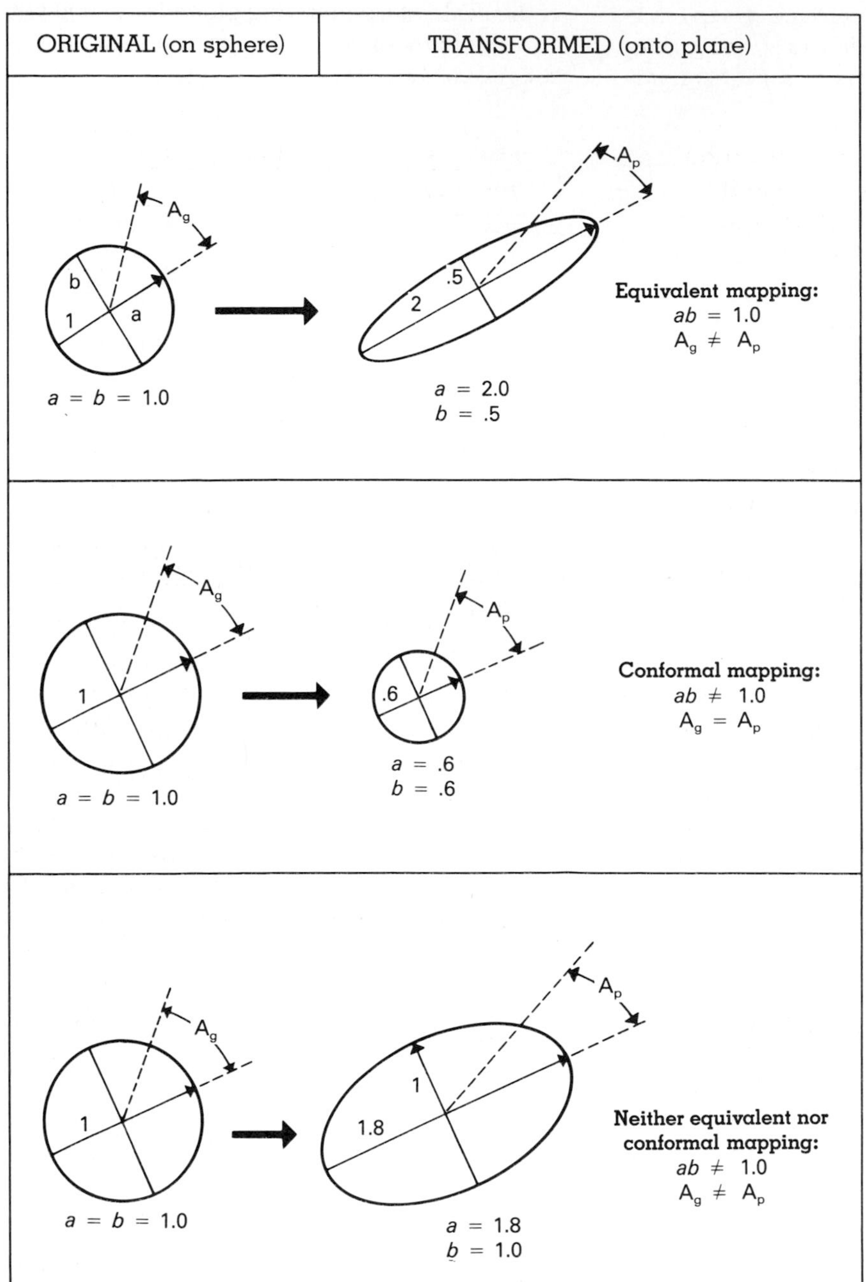

Figure 2.13. Hypothetical variants of the Tissot indicatrix. See the text for an explanation.

Every point on the new surface (map) has values computed for 2ω, *a, b,* and S. These conditions inform us of the distortion characteristics:

- If S = 1.0, there is no areal distortion, and the projection is equal-area. If S = 3.0, for example, *local* features are three times their proper areal size.
- If $a = b$ at all points on the map, it is a conformal projection. S varies, of course.
- If $a \neq b$, it is not conformal, and the amount of angular distortion is represented by 2ω.

Values of S or 2ω are usually mapped on projection graticules to illustrate the patterns of distortion. These will be discussed more fully below. Values of 2ω have been plotted on the sinusoidal projection in Figure 2.12.

Standard Lines and Points, Scale Factor

Standard lines are lines (usually meridians or parallels, especially the equator) on a projection that have identical dimensions to their corresponding lines on the nominal or reference globe. For example, if the circumference of the generating globe is 15 inches, the equator on the projection is drawn 15 inches long if it is a standard line. Scale can be thought to exist at points, so every point along a standard line has an unchanging or true scale when compared to the scale of the generating globe. On standard lines, scales are true. At all points on a standard line, using the indicatrix as a guide, $a = b$, $S = 1.0$, and $2\omega = 0°$.

In practice, the lines on the projection can be compared to their corresponding lines on the reference globe by a ratio called **scale factor** (S.F.):

$$\frac{\textbf{projection scale fraction}}{\textbf{nominal scale fraction}}$$

If the scale of a generating globe is 1:3,000,000 and the equator on the projection is drafted as a standard line, then

$$\frac{\textbf{projection scale fraction}}{\textbf{nominal scale fraction}} = \frac{\frac{1}{3{,}000{,}000}}{\frac{1}{3{,}000{,}000}} = \frac{3{,}000{,}000}{3{,}000{,}000} = 1.0 \text{ (S.F.)}$$

The scale factor of the equator will have a value of 1.0. For another example, suppose the nominal scale is 1:3,000,000 and a line on the projection has a linear scale of 1:1,500,000:

$$\frac{\textbf{projection scale fraction}}{\textbf{nominal scale fraction}} = \frac{\frac{1}{1{,}500{,}000}}{\frac{1}{3{,}000{,}000}} = \frac{3{,}000{,}000}{1{,}500{,}000} = 2.0 \text{ (S.F.)}$$

In this example, the line is two times *longer* than it should be, due to stretching. In our final example, the nominal scale is 1:3,000,000 and a line on the projection is drafted at 1:6,000,000:

$$\frac{\textbf{projection scale fraction}}{\textbf{nominal scale fraction}} = \frac{\frac{1}{6{,}000{,}000}}{\frac{1}{3{,}000{,}000}} = \frac{3{,}000{,}000}{6{,}000{,}000} = 0.5 \text{ (S.F.)}$$

Here we find the line to be one-half as long as its corresponding line on the globe. Compression has taken place in the transformation process.

Standard lines and scale factors are important to the overall understanding of projection distortion. The idea of scale factor can also be used in the assessment of areal scales. In this instance, small quadrilateral areas on the projection can be compared to the corresponding areas on the globe to determine the amount of *areal exaggeration* occurring on the projection.

Patterns of Deformation

Patterns of deformation on projections can best be approached by first looking at the different *projection families.* There are literally hundreds of individual projections. Some are quite old, such as the Mercator—dating back hundreds of years. Some have been devised within the last several years, most notably the Robinson pseudocylindrical.[9] Although there is great diversity among them, similarities in construction and appearance yield enough common elements to classify them into a few groups. The approach presented here is a conventional one.

Some map projections can be produced by descriptive, projective geometry. They can be developed with T-square and triangle on a drafting table. These **developable projections** form the basis for the classification presented here. There are three basic forms of developable surfaces: plane, cylinder, and cone. These yield the azimuthal, cylindrical, and conic projection families. (See Figure 2.14.) The purely mathematical projections (those that cannot be developed by projective geometry) are classified into these families on the basis of their appearances. There are also a few projections that bear striking resemblance to the developable ones but are enough different to be classed as pseudo-cylindrical, pseudo-conic, and pseudo-azimuthal.

Azimuthal Family

In the plane or azimuthal class, the spherical grid is projected onto a plane. This plane can be tangent to the sphere at a point (simple form), or pass through the sphere, making it tangent along a small circle (secant form). There are numerous versions of this group, distinguished from one another by the location of the assumed light source for the projection. In Figure 2.14, the light is shown emanating from the center of the globe; this is a gnomonic projection. If the light source is at the point opposite the point of tangency, the projection is stereographic; if it is at infinity (outside the generating globe), an orthographic projection results.

The plane may be tangent, of course, at any point on the spherical grid, depending on the **projection aspect.**[10] Tangency at the pole is a *polar aspect;* at mid-latitude, an *oblique aspect;* and at the equator, an *equatorial aspect.* The *normal aspect* is the position that produces the simplest graticule. Normal aspect for this family is the polar position when the plane is tangent at one of the poles. In this case, the meridians are straight lines intersecting the pole, and parallels are concentric circles having the pole as their centers. Directions to any point from the point of tangency (pole) are held true. All lines drawn to the center are great circles, as is also the case for equatorial and oblique aspects. Normally, only one hemisphere is shown on these projections. Some versions, for example the azimuthal equidistant map of the world, do indeed map the entire globe, but the grid departs radically from what we are accustomed to seeing. The point of tangency is a standard point; in the secant case, the standard line is the line of tangency. Scale deformation is nonexistent at these locations. Azimuthal (also called zenithal) projections became quite popular during World War II, when there was considerable circumpolar air navigation. They have remained so today.

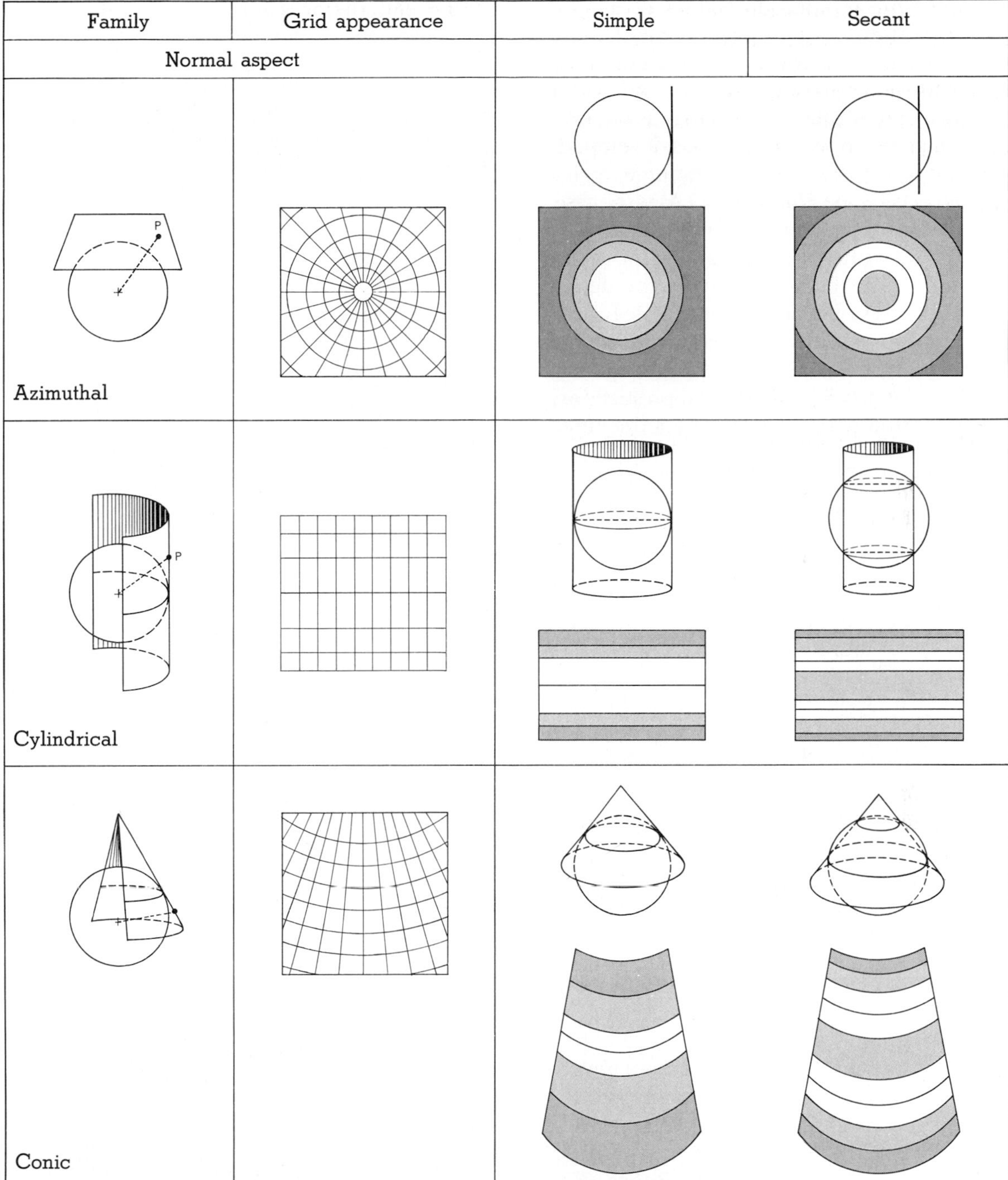

Figure 2.14. Projection families and patterns of deformation.
The patterns are clearly related to the way the projection is devised.

It is possible to compute values of *a, b,* S, and 2ω, plot these on a projection, and thereby see the pattern of deformation. Repetition of this procedure for many projections shows that the pattern of deformation is closely associated with the manner in which each was developed. Patterns of deformation begin to emerge for the azimuthal class. (See Figure 2.14.) As is the case for all projections, deformation increases with distance from either the standard point or the standard line. In the present case, the deformation increases outward in concentric bands. On conformal projections, areal exaggeration is extreme at greater distances from tangency. Likewise, for equivalent azimuthal projections, shape distortion reaches extremes at the outer edges.

Cylindrical Family

Cylindrical or retangular projections are common forms, frequently seen in atlases and other maps portraying the whole world. They are developed (graphically or mathematically) by wrapping a flat plane (sheet) into a cylinder and making it tangent along a line or lines on the sphere. Points on the spherical grid can be transferred to this cylinder, which is then unrolled into a flat map. The normal aspect for these projections is the equatorial aspect, with the equator as the standard line. In this case, the standard line is also a great circle. Oblique cases of this projection, notably the oblique Mercator, also have the tangent line as a great circle.

This class of projection is often used to show the worldwide distribution of a variety of geographical phenomena. A great deal of effort has therefore gone into the development of equivalent cylindrical projections. Most of the common ones used today are in fact pseudocylindrical versions. The popular ones include the Eckert family, Mollweide, Boggs Eumorphic, and Goode's Interrupted Homolosine. Two recent developments are the Robinson and Tobler pseudo-cylindricals.

The patterns of deformation for all rectangular projections depend on their method of development. (See Figure 2.14.) Areas of least distortion are bands parallel to the line(s) of tangency, with increasing exaggeration toward the outer edges of the map plane. Here, as with the azimuthal class, distortion may appear in area, angle, distance, or direction.

Conic Family

Conic projections are constructed by transferring points from the generating globe grid to a cone enveloped around the sphere. This cone is then unrolled into a flat plane. In the normal aspect, the axis of the cone coincides with the axis of the sphere. This aspect yields either straight or curved meridians that converge on the near pole and parallels that are arcs of circles. In the simple conic projection (normal aspect), the cone is tangent along a chosen parallel, along which there is no distortion. In the secant case, the cone intersects the sphere along two parallels. This reduces distortion.

One special form of the conic family is the polyconic. This projection uses several cones of development and consequently has several standard lines. Theoretically, each parallel is the base of a tangent cone. This form of polyconic projection is not conformal. It is often used for mapping areas of great latitudinal extent; the polyconic projection was used by the USGS in its topographic mapping program until being replaced in the 1950s.

The pattern of deformation includes concentric bands parallel to the standard parallels of the projection. (See Figure 2.14.) Secant conics tend to compress scale in areas between the standard lines and to exaggerate scale elsewhere. There are a number of very useful conic projections, one of which will be discussed at length below. Conic projections, simple or secant, are best for mapping earth areas having greater east–west extent than north–south.

Analysis of a variety of conic projections suggests that the **Albers conical equal-area**

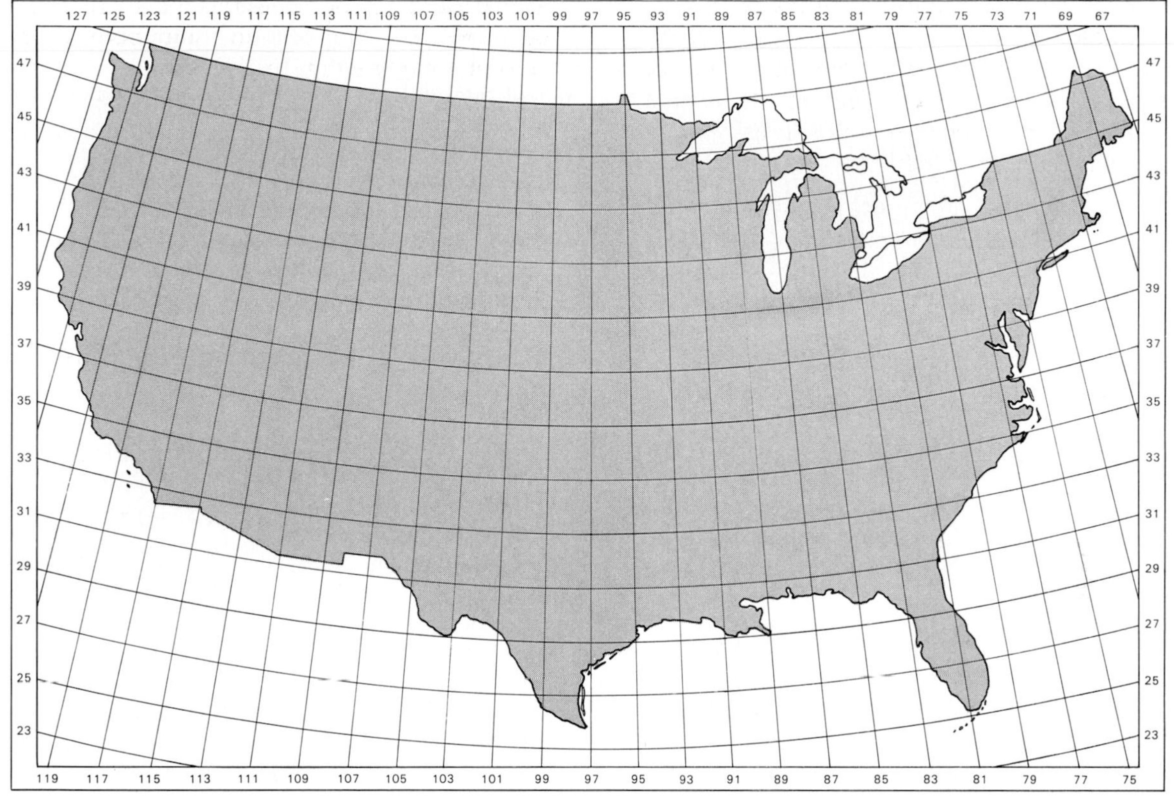

Figure 2.15. The United States, plotted on an Albers conical equal-area projection. Angular distortion is minor, nowhere exceeding 2°. This projection is frequently used for mapping the United States.

projection, with standard parallels at 29½° and 45½°, is a good choice for mapping the United States. (See Figure 2.15.) This projection, at scales of 1:7,500,000, 1:17,000,000, and 1:34,000,000, was used exclusively for mapping the United States in the *National Atlas of the United States of America.*[11] The reasons for its use, as stated by the atlas design team, were as follows:

1. Equivalent projection.
2. Simple to construct.
3. Easy to segment and reassemble.
4. Contains small errors in scale.
5. Long been popular in mapping the United States, with its large east–west expanse.

Indicatrix values for this projection suggest very little distortion in an area of the size of the United States. Values of 2ω nowhere exceed 2°, and linear values (a,b) are very nearly 1.0. S, the index value for areal change, is 0.0, as this is an equivalent projection. Thematic cartographers seeking a projection suitable for mapping the continental United States will find a most useful projection in the Albers equal-area.

Notes

1. Lloyd A. Brown, *The Story of Maps* (New York: Bonanza Books, 1959), pp. 28–29.
2. Desmond King-Hele, "The Shape of the Earth," *Science* 183 (1967): 67–76; and Desmond King-Hele, "The Shape of the Earth," *Science* 192 (1976): 1293–1300.
3. John P. Snyder, *Map Projections—A Working Manual.* U.S. Geological Survey, Professional Paper 1395 (Washington, D.C.: USGPO, 1987), p. 13.
4. Mark S. Monmonier, *Computer-Assisted Cartography, Principles and Prospects* (Englewood Cliffs, NJ: Prentice-Hall, 1982), p. 195.
5. Brown, *The Story of Maps,* pp. 209–40.
6. F. J. Marschner, "Structural Properties of Medium- and Small-Scale Maps," *Annals* (Association of American Geographers) 34 (1944): 44.
7. Borden D. Dent, "Continental Shapes on World Projections: The Design of a Poly-Centred Oblique Orthographic World Projection," *The Cartographic Journal* 24 (1987): 117–24.
8. A. Tissot, *Memoire sur la Representation des Surfaces et les Projections des Cartes Geographiques* (Paris: 1881).
9. Arthur H. Robinson, "A New Map Projection: Its Development and Characteristics," *International Yearbook of Cartography* 14 (1974): 145–55.
10. James A. Hilliard, Umit Bosoglu, and Phillip C. Muehrcke, *A Projection Handbook* (Madison, WI: University of Wisconsin at Madison, Cartographic Laboratory, 1978), pp. 3–5.
11. United States Department of the Interior, Geological Survey, *National Atlas of the United States of America* (Washington, D.C.: USGPO, 1970).

Glossary

Albers conical equal-area projection customarily used for small-scale mapping of the United States; a good selection for mapping any earth area at mid-latitude with considerable east–west extent

antipodal point point opposite; on the earth, the north pole is antipodal to the south pole; 20°S, 60°E is antipodal to 20°N, 120°W

axis of rotation imaginary line around which the earth rotates

azimuthal projection directions from the projection's center to all points are correct; also called a zenithal projection

cartesian coordinate geometry system of intersecting perpendicular lines in plane space, useful in analytic geometry and the precise specification of location

Clarke ellipsoid of 1866 a reference ellipsoid used by the United States and other countries of North America

conformal projection preserves angular relationships at points during the transformation process; cannot be equal-area; also called an orthomorphic projection

Descartes French mathematician whose early studies of algebra and geometry led to analytic geometry

developable projection can be constructed using the ordinary means of the draftsperson and the principles of projective geometry

digitizing transforming spatial elements of a map—or other two dimensional images—into *x*- and *y*-coordinates of cartesian space

ellipsoid a geometrical solid developed by the rotation of a plane ellipse about its minor axis

equal-area map projection no areal deformation; cannot be conformal; also called an equivalent projection

equator imaginary line of the earth's coordinate system that is formed by passing a plane through the center of the earth perpendicular to the axis of rotation, midway between the poles

equidistant projection preserves correct linear relationships between a point and several other points, or between two points; cannot show correct linear distance between all points to all other points

Eratosthenes (276–194 B.C.) Greek scholar living in Alexandria who first accurately measured the size of the earth

geodesy the science that measures the size and shape of the earth; often involves the measurement of the external gravitational field of the earth

geographic grid spherical coordinate system used for the determination of location on the earth's surface

geographic north and south the imaginary line forming the earth's axis of rotation intersects the earth's surface at two locations, the North and South Poles, referred to as geographic north or south

geoid term used to describe the shape of the earth; means "earth-shaped" and does not refer to a mathematical model

graticule meridians and parallels on a map projection

great circle arc segment of a great circle that is the shortest distance between two points on the spherical surface

GRS80 Geodetic Reference System reference ellipsoid used by the United States in the NAD83 datum.

latitude position north or south of the earth's equator; designation is by identifying the parallel passing through the position; determined by the angle subtended at the center of a sphere by a radius drawn to a point on the surface

longitude position east or west of the prime meridian; designation is by identifying the meridian passing through the position; determined by angular degrees subtended at the center of a sphere by a radius drawn to the meridian and the position in question

loxodrome a line on the earth that intersects every meridian at the same angle; because of meridional convergence, a loxodrome theoretically never reaches the pole; also called a rhumb line

map projection the systematic arrangement of the earth's spherical or geographic coordinate system onto a plane; a transformation process

map projection deformation results from the transformation of the spherical surface to the plane surface, and may include tearing, shearing, or compression

marine chronometer extremely accurate timepiece used to determine longitude; perfected by John Harrison in 1761

meridian great circle of the earth's geographical coordinate system formed by passing a plane through the axis of rotation; meridional number designation ranges from 0°–180° E or W of the prime meridian

minimum error projection no equivalency, conformality, azimuthality, or equidistance; chosen for its overall utility and distinctive characteristics

nominal scale the scale of the reference globe, expressed as a representative fraction; also called the defined scale

parallel small circle of the earth's geographical coordinate system formed by passing a plane through the earth parallel to the equator; parallel number designation ranges from 0° at the equator (a great circle) to 90° at the pole (either north or south)

pattern of deformation distribution of distortion over a projection; customarily increases away from point or line(s) of tangency of plane to sphere

prime meridian meridian adopted by most countries as the point of origin (0°) for determination of east or west longitude; passes through the British Royal Observatory at Greenwich, England

projection aspect the position of the projected graticule relative to the ordinary position of the geographical grid on the earth

reference globe the reduced model of the spherical earth from which projections are constructed; also called a nominal or generating globe

scale factor ratio of the scale of the projection to the scale of the reference globe; 1.0 on standard lines, at standard points, and at other places, depending on the system of projection

sexagesimal system of numbering that proceeds in increments of 60; for example, the division of a circle into 360 degrees, a degree into 60 minutes, and a minute into 60 seconds

small circles any circles on the spherical surface that are not great circles; parallels are small circles

standard lines and points transformed from the spherical surface to the plane surface without distortion; scale factors on these lines or points are 1.0

Tissot's indicatrix mathematical construct that yields quantitative indices of distortion at points on map projections

Readings for Further Understanding

Brown, Lloyd A. *The Story of Maps.* New York: Bonanza Books, 1959.

Burkard, R. K. *Geodesy for the Layman.* St. Louis: Geophysical and Space Sciences Branch, Aeronautical Chart and Information Center, 1964.

Cotter, Charles H. *The Astronomical and Mathematical Foundations of Geography.* New York: American Elsevier, 1966.

Deetz, Charles H., and Oscar S. Adams. *Elements of Map Projection.* United States Department of Commerce, Special Publication No. 68. Washington, D.C.: USGPO, 1945.

Defense Mapping Agency, Hydrographic Center. *American Practical Navigator.* Washington, D.C.: Defense Mapping Agency, 1977.

Dent, Borden D., "Continental Shapes on World Projections: The Design of a Poly-Centred Oblique Orthographic World Projection." *The Cartographic Journal* 24 (1987): 117–24.

Hilliard, James A.; Umit Bosoglu, and Phillip S. Muehrcke (compilers). *A Projection Handbook.* University of Wisconsin at Madison, Cartographic Laboratory, Paper No. 2, 1978.

Hsu, Mei-Ling. "The Role of Projections in Modern Map Design." *Cartographica* 18 (1981): 151–86.

Jackson, J. E. *Sphere, Spheroid and Projections for Surveyors.* New York: Halsted Press, 1980.

King-Hele, Desmond. "The Shape of the Earth." *Science* 183 (1967): 67–76.

———. "The Shape of the Earth." *Science* 192 (1976): 1293–1300.

Marschner, F. J. "Structural Properties of Medium- and Small-Scale Maps." *Annals* (Association of American Geographers) 34 (1944): 1–46.

Morgan, J. G. "The North American Datum of 1983." *Geophysics: The Leading Edge of Exploration* (January, 1987): 27–32.

Richardus, Peter, and Ron K. Adler. *Map Projections.* New York: American Elsevier, 1972.

———. "A New Map Projection: Its Development and Characteristics." *International Yearbook of Cartography* 14 (1974): 145–55.

Robinson, Arthur H. "An Analytical Approach to Map Projections." *Annals* (Association of American Geographers) 39 (1949): 283–90.

Steers, J. A. *An Introduction to the Study of Map Projections.* London: University of London Press, 1962.

Synder, John P. *Map Projections—A Working Manual.* U.S. Geological Survey, Professional Paper 1395. Washington, D.C.: USGPO, 1987.

———. "The Perspective Map Projection of the Earth." *American Cartographer* 8 (1981): 149–60.

Thompson, Morris M. *Maps for America* 2nd ed. United States Department of the Interior, Geological Survey. Washington, D.C.: USGPO, 1982.

Vogel, Steven A. "Network for the 21st Century." *ACSM Bulletin* (August, 1988): 21–23.

Chapter 3

Employment of Projections and Thematic Base Map Compilation

Chapter Preview

Designers of thematic maps have a choice of hundreds of projections, but the range is reduced by differences in the suitability of projections for mapping the desired part of the globe. Since the equal-area property is important in mapping most themes, the choice is further restricted. For world mapping on one sheet, only a handful of projections have become widely used. Only three are discussed in detail: Mollweide, Hammer, and Boggs. Projections suitable for continent- and country-scale mapping are generally different from those employed at world scales. Unique solutions to projections may also need to be explored.

Base map compilation, involving the generalizing activities of simplification, selection, and emphasis, is generally the next step after selection of the map projection. The compiler's chief concern is evaluation of the accuracy and reliability of source maps and materials. Good judgment comes with care and experience. Actual plotting may be done by direct tracing or by the grid squares method. The designer needs to be cautious when using other source maps and materials to avoid infringement of others' legal rights.

Chapter 2 introduced the earth's spherical grid, the development of projections, and patterns of deformation in projections. This chapter addresses two additional topics of importance: employment of projections and base map compilation. A thematic map is comprised of a base map and a thematic overlay. Only topics central to the preparation of the base map are discussed in this chapter; subsequent chapters deal with the operations that go into the production of various thematic overlays.

Thematic Map Compilation

Compilation is composition using materials from other documents. **Map compilation** has been given considerable treatment in the cartographic literature. The International Cartographic Association, in its Multilingual Dictionary, defines it as follows:

> The selection, assembly and graphic presentation of all relevant information required for the preparation of a map. Such information may be derived from other maps and from other sources.[1]

Another view is that compilation includes all "aspects of collection, evaluation, interpretation and technical assembly of cartographic material."[2]

Thematic map compilation refers to the construction of a special-purpose map from a variety of previously existing sources: base maps, other thematic maps, or both, and involves numerous steps. (See Figure 3.1.) In compilation, it is important to use only good sources.[3] Compilers of small-scale thematic maps can use either large-scale maps, usually topographic survey maps, or derived medium- to small-scale maps as source maps. **Derived maps** are all made from large-scale maps. United States Geological Survey maps at large scales are used as source maps for the production of smaller-scale maps in this country. Compilers of thematic maps use derived maps more frequently than maps at topographic scales. The availability, accuracy, reliability, and currency of derived maps are of great concern to the designer. Generalization, selection, scale changing, and plotting methods are also important aspects of compilation.

If a thematic map is to be compiled without using a previously existing base map, a new base map must be constructed. A projection must be selected, the coastline compiled, and major cities, political boundaries, and physical features located. Developing a new base map requires reference to previously produced maps, from which locational information about the mapped features can be taken. Consideration must be given to the accuracy and reliability of all previously produced source maps.

Map compilation works differently in larger map-making organizations. Specialists are employed to locate, annotate, and evaluate source materials used in the compilation of a new map. The designer then is concerned only with the production and design of the final camera-ready art and is not worried about the reliability of the source material. Designers may also find themselves compiling from digital data bases. There will be questions about reliability in these cases, too.

The Employment of Map Projections

The thematic cartographer must select the proper map projection for a given map design problem. Construction of projections by manual means is quite simple (though time

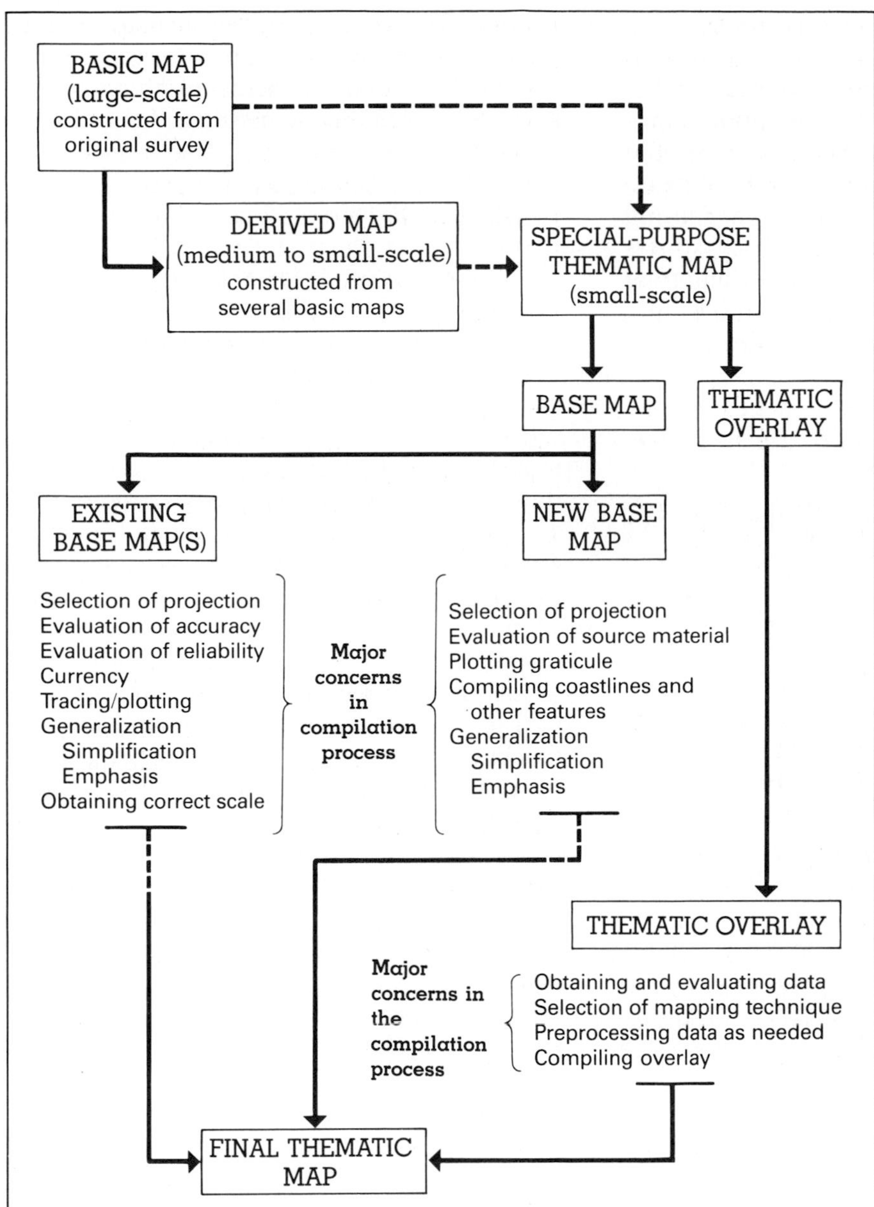

Figure 3.1. Thematic map compilation.
See the text for an explanation.

consuming), because of the numerous projection tables available. Today, the task has become even more simple and flexible because of the proliferation of computer programs that plot projections from simple instructions. These programs have eliminated the arduous hours of plotting points and fitting curves to them. For the first time, geographers and cartographers can spend more useful time in the proper *selection* of the projection, without concerning themselves with the routine of plotting it.

Several essential elements must be carefully considered in the selection of a map projection:

1. *Projection properties.* Are the properties of a particular projection suited to the design problem at hand? Are equivalency, conformality, equidistance, or azimuthality needed?
2. *Deformational patterns.* Are the deformational aspects of the projection acceptable for the mapped area? Is linear scale and its variation over the projection within the limits specified in the design goals? Do the characteristics of linear scale over the projection benefit the *shape* of the mapped area?
3. *Projection center.* Can the projection be centered easily for the design problem? Can the design accommodate experimentation with the recentering of the projection?
4. *Familiarity.* Will the projection and the appearance of its meridians and parallels be familiar to most readers? Will the form of the graticule detract from the main purpose of the map?
5. *Cost.* Is using a particular projection economical? Is there one already at hand, or will it need to be freshly plotted, by hand or by computer? Have land areas and political boundaries already been plotted on the graticule?

Although there are literally hundreds of projections from which the cartographer may choose, there are certain ones that have proven more useful in mapping particular places. (See Table 3.1.) The ones included in our discussion are offered only as a guide.

World Projections

Many world projections may be selected for thematic mapping. Equivalency is of overriding concern. Three projections, all equal-area, are presented here as good choices when mapping at the world scale on a single sheet: the Mollweide (or homolographic), Boggs Eumorphic, and the Hammer equal-area. None of these, nor any world equal-area projection on a single sheet, can avoid considerable shape distortion, especially along the peripheries of the map. These disadvantages must be accepted in order to represent the whole earth on one sheet.

The **Mollweide projection,** named after Carl B. Mollweide, who developed it in 1805, has become widely used for mapping world distributions.[4] The equator is a standard line, equally divided. (See Figure 3.2a.) The central meridian is one-half the length of the equator and drawn perpendicular to it. Parallels are straight lines parallel to the equator but are not drawn with lengths true to scale, except for the parallel at 40°40′. Each parallel, however, is equally divided along its length. The parallels are spaced along the central meridian to achieve equivalency.

Meridians on the Mollweide are curved, and the ones at 90° from the central meridian form a complete circle. Thus one hemisphere is illustrated by a whole circle. Other meridians are elliptical arcs with the shape of the whole projection an ellipse.[5]

Shape distortion on the Mollweide projection is consistent with the overall pattern of deformation for world pseudo-cylindrical equal-area projections. Maximum shape deformation occurs at the corners, where the intersections of the meridians and parallels are the most oblique. The principal use for the Mollweide is the plotting of world thematic distributions.

Table 3.1 Guide to the Employment of Projections for World, Continental, and Country-Scale Thematic Maps

Principle Use	Suitable Projections	Principal Features		
		Parallels	Meridians	Other
1. Maps of the world in one sheet				
Equal-area	Sinusoidal (Sanson-Flamstead)	Horizontal, spaced equally at true distances	Sine curves, spaced equally, true on each parallels	Awkward shape
Equal-area	Eumorphic (Boggs)	Horizontal, spaced equally	Curved	Arithmetic mean between sinsuoidal and Mollweide
Equal-area	Mollweide	Horizontal, spaced closer near poles	Ellipses, spaced equally	Pleasing shape
Equal-area	Hammer	Curved, spaced equally	Curved, spaced closer on central meridian	
Equal-area	Eckert (espc. IV and VI)	Horizontal, spaced closer near poles; poles are lines half length of equator	Ellipses, spaced equally. Sine curves, spaced equally.	Poles represented as lines; considerable distortion in poleward areas
Conformal	Mercator	Horizontal, spaced closer near equator	Vertical, spaced equally, true on equator	All bearings correct
2. Continental areas				
A. Asia and North America				
Equal-area	Bonne	Concentric circles, spaced equally at true distances	Curves, spaced equally, true on each parallel	Considerable distortion in NE and NW corners
Equal-area	Lambert azimuthal	Curves, spaced closer near poles on central meridian	Curves, spaced closer near sides of equator	Bearings true from center
B. Europe and Australia				
Equal-area	Lambert azimuthal Bonne			
Equal-area	Albers with two standard parallels	Concentric circles, spaced closer at N and S ends	Radiating straight lines, spaced equally, true on one or two standard parallels	Ideal for United States

Table 3.1 ***Continued***

Principle Use	Suitable Projections	Principal Features		
		Parallels	Meridians	Other
C. Africa and South America				
Equal-area	Mollweide			
Equal-area	Bonne			
Equal-area	Sinusoidal			
Equal-area	Lambert azimuthal			
3. Large countries in mid-latitudes				
A. United States, U.S.S.R., China				
Equal-area	Lambert azimuthal			
Equal-area	Albers equal area			
Equal-area	Bonne			
Conformal	Lambert conformal conic	Concentric circles, spaced wider at N and S ends	Radiating straight lines spaced equally true on one or two standard parallels	
4. Small countries in mid-latitudes				
Equal-area	Albers equal area			
Equal-area	Bonne			
5. Polar regions				
Equal-area	Lambert azimuthal			
6. Hemispheres and continents				
Visual	Orthographic	Ellipses, spaced closer near periphery	Ellipses, spaced closer near periphery	View of earth as if from space; neither equal-area nor conformal

Source of table: Compiled from a variety of sources; see especially Erwin Raisz, *Principles of Cartography* (New York: McGraw-Hill, 1962) 189; and J. A. Steers, *An Introduction to the Study of Map Projections* (London: University of London Press, 1962), pp. 222–225; and John P. Snyder. *Map Projections—A Working Manual.* U.S. Geological Survey Professional Paper 1395 (Washington, D.C.: USGPO, 1987), table "The Properties and Uses of Selected Map Projections," in pocket.

Its overall shape is rather pleasing, and if the distortion of shape at the peripheries is not at odds with the design purpose, this projection is an acceptable choice.

Very similar to the Mollweide are the Hammer and Boggs equal-area projections. The **Hammer projection,** developed in Germany in 1892, was for many years erroneously called the Aitoff.[6] The principal difference between this projection and the Mollweide is that the Hammer has curved parallels. (See Figure 3.2b.) This curvature results in less oblique intersections of meridians and parallels at the extremities, and thus reduces shape distortions in these areas. The outline (i.e., the ellipses forming the outermost meridians) is identical to the Mollweide.

Construction of the Hammer projection is somewhat more difficult than the Mollweide because of the curved parallels. This is a minor

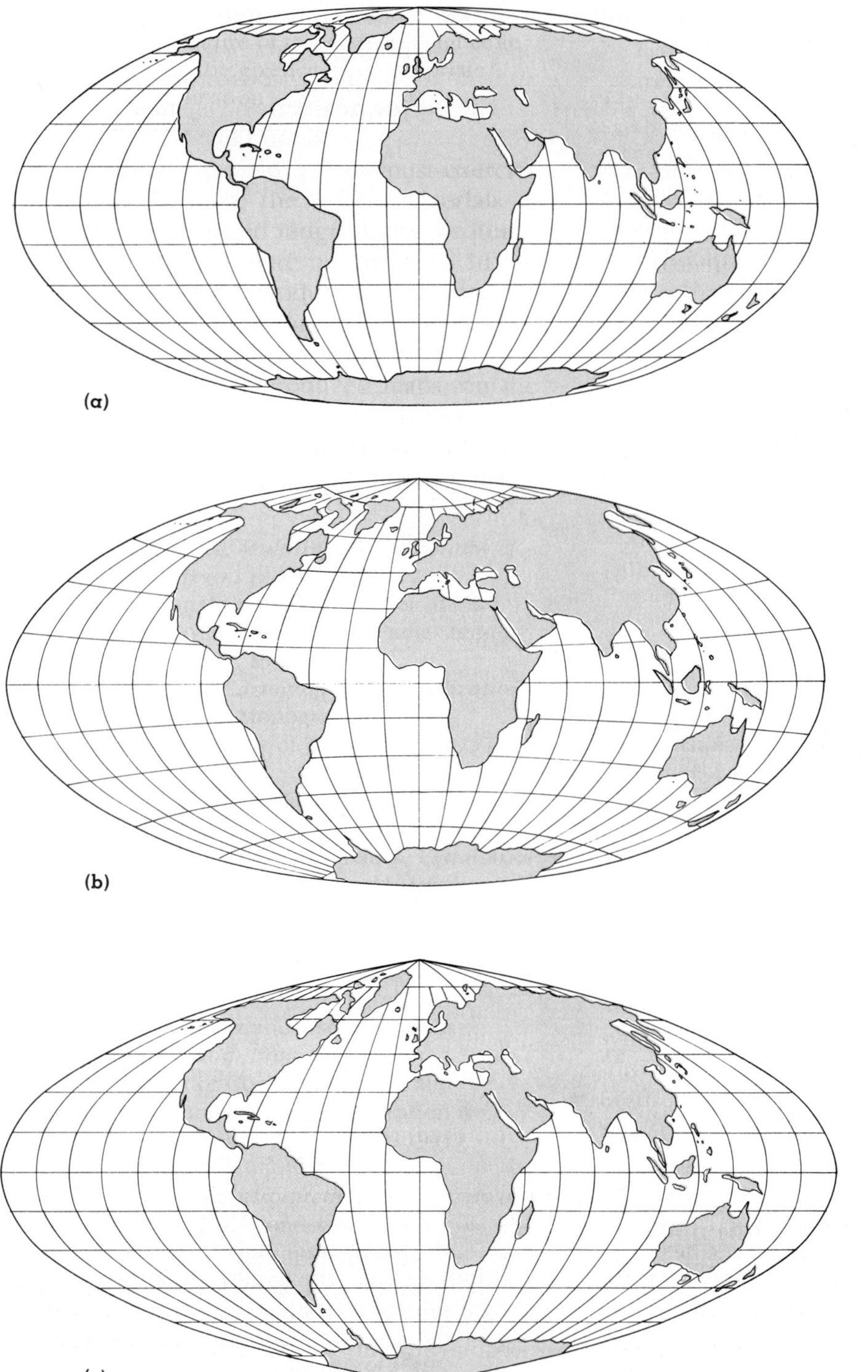

Figure 3.2. Three popular world projections on one sheet.
The ones illustrated here are (a) Mollweide, (b) Hammer, and (c) Boggs eumorphic.

inconvenience, especially if machine plotting is available. Because the parallels are unequally curved, they are not true to scale.

Hammer's projection also is quite acceptable for mapping world distributions. A comparison of it with the Mollweide shows little difference. Because the parallels are curved, east–west exaggeration at the poles is less on the Hammer than on the Mollweide. This is most notable when comparing the Antarctica land masses. Africa is less stretched along the north–south axis on the Hammer. Overall, however, these projections are very similar in appearance and attributes.

The **Boggs eumorphic projection** was developed by Whittemore Boggs in 1929.[7] In contrast to the Mollweide and the Hammer, the poles on the Boggs projection are accentuated by the converging meridians. (See Figure 3.2c.) This projection is a combination of the sinusoidal (Sanson–Flamstead) and the Mollweide. It is considered an arithmetic average between the two. Above 62° latitude, angular distortion is less than on sinusoidal projection. A chief advantage of the Boggs projection over the Mollweide is better shape preservation along the equator, brought about by greater equality of linear scales in both east–west and north–south directions. There is greater north–south stretching at the equator (e.g., Africa) on the Mollweide.

The number of other projections that can be chosen for world thematic mapping on one sheet is quite large. There are numerous publications that treat this subject in more depth than space allows here. These may be consulted for further details in selecting and constructing a map projection.

Projections for Mapping Continents

The projections that are suitable for world maps are generally not the best for mapping continental areas. Either equivalency or conformality can be better preserved at the larger scales by selecting other projections. As indicated in Table 3.1, the Bonne, Lambert azimuthal, Albers equal-area, and sinusoidal, in addition to the Mollweide, are wise choices when mapping continent-sized areas. We will look at the Bonne projection in detail.

The **Bonne projection** is named after its inventor, Rigobert Bonne (1727–1795).[8] It is an equal-area conical projection, with a central meridian and the cone assumed tangent to a standard parallel. (See Figure 3.3.) All parallels are concentric circles, with the center of the standard parallel the apex of the cone. The central meridian is divided true to scale. All parallels are drawn with their lengths true to scale, and each is divided truly. The meridians are drawn through the points of division along the parallels.[9] If the standard parallel selected is the equator, the projection becomes identical to the sinusoidal.

Map designers select the Bonne projection for a variety of continental mapping cases. It is commonly used to map Asia, North America, South America, Australia, and other large areas. Europe may also be adequately mapped with the Bonne. Caution should be exercised, however; although equivalency is maintained throughout, shape distortion is particularly evident at the NE and NW corners. Because of this, the Bonne projection is really best suited for mapping compact regions lying on only one side of the equator.[10] Because shape is best along the central meridian the distortion becomes objectionable at greater distances from it, the selection of the central meridian relative to the important mapped area is critical.

Mapping Large and Small Countries at Mid-latitudes

Mapping large countries at mid-latitudes can be handled in a variety of ways. The Bonne, the Lambert azimuthal equal-area, or the Albers equal-area projections may be used. If conformality is desired, the Lambert conformal conic can be selected. In general, a conic projection is usually adequate for mapping rather large countries lying in the mid-latitudes.

All simple azimuthal projections are developed onto a plane that is tangent to the generating globe at a point. The **Lambert equal-**

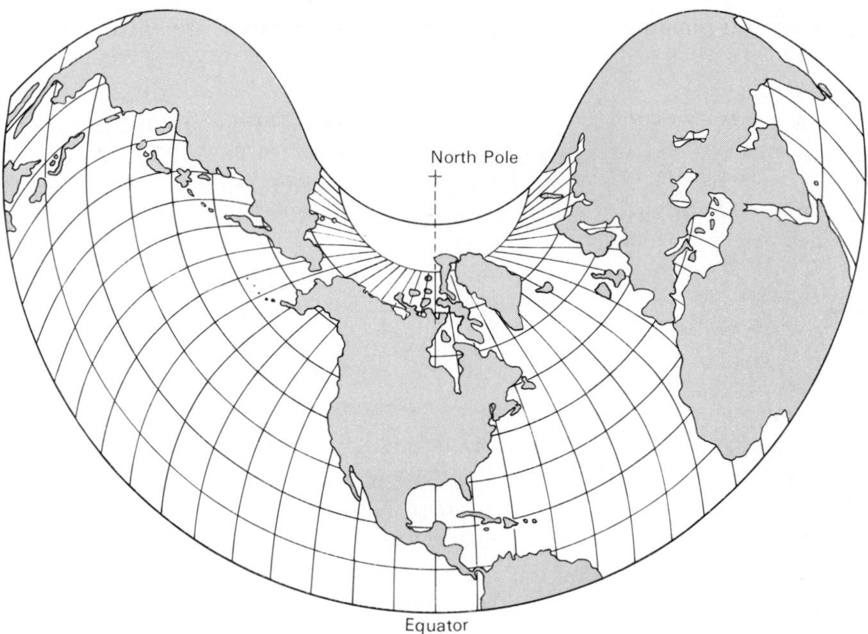

Figure 3.3. The Bonne projection.
This projection is suitable for mapping continents, but should never be used for mapping complete hemispheres. Notice the severe shape distortion, brought about by shearing, at the northeast and northwest corners of the projection.

area azimuthal projections follow this pattern. They may be tangent at any point, although the polar case is considered the normal aspect. It is referred to as the Lambert equal-area meridional (or central equivalent) projection when the point of tangency is the equator and some meridian.[11] It may also be centered at some latitude between the equator and the pole; this is is called the oblique case. (See Figure 3.4.) This projection is quite adequate for showing countries that have symmetrical shapes; little shape distortion is evident. In addition to the equivalency feature, because the Lambert is an azimuthal projection, the azimuth of any point on the map (as measured from the center of the projection) is correct. This makes this projection especially useful when mapping phenomena having an important directional relationship to the central point chosen for the map.

The **Albers equal-area projection** is a conic projection having two standard parallels. Its utility is not limited to mapping the United States; any land area having considerable east–west extent is well represented by the Albers. Overall scale distortion is nearly the lowest possible for an area the size of the United States. The desirable properties of this projection are as follows:[12]

1. It is an equal-area projection.
2. Maximum scale error is approximately 1.25 percent over an area the size of the United States.
3. Meridians are straight lines that intersect parallels at right angles.
4. Parallels are concentric circles, making construction relatively easy.
5. Because it is a single-cone conic projection, its properties do not deteriorate in an east–west direction.
6. It is suitable for a series of "section" maps, since adjacent sections of this map will fit together exactly.

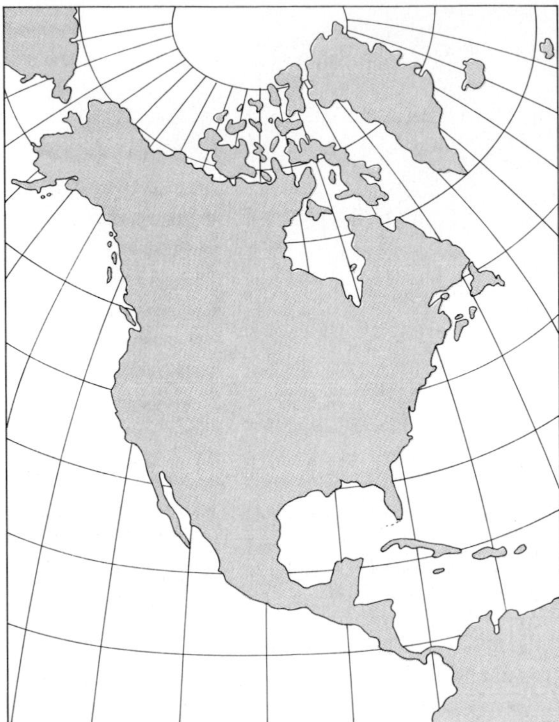

Figure 3.4. An oblique Lambert equal-area projection.
This projection is suitable for mapping both large and small countries at mid-latitudes.

Equal-area mapping of small countries at mid-latitudes (e.g., France, Spain) can be accomplished by using such projections as the Bonne or the Albers. As mapped areas become smaller in extent, the selection of the projection becomes less critical; potential scale errors begin to drop off considerably.

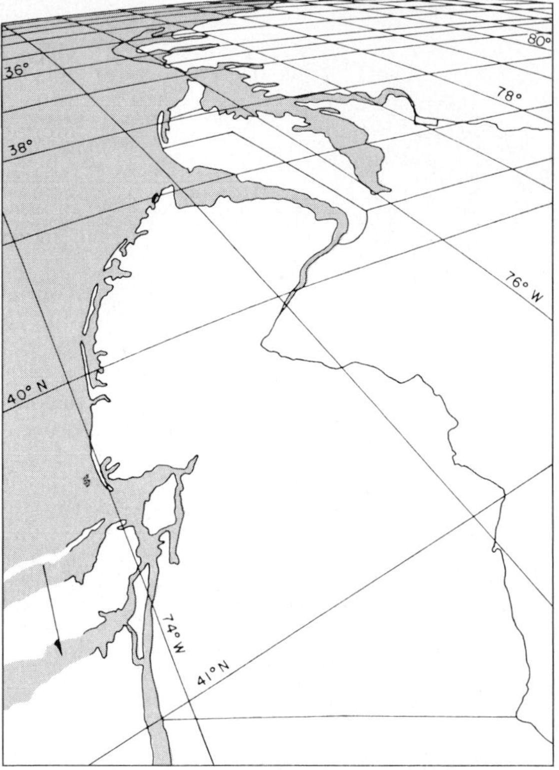

Figure 3.5. A tilted perspective or "space photo" projection.
This view is of the eastern seaboard of the United States, viewed from about 160 km above Newburgh, N.Y. Cartographers now have greater opportunity to select unique projections because of the proliferation of computer programs to plot them.

Redrawn by permission from John P. Snyder, "The Perspective Map Projection of the Earth," *American Cartographer* 8 (1981): 155.

Mapping at Low Latitudes

Many of the projections already discussed are suitable for mapping countries, large and small, on or near the equator. Most of the equal-area projections suited for world mapping (e.g., Mollweide, Hammer, sinusoidal) are cylindrical projections having relatively little angular or linear scale distortion at low latitudes. Coupled with a larger scale (showing smaller earth areas), these projections are quite satisfactory for most thematic map applications in equatorial regions.

The Selection of Projections for Individual States

For most large-scale thematic mapping, the selection of a projection on which to plot a state outline and its geographical features does not present any significant problem. In general, the projection should be equal-area, and other desired qualities can be chosen as appropriate (azimuthality from the center, for example). It is important to select a projection having the properties that are suited for the state. For example, a good choice for mapping Tennessee

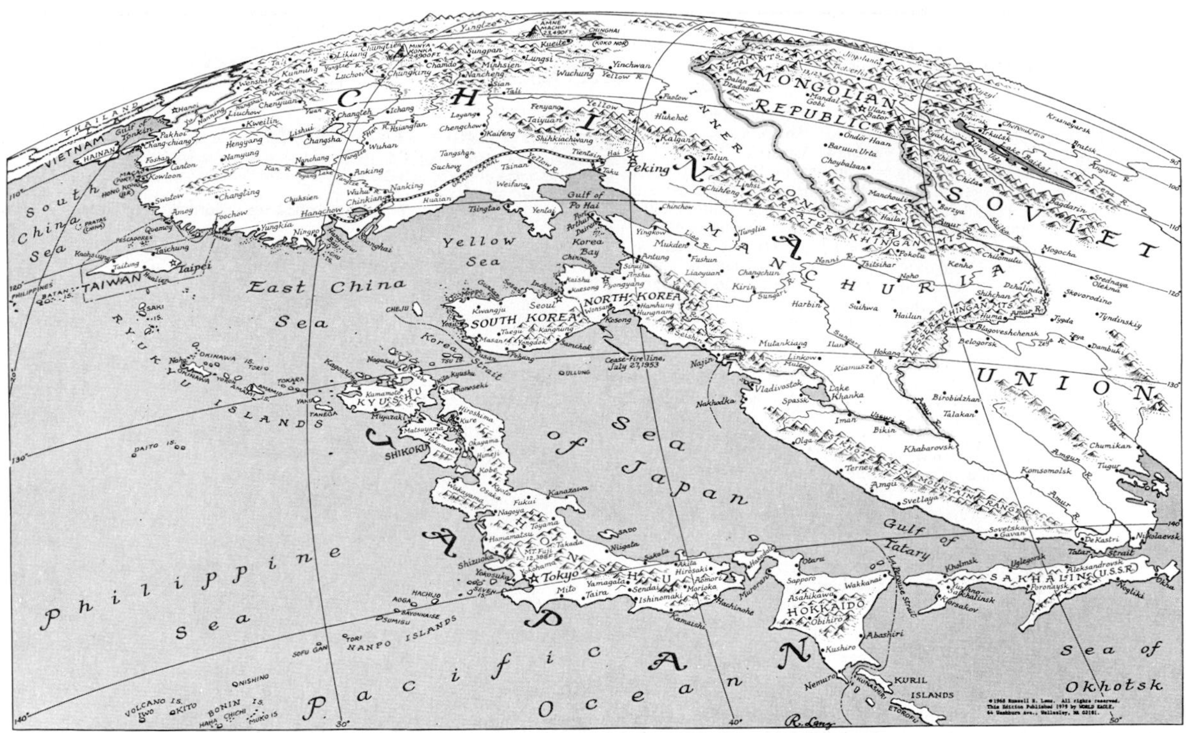

Figure 3.6. The People's Republic of China as seen from Japan.
The orientation of projection, which was developed by photographing the graticule on a globe, gives this map its unique quality. This map is one of a series done by cartographer Russell Lenz.

would be an equal-area conic projection whose standard parallel runs through the main east–west axis of the state.

Each state has been mapped with either the **Lambert conformal conic** or **transverse Mercator** projection, depending on suitability for the given state. The United States Coast and Geodetic Survey superimposes a rectangular grid over the geographical grid, and **state plane coordinates** are produced. The grids are marked in feet and are employed by highway engineers, utility companies, and regional planners. Topographic maps at the 1:24,000 scale include marks along the margin to locate the grid of the state plane system.

The United States Geological Survey's state base maps are produced on either of these projections. Because earth areas are relatively small and the mapping scale so large, areal distortion of these conformal projections is negligible. Therefore, the thematic map designer has a convenient source for state base maps on acceptable projections.

Unique Solutions to the Employment of Projections

Thematic cartographers should spend considerable time on the selection of a projection in furthering the communication effort of the map.[13] They should therefore be willing to investigate the impact of recentering the projection and to see the effects of new orientations on the map's message. Two such maps are reproduced here as Figures 3.5 and 3.6.

Unique solutions to the employment of projections have become more commonplace with the increasing use of computer mapping. Cartographers were previously discouraged from

New World Map Cuts Nations Down to Size

From Wire Reports

WASHINGTON — Pentagon officials are in for a pleasant surprise: The Soviet Union finally has been cut down to size — reduced by 63 percent, in fact.

But Canadians probably will be furious to see their country trimmed by two-thirds. And real estate agents all over the world will be alarmed to find the Earth's land surface has shrunk dramatically.

Blame it on the National Geographic Society.

"It's not every day that you get to change the world," National Geographic spokesman Robert B. Sims said Thursday as he unveiled the society's new official world map.

"The Robinson Projection more accurately portrays round Earth on a flat surface," said Gilbert M. Grosvenor, president and chairman of the society.

The map is named for Arthur H. Robinson, professor emeritus of geography and cartography at the University of Wisconsin at Madison, who came up with the design in 1963. National Geographic also honored him with a medal Thursday for his contributions to the field.

The new map, like the old — the Van der Grinten projection — has the inherent problem of trying to capture a three-dimensional world on paper.

It differs from the old primarily in the polar regions, which are less exaggerated. Greenland, for example, was 5½ times its actual size on the old map, which made it look about the size of South America. On the new map it is a mere 60 percent larger than life. Antarctica no longer is as big as Asia.

On Van der Grinten maps, the Soviet Union appears 223 percent too large and the United States seems 68 percent bigger than it actually is. On the Robinson maps, the Soviet Union is 18 percent larger than it should be, and the United States is 3 percent smaller.

By reducing the sizes of land masses relative to oceans, the Robinson Projection shows the Earth's surface more like what it is — 71 percent water.

Finally, because most people live in temperate and tropical zones, the new map includes more place names in those areas.

National Geographic's move to the new map coincides with the organization's centennial. The society, which has 11 million members, estimates 40 million people will use the new map.

Although the society fully endorses the map, it — and the rest of the world — still will use the Mercator Projection, a map designed in the 16th century, to show areas smaller than the whole world. The Mercator greatly distorts the size of the world's land masses

Even the creator of the new map admits it is not perfect.

"I hesitate to say this, but it is pretty much an artistic process rather than a scientific one. You try to get the most realistic view of the world," Mr. Robinson said. "I worked with the variables until it got to the point where, if I changed one of them, it didn't get any better. And so I stopped."

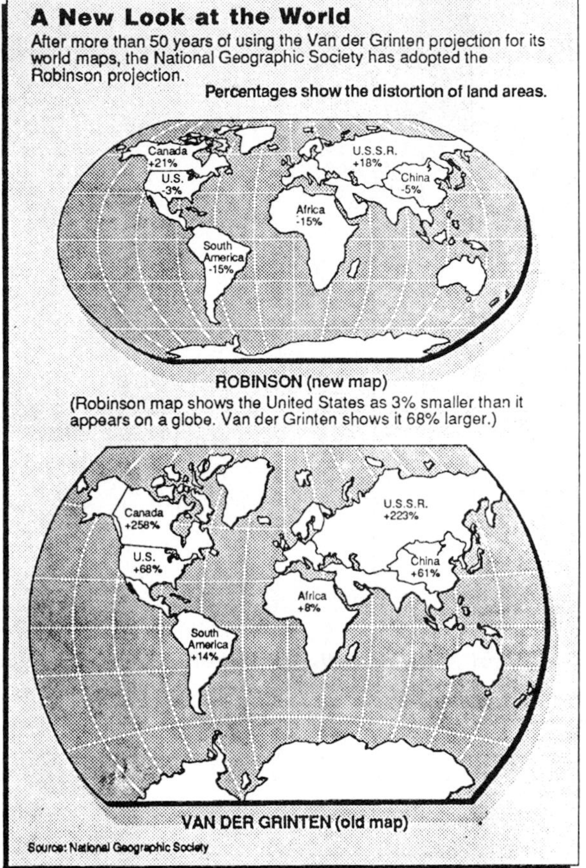

experimentation by the drudgery of computing and plotting projections. Now the emphasis can be on selection.

In sum, the projection provides the geometrical framework for all the points and areas compiled on the final map. Projection is thus a key element in the overall design. It controls the locational accuracy and final appearance of the mapped area and can be used to provide a focus for the map's message. The designer must have a firm grasp of the earth's spherical coordinate system in order to understand projections and to use them effectively in the design effort. And finally, the cartographer should provide the map reader information about the essential projection properties. These include, where applicable: scale, central meridian, standard parallel(s), central parallel, scale factor on the central meridian, and any other unique attributes.[14]

Gall-Peters Projection

Special consideration is devoted here to what has become known as the **Gall-Peters projection,** primarily to stimulate the cartographic designer to look into the literature further, and

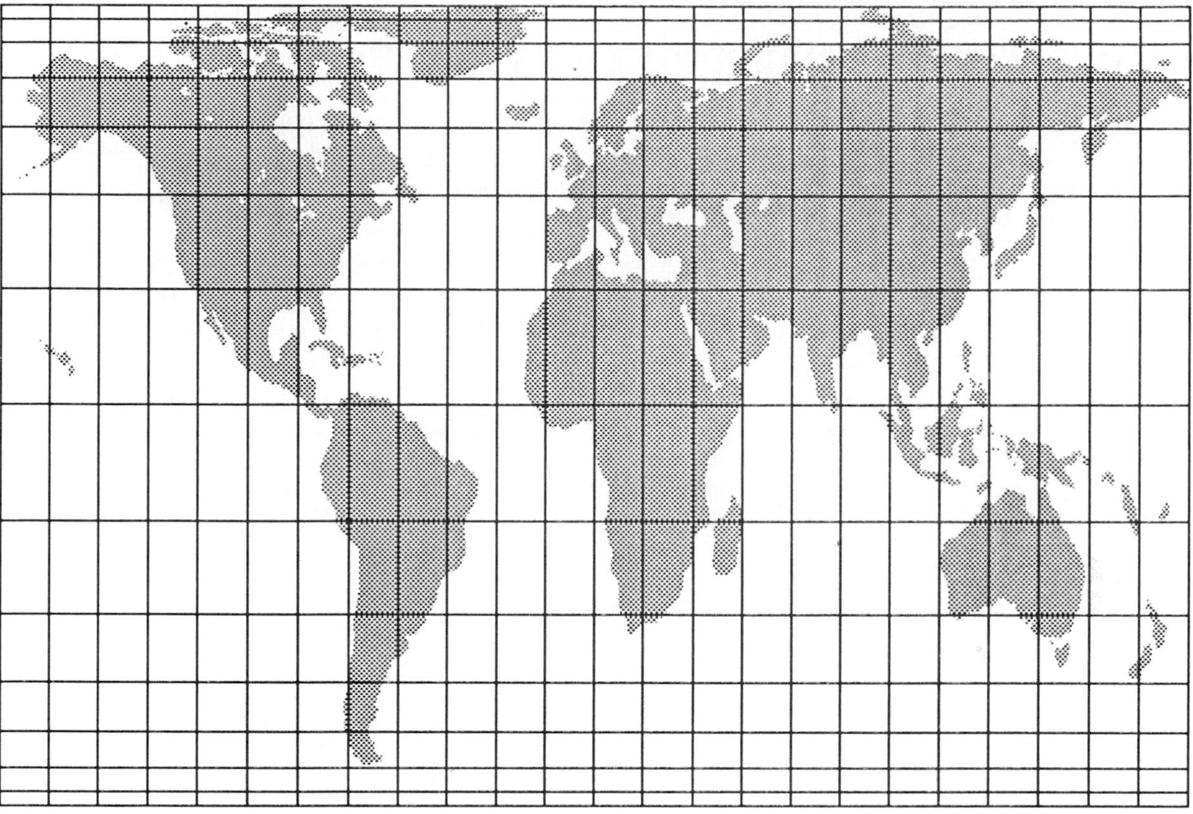

Figure 3.7. Gall-Peters projection.
Projection advanced by Arno Peters in 1972 to overcome area distortion of the Mercator projection. May be based on the Gall cylindrical orthographic.

partly because of the controversy surrounding its use. In recent times, probably no other map projection has received as much attention in both the scientific *and* popular literature.

Dr. Arno Peters, of Germany, in 1972 published what he called a new map projection—the Peters projection.[15] In fact, this projection had been devised earlier by Gall in the mid-1880s, who then called it the orthographic; a form of cylindrical projection (although it may be speculated that Peters had no knowledge of Gall's previous work).[16] In the conventional aspect, the Peters projection is an equal-area rectangular projection with standard parallels at 45° N and S. (See Figure 3.7.) Peters was reacting to the areal misrepresentation on the Mercator projection, which he felt showed European dominance over the Third World.[17] Shapes are terribly distorted on the Mercator projection. Because the projection is not new, it is referred to as the Gall-Peters projection.

The purpose behind the invention, as Peters argues, is that the Mercator projection, often used for mapping the world, is inadequate in that purpose because it so grossly distorts area, which is true. Peters, therefore, stressed that his projection be used exclusively and that it portrays distances accurately (which is false). The real objection to Peters is with the misconceptions he renders about the projection. It has been widely adopted by three United Nation's organizations (UNESCO, UNDP, and UNICEF), the National Council of Churches, and by Lutheran and Methodist organizations.[18]

One good thing has come from the controversy, namely that the inadequacies of using the Mercator projection for world thematic mapping have finally been noted to the general population. Snyder, for example, has said:

> Nevertheless, Peters and Kaiser (his agent) appear to have successfully accomplished a feat that most cartographers only dream of achieving. Professional mapmakers have been wringing their hands for decades about the misuse of the Mercator Projection, but, as Peters stresses, the Mercator is still widely misused by school teachers, television news broadcasters and others. At least Peter's supporters are rightly communicating the fact that the Mercator should not be used for geographical purposes, and numerous cartographers agree. The Gall-Peters Projection does show many people that there is another way of depicting the world.[19]

Base Map Compilation

The rest of this chapter treats base map compilation, previously defined in broad terms. We now turn to some specific aspects of the subject: compilation and generalization, compilation sources, methods of compilation, and copyright issues.

Compilation and Generalization

Compiling thematic maps involves **cartographic generalization,** which ordinarily means simplification.[20] Simplification requires the cartographer to use good judgement in the elimination of detail and to reduce the number of unnecessary features taken from source maps in arriving at the new map. Selection and emphasis are also important in generalization, which is considered one of the principal functions of cartographers.[21] In thematic map design, poorly performed generalization can cause the whole map effort to fail. Unfortunately, generalization and simplification in thematic mapping are not subject to quantification or strict rules, so the cartographer must become proficient in these through experience and good judgment.

However, guidelines have been suggested to help in dealing with problems in generalization and compilation.[22]

1. *Map purpose* must be kept clearly in mind.
2. The *reduction factor* of source materials to the new manuscript map scale must be considered in order to develop uniformity.
3. Use *objective evaluation:* utilize many sources in the selection process.
4. Be mindful of *local bias.* Avoid allowing biased local sources to undermine the consistent selection and application of detail.
5. *Personal bias* must be avoided. Become acquainted with details of the whole region, so that areas about which you have greater knowledge do not appear with inordinate precision and detail.
6. Determine what elements really typify the *character* of an area, and retain only these during the selection process.
7. Strive for *uniformity* of treatment over the new map. Uniformity applies to the *level* of generalization, as well as to the amount of detail chosen for the map.

These guidelines, although quite general, will serve you well and are worth learning.

What is Compiled?

The answer must be in terms of the purpose of the map, since we are concerned with the compilation of the base map portion of the thematic map. Map purpose will also dictate the level of generalization believed important for a particular message. First, let us look at the selection of features.

Base map information serves to help map readers orient the thematic information to a spatial or geographical frame of reference. The designer's task is thus to select only those base map features that will help the reader in the context of the map's purpose. For example, a map that shows urban employment in aircraft

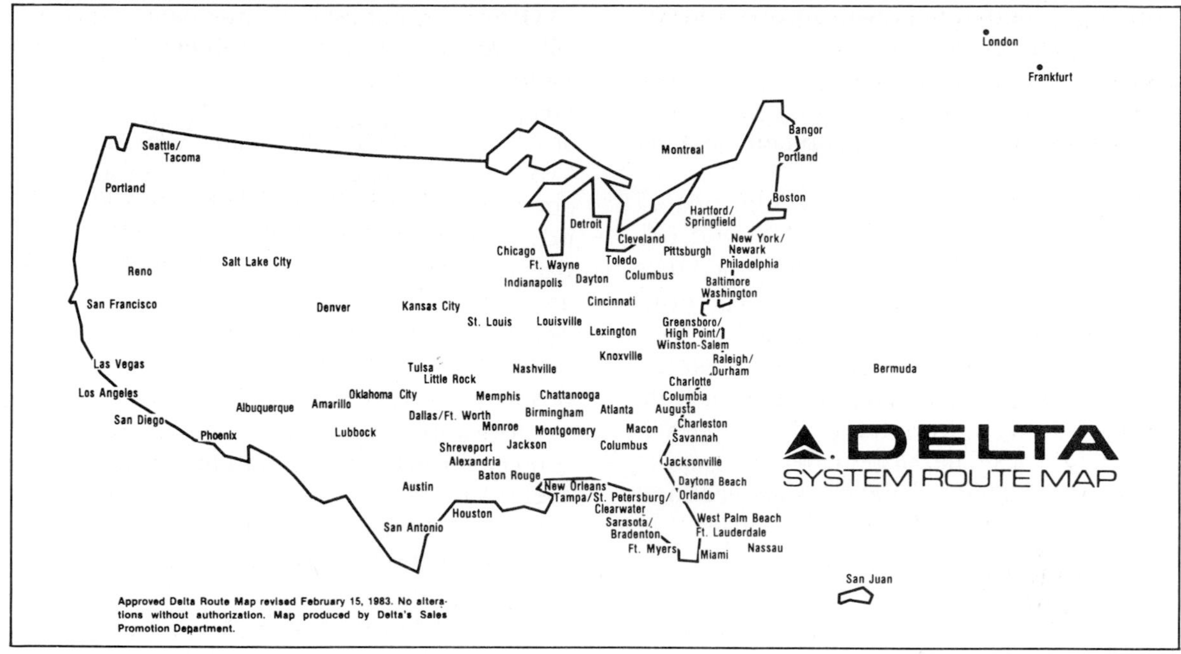

Figure 3.8. Advertising map of a commercial airline.
Notice the extreme generalization of the outline of the United States. Even though the irregularities of the outline have been simplified greatly, this map communicates effectively.
Delta Air Lines Route Map, effective February 15, 1983. Courtesy of Delta Air Lines. Reproduced by permission.

manufacturing will not need rivers on the base map. State boundaries, and perhaps other cities not having employment, should be shown. A map showing major cattle drives on the Great Plains in the 1870s, however, should probably include the major rivers. The designer should ask, "Does this information assist in orienting the reader? Is this information helpful to the map purpose?"

The kinds of base map information that often prove useful are the inclusion of coastlines, lakes, rivers and streams, major or minor physical features (e.g., mountains), and other special physical features. Other base data helpful for reader orientation are political boundaries and the locations of cities and towns. Of particular importance is the geographic grid (the graticule), especially if the mapped subject is dependent on its location on the earth. For example, a map showing annual temperature should include the graticule, or at least tic marks along the map border.

Many beginning cartographers are reluctant to reduce the amount of base map information for fear of producing an "unscientific" map. Even many experienced cartographers are reluctant to adopt a high level of generalization. This is unfortunate, since ideas can often be communicated with simple solutions. In Figure 3.8, for example, the overall message intended by the airline company is effectively conveyed even though the outlines (composed of national political boundaries and coastlines) are highly generalized. The map's purpose and the reader's preexisting knowledge about shapes must determine the level of generalization and the selection of base map information.[23]

Compilation Sources

Thematic base maps are compiled from other sources (usually other maps). Even if a new projection is drawn for a thematic map, certain features, notably coastlines, rivers, and lakes, must be obtained from previously drawn maps. Likewise, man-made features such as roads, railroads, towns, and cities, have to be obtained from other maps. Knowing about map sources that can be used in compilation is most helpful to the cartographer.

Fortunately, there are many high-quality source maps available. Many public and private agencies produce very useful base maps and other thematic map materials. The principal government source for maps in this country is the United States Geological Survey, which produces maps ranging in scale from 1:24,000 (topographic scale) to 1:1,000,000. (See Table 3.2.) The large-scale topographic maps include detailed physical and cultural features. The smaller-scale (1:250,000 or smaller) maps show a limited number of features, such as boundaries, state parks, airports, major roads, and railroads.

Over thirty federal agencies, states, counties, and municipalities are involved in producing maps of all kinds. It is estimated that more than 25,000 new maps are produced each year. These range from topographic maps to census maps to maps of the moon. Information concerning the products, sources, and addresses of government mapping activities is provided by this coordinating agency:

National Cartographic Information Center (NCIC)
United States Geological Survey
507 National Center
Reston, Virginia 22092
(703) 860–6045

Appendix C is a comprehensive list of map sources, including types, publishing agencies, and addresses.

Of special interest to the thematic map compiler are the Geographic Names Information System (GNIS) files produced by the Branch of Geographic Names, Office of Geographic Research of the USGS. These files contain nearly two million name entries, with information about each feature, its category (stream, town, ect.), and its location by geographic coordinates, county, and topographic map. The files are organized by state and are available in printed or microfiche forms from NCIC. They are extremely helpful in checking the location of small places.

Cartographic designers often require other sources for mapping foreign lands. Especially useful here are two sources: the International Map of the World (IMW) series, produced by the USGS at a scale of 1:1,000,000; and the Defense Mapping Agency (DMA) World Series (at varying scales). The United States Air Force Navigation Charts are also quite useful. All these source maps have good reliability.

Governmental agencies are not the only places to find map source materials. Private map and atlas producers are extremely valuable resources. The thematic map designer should gather a personal inventory of atlases and single-sheet maps for reference.[24] It is especially important to collect a variety of maps having different scales, projections, and content. A good gazetteer is useful, as well as governmental-produced national atlases.

Accuracy and Reliability

Having obtained a source map or maps, the designer must judge their accuracy and reliability. The newly compiled base map can be no

What's the map for? To direct visitors how to get to your house from the station? Then subordinate everything else to that, even scale. The point is to simplify your guest's way to your residence. The simplification may reduce your project from a map to a mere cartogram. No matter: it will still be cartography. If it works, that's something. A map's first business is to function. Utter simplification need not inhibit interest, and it may even improve the design. The simplest map can be fascinating, though obvious.

Source: David Greenhood, *Mapping* (Chicago: University of Chicago Press, 1964), p. 175.

Table 3.2 Scales of National Topographic Maps Produced by the United States Geological Survey

Series	Scale	1 inch represents	1 centimeter represents	Standard Quadrangle Size (latitude-longitude)	Quadrangle Area (sq mi)
7.5 min	1:24,000	2,000 ft	240 m	7.5 × 7.5 min	49–70
7.5 × 15 min	1:25,000	2,083 ft	250 m	7.5 × 15 min	98–140
Puerto Rico					
7.5 min	1:20,000	1,667 ft	200 m	7.5 × 7.5 min	71
15 min	1:62,500	1 mi	625 m	15 × 15 min	197–282
Alaska	1:63,360	1 mi	634 m	15 × 20 to 36 min	207–281
Intermediate	1:50,000	3.2 mi	2 km		(by county)
Intermediate	1:100,000	1.6 mi	1 km	30 × 60 min	1,568–2,240
United States	1:1,000,000	4 mi	2.5 km	1° × 2° or 3°	4,580–8,669
State maps	1:500,000	8 mi	5 km		
United States	1:1,000,000	16 mi	10 km	4° × 6°	73,734–102,759
Antarctica	1:250,000	4 mi	2.5 km	1° × 3° to 15°	4,089–8,336
Antarctica	1:500,000	8 mi	5 km	2° × 7.5°	28,174–30,462

Although the objective of the map must be kept constantly in view, freedom of expression must be permitted. . . . Practical experience combined with common sense and a flair for the subject is the essential requirement for successful generalization of information.

Source: O. M. Miller and Robert J. Voskuill, "Thematic-Map Generalization," *Geographical Review* 54 (1964): 13–19.

more accurate than the one on which it is based. The locational accuracy of features on USGS topographic maps is quite good; occasional errors, mostly due to omission, are found among names, symbols, and spellings, so they must be checked. Generally, however, federal mapping accuracy standards are maintained.

The word *accuracy* can only be applied to base map features: the location of features or the graticule. **Map accuracy** is the degree of conformance to an established standard. Since thematic maps are not devised relative to standards, their accuracy cannot be properly evaluated. Thus, when other thematic maps are used as sources for a new compilation, there is no accuracy standard by which to judge them. Only their base map component can be so evaluated; even locational accuracy is often irrelevant, because some thematic maps do not require planimetric precision. Because of this,

We are all aware that to produce a successful thematic map the summarizing or generalizing process rightly starts at the planning stage and continues on through the steps of design, compilation, and drafting. But before planning can even begin, the source material must be sought out and appraised for its reliability and significance.

Source: O. M. Miller, and Robert J. Voskuill, "Thematic Map Generalization," *Geographical Review* 54 (1964): 13–19.

United States National Map Accuracy Standards

With a view to the utmost economy and expedition in producing maps which fulfill not only the broad needs for standard or principal maps, but also the reasonable particular needs of individual agencies, standards of accuracy for published maps are defined as follows:

1. Horizontal accuracy. For maps on publication scales larger than 1:20,000, not more than 10 percent of the points tested shall be in error by more than 1/30 inch, measured on the publication scale; for maps on publication scales of 1:20,000 or smaller, 1/50 inch. These limits of accuracy shall apply in all cases to positions of well-defined points only. Well-defined points are those that are easily visible or recoverable on the ground, such as the following: monuments or markers, such as bench marks, property boundary monuments; intersections of roads, railroads, etc.; corners of large buildings or structures (or center points of small buildings); etc. In general, what is well-defined will also be determined by what is plottable on the scale of the map within 1/100 inch. Thus while the intersection of two road or property lines meeting at right angles, would come within a sensible interpretation, identification of the intersection of such lines meeting at an acute angle would obviously not be practicable within 1/100 inch. Similarly, features not identifiable upon the ground within close limits are not to be considered as test points within the limits quoted, even though their positions may be scaled closely upon the map. In this class would come timber lines, soil boundaries, etc.
2. Vertical accuracy, as applied to contour maps on all publication scales, shall be such that not more than 10 percent of the elevations tested shall be in error more than one-half the contour interval. In checking elevations taken from the map, the apparent vertical error may be decreased by assuming a horizontal displacement within the permissible horizontal error for a map of that scale.
3. The accuracy of any map may be tested by comparing the positions of points whose locations or elevations are shown upon it with corresponding positions as determined by surveys of a higher accuracy. Tests shall be made by the producing agency, which shall also determine which of its maps are to be tested, and the extent of such testing.
4. Published maps meeting these accuracy requirements shall note this fact in their legends, as follows: "This map complies with National Map Accuracy Standards."
5. Published maps whose errors exceed those aforestated shall omit from their legends all mention of standard accuracy.
6. When a published map is a considerable enlargement of a map drawing (manuscript) or of a published map, that fact shall be stated in the legend. For example, "This map is an enlargement of a 1:20,000-scale map drawing," or "This map is an enlargement of a 1:24,000-scale published map."
7. To facilitate ready interchange and use of basic information for map construction among all Federal mapmaking agencies, manuscript maps and published maps, wherever economically feasible and consistent with the use to which the map is to be put, shall conform to latitude and longitude boundaries, being 15 minutes of latitude and longitude, or 7½ minutes, or 3¾ minutes in size.

Source: Morris M. Thompson, *Maps for America.* United States Geological Survey (Washington, D.C.: USGPO, 1982), p. 104.

thematic map compilers need to familiarize themselves with all aspects of the mapped area, so that the accuracy and reliability of a source map can better be judged.[25] The key to source evaluation is comparison and an accumulated sense of the reliability of the map-producing agency. There are few substitutes for experience and knowledge.

A source map should be checked to determine if it is a *basic* or a *derived* map. Basic maps result from original survey work; derived maps have been compiled from basic maps. The potential for error is greater on derived maps.

Methods of Compilation

Compilation of the base map component of a thematic map involves selecting and constructing a projection at the proper manuscript map scale, plotting relevant coastlines, rivers, lakes, and other physical features, and plotting any man-made features to be included on the final map. There are three compilation methods often used: the reduction method, the common-scale method, and the one-to-one method.

Reduction methods involve "lifting" or tracing the selected objects from one source map onto a work sheet. The work sheet is then reduced, by one of several means discussed below, to the desired manuscript or artwork map size. Locational accuracy is the chief advantage of the reduction method. The compiler makes selection and generalization decisions while tracing. False information should not be introduced, and only relevant information is traced.[26] Consideration must be given to the effects of reduction. (See Figure 3.9.) The reduction method is *not* a simple tracing operation.

Common-scale methods involve compilation from various sources onto one work sheet. A political base outline may be traced from one source map, and another feature (e.g., railroads) traced from another. The different source maps are likely to be at different scales—perhaps even on different projections—because they have been produced for different purposes. Through reduction methods, generalization, and selection, the compiler develops a single-scale manuscript map that is homogeneous in detail.[27] Extreme caution must be exercised at this stage because of the danger that variations in scale and level of generalization may be introduced into the new map. Expecially careful treatment is needed in order to assure the locational accuracy of features compiled from maps having different projections.

In **one-to-one methods,** the cartographer simply traces, at the same scale, the features from the source map onto the new work sheet. Locational accuracy is very good. As with other compilation methods, generalization and selection are performed during the tracing process. It is unlikely that a source map will have the exact scale needed for the new compilation.

Plotting Projections

One of the most time-consuming of compilation activities is the plotting of projections. If the compiler has not been fortunate enough to obtain a graticule of the desired projection, one must be drawn. Computer plotting is the easiest and most flexible, if the facilities are available. Drawing the graticule from projection tables is next best, and developing projections from scratch is the least desirable. With each of these three methods, the projection can be custom-designed and developed at the desired work sheet size. The graticule of the projection should include only the number of meridians

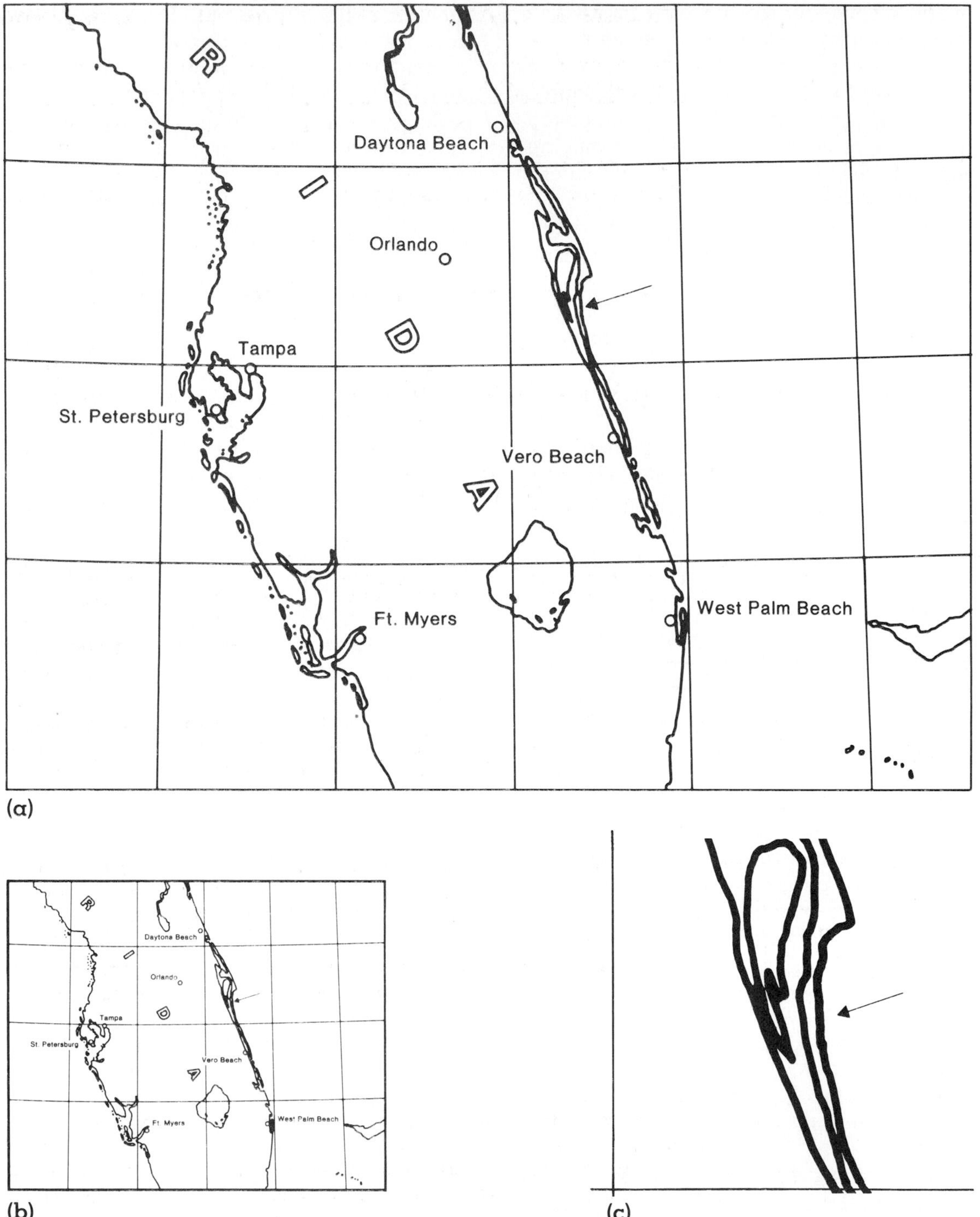

Figure 3.9. The effects of changing map scale photographically.
Source maps should be reduced with caution and never enlarged. The original compilation (a), when photographically reduced (b), may close up and detail can be lost or confused. Enlargement of the map (c) creates an overgeneralized picture of the same features. Notice the coastline at the arrows.

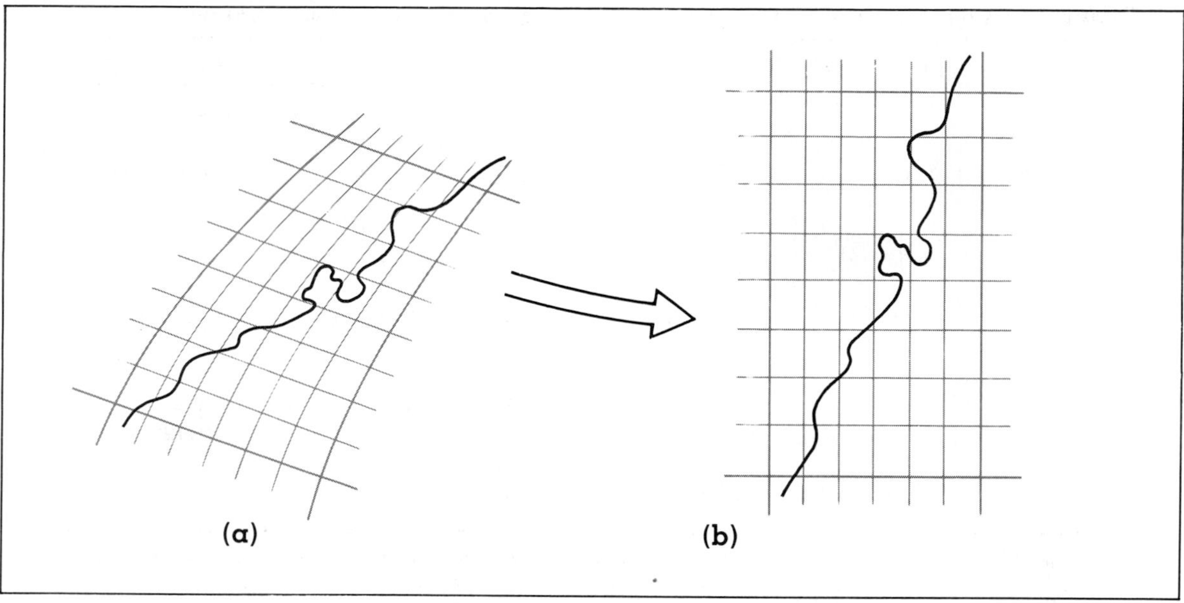

Figure 3.10. Compilation using grid squares.
Features from the original or source map, as in (a), can be transferred to the new map (b) by gridding mapped space. A similar pattern of cells is chosen. Good accuracy can be achieved and increases with smaller cell sizes. A change in scale may be also accomplished by this method.

and parallels suitable for the map purpose. Too dense a graticule detracts from the thematic map.

Compiling Coastlines and Other Features

Coastlines and other features are placed on the new base map by reference to source maps. The source maps need not be on the same projection, but their date, reliability, and accuracy must be evaluated. This is especially important for coastlines and political boundaries. If these are acceptable, compiling may begin.

An acceptable way of compiling coastlines and other features is the **grid squares** method. (See Figure 3.10.) This is especially useful when the new map has a projection and scale differing from the source map's. The cartographer grids the source map and the new map with a similar pattern of grid cells. Details from each cell on the source map are then transferred to their corresponding cell on the new map. Any type of feature can be transferred this way. If care is exercised, good locational accuracy can be achieved. Smaller grid cell sizes yield better precision.

Changing Scales

Scale changing may be accomplished by three methods: *photography, optical enlarger/reducer,* or the *grid square* method. The most accurate and efficient method is to use photography. The cartographer simply submits the artwork (source map or new work sheet) to a professional photographer, or photographs the art in the cartographic darkroom. The camera, however, is not selective; the designer must generalize and select, either before or after photography. Exact scales can be specified with this method.

The rule is to go from larger to smaller scale for any of these methods of scale changing. In this way, the compiler can select for simplicity, whereas in the opposite direction, information must be added, which may result in error.

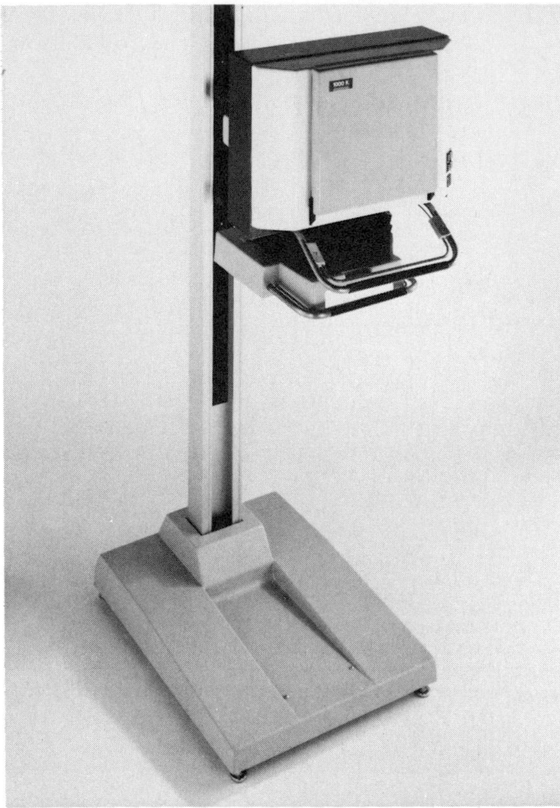

Figure 3.11. An enlarging and reducing optical projector.
This kind of projector is used to change scales during compilation.
Photo courtesy of Map-O-Graph, Inc.

The next most popular method is to use the optical enlarger/reducer to achieve the desired scale. The source maps or work sheets are placed in a special optical machine that projects an image of the original artwork onto a drafting surface. (See Figure 3.11.) The machine is adjusted to achieve the desired scale, and the image is traced onto the new work sheet. These optical machines must be used with caution. The accuracy of the projected image often drops off at the peripheries, causing considerable distortion. This facet may be checked with sample grids produced and projected onto the drawing surface. If possible, it is best to use only the center of the image area, where accuracy is greatest.

Grid squares, discussed above, may also be used for scale changing. The grid cell sizes are simply reduced on the new work sheet, but similar proportions are maintained.

Copyright

Map copyright protects against infringement.[28] The category of maps includes charts, globes, and atlases that deal with terrestrial, marine, planetary, or celestial subjects. Copyright infringement is any violation of the rights of the copyright holder. The rights most applicable to the map case are as follows:

1. to reproduce the copyrighted work in copies,
2. to prepare derivative works based on the original copyrighted work, and
3. to distribute copies of the copyrighted work to the public by sale or other transfer of ownership or by rental, lease, or lending.

The rules governing the copyright of maps are vague at best. In general, a map must display an "appreciable" amount of original cartographic material to be copyrightable. A reprint of a map that is already in the public domain cannot be copyrighted. Generally, outline or base maps are not copyrightable. For example,

> A revised version of a previously published map may be copyrighted as a "new work" if the additions and changes in the new version are copyrightable in themselves. The preparation of most present-day maps involves the use of previously published source material to a significant degree, and the copyrightable authorship therefore is generally based upon elements such as compilation and drawing rather than upon original surveying, aerial photography, and field work alone. Where any substantial portion of the work submitted for registration is

based upon previously published sources, a statement of the nature of the new, copyrightable authorship should be given at the appropriate space on the application submitted to the Copyright Office.[29]

Compilers of thematic maps must exercise caution when using the source materials of others. Care and good judgment are required so that others' rights are not violated. In the United States, maps produced by the federal government are in the public domain and are therefore free of potential copyright problems. The use of privately produced maps, on the other hand, may require permission from the copyright holder.

Notes

1. E. Meynen, ed., *Multilingual Dictionary of Technical Terms in Cartography* (International Cartographic Association, Commission II—Wiesbaden: Franz Steiner Verlag, 1973), p. 137.
2. H. V. Steward, *Cartographic Generalization* (Cartographica, Monograph No. 10—Toronto: University of Toronto Press, 1974), p. 20.
3. David Greenhood, *Mapping* (Chicago: University of Chicago Press, 1964), p. 176.
4. F. McBryde and Paul D. Thomas, *Equal Area Projections for World Statistical Maps* (United States Department of Commerce, Coast, and Geodetic Survey, Special Publication No. 245—Washington, D.C.: USGPO, 1949), p. 7.
5. Charles H. Deetz and Oscar S. Adams, *Elements of Map Projections* (United States Department of Commerce, Coast and Geodetic Survey, Special Publication No. 68—Washington, D.C.: USGPO, 1944), pp. 163–64.
6. J. A. Steers, *An Introduction to the Study of Map Projections* (London: University of London Press, 1962), pp. 161–63.
7. Ibid., p. 182.
8. Erwin Raisz, *Principles of Cartography* (New York: McGraw-Hill, 1962), p. 177.
9. Deetz and Adams, *Elements of Map Projections,* p. 67.
10. Porter W. McDonnell, Jr., *An Introduction to Map Projections* (New York: Marcel Dekker, 1979), p. 61.
11. Deetz and Adams, *Elements of Map Projections,* p. 75.
12. Ibid., pp. 94–96.
13. Mei-Ling Hsu, "The Role of Projections in Modern Map Design," *Cartographica* 18 (1981): 151–86. This informative contribution to the literature of thematic mapping is highly recommended.
14. John P. Snyder, "Labeling Projections on Published Maps," *American Cartographer* 14 (1987): 21–27.
15. Arthur H. Robinson, "Arno Peters and His New Cartography," Views and Opinions, *American Cartographer* 12 (1985): 103–11; John P. Snyder, "Social Consciousness and World Maps," *The Christian Century* (February 24, 1988): 190–92.
16. John Loxton, "The Peters Phenomenon," *The Cartographic Journal* 22 (1985): 106–108.
17. Marshall Faintick, "The Politics of Geometry," *Computer Graphics World* (May, 1986): 101–104.
18. Loxton, "The Peters Phenomenon," pp. 106–108; A multicolored, 35 × 52 in version may be obtained from the Friendship Press in Cincinnati, Ohio.
19. Snyder, "Social Consciousness," p. 192.
20. O. M. Miller and Robert J. Voskuil, "Thematic Map Generalization," *Geographical Review* 54 (1964): 13–19.
21. A. J. Pannekoek, "Generalization of Coastlines and Contours," *International Yearbook of Cartography* 2 (1962): 55–73.
22. Gösta Lundquist, "Generalization—A Preliminary Survey of an Important Subject," *Canadian Survey* (1959): 466–70.
23. Steward, *Cartographic Generalization,* p. 30.
24. Alan G. Hodgkiss, *Maps for Books and Theses* (New York: Pica Press, 1970), pp. 29–32.
25. Barbara A. Bond, "Cartographic Source Material and Its Evaluation," *Cartographic Journal* 10 (1973): 54–58.
26. David Cuff and Mark T. Mattson, *Thematic Maps, Their Design and Production* (New York: Methuen, 1982), p. 88.

27. J. S. Keates, *Cartographic Design and Production* (New York: John Wiley, 1973), p. 170.
28. James W. Cerney, "Awareness of Maps as Objects of Copyright," *American Cartographer* 5 (1978): 45–56.
29. Copyright Office, *Copyright for Maps,* Circular 40F, (Washington, D.C.: USGPO, 1974), p. 2.

Glossary

Albers equal-area projection secant conical projection having equal-area properties; useful for mapping continent-size areas on the earth; used most frequently for medium-scale maps of the United States

Boggs projection equal-area projection useful for mapping the world on one sheet; adequate for small-scale thematic maps containing a world theme

Bonne projection simple conical equal-area projection, useful for mapping continent-size areas of the earth; should not be used for areas of considerable east-west extent

cartographic generalization a central function in cartographic compilation; involves simplification, selection, and emphasis

common-scale method compilation of a new manuscript map from more than one source map, often at different scales, each containing unique information useful to the new map

derived maps maps, usually at intermediate to small scale, that have been compiled from topographic scale maps

Gall-Peters projection based on the earlier work of James Gall in 1855 and adopted by several UN organizations

grid squares method of plotting features from one map to another; can also be used to change scales

Hammer projection equal-area projection useful for mapping the world on one sheet; adequate for small-scale thematic maps containing a world theme

Lambert azimuthal equal-area projection has equal-area properties useful for mapping continent-size areas on the earth

Lambert conformal conic projection has conformal properties; used as the standard reference projection for the plane grid system for many states

map accuracy degree of conformance to established standards; has particular significance with reference to maps at topographic scales but loses relevance when applied to thematic maps

map compilation includes all aspects of collection, evaluation, interpretation, and technical assembly of cartographic material

map copyright legal protection against the infringement of the rights of the copyright holder of a map

Mollweide (or homolographic) projection equal-area projection useful for mapping the world on one sheet; adequate for small-scale thematic maps containing a world theme

National Cartographic Information Center (NCIC) a centralized source for obtaining information about national cartographic products and ordering them

one-to-one method compilation by simple tracing of information from a source map at the required scale onto the new manuscript map

reduction method compilation by tracing of information from one source map onto the new manuscript map being compiled; the new map is then reduced to final size

state plane coordinates rectangular plane coordinate system applied to states; used especially by highway engineers, utility companies, and planners; grids labelled in feet

thematic map compilation construction of a special-purpose map from a variety of previously existing sources, including maps and non-map materials

transverse Mercator projection has conformal properties; used as the standard reference projection for the plane grid system for many states

Readings for Further Understanding

Birch, T. W. *Maps, Topographical and Statistical.* London: Oxford University Press, 1964.

Bond, Barbara A. "Cartographic Source Material and its Evaluation." *Cartographic Journal* 10 (1973): 54–58.

Cerny, James W. "Awareness of Maps as Objects of Copyright." *American Cartographer* 5 (1978): 45–56.

———. *Choosing a World Map.* Falls Church, VA: American Congress on Surveying and Mapping, 1988.

Committee on Map Projections. *Which Map is Best?* Falls Church, VA: American Congress on Surveying and Mapping, 1986.

Copyright Office. *Copyright for Maps* (Circular 40F). Washington, D.C.: USGPO, 1974.

Cuff, David, and Mark T. Mattson. *Thematic Maps, Their Design and Production*. New York: Methuen, 1982.

Deetz, Charles H., and Oscar S. Adams. *Elements of Map Projections.* United States Department of Commerce, Coast and Geodetic Survey, Special Publication No. 68. Washington, D.C.: USGPO, 1944.

Dent, Borden D. "Continental Shapes on World Projections: The Design of a Poly-Centred Oblique Orthographic World Projection." *The Cartographic Journal* 24 (1987): 21–27.

Dickinson, G. C. *Maps and Air Photographs.* New York: John Wiley, 1979.

Espenshade, Edward B., ed. *Goode's World Atlas.* 16th ed. Chicago: Rand McNally and Co., 1983.

Faintich, Marshall. "The Politics of Geometry." *Computer Graphics World* (May, 1986): 101–104.

Gilmantin, Patricia P. "Aesthetic Preferences for the Proportions and Forms of Graticules." *The Cartographic Journal* 20 (1983): 95–100.

Greenhood, David. *Mapping.* Chicago: University of Chicago Press, 1964.

Hazelwood, L. K. "Semantic Capabilities of Thematic Maps." *Cartography* 7 (1970): 69–75, 87.

Hodgkiss, Alan G. *Maps for Books and Theses.* New York: Pica Press,1970.

Hsu, Mei-Ling. "The Role of Projections in Modern Map Design." *Cartographica* 18 (1981): 151–86.

Keates, J. S. *Cartographic Design and Production.* New York: John Wiley, 1973.

Kelley, Philip S. *Information and Generalization in Cartographic Communication.* Unpublished Ph.D. Dissertation, University of Washington, Department of Geography, 1977.

Lawrence, G. R. P. *Cartographic Methods.* London: Methuen, 1971.

Loxton, John. "The Peters Phenomenon." *The Cartographic Journal* 22 (1985): 106–108.

Lundquist, Gösta. "Generalization—A Preliminary Survey of an Important Subject." *Canadian Survey* (1959): 466–70.

Mainwaring, James. *An Introduction to the Study of Map Projection.* London: Macmillan and Company, 1942.

McBryde, F., and Paul O. Thomas. *Equal Area Projections for World Statistical Maps.* United States Department of Commerce, Coast and Geodetic Survey, Special Publication No. 245. Washington, D.C.: USGPO, 1949.

McDonnell, Porter W., Jr. *Introduction to Map Projections.* New York: Marcel Dekker, 1979.

Meynen, E., ed. *Multilingual Dictionary of Technical Terms in Cartography.* International Cartographic Association, Commission II. Wiesbaden: Franz Steiner Verlag, 1973.

Miller, O. M., and Robert J. Voskuill. "Thematic Map Generalization." *Geographical Review* 54 (1964): 13–19.

Pannekoek, A. J. "Generalization of Coastlines and Contours." *International Yearbook of Cartography* 2 (1962): 55–73.

Pearson, Frederick, II. *Map Projection Methods.* Blacksburg, VA: Sigma Scientific, Inc., 1984.

———. *Map Projection Software.* Blacksburg, VA: Sigma Scientific, Inc., 1984.

Raisz, Erwin. *Principles of Cartography.* New York: McGraw-Hill, 1962.

Robinson, Arthur, Randall Sale, and Joel Morrison. *Elements of Cartography* 4th ed. New York: John Wiley, 1978.

Robinson, Arthur H. "Arno Peters and His New Cartography." Views and Opinions, *American Cartographer* 12 (1985): 103–11.

Snyder, John P. *Map Projections Used by the U.S. Geological Survey* 2nd ed. Geological Survey Bulletin 1532. United States Geological Survey. Washington, D.C.: USGPO, 1983.

———. *Map Projections—A Working Manual.* U.S. Geological Survey, Professional Paper 1395. Washington, D.C.: USGPO, 1987.

———. "Labeling Projections on Published Maps." *American Cartographer* 14 (1987):21–27.

———. *Map Projections Used for Large-Scale Quadrangles by the U.S. Geological Survey.* U.S. Geological Survey, Circular 982. Denver: U.S. Geological Survey, 1987.

———. "Social Consciousness and World Maps." *The Christian Century* (February 24, 1988): 190–92.

Steers, J. A. *An Introduction to the Study of Map Projections.* London: University of London Press, 1962.

Steward, H. J. *Cartographic Generalization.* Cartographica, Monograph No. 10. Toronto: University of Toronto Press, 1974.

Thompson, Morris M. *Maps for America.* Untied States Geological Survey. Washington, D.C.: USGPO, 1982.

Thrower, Norman J. W. *Maps and Man.* Englewood Cliffs, NJ: Prentice-Hall, Inc., 1972.

Chapter

4

The Nature of Geographical Phenomena and the Selection of Thematic Map Symbols

Chapter Preview

Cartographers are often faced with decisions about the presentation and symbolization of geographical phenomena and data. They must therefore learn the fundamental concepts of geography: direction, distance, scale, location, distribution, functional association, spatial interaction, and the concept of region. Geographic phenomena are on four levels: nominal, ordinal, interval, and ratio. It is entirely possible to symbolize geographic data at each of these levels of measurement. Cartographic symbol schemes are examined by looking at map types (quantitative or qualitative), symbol dimensions, the forms of geographical distributions, and symbols at the different measurement levels. The designer must know the concepts of geography so that appropriate symbols can be chosen to function with their referents.

Cartography involves the graphic presentation of some aspect of the real world. A map maker usually works in one of two ways: as content-author of the map's presentation of reality or on commission to render someone else's version of reality in graphic form. In either case, the cartographer deals with a wide variety of geographical phenomena. Understanding and interpretation of geographical analyses is often necessary to the selection of the appropriate mapping solution.

Geography and Geographical Phenomena

Designers are increasingly faced with perplexing graphic design problems as geographers become more sophisticated in their analytic methods. Indeed, the types of presentation problems brought about by complex modelling and quantitative approaches used today are forcing cartographers to be more innovative than ever before. Familiarity with basic geographical concepts, trends, and data forms is important for the student designer.

Geography Defined

Our present concern is with modern geography only, not the many changes and philosophical viewpoints in the history of the discipline. Even for present and future geographers, the following integrating materials are useful in illuminating geography's relationship to modern thematic cartography.

In the words of one prominent scholar on the philosophy of geography, Professor Richard Hartshorne, "The intrinsic characteristics of geography are the product of man's effort to know and understand the combinations of phenomena as they exist in areal interrelation in his world."[1] Hartshorne also pointed out that geography does not contain its own particular set of phenomena but studies the integration of heterogeneous phenomena over area, and that geography is both a social and a physical science. The discipline is so diverse that geographers are equally at home studying the distribution of climates throughout the world as they are analyzing transport between large urban centers. A wide range of research interests is one of the hallmarks of geography.

Geography is the study of areal interrelation; it can be called the *science of spatial analysis.* The distinctive thrust of the discipline is its attempt to answer the question, "Why are spatial distributions structured the way they are?"[2] Geographers attempt to reveal the underlying processes that are causally related to spatial patterns or structures. Its central focus on the spatial dimension sets the field of geography apart from other sciences.

In asking questions about the spatial world, geographers look at a wide variety of phenomena: drugstore locations in urban commercial districts, the distribution of rat bites in slum areas, patterns of land ownership in Third World countries, and worldwide language regions. Any phenomenon that has or can be thought to have a spatial attribute is subject to geographical inquiry.

Certain limits, however, appear to restrict the field of geography. Geographers do not normally pay attention to spaces smaller than architectural or larger than terrestrial. As a matter of fact, room geography (the architectural scale) has become the subject of some geographical researchers. The earth and its varying phenomena, at the other extreme, are often studied in geographic research. Worldwide distributional patterns (e.g., climates, soils, or religions) have been of interest to geographers for many years. Within the range between the architectural and terrestrial scales, geographical inquiry has been free to operate.

What? Where? Why? These are the typical questions geographers ask. Answers to "what" questions have traditionally been descriptive. For much of the history of geography, such questions have attracted the most attention, as is natural for an emerging science. For most of our history, we did not know "what" because most of the world was unknown. "Where"

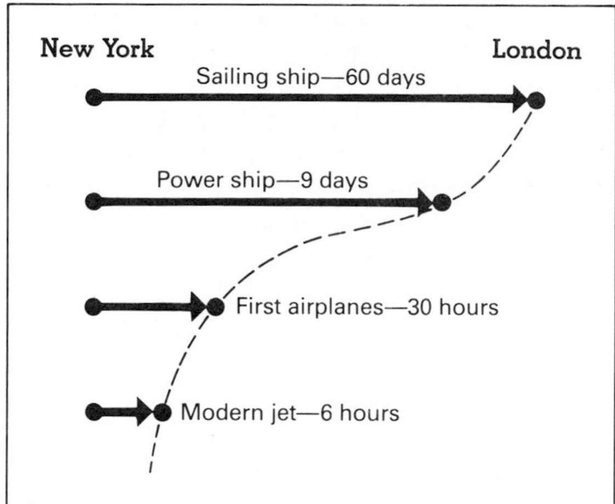

Figure 4.1. The shrinking earth.
As travel times decrease, relative distances diminish. What will it be like in 2025? What effect do shrinking distances have on the way we perceive different peoples?

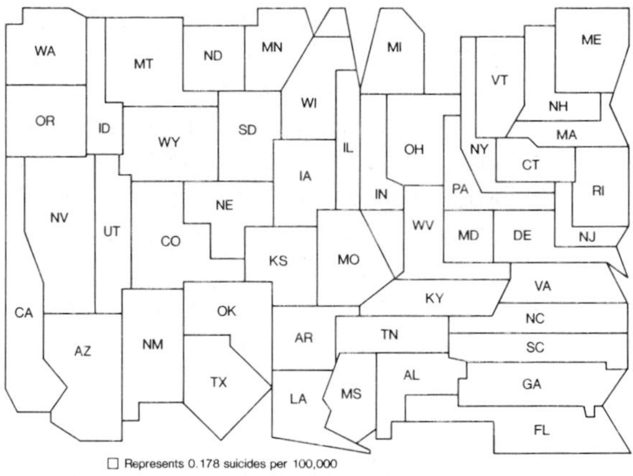

Figure 4.2. Value-by-area cartogram of suicide in the United States.
Note the overwhelming regularity in the spatial pattern of suicide. Mapping suicide rate in this manner reveals this pattern easily.
Map compiled by Monica Hendrix, a student at Georgia State University. Used by permission.

questions have also concerned geographers for a long time, and the answers have been largely descriptive until recent years. **Absolute location** with reference to a fixed coordinate system, such as the earth's geographic reference system, is characteristic of the answers to "where" questions. Answering questions in terms of absolute location allowed geographers and cartographers to "fill in" the world map.

More recently, "where" questions have taken on more dynamic qualities as **relative location** has gained prominence.[3] The traditional view of location is based on a rigid grid system—an arbitrary cartesian one or the geographic grid. Regardless of the reference grid adopted, location was viewed as absolute and unchanging. In the new view, location is a relative matter, dynamic and ever-changing; the distance between any two places may be stretching, or more likely shrinking. For example, London is closer to New York now than it was 100 years ago because travel time has decreased. (See Figure 4.1.) Remote places on the earth have become more accessible because transportation technology has led to more rapid travel and telecommunications to more rapid exchange of ideas.

Geographical exploration of these new spatial dimensions has resulted in different ways of mapping. In the conventional geography of absolute location, the mapping activity is based on Euclidean geometry. True and predictable distances and angles are the main features of Euclidean space. In order to map spaces that reflect relative perspectives (e.g., social, psychological, cost, or time), new forms of maps have emerged. For example, special transformations (cartograms) show content spaces and illustrate relative perspectives better than traditional maps and are often useful in showing the underlying spatial structure of dynamic distributions. (See Figure 4.2.) Nonconventional mapping often employs logarithmic projections (Figure 4.3), which tend to enlarge places closer to the center and diminish peripheral spaces. These projections are useful in showing geographical distributions or processes whose magnitudes decrease outward with distance. For example, migration patterns often show numerous short moves closer to home and fewer long-distance moves.

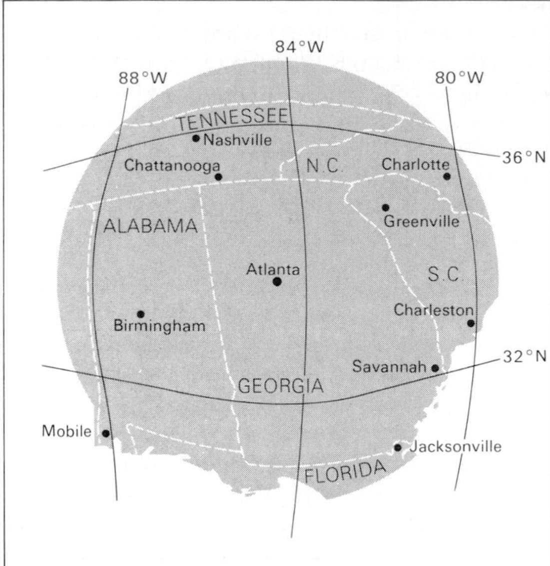

Figure 4.3. Logarithmic projection is suitable for showing activities in relative space.
Here we see that the nature of the projection enlarges places closer to the projection center and reduces places farther away.

Major Themes in Geographical Research

Certain major themes have dominated geographical inquiry over the past several decades.[4]

Areal or Spatial Association

The geographer studies the spatial location of a variable with respect to another or others functionally or causally related to it. An example at the micro-scale would be the locational linkage of florist shops to urban high-rise office buildings.

Forms and Processes

In this approach, the researcher examines the processes that result in random, clustered, or regular spatial distributions. For example, some studies show the location, extent, and spacing of central places (towns); others attempt to describe the tectonic processes that yield landform types and their distributions.

Spatial Interaction

Investigations of this type seek explanations for the flow (interaction) of people or goods from place to place. Forces of attraction and friction over distance are integrating ideas. An example of this theme would be the development of road networks in response to regional traffic patterns.

Distance Decay

These investigations examine the decreasing occurrence of certain events from a central point. Nodal area development (the influence of a city in its region) or "action spaces" (places where daily activities take place) can be determined by distance-decay studies.

A geographer may conduct research in many of these theme areas, and sometimes a particular study may overlap several areas. Geography is an eclectic field of study—one reason for its richness and attraction.

Key Concepts in Geography

Regardless of the approach taken or the nature of the content studied, certain key concepts bind geography together. The cartographer deals with geographical items and must therefore be thoroughly familiar with the concepts of geography.

Direction

Direction is a geometric property that can be used to describe relative location. True direction (azimuth) is measured by reference to a meridian. Direction can also be used to describe the location of a point in other spaces (such as on cartograms), which are more difficult because a different system of reference must be established.

Geography is a very old field of study. Writers of geography were among the earliest scholars of antiquity, and spoken geography must have been widely practised long before the invention of writing. During the long history of man's effort to gain further understanding of the forces and objects of his environment, the essential nature of geography remained unchanged. Today as in the past, geography is concerned with the arrangement of things on the face of the earth, and with the associations of things that give character to particular places. Those who face problems involving the factor of location, or involving the examination of conditions peculiar to specific locations, are concerned with geography, just as those who must be informed about a sequence of events in the past are concerned with history.

Source: Preston E. James, "Introduction: The Field of Geography," in American Geography, Inventory and Prospect, Preston E. James and Clarence F. Jones, eds. (Washington, D.C.: Association of American Geographers, 1954), p. 4.

Distance

The property of **distance** has long been instrumental in geographical inquiry. Distance is conventionally expressed in *units of length,* based on Euclidean geometry, but it can be expressed in terms of time, cost, stress, or any number of other units. As geographical inquiry becomes more and more behaviorally oriented, a greater variety of units is likely to be used to measure distance.

Scale

Scale relates to the size of the area being studied and determines the level of precision and generalization applied in the study. Micro-scale studies are done over small earth areas; macro-scale studies deal with larger areas. The nature of the inquiry sets the scale, and the scale in turn determines the degree of generalization. The study of the distribution of manufacturing employment in United States cities (macro-scale) usually dictates that cities be perceived as points; therefore, the intracity variation in employment areas becomes unimportant. Studying this same phenomenon at the intracity level (micro-scale), however, leads the researcher to focus on the distribution of employment areas within the city. Also, human interaction varies with scale, as shown in Table 4.1. Voice, touch, taste, and smell can link people at the personal-space scale but are impossible at greater distances. At the global-space scale, whole new networks are formed to facilitate interaction.

Location

A geographical feature can be evaluated in terms of either absolute or relative location. Absolute location is restricted to a reference or fixed-grid system. An object's location relative to another (or to all others) involves dynamic aspects of the elements and reflects cultural and technological change.

Distribution

Distribution is the spatial property of areal *pattern.* The chief pattern types are *dispersed* and *agglomerated.* A densely settled area is an example of agglomerated pattern, as is clustering of manufacturers in a given area.

Localization

This term applies to spatial clustering; it measures the concentration of an activity in a limited area. The concentration of petrochemical plants along the Gulf Coast is an example of localization of industry.

Functional Association

The concept of **functional association** involves the *occurrence* of objects or activities in space as a response to the presence of other elements. One feature is located as it is because another is close by. Medical offices tend to locate near hospitals. *Connectivity* is a special case of functional association; it implies actual ties.[5] Functional association can also be

Table 4.1 Human Interaction at Varying Scales

Type of Space	Spatial Range	Some Interaction Characteristics	
	Radius from person (representative fraction)	**Primary modes of interaction**	**Number of people involved**
Personal	Arm's length (1:1)	Voice, touch, taste, smell	1–3
Living or working (private space)	10–50 ft (1:50)	Audio and visual (sharp focus on facial expression, slight movement, or tonal changes)	50–400
House and neighborhood	100–1000 ft: line of sight (1:500)	Audio-visual (sound amplification)	100–1000
		Limit of proxemic interaction	
City–hinterland	4–40 mi: 60-minute one-way commute (1:50,000)	Local news media, TV, urban institutions, commuter systems	50,000–10,000,000
Regional–national	200–3000 mi (1:500,000)	National network of news media, national organizations (common language desirable)	200,000,000+
Global	12,000 mi (1:50,000,000)	International communication networks, translation services, world organizations, international blocs	3,000,000,000+

Source: Modified from John F. Kolars and John D. Nystauen, Geography: The Study of Location, Culture, and Environment (New York: McGraw-Hill, 1974), p. 29.

called *areal coherence.* Much of geographical explanation results from the discovery of functional associations.

Spatial Interaction

Chief among the geographical concepts is **spatial interaction.** It involves the movement of people, ideas, or goods from place to place over the surface of the earth. A special case of interaction is *spatial diffusion*—the spread of people (migration), things (e.g., agricultural tools), or ideas.

The Regional Concept

A **region** is defined by the researcher to study the likenesses and differences of areas on the earth. Regions may be cultural or physical. A region has an *internal homogeneity* that sets it apart from others. Defining regions often provides insight into geographical distributions and the explanations for them.

Concept of Change

Last but not least is the **concept of change.**[6] There is really no such thing as equilibrium in geography—only temporary phases in the process of transformation. If change is accepted as a general principle, then rate, regularity, and direction of change become important. One of the most dynamic elements of a city is its land-use map. It should never be viewed as a static picture of conditions, but rather an inventory of conditions for a small slice of time. Tomorrow it will be different. To what degree the land-use map changes, how fast, and with what regularity and direction, are important in explaining the economic forces in urban land-use theory.

Measurement in Geography

Geographers do not study a unique set of things that they can call their own. Geography is eclectic; the geographer uses techniques of observation in seeking to explain images of reality.[7] Geographers can study any phenomenon that has variability of spatial dimension, be it from the cultural world (people-oriented) or the physical (nature-oriented). Geography can be applied to such diverse topics as human poverty and drifting sand on a beach.

A distinction must be made between **geographical phenomena** and **geographical data.** Data are facts from which conclusions can be drawn. *Information* is another word that can be used in place of *data.* Geographical data are selected features (usually numerical) that geographers use to describe or measure, directly or indirectly, phenomena that have a spatial quality. For example, the phenomenon of climate can be described in part by looking at precipitation data. From the cultural world, the phenomenon of spatial interaction can be described and measured by collecting and analyzing data on telephone calls. In a recent study ranking cities, the phenomenon of stress brought about by psycho-social pathology was recorded with rates of alcoholism, suicide, divorce, and crime *data.*[8] Cartographers often map data when they think they are mapping phenomena.

Spatial Dimensions of Geographical Phenomena

Various **forms of geographical phenomena** can be considered to exist:

1. *Point* (zero-dimensional)
2. *Line* (one-dimensional)
3. *Area* (two-dimensional)
4. *Volume* (three-dimensional)
5. *Space-time* (four-dimensional)[9]

Most geographical study involves explanation of phenomena of the first four kinds, and to some degree the fifth.

An interesting feature of phenomena is that their form is intricately related to scale and may change with the level of inquiry. A city, for example, may be a point phenomenon at the macro-scale but can be considered to have two-dimensional qualities when examined at micro levels. At one level of investigation, a road may be a link between points (one-dimensional), but at micro levels the road can have the two-dimensional, areal qualities of length and width.

Examples of volume phenomena include landforms, oceans, and the atmosphere. Other geographical phenomena that are conventionally treated as three-dimensional because of their similarity to volumes are rainfall, temperature, growing-season days, and such derived ratios as population density and disease-related deaths. Rainfall collects in a glass and is volumetric; people piled close together make up a volume. Space-time phenomena are best exemplified by succession (for example, human settlement over time) and migration/diffusion.

Thematic cartographers are faced with a dilemma: they do not have a unique way of symbolizing each different class or form of geographical phenomena.

Discrete, Sequential, and Continuous Phenomena

Besides the above dimensional classification of geographical phenomena, they can be classed as *discrete* (or discontinuous), *sequential,* or *continuous.* **Discrete geographical phenomena** are those that do not occur between spatial observations, that form a separate entity, and that are individually distinguishable from place to place. Cows on the landscape are discrete phenomena because there are no other cows between the observed ones. Discrete phenomena can be either dispersed or concentrated; this is largely determined by scale. The important point is that the counting unit is the distinguishing characteristic of discrete phenomena.[10] Discrete scales used by geographers involve whole numbers (numbers of people, rivers, cities, cases of disease, and so forth).

In special cases lines, or linear phenomena, can be referred to as *sequential.* A series of points or discrete elements makes up a sequential phenomenon. Roads, telephone lines, and boundaries are examples—they can be redefined as points appearing close together.

Continuous geographical phenomena, either two- or three-dimensional, exist spatially between observations. For example, landform elevation is a continuous phenomenon, because everywhere between observations there are other elevations. Temperature is another example: Between spatially arranged weather stations where observations are taken, temperature exists. With continuous phenomena any number of possible values may occur between observations, and division of whole numbers is likely (45.22 m for example).

The cartographic process of symbolization has traditionally mapped a variety of geographical phenomena by symbol schemes that are based on the concept of discrete and continuous forms. For example, although population is discrete (people do not exist between people), population density is usually considered to be continuous. The use of an areal unit of land whose ratio is continuous (people/sq km) to express density assumes continuity from place to place. To confuse the issue even more, population density is often mapped using data aggregated by discrete real units (e.g., states or counties). Although these areal units are discrete, the data are assumed to be uniform or continuous within units and change only between units.

Measurement Scales

S. S. Stevens, a noted scientist and psychologist, has said that measurement is the "assignment of numerals to things so as to represent facts and conventions about them."[11] It has been customary in recent years for geographers to classify the ways they measure events into **measurement scales.** Measurement is an attempt to structure observations about reality. Ways of doing this can be grouped into four levels, depending on the mathematical attributes of the observed facts. A given measurement system can be assigned to one of these four levels: nominal, ordinal, interval, and ratio, listed in increasing order of sophistication of measurement. Methods of cartographic symbolization are chosen specially for representing geographical phenomena or data at these levels of measurement.

Nominal Scaling

Nominal scaling is the lowest level of measurement scale,[12] sometimes considered to be no more than an 0-or-1 classification. In this system, objects are classed into groups that are often labeled with letters or numbers. At this level of measurement, we cannot perform any mathematical operations between classes. We can only ascertain equality or inequality between groups (region A = region B; region A $\neq$ region C). *Identification* is the key word in the use of this measurement scale. An example would be the identification of wheat regions, corn regions, and soybean regions. Each crop region is distinct; arithmetic operations between the regions are not possible at this level.

Ordinal Scaling

The underlying structure of ordinal scales is a hierarchy of rank. Objects or events are arranged from least to most or vice-versa, and the information obtainable is of the "greater than" or "less than" variety. Ordinal scaling provides no way of determining how much distance separates the items in the array.

There are several types of ordinal measurement. In *complete* ordering, every element in the array has its own position, and no other element can share this position. This kind of ordinal scale is considered relatively strong because *statistical* observations about the ranking are possible. This is not the case with the second main type, *weak* ordering, in which elements can share positions (called *paired ranks*) along the ordinal continuum.

One interesting feature of ordinal scales is that we can attach any numerical scale to the ranking without violating the underlying structure. If we know that the order is K L M, we can have either $K = 5$, $L = 3$, $M = 1$ or $K = 500$, $L = 300$, $M = 1$ while retaining the original scale. Remember that in ordinal scaling we do not know how much difference separates the events in the array. Several geography students spend a summer touring fifty large cities throughout North America. After they return, their professor asks them to rank the cities based on their appeal, from best-liked to least-liked. In this array, a city's position is known in the overall ranking, but it is not possible to discern how much it differs from those ranking above and below it.

Interval Scaling

At this measurement scale, we can array the events in order of rank and know the distance between ranks. Observations with numerical scores at the interval scale are important in geographical analysis because data at this level are needed to perform fundamental statistical tests having predictive power. From interval scales, we can ascertain equality or inequality of interval.

Magnitude scales at the interval level have no natural origin; any beginning point may be used. The classic example is the Fahrenheit temperature scale. It cannot be said that 50° F is twice as hot as 25° F. There are no absolute values associated with interval scales; they are *relative*.

Ratio Scaling

Like interval scaling, ratio scaling involves ordering events with known distances separating the events. The difference is that ratio-scale magnitudes are absolute, having a known starting point. For example, 100 miles is twice as far as 50 miles. Ratio scaling is important to geography because more sophisticated statistical tests can be performed using this level of measurement.

The different levels of measurement are summarized in Table 4.2 and discussed in more detail in Chapter 5. The amount of information obtainable, statistical confidence, and predictive power increase as one progresses from nominal to ratio scaling.

Table 4.2 Measurement Scales and Methods of Studying Them

Scale	Relations	Appropriate Statistics		
		Central tendency	**Dispersion**	**Tests**
Nominal	Equivalence	Mode	Variation ratio	Chi-square contingency coefficient
Ordinal	Equal to; greater than	Median	Percentiles	Spearman's rho; Kendall's Tau
Interval	Equal to; greater than; difference of interval	Mean	Standard deviation	*F*-test; *t*-test; Pearson's *V*
Ratio	Equal to; greater than; difference of interval; difference of ratio	Geometric mean	Coefficient of variation	*F*-test; *t*-test; Pearson's *V*

Source: Adapted from Peter J. Taylor, *Quantitative Methods in Geography: An Introduction to Spatial Analysis* (Boston: Houghton Mifflin, 1977), p. 41; and David Harvey, *Explanation in Geography* (New York: St. Martin's Press, 1969), p. 312; originally published in S. S. Stevens, "On the Theory of Scales of Measurement," Science 103 (1946): 677–680.

Measurement Error

It is not usually possible to measure without some error, of one or all the following types: observer error, instrument error, environmental error, or error due to the observed material. Cartographers are faced with these types of error to a lesser or greater degree and should be sufficiently aware of their dangers to develop methods of minimizing them.

Observer error relates to the performance of the investigator, resulting in biased measurement. This type of error usually stems from the inability of observers to remove themselves from the measurement process. For example, an interviewer may consistently word questions in such a way as to favor a particular kind of response; an instrument reader may misread a dial because of visual parallax.

Errors caused by poorly designed or malfunctioning recording instruments belong to a class called *instrument error*. These errors may be biased (i.e., in one direction) or random. The classic example of the former is a thermometer that consistently reads one-half degree above the true temperature. If at all possible, the researcher should state the degree of instrument error in order to alert the reader to the measurement's reliability.

Conditions extraneous to the measurement can lead to *environmental error*. Errors in instrument recording result if the conditions deviate excessively from those in which the instrument was designed to operate. Observer error occurs, for example, when excessively cold temperatures cause the observer to lose concentration. Errors in the observed material would result from such environmental factors as high winds causing freak variations in the object of the measurement. The researcher must provide adequate environmental control in the context of the nature of the study. Environmental conditions may need to be stated as part of the study.

Errors due to changes in the *observed material* can lead to imperfect results. People interviewed for a psychological survey may interact with the researcher, causing the observer to measure the wrong behavior. These forms of error may be random or persistent (noncompensating). Observers need to watch out for such situations and inform readers of any potential sources of error they recognize.

Of special concern today are errors that may lurk in computerized data bases. MacEachren, for example, warns us:

> Because of the precision that computer systems are capable of, there is a disturbing tendency to accept solutions arrived at by these systems as being accurate as well. In order for students to become successful makers or users of maps, they must learn to critically evaluate the potential for error inherent in mapping systems and geographic data bases that are increasingly being used to generate maps.[13]

In cartography, error sources range from false images of reality to tricky reproduction techniques. Not all errors can be eliminated, but every conceivable means must be employed to reduce error to tolerable levels. Each mapping method discussed here has its own error sources, based on its unique way of mapping and symbolization. The thematic map designer must learn to take these error sources into account.

Data Sources

Geographical data are of three kinds: those obtained from direct field observation, those obtained from archival sources, and those generated from theoretical work.[14] Field observations may be either quantitative or qualitative in nature. Archival sources may be areal (maps, air photographs), or essentially nonareal. Theoretical work can involve modelling in a variety of forms. Cartographers are called on from time to time to treat data from each of these sources. Of particular interest, because of their extent of coverage and easy accessibility, are enumeration (census) data aggregated for areal units.

The Geographical Areal Unit

Geographical areal units may be of two kinds, natural and artificial.[15] *Natural areal units* are exemplified by such phenomena as lakes, countries, farms, cities, and other geographical elements having two dimensions. The boundaries are intimately tied to the feature. *Artificial areal units* are imposed by researchers in their attempt to organize reality and facilitate data collection. Examples are regions, Metropolitan Statistical Areas (MSAs), and census tracts.

It is likewise possible to work with either singular or collective areal units. *Singular areal units* correspond to only themselves, whereas one *collective unit* is composed of several singular units. The important distinction is that data collected at the singular level can be used in combination to make inferences about collective units. However, it is difficult to make inferences about singular units from collective-unit data. Suppose you have collected data at the singular level. It is possible to compute a mean and use it to describe the nature of the region composed of the individual units. By contrast, if the mean is all you have, you cannot describe the nature of each singular unit—the process of data aggregation can actually yield a *loss* of information.

The Multidimensional Characteristics of Thematic Map Symbols

The selection of symbols is one of the most difficult tasks for the cartographic designer. The choice is wide and the rules loosely defined.

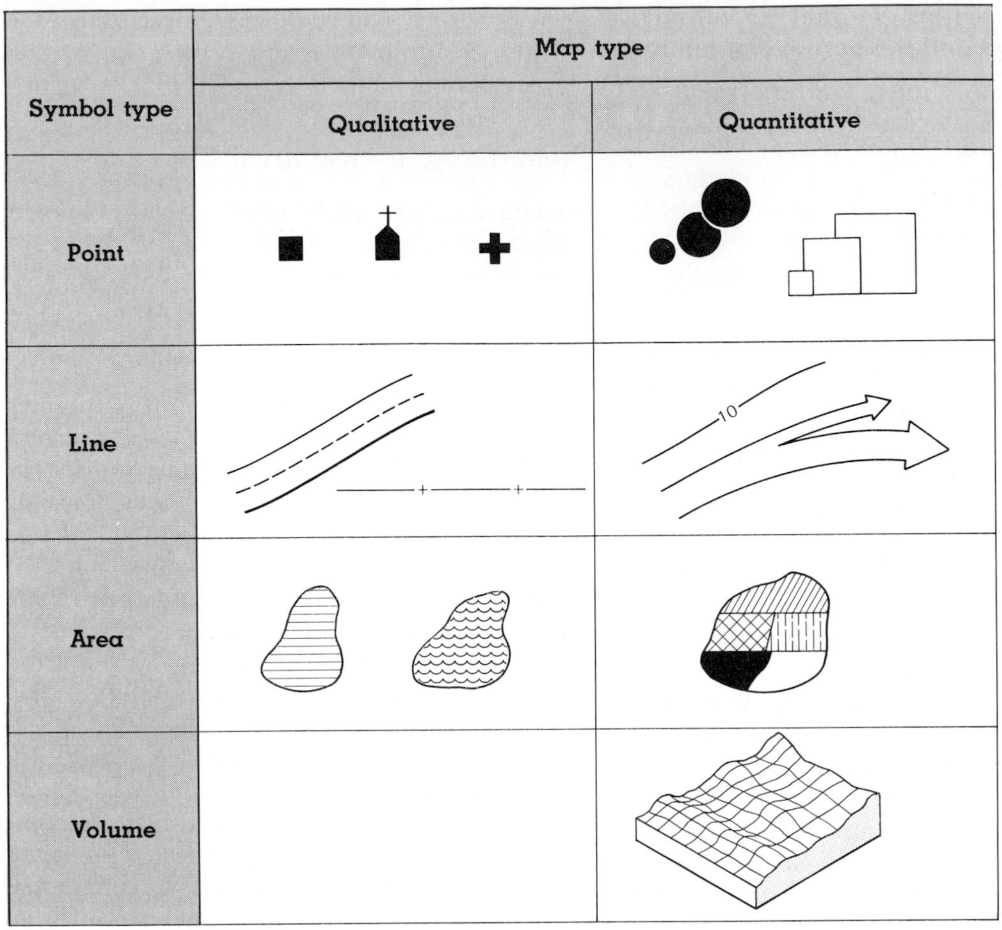

Symbol type	Map type	
	Qualitative	Quantitative
Point		
Line		10
Area		
Volume		

Figure 4.4. Various map symbols and map types.
Some common symbols used in both map types, qualitative and quantitative, are illustrated here. These are only a few of the large number of individual symbols that may be chosen. Point, line, and area symbols have been used for many years. To this group we may add the three-dimensional diagram made popular by the use of computer graphics (for quantitative mapping).

Some designers view this as a plus, since more experimentation and self-expression is possible without rules. Others would prefer to have a "cookbook." Symbol selection between these two extremes is increasingly based on a compelling system of logic.

Map design was earlier described as a process using various filters to select the design elements of the final map. One important filter involves the selection of quantitative and qualitative symbols. As explained in Chapter 1, symbols are the graphic marks used to encode the thematic distribution onto the map. From a vast array of symbols having different dimensions, the designer selects the symbol that best represents the geographical phenomena. Fortunately, the task is reduced somewhat by controlling factors, such as cartographic convention and the inability of most map readers to easily understand the more complex symbols that might be chosen.

Symbol Types, Symbol Visual Dimensions; Qualitative and Quantitative Mapping

The three generally recognized cartographic **symbol types** are *point, line,* and *area.*[16] To this list might be added the three-dimensional diagram so commonly used to describe volumetric distributions. (See Figure 4.4.) The first three symbol types have been virtually standard in thematic mapping.

The design of cartographic symbols requires attention to their visual qualities. These **symbol dimensions** are *shape, size, color hue, color value, color intensity, pattern orientation, pattern arrangement,* and *pattern texture.*[17] The cartographic designer makes selections on the basis of visual dimensions, such that the symbol is functionally tied to the level of measurement of the data. For example, we can draw circles sized proportionally to values in a data set because size variations correspond to intervals in the array. Size, however, cannot be the visual dimension when we scale symbols to nominal data because we do not know the numerical distance between nominal classes.

Thematic maps can be either qualitative or quantitative. The chief purpose of qualitative mapping is to show the locational characteristics of phenomena or data. Amounts, other than the "eyeballing" possible by comparing map areas, are not the content communicated by such mapping. Quantitative maps, on the other hand, stress how much of something is present at various locations. Not all of the symbol dimensions can be used in both qualitative and quantitative maps. (See Table 4.3.) The dimensional qualities of each symbol type (point, line, and area) have unique graphic expression for each map type. A point symbol, for example, can be used in either qualitative or quantitative mapping. In the former map type, shape is the quality of visual dimension, while size is the dimensional quality in the latter map type.

Table 4.3 Symbol Dimensions and Map Type (applicable to point, line, and area symbols)

Symbol Dimension	Map Type	
	Qualitative	Quantitative
Shape	X	
Size		X
Color		
Hue	X	
Value		X
Intensity		X
Pattern		
Orientation	X	
Arrangement	X	
Texture		X

Source: Adapted from Phillip C. Muehrcke, *Map Use, Reading, Analysis and Interpretation* (Madison, Wis.: JP Publications, 1978), pp. 62–63.

The design of thematic map symbols takes place within the limits of these visual dimensions and is governed by the type of thematic map. There are literally hundreds of kinds of symbols devised, even within these constraints, although certain ones are standard. Figure 4.5 presents a number of unique symbols that fit each of the categories listed by symbol and map type.

Characteristics of Geographical Phenomena and Symbol Selection

There is a traditional correspondence between geographical phenomena (point, line, area, and volume) and the employment of symbol types

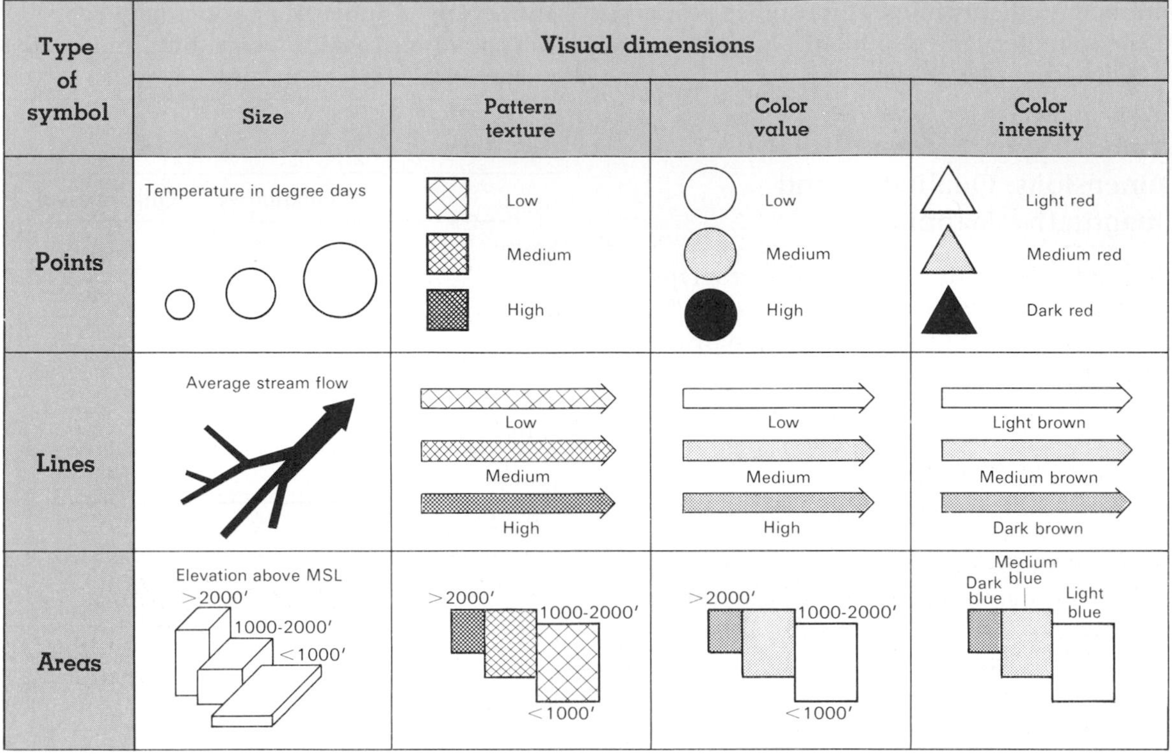

Figure 4.5. Types and symbols useful in quantitative mapping and their visual dimensions.
Courtesy JP Publications.

(point, line, area). Of course, the match is not a convenient one-to-one correspondence. Point data (e.g., cities at the appropriate scale) are customarily mapped by point symbols such as dots, or scaled circles. Roads, which are linear phenomena, can be mapped by line symbols; geographical phenomena having areal extent (lakes, countries, nations) can be mapped with areal symbolization—patterns or tints. Geographical landforms have been mapped by a form of area symbolization called *hypsometric tinting:* Areas between selected elevation boundaries are rendered in various color shades. Elevation is also mapped by *contours of elevation,* which are line symbols. The use of three-dimensional *diagrams* may become a standard form of symbol to represent volume phenomena.

The three symbol types may be matched to the four measurement scales to produce a valuable **typology of map symbols.** (See Figure 4.6.) Cartographic presentations are thus possible even at the highest (interval/ratio) levels of measurement.[18] Choosing to symbolize the data at a level below that which is possible for their level of precision is unwise and results in loss of information.[19] Symbolization above the level of precision of the data is a questionable practice because it forces the cartographer to make assumptions not warranted by the data.

Another consideration in symbol selection is the nature of the geographical phenomenon

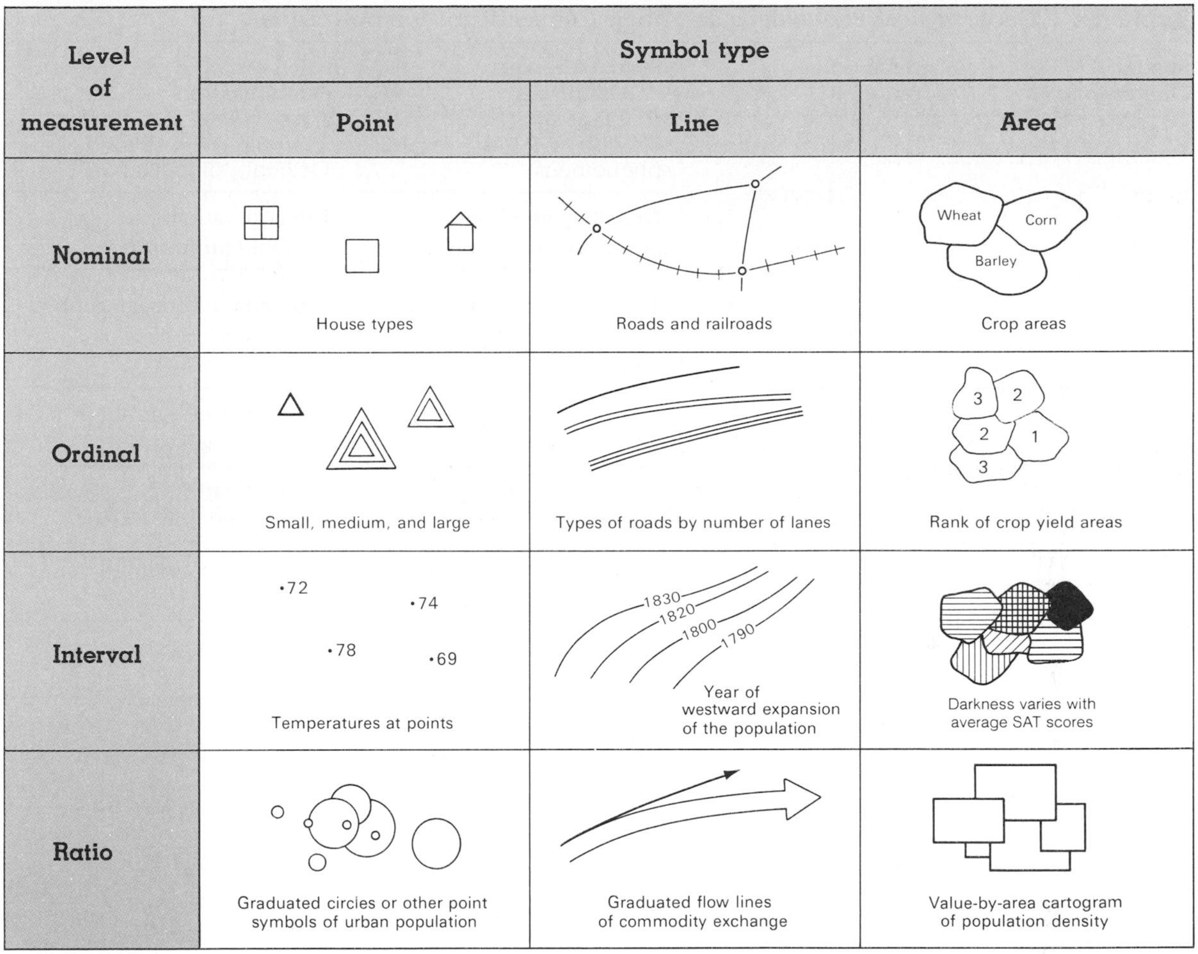

Level of measurement	Symbol type		
	Point	**Line**	**Area**
Nominal	House types	Roads and railroads	Wheat Corn Barley Crop areas
Ordinal	Small, medium, and large	Types of roads by number of lanes	3 2 2 1 3 Rank of crop yield areas
Interval	•72 •74 •78 •69 Temperatures at points	1830 1820 1800 1790 Year of westward expansion of the population	Darkness varies with average SAT scores
Ratio	Graduated circles or other point symbols of urban population	Graduated flow lines of commodity exchange	Value-by-area cartogram of population density

Figure 4.6. Typology of map symbols.
Classification of thematic map symbols is possible by examining their symbol type and the level of measurement being used.

being mapped. The designer should select a symbol that will best show its distributional form: discrete, sequential, or continuous. Table 4.4 illustrates several ways of incorporating the different levels of measurement; it can be used as a comprehensive guide to symbol selection.

In practice, a large portion of the thematic map designer's role is to select a symbolization plan that best portrays the appropriate concept of geography. Regardless of the nature or level of sophistication of the geographical study, the cartographic communication of the findings can usually be reduced to one of the key concepts. It is therefore important for cartographers to be familiar with the nature of geography. Matching the symbolization plan to the geographical concept is part of the challenge and enjoyment of cartographic design.

Table 4.4 Geographical Phenomena and Their Conventional Symbolization

Spatial Dimension	Distributional Form		Spatial Message and Emphasis	Symbolization
Point	Discrete	Dispersed	Patterns of point phenomena	Nominal point symbol Ordinal point symbol
			Density patterns of point phenomena	Interval/ratio point symbol (uniform dots)
		Concentrated	Total quantities at points or centers of unit areas	Interval/ratio point symbol (graduated symbols)
Line	Sequential		Pattern of linear phenomena	Nominal line symbol Ordinal line symbol
			Interactions among places	Interval/ratio line symbol (graduated flow lines)
Area	Continuous		Patterns of areal phenomena	Nominal area symbol Ordinal area symbol
			Patterns of quantities in areal units	Interval/ratio area symbols
Volume	Continuous	Smooth	(Volumetric phenomena cannot be mapped with nominal symbols)	
			Relative variations of volumetric phenomena	Ordinal area symbol pattern (to show slopes)
			Spatial form and gradient of volumetric phenomena	Interval/ratio line or area symbol (isarithms with or without shaded patterns)
		Stepped	Variations of volumetric phenomena	Interval/ratio area symbol (choropleth)

Source: Adapted from Mei-Ling Hsu, "The Cartographer's Conceptual Process and Thematic Symbolization," American Cartographer 6 (1979): 121.

In short, maps and other graphics comprise one of three major modes of communication, together with words and numbers. Because of the distinctive subject matter of geography, the language of maps is the distinctive language of geography. Hence sophistication in map reading and composition, and ability to translate between the languages of maps, words, and numbers are fundamental to the study and practice of geography.

Source: John R. Borchert, "Maps, Geography, and Geographers, *The Professional Geographer.* 39(1987): 387–89.

Enumeration Data, Geographical Units, and Census Definitions

Geographers and cartographers frequently use enumeration data provided by federal government census. Enumeration data are aggregated data, tabulated by specific geographic areas, and are often the only information available. The geographical units for which these data are published are important background for the thematic cartographer.

The United States Census

One of the principal sources of geographical data (information associated with a particular areal unit) is the United States decennial *Census of Population and Housing.* Census data have been collected for the United States since 1790. The purpose of the census of population is to take a "head count" for determining representation in the House of Representatives. Its scope, thoroughness, and reliability have increased over the years; it now constitutes the single best source of broad-based demographic and socioeconomic data available to geographic researchers.

Census Geography

Of particular interest to the cartographer is the nature of the **census geography** used in the decennial censuses. The Census Bureau tabulates data into more than 40 types of geographic areas. (See Table 4.5 for some types.) Other censuses are also conducted by the Census Bureau, including the Census of Governments, six Economic Censuses, and the Census of Agriculture. As Table 4.5 shows, more geographic detail is provided by the Census of Population and Housing.

The larger the geographic area, the greater the data detail available. At the block level, for example, the only 1980 population data published are total population, number Black, and counts of Asian/Pacific Islanders and those of Spanish origin, number under 18-years old, and number 65 years or older. A few items of housing are also available at the block level. At the census tract level, numerous socioeconomic data are published, in addition to expanded population characteristics. This is partly because of disclosure rules. The Census Bureau will not publish data that permit violation of confidentiality; this could be a problem in small geographic areas having very small populations.

The boundaries of census geographic areas are determined by a variety of agencies. Governmental authorities provide the Census Bureau with information on the boundaries of such units as states, congressional districts, Indian reservations, election districts, counties, minor civil divisions, incorporated places, and city wards. Metropolitan areas are statistical divisions defined by the Office of Management and Budget. Local government agencies or public committees and the Census Bureau determine other statistical geographic areas, including urbanized areas (UAs), census county divisions (CCDs), census designated places (CDPs), census tracts, enumeration districts (EDs), blocks, and block groups.

There is a major distinction between governmental area units and statistical area units. The boundaries of governmental units are established by law; they really exist and are often marked on the landscape. Census statistical units, on the other hand, are established merely for convenience in enumeration and tabulation. They have been devised to satisfy the needs of governmental agencies and public users for analyzing population characteristics or change. For example, urbanized area boundaries are used to measure the areal extent of larger American cities, but they do not exist on the landscape—only on maps provided by the Census Bureau.

Table 4.6 lists the governmental and statistical area units for which the census, currently or in the past, has provided data, and the or-

Table 4.5 Major Geographic Areas and Census Bureau Programs

	Population and Housing Censuses		Census of Govern-ments	Economic Censuses						
	Area reports	Subject reports		Retail trade	Whole-sale trade	Selected services	Manufac-tures	Mineral indus-tries	Con-struc-tion indus-tries	Agri-culture
United States	a	a	a	a	a	a	a	a	a	a
Census Regions	a	s	a	a	a	a	a	a	a	a
Census Divisions	a	s	a	a	a	a	a	a	a	a
States	a	s	a	a	a	a	a	a	a	a
Metropolitan areas	a	s	a	a	a	a	a			a
Urbanized areas	a									
Counties	a		a	a	a	a	a	a		a
Incorporated places	a		a	s	s	s	s			
Census designated places (CDPs)	a									
Minor Civil Divisions (MCDs)	a		a							

Note: Other areas unique to the population and housing census are urbanized areas, urban/rural, and congressional districts.

a All areas

c All, by addition of components

s Selected areas—larger or with more activity

Table 4.6 Governmental and Statistical Geographical Units Used by the United States Bureau of the Census

Governmental Areas	Statistical Areas
United States	Geographic Regions of the United States
States (and outlying areas), American Indian Reservations, Native Villages	Geographic Divisions of the United States
Counties (and county equivalents)	Metropolitan Statistical Areas (MSAs)
Minor civil divisions (MCDs)	Urbanized areas (UAs)
Incorporated places	Urban and rural areas
Congressional districts	Census designated places (CDPs)
Election precincts (in selected areas)	Census county divisions (CCDs)
Zip Code areas (a governmental administrative area)	Census tracts
	Block numbering areas (BNAs)
	Block groups (BGs)
	Census blocks
	County groups (in the Public—Use Samples)

Source: United States Bureau of the Census, Census Geography—Data Access Description No. 33 (Washington, D.C.: USGPO, 1979), pp. 2, 6; and United States Bureau of the Census, "Census Geography—Concepts and Products," *Factfinder* No. 8 rev. (August 1985), p. 2.

Area reports / Subject reports	Population and Housing Censuses	Census of Governments	Economic Censuses: Retail trade	Wholesale trade	Selected services	Manufactures	Mineral industries	Construction industries	Agriculture
Census County Divisions (CCDs)	a								
Census Tracts	a								
Enumeration Districts (EDs) and block groups*	a								
ZIP Code areas*	a								
Blocks*	a								
Central business districts**	c				a				
Major retail centers**					a				

* Not in printed reports

** These areas are dropped beginning with the 1987 Census of Retail Trade.

Source of Table: United States Bureau of the Census, Census Geography—Data Access Description No. 33 (Washington, D.C.: USGPO, 1979), p. 2; United States Bureau of the Census, "Census Geography—Concepts and Products," *Factfinder* No. 8 rev. (August 1985), p. 6.

ganizational and hierarchical arrangements of these are illustrated in Figure 4.7. Figure 4.8 presents the geographical relationships of the principal metropolitan areal units. The Athens, Georgia, urbanized area map is reproduced as an example in Figure 4.9. Definitions of many of the geographical units are included in Appendix B.

Caution is necessary when dealing with census data taken from decennial censuses. Definitions often change from one census to the next, so they must be checked before using these data sources. Comparability is doubtful in many instances. Furthermore, census geography may likewise change from time to time—urbanized area boundaries, for example, are usually different from one census to the next. In fact, the urbanized area concept was not even used prior to 1950. Census tract boundaries may change because boundary features may disappear (perhaps because of urban renewal). A thorough examination of all data definitions and geographical descriptions is recommended before using these data.

The Census of Canada

The Census of Canada is conducted decennially and published in the first year of the decade; the current census is from 1981. It is taken by law for the purpose of establishing representation in the Federal House of Commons. Like the United States *Census of Population and Housing,* the *Decennial Census of*

Figure 4.7. Hierarchical arrangement of census geographical units.
Note that some hierarchies overlap. For example, counties may be subdivided into minor civil divisions (MCDs) or census county divisions (CCDs).

Source: Redrawn from United States Bureau of the Census, "Census Geography—Concepts and Products," *Factfinder,* No. 8 rev. (August 1985), p. 2.

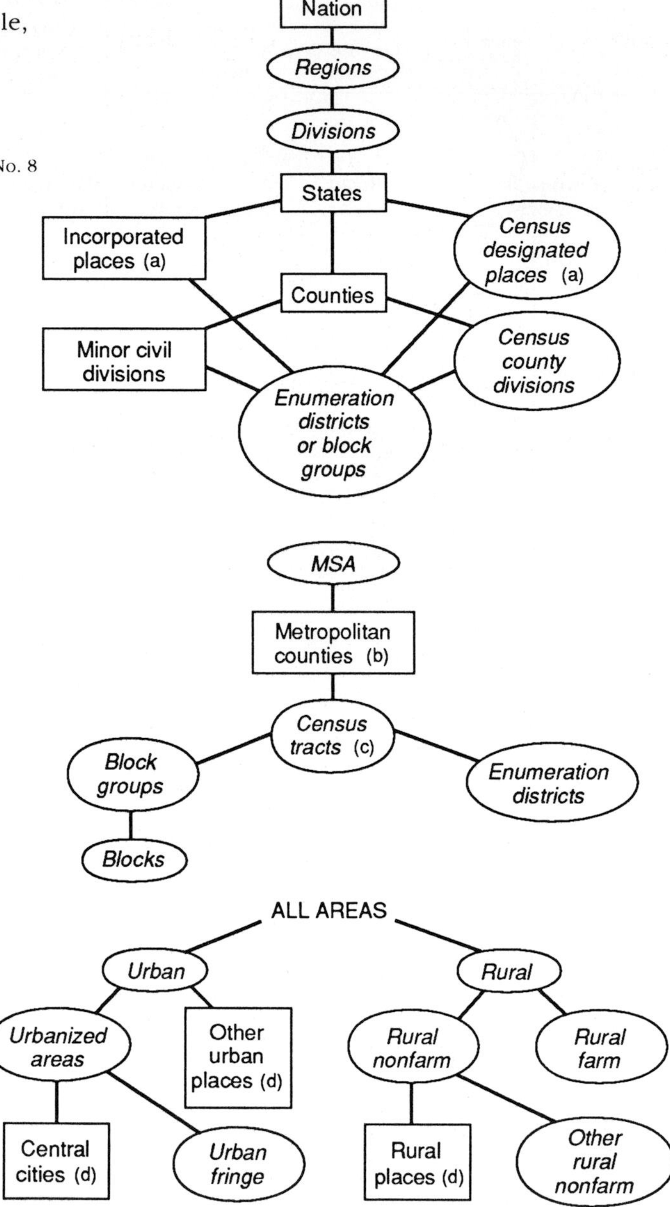

Notes:

(a) Note that places (incorporated and census designated) are not shown within the county and county subdivision hierarchy, since places may cross the boundries of these areas.

(b) In New England, MSA's are defined in terms of towns and cities, rather than counties (as in the rest of the country).

(c) Census tracts subdivide most MSA counties as well as about 200 other counties. As tracts may cross MCD's and place boundries, MCD's and places are not shown in the hiearachy.

(d) Includes both governmental and statistical units.

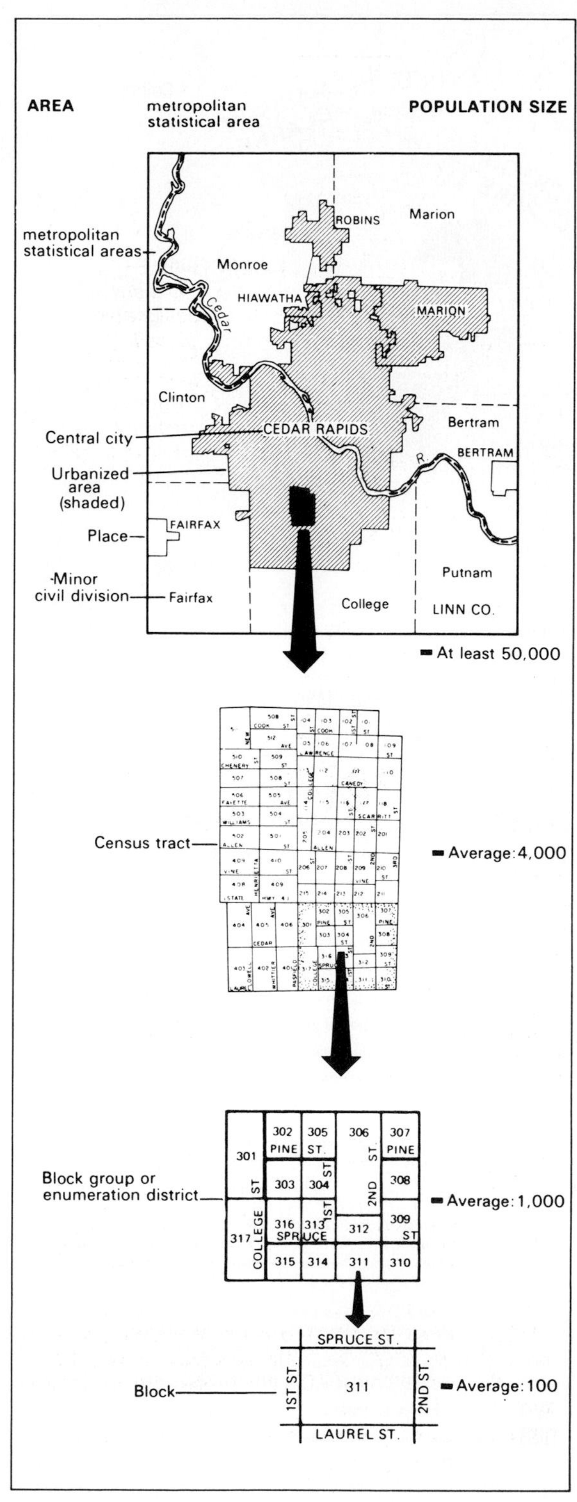

Figure 4.8. Metropolitan census geographic areas.

Source: United States Bureau of the Census, *Census Geography,* Data Access Description No. 33 (Washington, D.C.: USGPO, 1979), p. 4.

Figure 4.9. **Athens, Georgia urbanized area (UA) map.**
Note the areas that go to make up the urbanized area.

Source: United States Bureau of the Census, *1980 Census of Housing.* General Housing Characteristics: Georgia (Washington, D.C.: USGPO, 1982), p. 298.

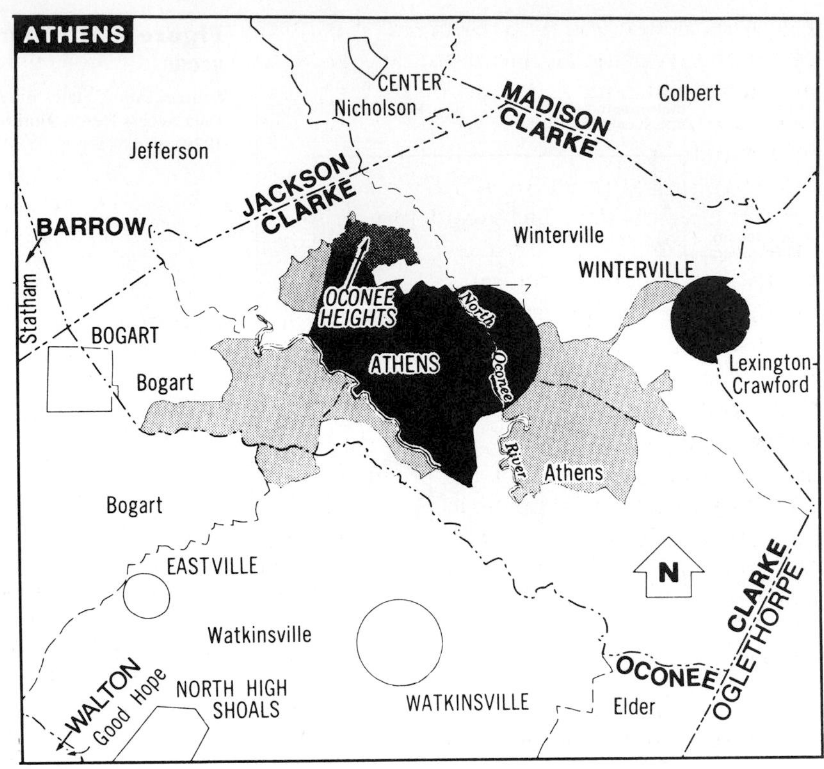

SYMBOLS	TYPE STYLES	GEOGRAPHIC AREAS
	MEXICO	Foreign country
	IOWA	State
	DANE	Subject SMSA county
	POWER	County not part of subject SMSA
	Locust	County subdivision
	SILAS	Incorporated place
	PERDIDO	Census designated place
	Pyramit	American Indian reservation
	Lake Wingra	Major water feature

Note: All political boundaries are as of January I, 1980. Boundaries of small areas may not be depicted exactly due to scale of map. Where boundaries coincide, boundary symbol of higher level geographic area is shown.

Open six-spoked asterisk following place name indicates the place is coextensive with a county subdivision. The county subdivision name is shown only when it differs from that of the place.

Solid eight-spoked asterisk following an incorporated place name indicates the place is treated as a county subdivision for census purposes.

COMPONENTS OF URBANIZED LAND AREA

Incorporated place

Census designated place

Other area

Canada has become more complex. It is used by federal, provincial, and municipal governments for planning and decision making. Private business and industry also use the information it provides.

Geographical units for which data are tabulated are as follows: Canada, provinces, census divisions, census subdivisions, unorganized townships, federal electoral districts, census metropolitan areas, census agglomerations, and census tracts.[20] As is the case in the United States census, the geographic divisions of the Census of Canada are hierarchical and can be divided into both governmental and statistical area units. A perusal of a current census of Canada will demonstrate that using its documents is similar to using those of the United States census.

Notes

1. Richard Hartshorne, "The Concept of Geography as a Science of Space, from Kant and Humboldt to Hettner," in *Introduction to Geography: Selected Readings,* Fred E. Dohrs and Lawrence M. Sommers, eds. (New York: Thomas Y. Crowell, 1967), pp. 89–90.
2. Ronald Abler, John S. Adams, and Peter Gould, *Spatial Organization: The Geographer's View of the World* (New York: Prentice-Hall, 1971), p. 56.
3. *Ibid.,* p. 59.
4. Douglas Amedea and Reginald G. Golledge, *An Introduction to Scientific Reasoning in Geography* (New York: John Wiley, 1973), pp. 4–6.
5. John F. Kolars and John D. Nystuen, *Geography: The Study of Location, Culture, and Environment* (New York: McGraw-Hill, 1974), pp. 14–16.
6. Jan O. M. Broek, *Geography: Its Scope and Spirit* (Columbus, OH: Charles E. Merrill, 1965), pp. 72–76.
7. David Harvey, *Explanation in Geography* (New York: St. Martin's Press, 1969), pp. 293–98.
8. Robert Levine, "City Stress Index," *Psychology Today* (November, 1988): 53–58.
9. Harvey, *Explanation in Geography,* p. 351.
10. Kirk W. Elifson, *Fundamentals of Social Statistics* (Reading, MA: Addison-Wesley, 1982), p. 31.
11. Harvey, *Explanation in Geography,* p. 306.
12. Peter J. Taylor, *Quantitative Methods in Geography: An Introduction to Spatial Analysis* (Boston: Houghton Mifflin, 1977), p. 39.
13. Alan M. MacEachren, "Map Use and Map Making Education: Attention to Sources of Geographic Education," *The Cartographic Journal* 23 (1986): 115–22.
14. Peter Haggett, *Locational Analysis in Human Geography* (New York: St. Martin's Press, 1966), p. 186.
15. Harvey, *Explanation in Geography,* pp. 351–52.
16. Kang-tsung Chang, "Data Differentiation and Cartographic Symbolization," *Canadian Cartographer* 13 (1976): 60–68.
17. Joel Morrison, "A Theoretical Framework for Cartographic Symbolization," *International Yearbook of Cartography* 14 (1974): 115–27.
18. David Unwin, *Introductory Spatial Analysis* (London: Methuen, 1981), p. 25.
19. Phillip Muehrcke, *Thematic Cartography* (Commission on College Geography, Resource Paper No. 19—Washington, D.C.: Association of American Geographers, 1972), p. 15.
20. Statistics Canada, *1971 Census of Canada,* vol. 1, pt. 1 (Population) (Ottawa: Ministry of Industry, Trade, and Commerce, 1974).

Glossary

absolute location specifies position relative to a fixed grid system

census geography the divisions and descriptions of areal entities, both political and statistical; used to tabulate census (enumerated) data

concept of change fundamental idea in geography; both space and time events are merely brief states of equilibrium separating periods of change

continuous geographical phenomena extend unbroken and without interruptions

direction a line going from one place to another; azimuth is a formal way of specifying direction

discrete geographical phenomena do not occur between spatial observations; distinguishable individual entities

distance the number of linear units separating two objects; in relative location, distance can be measured in units of time, cost, psychological factors (e.g., stress), or other units

distribution characteristic of the spatial pattern of geographical phenomena; dispersed and agglomerated are two kinds

forms of geographical phenomena point, linear, areal, volumetric, and space-time; sometimes called the *spatial dimensions* of geographical phenomena

functional association one object is located relative to another because of a link (e.g., physical or economic) between the two; also called *areal coherence* or *connectivity*

geographical areal units entities either natural (e.g., lakes, countries, oceans) or artificial (e.g., winter wheat region, urbanized areas); often used to tabulate aggregated data

geographical data facts about which conclusions can be drawn; chosen to describe geographical phenomena; associated with a spatial dimension

geographical phenomena elements of reality that have spatial attributes; any spatial phenomena can be the subject of geographical analysis within the limits of scale

geography science that deals with the analysis of spatially distributed phenomena

measurement scales definite rules established for each of four levels of describing numerical events; the nominal, ordinal, interval, and ratio scales; precision increases from nominal to interval/ratio

region mental constructs formed for the purpose of understanding spatial patterns of geographical phenomena; regions are based on the idea of internal homogeneity

relative location position of an object relative to another object; dynamic in nature

scale level of inquiry, ranging from micro (small earth areas) to macro (large earth areas); scale determines precision and generalization

spatial interaction movement of people, goods, or ideas between places

symbol dimensions shape, size, color hue, color value, color intensity, pattern orientation, pattern arrangement, and pattern texture

symbol types cartographic symbols are classed as point, line, or areal symbols

typology of map symbols description of thematic map symbols based on a cross-tabulation of measurement scales and symbol types

Readings for Further Understanding

Abler, Ronald; John S. Adams, and Peter Gould. *Spatial Organization: The Geographer's View of the World.* New York: Prentice-Hall, 1971.

Ackerman, Edward A. *Geography as a Fundamental Research Discipline.* University of Chicago, Department of Geography, Research Paper No. 53. Chicago: University of Chicago, 1958.

Amedeo, Douglas, and Reginald G. Golledge. *An Introduction to Scientific Reasoning in Geography.* New York: John Wiley, 1975.

Borchert, John R. "Maps, Geography, and Geographers." *The Professional Geographer* 39 (1987): 387–89.

Broek, Jan O. M. *Geography: Its Scope and Spirit.* Columbus, OH: Charles E. Merrill, 1965.

Chang, Kang-tsung. "Data Differentiation and Cartographic Symbolization." *Canadian Geographer* 13 (1976): 60–68.

Dent, Borden D., ed. *Census Data: Geographic Significance and Classroom Utility.* NCGE Pacesetter Series. Chicago: National Council for Geographic Education, 1976.

Elifson, Kirk W. *Fundamentals of Social Statistics.* Reading, MA: Addison-Wesley, 1982.

Haggett, Peter. *Locational Analysis in Human Geography.* New York: St. Martin's Press, 1966.

Hartshorne, Richard. "The Concept of Geography as a Science of Space, from Kant and Humboldt to Hettner." In *Introduction to Geography: Selected Readings.* Fred E. Dohrs and Lawrence M. Sommers, eds. New York: Thomas Y. Crowell, 1967.

Harvey, David. *Explanation in Geography.* New York: St. Martin's Press, 1969.

Hsu, Mei-Ling. "The Cartographer's Conceptual Process and Thematic Symbolization." *American Cartographer* 6 (1979): 117–27.

Jenks, George. "Conceptual and Perceptual Error in Thematic Mapping." *Proceedings* (American Congress on Surveying and Mapping, March, 1970): 174–88.

Kolars, John F., and John D. Nystauen. *Geography: The Study of Location, Culture and Environment.* New York: McGraw-Hill, 1974.

Levine, Robert. "City Stress Index." *Psychology Today* (November, 1988): 53–58.

MacEachren, Alan M. "Map Use and Map Making Education: Attention to Sources of Geographic Education." *The Cartographic Journal* 23 (1986): 115–22.

Monmonier, Mark S. *Maps, Distortion and Meaning.* Association of American Geographers, Resource Paper No. 75–4. Washington, D.C.: Association of American Geographers, 1977.

Morrison, Joel. "A Theoretical Framework for Cartographic Generalization with Emphasis on the Process of Symbolization." *International Yearbook of Cartography* 14 (1974): 115–27.

Muehrcke, Phillip. *Thematic Cartography.* Commission on College Geography, Resource Paper No. 19. Washington, D.C.: Association of American Geographers, 1972.

Statistics Canada. *1971 Census of Canada,* Vol. 1, pt. 1 (Population). Ottawa: Ministry of Industry, Trade and Commerce, 1974.

Stevens, S. S. "On the Theory of Scales of Measurement." *Science* 103 (1946): 677–80.

Taaffe, Edward J. "The Spatial View in Context." *Annals* (Association of American Geographers) 64 (1974): 1–16.

Taylor, Peter J. *Quantitative Methods in Geography: An Introduction to Spatial Analysis.* Boston: Houghton Mifflin, 1977.

Unwin, David. *Introductory Spatial Analysis.* London: Methuen, 1981.

Chapter 5

Preprocessing Geographical Data: Common Measures Useful in Thematic Mapping

Chapter Preview

Preprocessing involves both mathematical and statistical measures designed to develop single-value statistics that describe nominal, ordinal, or interval/ratio data distributions: notably the mode, median, and mean. Dispersion characteristics are also part of the description of distributions: the variation ratio, percentiles, and the standard deviation. Skewness and Kurtosis are descriptors of distribution shapes and prove useful in classing activities. The useful measures of areal concentration are the coefficient of areal correspondence, areal means, and the location quotient. The mathematical measurement of association between spatial distributions is another important preprocessing activity. Regression, correlation, and mapping residuals from regression are very useful data preprocessing techniques. Data classification, especially numerical classification, is a most important preprocessing task in which groups are developed, based on similarity criteria. Graphic methods and methods of one-way analysis of variance are borrowed from statistics for use in numerical classification.

Data preprocessing—analyzing and preparing geographical data for mapping—has become increasingly important in the broader contexts of thematic mapping. Cartographers often work with professionals, other than geographers, who customarily deal with statistics or who need data summarized before portraying them in graphic form. This chapter introduces the more common measures used in preprocessing data for mapping.

The Need for Data Preprocessing

The goal of cartographic design is to convey thought in graphic form as simply as possible. The contents may be simple, based on routinely derived measures, or complex, based on lengthy and sophisticated data compilations. With the widespread use of computers for data storage and processing, cartographers are finding it increasingly difficult to portray the data simply. Preprocessing data is one way to reduce them into forms more suitable for straightforward communication. It provides mechanisms for data reduction, enhancement, retention of key features, and simplification of spatial patterns.[1]

Mathematical and Statistical Methods

The principal methods of preprocessing data for mapping are mathematical and statistical. Mathematical measures are usually quite simple and involve the production of common measures such as ratios, proportions, and percentages, including such activities as area and shape determination, calculating trends, or indices of comparison. For our purposes, *mathematics* will be considered to mean arithmetic, geometry, algebra, and other methods dealing with magnitudes and relationships that can be expressed in terms of numbers and symbols.

Statistics is a separate discipline and has been defined as follows:

> Statistics is a body of *concepts* and *methods* used to collect and interpret data concerning a particular area of investigation and to draw conclusions in situations where uncertainty and variation are present.[2]

Statistics is *not* the displaying of large numerical arrays such as census enumeration tables, although statisticians have historically spent much time compiling such data.
Three goals of statistics may be listed:[3]

1. Summarize observations
2. Describe relationships between two variables
3. Make inferences, both estimations and tests of significance

The first two of these will be treated in detail below.

Ratio, Proportion, and Percent

Three of the simplest *derived* indices used by geographers and cartographers are ratios, proportions, and percents. The first two are often used interchangeably, but in fact they differ. The **ratio** is a good way of expressing the *relationship* between data. It is expressed as

$$\frac{f_a}{f_b}$$

where f_a is the number of items in one class and f_b the number in another class. The number of items is referred to as the *frequency,* a term used in most statistical work.

As an example, consider the ratio of the sexes in the population of a small town. If we discover that there are 3000 males, this is one thing, but if we discover that there are also 1500 females, we learn something else. The ratio is expressed this way:

$$\frac{3000}{1500} = 2$$

The 2 really means

$$\frac{2}{1}$$

Thus, there are two males in the town for every female.

A familiar ratio in geography is *population density,* defined as the number of people per square kilometer or other areal unit. Thus, if the same small town has a total area of ten square kilometers, the density would be:

$$\frac{4500}{10}$$

We must reduce this ratio so that unity (1 sq km) is in the denominator:

$$\frac{4500}{10} = \frac{x}{1};\ 10x = 4500;\ x = 450;$$

$$\frac{4500}{10} = \frac{450}{1}$$

or, simply, the population density is 450.

Proportion is the ratio of the number of items in one class to the total of all items. It is written

$$\frac{f_a}{N}$$

where f_a is the number of items (frequency) in a class and N is the total number of items or total frequency. In our small town, the proportion of males would be:

$$\frac{3000}{4500} = \frac{30}{45} = \frac{2}{3} = .66$$

Typically, proportions are multiplied by 100, yielding **percentage.** In this case,

$$.66 \times 100 = 66\%$$

It is easier to describe the proportion of males in the small town as 66 percent than as .66.

Percentage is a familiar and useful index in many geographical analyses. It is simply derived and, when coupled with location, can yield information about geographical concentration.

Variables, Values, and Arrays

A distinction must be made between values and variables. **Variables** are the subjects used in statistical analyses: for example, height, rainfall, elevation, and income. Individual observations of the variable are called **values.** Values of the rainfall variable could be 16, 18, 19, 25, 34, and so on. It is typical in statistics to name a variable with a capital letter (e.g., X or Y) and the values by subscripts to the lower case form of the letters: x_1, x_2, x_3, x_4, x_5, and so on. To complete the symbology, the subscript is usually referred to simply as i. Thus x_i refers to any unspecified value of the variable X.

After an examiner completes the process of collecting measures, the values are arranged in an order called an **array.** Typically, an array arranges the values in ascending or descending order of magnitude or else lists nominal observations in similar groups. The following is a nominal array:

apple apple apple orange orange
banana banana banana
peach peach peach peach

Arrays may also be developed for ordinal data:

city city city town town town town
town village village village village village

Here is an example of an interval array:

1500 1450 1400 1395 1200 1195 1100
1050 1025 955 905 859 825 795 750

Table 5.1 Typical Frequency Distributions in Nominal, Ordinal, and Interval Scales

Nominal		Ordinal		Interval	
Crop	Frequency	Hardiness	Frequency	Yields	Frequency
Wheat	10	Most	12	400	18
Barley	8	Intermediate	15	300	22
Corn	6	Least	8	250	23
				175	15
				125	10

Frequency Distributions and Histograms

A *frequency distribution* is an ordered array that shows the frequency of occurrence of each value. The distribution is simply a method of portraying the data in a way that allows for easy inspection. The simple array can be placed in a table, as in Table 5.1.

When the number of observations is quite large, a simple table showing the frequency of each class or value becomes quite cumbersome. A common solution is to develop a frequency table from which a histogram can easily be drawn. The frequency table is constructed by dividing the total range of the data (the difference between the lowest and highest values in the array) into non-overlapping subgroups called *classes.* The number of values (*cell* or *class frequencies*) belonging to each class is then counted. Usually, the relative frequency in each cell is determined. **Relative frequency** is that proportion of the observations belonging to a particular class:

$$\text{Relative frequency of a class} = \frac{\text{class frequency}}{\text{total number of observations}}$$

When added together, all relative frequencies should total 1.00. (See Table 5.2.)

Table 5.2 Classed Frequency Distribution and Relative Frequency Table

Class Interval	Frequency	Relative Frequency
0–10,000	39	.204
10,000–20,000	51	.267
20,000–30,000	33	.172
30,000–40,000	25	.130
40,000–50,000	20	.104
50,000–60,000	18	.094
60,000–70,000	5	0.26
Total	191	1.000

Histograms are special graphs drawn using the information from a relative frequency table. Most often, a *relative frequency histogram* is drawn by plotting the class boundaries on the horizontal axis and erecting rectangles over each class as a base. The vertical axis is scaled in divisions of relative frequency. If the horizontal axis is laid out with the widths of the rectangle bases equal, their heights are simply scaled to the relative frequencies shown on the vertical axis. (See Figure 5.1.)

Relative frequency histograms with proportional areas are more useful than simple histograms that have their vertical axes scaled to frequency. Since the total height of all rectangles equals one, a quick visual inspection will reveal how much of the total each class represents.

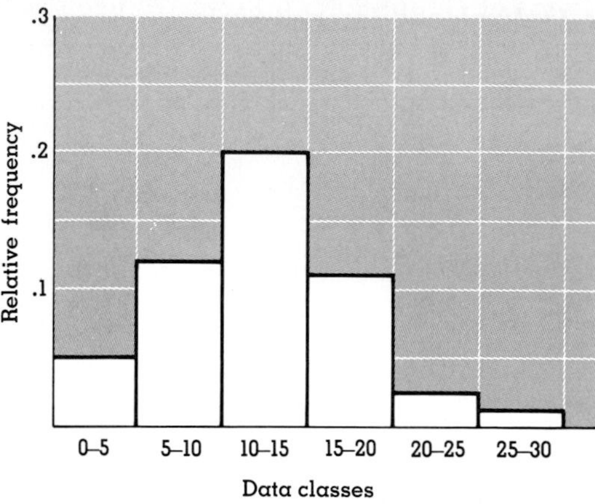

Figure 5.1. Relative frequency histogram. Heights of bars are drawn proportional to their relative frequencies. The sum of the bar heights equals one.

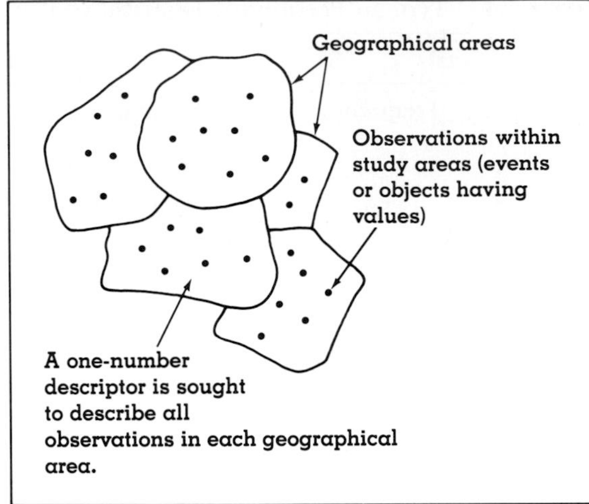

Figure 5.2. Summarizing geographical data. Summarizing geographical data usually involves developing a one-number descriptor from several numerical observations occurring in geographical study areas.

Summarizing Data Distributions

The principal purpose of most descriptive statistics is to develop one best numerical description for a data set. (See Figure 5.2.) In many mapping cases, events or observations are found at specified locations within a bounded area. The area may be a specially created study region, a political subdivision, or a statistical division defined by the Census Bureau. It is advantageous to work with the latter areas because they function as enumeration areas for which numerous data are available.

Each of the events or observations conceptualized in Figure 5.2 has values associated with it. Because mapping the values at each location is usually impracticable, one best numerical descriptor is often computed for all observed values within the bounded area. The values are *aggregated*—this has already been done for us in many census enumerations. For example, the median family income in the Little Rock, Arkansas, MSA in 1969 was $8285; this is an aggregated and summarized descriptor for all the individual incomes reported in the MSA.

Several methods exist for determining the one best descriptor for a data set, depending on the scale of measurement. It is often desirable to know how values are arrayed around the descriptor, since this can show how effective the descriptor is. Are values closely clustered around the descriptor or uniformly spread throughout the data range?

Nominal Scale: The Mode and the Variation Ratio

For nominal scale measurement, the *mode* is the one best numerical descriptor, or summary statistic, and the *variation ratio* is the index that tells us how the values are arrayed in the distribution and defines a distribution in terms of its dominant class. The **mode** is the name or number of the class in a nominal distribution having the greatest frequency.

A market researcher conducted a study to determine shoppers' opinions of the types of apparel sold in a large women's apparel store. Ten Zip Code areas immediately surrounding the store were identified as the study area, and 200 people from each Zip Code area were asked to choose among the following alternatives:

A. Prefer the better-quality merchandise, and am willing to pay for it

Table 5.3 Results of Marketing Opinion Research on Types of Goods Sold at Large Women's Apparel Store
(m = modal class)

Zip Code area 1	Frequency	Zip Code area 6	Frequency
A	10	A	0
B	147 m	B	172 m
C	32	C	26
D	11	D	2
Zip Code area 2	Frequency	Zip Code area 7	Frequency
A	125 m	A	38
B	12	B	70 m
C	33	C	42
D	30	D	50
Zip Code area 3	Frequency	Zip Code area 8	Frequency
A	137 m	A	24
B	10	B	20
C	25	C	24
D	28	D	132 m
Zip Code area 4	Frequency	Zip Code area 9	Frequency
A	153 m	A	36
B	5	B	40
C	5	C	46
D	37	D	78 m
Zip Code area 5	Frequency	Zip Code area 10	Frequency
A	5	A	32
B	135 m	B	17
C	50	C	9
D	10	D	142 m

Zip Code Area	Variation Ratio
1	.265
2	.375
3	.315
4	.235
5	.325
6	.140
7	.650
8	.340
9	.610
10	.290

B. Always seek lower-quality merchandise, will usually not pay for high-priced goods
C. Prefer the better-quality merchandise but will only pay it if on sale
D. Like the current mix of merchandise and will buy both the better-quality and lower-quality merchandise

Table 5.3 lists the final results of the study. The modal class is identified for each Zip Code area.

In this example, people in Zip Code areas 2, 3, and 4 clearly prefer the better-quality merchandise and are apparently willing to pay for it; people in Zip Code areas 8, 9, and 10 have no clear-cut preferences. The results of this study could be used by the store manager to direct advertising to certain geographical locations.

It is clear from these results that the frequencies tallied in Zip Code areas 7 and 9 do not indicate modes as strong as for the others. The variation ratio (v) is a quantification of how well the mode describes the distribution. The **variation ratio** is the proportion of occurrences not in the modal class. It is computed this way:

$$v = 1 - \frac{f_{\text{modal}}}{N}$$

where f_{modal} is the frequency of the modal class and N equals the total number of occurrences in all classes. The smaller the v is, the better it summarizes the distribution; the higher the v, the less satisfactory it is.

The variation ratio has been computed for each Zip Code area in our example. It is clear that areas 7 and 9 contain less definite modal classes than in the others ($v = .65$ and $.61$, respectively). This indicates to the advertiser that a different approach might be used in these areas.

Modes and variation ratios are the most important indices of central tendency and dispersion for nominally scaled data. They can also be used for interval/ratio scales, although other statistics are preferable at such scales.

Table 5.4 **Arkansas: Health System Area I**

Counties Ranked by Birth Rates (per 1000 population)

County	Rank		Birth Rate
Boone	1		26.2
Perry	2		16.1
Washington	2		16.1
Newton	3		16.0
Crawford	4		15.8
Polk	5		15.6
Sebastian	6		15.5
Madison	6		15.5
Hot Spring	7		15.0
Conway	8		14.7
Johnson	9		14.5
Montgomery	9		14.5
Pope	10	Median	14.3
Logan	11		13.7
Franklin	11		13.7
Clark	12		13.3
Pike	13		13.2
Carroll	14		12.3
Learcy	14		12.3
Garland	15		12.2
Benton	16		11.5
Yell	16		11.5
Marion	17		11.2
Scott	18		10.9
Baxter	19		9.8

Source: *Arkansas Division of Health Statistics,* Arkansas Vital Statistics, 1980, p. 61.

Ordinal Scale: The Median and Percentiles

When data are collected at the ordinal scale, the statistic that describes central tendency is the *median,* and one of several *decile ranges* describes the dispersion of observations around the median. The **median** can be defined as the point that neither exceeds nor is exceeded in rank by more than 50 percent of the observations in the distribution. Its determination is straightforward when there are an odd number of ranks; special rules apply when the number of values is even.

Just as the mode can be determined for interval/ratio observations, the median can be calculated for such higher-level data. It is probably more common for the median and decile ranges to be used in this way.

Table 5.4 illustrates an array of 1980 birth rates for an odd number of Arkansas counties. The median birth rank is 10; it is the typical rank. Its value of 14.3 is the middle value of the array. For this geographic area, it might be said that 14.3 births per 1000 people is the typical

Table 5.5 Arkansas: Health Systems Area III

Counties Ranked by Birth Rates (per 1000 population)

County	Rank		Birth Rate
Monroe	1		19.4
Pulaski	1		19.4
Lonoke	2	Median	16.8
Saline	3		13.8
Faulkner	4		13.6
Prairie	5		12.1

Median rank = 2½

$$\text{Median value} = \frac{16.8 + 13.8}{2} = 15.3$$

Source: *Arkansas Division of Health Statistics,* Arkansas Vital Statistics, 1980, p. 73.

Table 5.6 Two Hypothetical Ranked Distributions with Identical Medians

Distribution A		Distribution B
69		722
67		341
66		250
52		99
49		98
37	Median	37
36		30
32		28
32		14
32		1
30		1

birth rate. We judge the other ranks, or counties, as being high or low *relative to the median value.*

Special problems arise when the number of ranks is even. When dealing solely with ordinal data, we can say only that the median *position* is halfway between the two middle ranks. We *cannot* say that it is *exactly* halfway in terms of value—with ordinal data, we have no idea of values. However, when dealing with ratio data with an even number of ranks, the median is the average of the values of the two middle ranks. (See Table 5.5.)

Because the concept of the median is related to position, a median value may be the same for two distributions but fail to provide adequate comparison between the two data sets. (See Table 5.6.) When comparing distributions, it is best not to attach too much significance to the median values unless additional measures of dispersion are computed.

One method of looking at the dispersion of ranks or scores for ranked observations is to compute **quartile values,** which are obtained in a manner similar to finding the median. The lower (first) quartile separates the lower 25 percent of the observations from the upper 75 percent and is called the 25th **percentile.** The second quartile is the median, and the upper (third) quartile marks the division between the lower 75 percent of the observations from the upper 25 percent and is referred to as the 75th percentile. Any percentile can be computed, but it is not customarily done unless there are at least 25 observations in the array.[4]

The *quartile deviation* is defined as

$$\frac{\textbf{upper quartile} - \textbf{lower quartile}\ \textbf{(inner-quartile range)}}{\textbf{2}}$$

and is the average expectation of how the occurrences will be distributed on either side of the median.[5] The employment of quartile measure or the quartile deviation is most useful in comparing distributions.

The inner-quartile range identifies the 50 percent of all values that are closest to the median value. This is useful in preprocessing data before mapping, because mapping those geographical areas within the inner-quartile range reveals the geographical pattern of values nearest the typical (median) value. (See Table 5.7 and Figure 5.3.)

Table 5.7 1980 Arkansas Infant (Age 0–1) Mortality by County of Residence—Ranked

County	Rate	Rank	County	Rate	Rank
Lawrence	3.7	1	Desha	12.0	35
Bradley	4.5	2	Clay	12.2	36
Franklin	5.0	3	Lafayette	13.2	37
Lincoln	5.1	4	Drew	13.3	38
Woodruff	5.2	5	Conway	13.9	39
Madison	5.7	6	Faulkner	14.3	40
Boone	5.9	7	Jefferson	14.4	41
White	6.0	8	Mississippi	14.9	42
Nevada	6.4	9	Calhoun	14.9	43
Sharp	6.4	10	Carroll	15.1	44
Poinsett	6.6	11	Crawford	15.4	45
Jackson	7.0	12	Independence	15.4	46
Pike	7.3	13	Grant	15.5	47
Baxter	7.4	14	Clark	16.1	48
Fulton	8.8	15	Hempstead	17.1	49
Greene	9.1	16	Union	17.1	50
Randolph	9.2	17	Crittenden	17.2	51
Little River	9.3	18	Hot Spring	17.4	52
Sebastian	9.5	19	Polk	18.8	53
Ashley	9.9	20	Benton	18.9	54
Lonoke	10.3	21	Scott	18.9	55
Craighead	10.5	22	Saline	19.2	56
Cleburne	10.6	23	Howard	20.2	57
Pope	10.8	24	St. Francis	20.3	58
Dallas	11.0	25	Cleveland	20.8	59
Pulaski	11.0	26	Monroe	22.1	60
Miller	11.1	27	Van Burden	22.4	61
Ouachita	11.1	28	Marion	23.6	62
Washington	11.2	29	Prairie	24.4	63
Garland	11.7	30	Cross	25.9	64
Phillips	11.7	31	Lee	31.7	65
Johnson	11.9	32	Newton	32.3	66
Arkansas	12.0	33	Chicot	37.5	67
Columbia	12.0	34			

Median rank = 34, Q_1 (25th percentile or lower quartile) = 9.2
Median value = 12.01, Q_3 (75th percentile or upper quartile) = 17.2

Inner-quartile range = 17.2 − 9.2 = 8.0
Quartile deviation = 8.0 ÷ 2 = 4.0

Note: The following counties had no infant mortalities (their rates were 0): Izard, Logan, Montgomery, Perry, Searcy, Sevier, Stone, and Yell. They are not listed above but are mapped in Figure 5.3.

Source: *Arkansas Division of Health Statistics,* Arkansas Vital Statistics, 1980, p. 41.

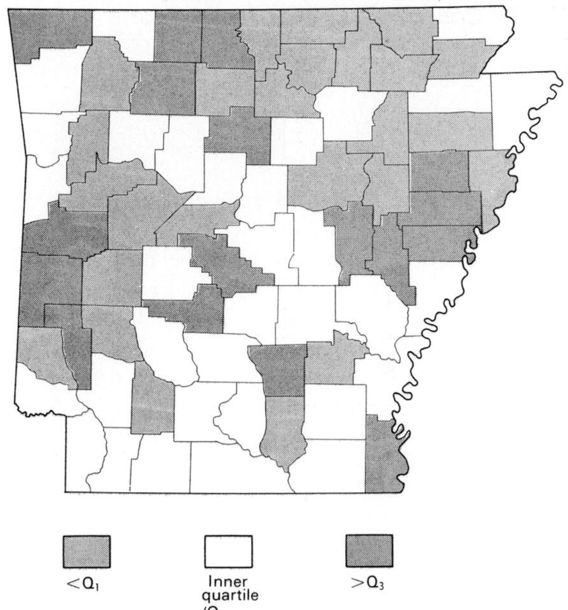

Figure 5.3. Mapping the inner-quartile range.
Counties whose values are in the inner-quartile range are mapped, as are those above Q_3 and below Q_1, with distinct area patterns. There appears to be an absence of spatial clustering of countries in the quartile ranges. This map presents the data of Table 5.7.

Geographical patterns of infant mortality in Arkansas reveal an absence of spatial clustering. Perhaps somewhat more interesting than looking at the counties in the inner-quartile range is examining those above Q_3 (upper quartile), or below Q_1 (lower quartile), since their values are dispersed farther from the median. Especially important to health care planners would be to examine the health care delivery systems in these counties; infant mortality is frequently used as a measure of overall health care. The geographical pattern revealed by Figure 5.3. suggests that local health delivery systems vary widely in quality throughout the state, and that infant mortality in Arkansas may not be the result of statewide health policies.

Interval and Ratio Scales: The Arithmetic Mean and the Standard Deviation

Data that exist at the interval or ratio scales are sometimes more useful in geographical analysis because more statistical measures can be applied to them. When we summarize a distribution at these scales, the goal is to obtain a one-number index to describe the array of values. The statistic most frequently used for this purpose is the *arithmetic mean,* and the statistic to describe dispersion around the mean is the *standard deviation.*

The **arithmetic mean** is derived by adding all the values and dividing by the number of values:

$$\overline{X} = \frac{\Sigma x}{N}$$

where $\overline{X}$ is the mean; Σ is a summation sign; x is a value in the array (Σx means to sum all xs), and N is the total number of x-values.

It may be easier to see the idea of the mean, and later the standard deviation, by looking at Figure 5.4, a distribution of values on a number line. The mean is a point on the number line around which all values can be thought to balance. The arithmetic mean is considered the best one-number description of all values in a data array.

Like the other measures of central tendency (the mode and median), the mean does not show *how* the values are arranged around the mean. Two quite different numerical arrays can have identical arithmetic means. In Table 5.8,

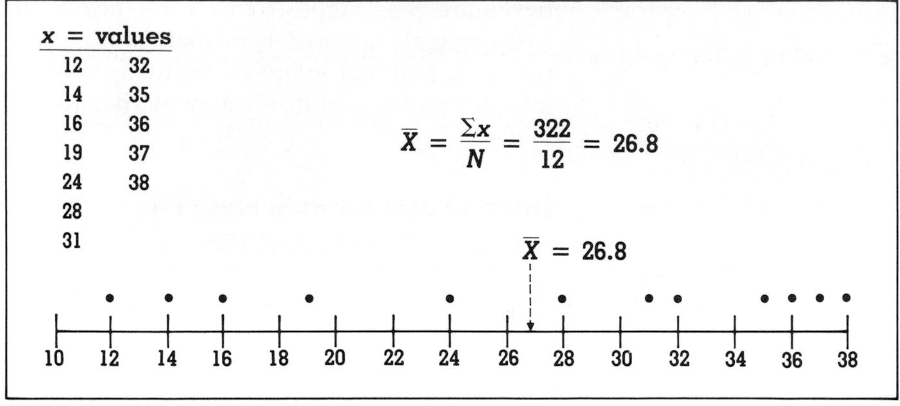

Figure 5.4. Plotting the arithmetic mean on a number line.
The arithmetic mean can be thought of as a balance point for the values on the line.

Table 5.8 Two Data Sets with Identical Means
(Units = Tons)

Variable X	Variable Y
6	28
39	24
11	26
22	29
23	23
44	27
49	21
1	20
31	21
19	29
28	30
27	22
$\Sigma x = 300$ $N = 12$ $\overline{X} = 25$ range = $\lvert 49 - 1 \rvert = 48$ $\overline{X}^2 = 625$ $\Sigma x^2 = 9864$ σx^2 (variance) = 196.84 σ (standard deviation) = 14.03 tons	$\Sigma y = 300$ $N = 12$ $\overline{Y} = 25$ range = $\lvert 30 - 20 \rvert = 10$ $\overline{Y}^2$ 625 Σy^2 7642 σy^2 (variance) = 11.76 σ (standard deviation) = 3.43 tons

for example, the range of the values (difference between highest and lowest values) for variable X is 48, whereas it is only 10 for variable Y. The identical means do not reflect these quite different deviation patterns. We may also call this the *variation* around the mean.

One might think that a good way to express this deviation is to calculate the average deviation. Doing this, however, yields no measure at all! If we calculate each x-value's deviation from the mean and sum these (as we must do in computing an average), they would total 0. The positive deviations will be offset by the negative deviations.

The standard deviation is the solution to this dilemma. We begin by first squaring each deviation. This eliminates the negative signs (minus times minus equals a plus). Now we can sum them, divide by N, and compute the mean squared deviation. This is also called the **variance** and is symbolically written

$$\sigma^2 = \frac{\Sigma(x - \overline{X})^2}{N}$$

where σ^2 is the variance; $(x - \overline{X})^2$ means to compute each x's deviation from the mean $\overline{X}$, square this deviation and sum them all; N is the number of values.

The large Greek sigma Σ is used in statistical notation to denote summation. It may be $\sum_{i=1}^{n}$. This means that all values from the first ($i = 1$) to the nth are summed in the operation. Hence, $\sum_{i=1}^{n} x$ directs you to add all x-values from the first to the nth.

It is sometimes necessary to compute the mean of grouped data. In these cases, we use this formula:

$$\overline{X} = \frac{\Sigma fx}{\Sigma f}$$

The *x*s are now class midpoint values, and the *f*s are frequencies in each class. The denominator is the sum of all frequencies.

The formula also may be written this way:

$$\sigma^2 = \frac{\Sigma x^2 - N\overline{X}^2}{N}$$

This version of the formula is preferred because it is easy to use, especially when using electronic calculators. First, each x is squared and these are summed. The average is squared, multiplied by N, and this product is subtracted from the sum of xs squared. Finally, this is divided by the number (N) of observations.

There are two main advantages to using this method of determining variation. First, it is relatively simple; many hand-held calculators perform these computations automatically. Second, the units of the variance are identical to the original units of the variable. If variable X in Table 5.8 is in tons, the variance is expressed in (tons)^2. We take the square root of the variance to obtain the measure of deviation in the same units as the data, and this is called the **standard deviation:**

$$\sigma = \sqrt{\text{variance}} = \sqrt{\frac{\Sigma x^2 - N\overline{X}^2}{N}}$$

The standard deviation is especially useful when comparing two variables. The smaller of two standard deviations indicates values more closely packed around the mean than a larger standard deviation. An examination of Table 5.8 reveals this. Variable X's σ of 14.03 indicates a wider dispersion than variable Y's 3.43. Their ranges (48 and 10 respectively) can also be compared. Of course, comparison of standard deviations is only meaningful with variables in the same units.

Using the standard deviation as a measure of dispersion in one distribution is more subtle, but there are certain relationships that exist between $\overline{X}$ and the standard deviation that assist in this analysis. Using Chebyshev's rule, these summaries may be drawn *for all data sets:*[6]

1. The space between $\overline{X} - 2\sigma$ and $\overline{X} + 2\sigma$ contains *at least* 75 percent of the observations.
2. The space between $\overline{X} - 3\sigma$ and $\overline{X} + 3\sigma$ contains *at least* 88 percent of the observations.
3. In general cases, for any multiplier K (>1), the space between $\overline{X} - K\sigma$ and $\overline{X} + K\sigma$ contains at least $1 - \frac{1}{k^2}$ fraction of the observations.

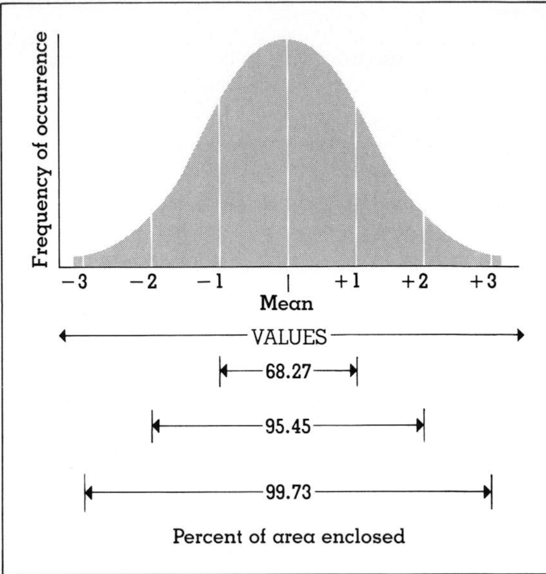

Figure 5.5. The normal distribution. This figure illustrates the percentage of observations that fall within the standard deviations ±1, ±2, and ±3. Note that values above or below three standard deviations are very rare (less than 1 percent).

Frequency or relative frequency distributions having certain specified shapes represented by bell-shaped curves are known as **normal distributions.** It has long been thought that a great many naturally occurring phenomena display such a distributional shape. This is not necessarily true, but the normal distribution remains at the heart of much statistical analysis. If data are distributed normally, *specific* percentages of observations occurring within spaces between $\overline{X}$ and σ can be specified. (See Figure 5.5.)

Central to the use of normal distributions is the calculation of the *probability* that certain observed values will occur. For data distributed normally, events that occur at further distances from the mean are less likely to occur.

The arithmetic mean and the standard deviation are frequently used in analysis of geographical data. They have been introduced here as a way of describing data observations, but they can also be used in other areas of statistical analysis involving hypothesis testing. The arithmetic mean and the standard deviation are employed in a great many cartographic preprocessing activities. (See Table 5.9 and Figure 5.6.)

Skewness

When the peak or mode of a distribution is displaced to either side of the mean in a distribution it is referred to as skewed. If the bulk of the frequencies are found less than the mean (and a long tail of low frequencies to the right), it is positively skewed. If the long tail is to the left of the mean, and the greater portion of frequencies greater than the mean, it is negatively skewed. **Skewness** is the measure of the displacement, and is calculated this way:[7]

$$\text{Skewness} = \frac{\Sigma(x - \overline{X})^3}{n\,\sigma^3}$$

Negative values indicate negative skewness and positive values positive skewness. The higher the value the greater the difference away from a normal distribution. Normal distributions have a skewness of 0.0.

Kurtosis

Kurtosis is a measure which describes the "peakiness" of a distribution. A flat distribution is one in which there is a nearly equal number of observations in each cell in the histogram. A distribution having a high "peaki-

Table 5.9 Percentage of Families below Poverty Level, by States, 1975

State	Percent	State	Percent
Alabama	12.9	Montana	8.9
Alaska	5.2	Nebraska	7.1
Arizona	10.8	Nevada	7.0
Arkansas	14.1	New Hampshire	5.9
California	8.5	New Jersey	6.9
Colorado	6.3	New Mexico	15.5
Connecticut	5.6	New York	7.6
Delaware	6.6	North Carolina	12.1
Washington, D.C.	11.4	North Dakota	8.0
Florida	11.0	Ohio	7.3
Georgia	14.6	Oklahoma	11.1
Hawaii	6.4	Oregon	6.7
Idaho	8.2	Pennsylvania	7.4
Illinois	8.3	Rhode Island	6.9
Indiana	6.0	South Carolina	12.9
Iowa	5.8	South Dakota	10.6
Kansas	6.1	Tennessee	12.6
Kentucky	14.9	Texas	11.7
Louisiana	15.0	Utah	7.0
Maine	9.3	Vermont	10.8
Maryland	6.2	Virginia	8.3
Massachusetts	6.1	Washington	6.6
Michigan	7.6	West Virginia	11.5
Minnesota	6.4	Wisconsin	5.8
Mississippi	20.4	Wyoming	7.0
Missouri	9.5		

Source: *United States Bureau of the Census,* Statistical Abstract of the United States: 1980, 101st ed. (Washington, D.C.: USGPO, 1980), p. 467.

ness" is one in which a greater bulk of the observations are in one cell. Kurtosis is calculated using the formula:[8]

$$\text{Kurtosis} = \frac{\Sigma(x - \overline{X})^4}{n\,\sigma^4}$$

A normal distribution has a kurtosis of 3.0. Values greater than 3.0 indicate "peakiness," and values below 3.0 a flatness in the histogram.

Measures of Areal Concentration and Association

Since cartographers and geographers are concerned with the analysis of spatial patterns, they naturally examine data sets during preprocessing to discover any significant patterns of spatial variation. Some of the more useful ways of describing spatial patterns are discussed below.

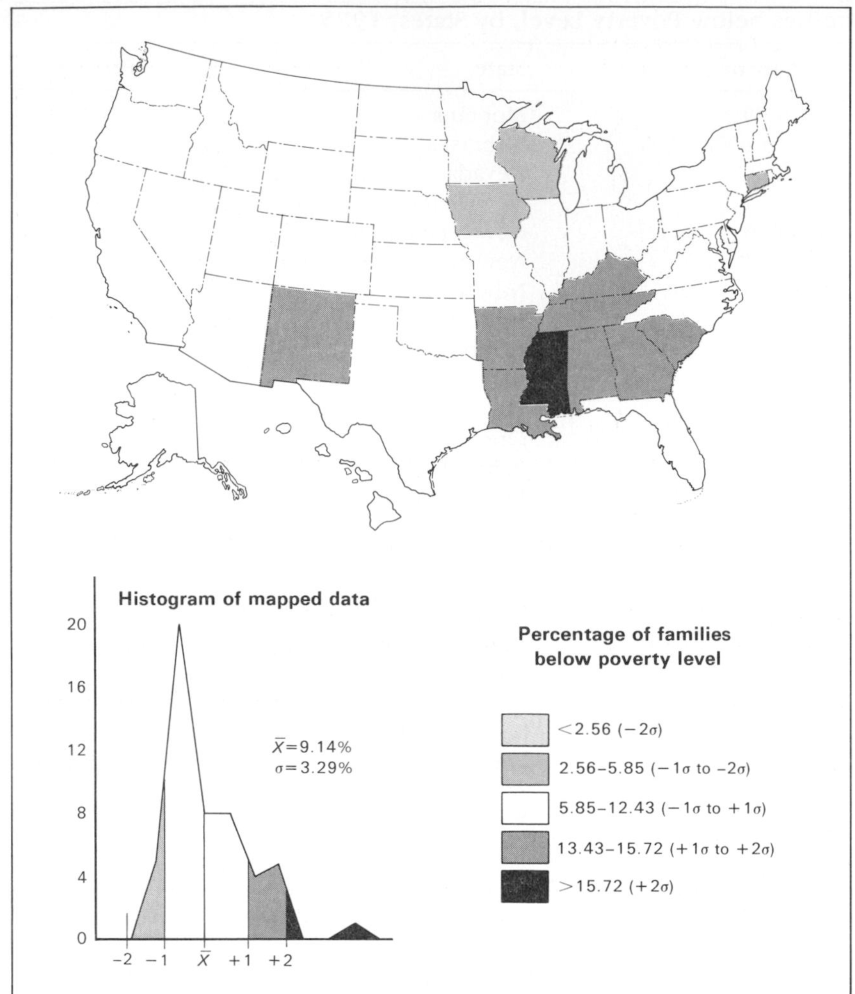

Figure 5.6. Mapping data using means and standard deviations.
Preprocessing data can involve using means and standard deviations and may be used to present final geographical analyses or simply as work-maps. In final presentations, it is helpful to include as part of the legend material a histogram of the data distribution.

Source: United States Bureau of the Census, *Statistical Abstract of the United States:* 1980, 101st ed. (Washington, D.C.: USGPO, 1980), p. 467.

The Coefficient of Areal Correspondence

Researchers often need to assess the degree of correspondence between two areal distributions, usually to discover or demonstrate causal relationships. Figure 5.7, for example, shows two maps that present nominal data. One way to examine the areal correspondence—a measure of congruence—is to inspect them visually, but inconsistency will result because of the individual differences and biases of the observers. A *numerical* measure is preferable.

One such numerical measure is the **coefficient of areal correspondence,** computed as follows:[9]

$$C_a = \frac{\text{area covered jointly by both phenomena}}{\text{total area covered by the two phenomena}}$$

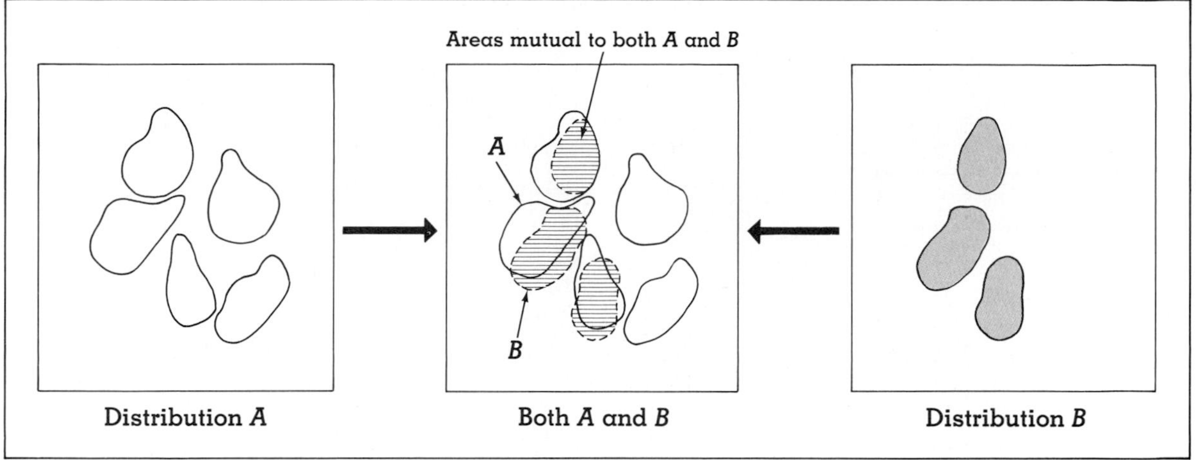

Figure 5.7. Coefficient of areal correspondence.
This index is a simple proportion measuring the amount of congruence between two areal distributions.

C_a varies from 1.0 (perfect correspondence) to 0 (no areal correspondence). Multiplication by 100 will yield a percentage. Table 5.10 contains the areas applicable to the example used in Figure 5.7. For this example, $C_a = .21$.

This index is descriptive and does not necessarily suggest causal relationships. However, values approaching 1.0 may indicate that the areal variables are somehow related. Further inquiry would certainly be justified.

Table 5.10 Areas in Figure 5.7, the Example of the Coefficient of Areal Correspondence

Distribution *A* (sq km)
63
26
23
17
20
149
Distribution *B* (sq km)
18
21
12
51
Total of areas mutual to *A* and *B* (sq km)
16
17
9
42

Areal Means

Cartographers often deal with point patterns and have to describe them in some meaningful way. The *areal mean* is a useful way of illustrating the spatial balance point, in much the same way the mean describes balance along a number line. A plot of several areal means representing the same phenomenon over time can be useful in clarifying spatial change.

Areal Mean

In the simplest version of computing an **areal mean,** a cartesian coordinate system is superimposed over the distribution of scattered points. (See Figure 5.8a.) The axes are scaled in convenient units with arithmetic intervals; the smaller the numbers, the easier the computations. Vertical and horizontal ordinates are then constructed from each point to each axis, and *X*- and *Y*-values are obtained for each point. After all *X*- and *Y*-values are noted, an arithmetic mean is computed for each. The pair of means is called the *areal mean.* It is plotted by using the same cartesian coordinates.

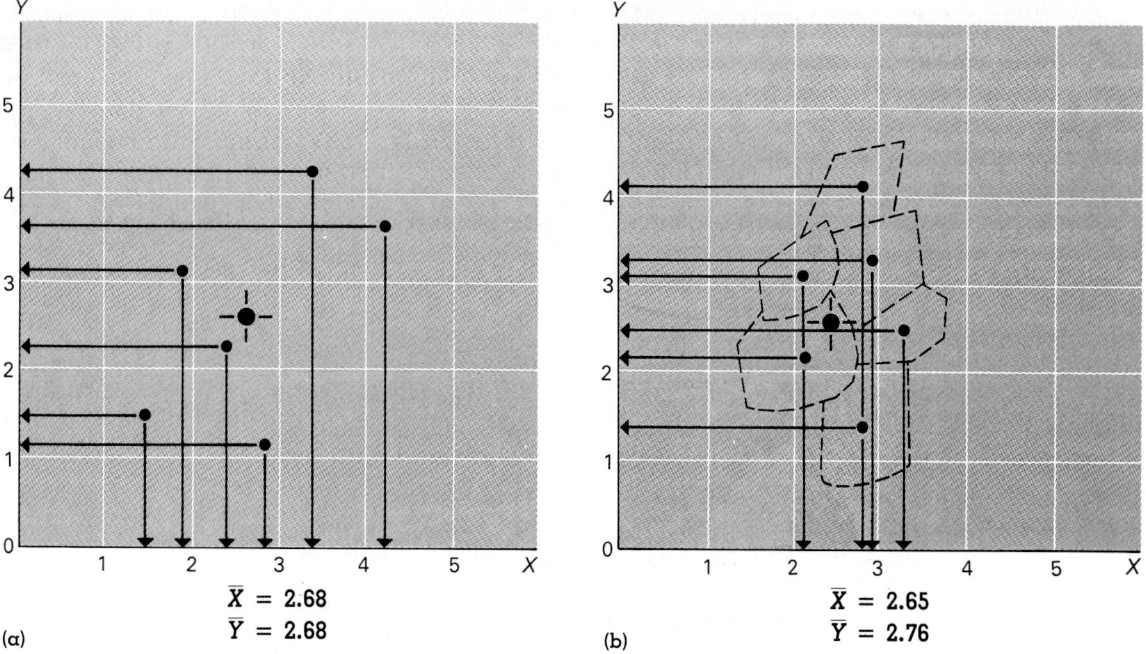

Figure 5.8. Plotting areal means.
Areal means for point distributions (a), or for centers of areas (b), are computed with equal ease. In each case, the points may be weighted by geographical values associated at each point.

Most geographical analyses involve point patterns (or centers of areas) that have weights attributed to them. Except at the simplest nominal scales, geographical point phenomena do not usually occur everywhere with equal value. An especially interesting study is to plot weighted means over time to discover a spatially dynamic pattern. Weights are most often socioeconomic data such as income, production, sales, or employment data. Sometimes, data are available only for larger enumeration areas such as counties, census tracts, or blocks. In these cases, the data can be assumed to be aggregated at a central point within the area. (See Figure 5.8b.) The central point is then used as before in computing the weighted areal mean.

Advanced techniques, too lengthy to introduce here, are available to study other aspects of areal means—for example, centrographic analysis, including confidence plotting and hypothesis testing.

Standard Distance

Another useful statistic related to the areal mean is the **standard distance** (SD). For nonareal data, the standard deviation is the statistic ordinarily used to describe the dispersion of values around the mean. In the areal case, the equivalent is the standard distance, which is obtained by measuring the distance from each point to the areal mean, squaring each distance, summing these, dividing by *N*, and obtaining the square root.[10]

$$\text{SD} = \sqrt{\frac{\Sigma d^2}{N}}$$

where SD = standard distance, d is the distance from each point to the center, and N is the total number of points.

In practice, this involves considerable calculation—the distance from each point to center can be obtained using analytic geometry, which is time consuming—and there is much potential for error. Other formulas have therefore been worked out, notably the following:

$$SD = \sqrt{\sigma y^2 + \sigma x^2}$$

where σy^2 is the variance of Y and σx^2 is the variance of X. The values of X and Y are obtained from a grid placed over the distribution of points (as in the areal mean).

Standard distances become useful when compared to others computed for identical data at different time periods.[11] Together with areal means, standard distances give very good quantitative assessment of point patterns. Standard distances may be plotted on the original point distributions using a circle, centered on the areal mean, with a radius equal to the standard distance. Used by themselves, however, these plotted circles have little utility.

A summary of the most important properties of areal means and standard distances includes the following:[12]

1. Both include all events (points) used in the analysis.
2. Because all observations are used, both are extremely sensitive to change in any one observation.
3. Because squares of distance are used in computing the standard distance, its value is strongly affected by points at extreme distances from the areal mean.

Location Quotient

A simple measure to illustrate geographical localization is the **location quotient.** The location quotient is a ratio of proportions and provides an index for determining geographical share. One normally assumes at the outset that the variable being measured is uniformly distributed. It is not difficult to calculate and can be used to determine if more sophisticated measures should be pursued. Moreover, quotients can be mapped easily and show at a glance the idea of geographical concentration.

The location quotient concept can be illustrated by example. A geographer wishes to determine if there is anything unusual about the distribution of food store sales in Maryland. Data are collected on food store sales, as a percent of all retail sales, for the state and for each county. (See Table 5.11.)

First, the proportion of food sales to total retail sales in the state is determined (expressed as a percent):

$$\frac{5{,}718{,}377}{30{,}205{,}991} \times 100 = 18.9\%$$

Next, the food sales to total retail sales in each county is calculated. Each county's location quotient is then determined by a ratio between the county's proportion and that of the state's. For example,

$$\text{Caroline } \frac{36.5}{18.9} = 1.93$$

If the proportion of food sales to total sales were the same in a county as it is in the entire state, the location quotient would be 1.0, and the distribution of the proportion of food sales would be *uniform* throughout the state. In counties having a proportion less than the state's, a location quotient of less than 1.0 results, suggesting less food sales than would be

Areal means can reflect weighted values. A weighted average is computed in the same way as for grouped data:

$$\overline{X} = \frac{\Sigma fx}{\Sigma f} \qquad \overline{Y} = \frac{\Sigma fy}{\Sigma f}$$

The f is the weight at each x, y location. The denominator is the total of all the weights in the study areas.

Table 5.11 Maryland: Food Store, and Total Retail Sales by County and for the State, 1986

County	Total Retail Sales (x 1000 dollars)	Food Store Sales (x 1000 dollars)	Food Sales as Percent of Total Sales	Location Quotient
Allegany	481,668	95,513	19.8	1.05
Anne Arundel	2,534,556	427,237	16.8	.88
Baltimore	5,688,127	1,039,909	18.2	.96
Baltimore City	3,775,658	732,202	19.3	1.02
Calvert	107,384	32,754	30.5	1.61
Caroline	58,642	21,440	36.5	1.93
Carroll	498,542	120,438	24.1	1.28
Cecil	319,283	57,096	17.9	.95
Charles	318,420	85,140	26.7	1.41
Dorchester	182,872	33,311	18.2	.96
Frederick	771,251	169,330	21.9	1.16
Garrett	136,742	23,659	17.3	.92
Harford	905,157	186,551	20.6	1.09
Howard	752,198	141,776	18.8	.99
Kent	94,276	30,254	32.0	1.69
Montgomery	6,053,328	1,051,753	17.4	.92
Prince George's	5,253,177	1,024,023	19.5	1.03
Queen Anne's	109,947	33,444	30.4	1.61
St. Mary's	236,175	60,927	25.8	1.37
Somerset	61,924	27,387	44.2	2.34
Talbot	242,648	46,783	19.2	1.02
Washington	691,825	116,915	16.9	.89
Wicomico	581,718	88,810	15.3	.81
Worcester	350,473	71,725	20.4	1.08
State	30,205,991	5,718,377	18.9	1.00

Source of data: Rand McNally. *Commercial Atlas and Marketing Guide*. 119th Ed. Chicago: Rand McNally, 1988; location quotients calculated by author.

expected. A location quotient greater than 1.0 suggests that a county has a greater share of food sales than expected. The quotient cannot be less than 1.0. Of course, it is not likely for all counties to have a quotient of 1.0. What is unusual and bears checking is when all the counties having greater or lesser than 1.0 are grouped together geographically.

In the case of Maryland, food sales as a proportion of all retail sales is hardly uniform, and can be seen by mapping the quotient. (See Figure 5.9.) The counties of Calvert, Caroline, Charles, Kent, Queen Anne's, and especially Somerset, considerably exceed the quotient of 1.0 (all are in the top quartile). It may or may not be geographically significant that these counties (except Caroline and Charles) border the Chesapeake Bay. Somerset county deserves closer scrutiny at least. Food distributors and marketing teams may use this simple technique to guide them in any future or more sophisticated study.

Performing data calculations such as the location quotient is very nearly out of the realm of preprocessing geographical data for mapping. It is in fact bordering on data analysis.

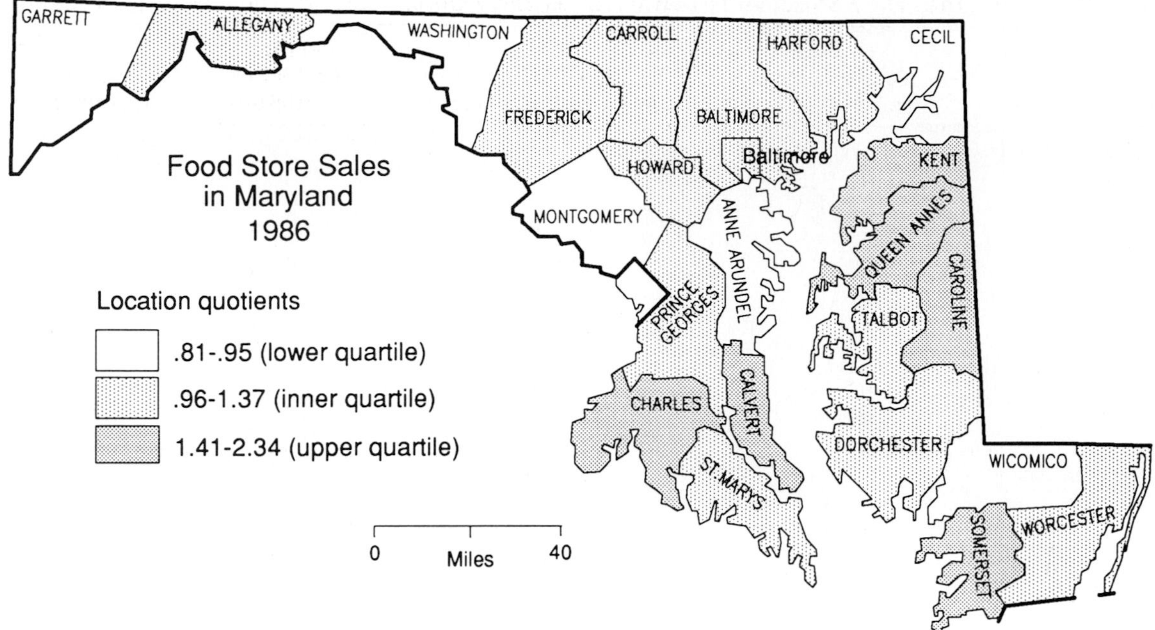

Figure 5.9. Location quotient comparing food sales to total retail sales in Maryland. Counties having a location quotient greater than 1.0 are clustered in the east and south of the state, and four of those in the top quartile border the Chesapeake Bay.

Nonetheless, if measures such as these are useful for the cartographer or the client, they should indeed be attempted.

Measuring Spatial Association by Regression Methods

An especially useful method for examining spatial association is **residuals from regression.** It is more complex than the foregoing methods but yields extremely valuable results.

The essential idea behind regression is to measure the degree of association between two magnitude (interval/ratio) variables. One variable is normally considered *independent* and the other *dependent,* because we are looking for causal relationships. A classic example is the relation of crop yield (dependent variable) to fertilizer application (independent variable). Regression methods in themselves do not establish causality and are useful only when applied to related variables.

Each pair of corresponding magnitudes for each variable is plotted on suitable graph paper to form a **scattergram.** (See Figure 5.10.) Normally, the independent variable is plotted on the horizontal axis and the dependent variable on the vertical axis. A **regression line** is a "best fit" line plotted on the graph, summarizing the mathematical relationship between the two variables. Numerous regression lines may be drawn; the simplest and most common is the straight, or linear, regression line.

Regression lines are plotted so that they minimize all deviations of the points from the line. (See Figure 5.10.) This is why they are termed *lines of best fit.* By plotting the line at different locations or at different slants, one can

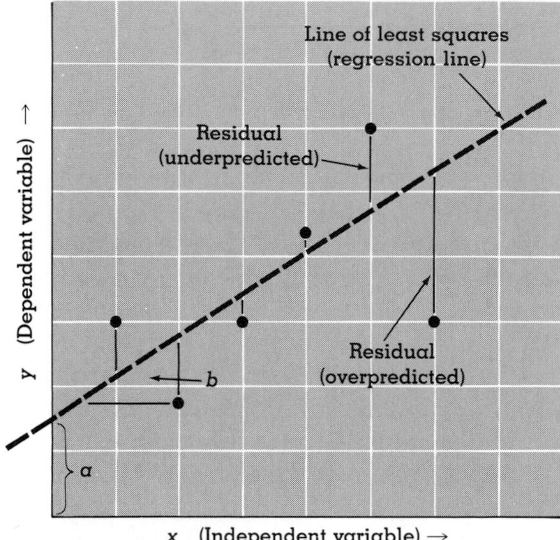

Figure 5.10. Components of the linear regression line.
The a in the equation of a straight line ($y = a + bx$) is the intercept of the line with the y-axis. The b refers to the slope of the line. Notice that both underpredicted and overpredicted residuals may be recorded. When a computed (expected) value of y is greater than the observed value, it is overpredicting.

easily see how its position could affect the minimizing operation. The general **equation for a straight line** is

$$y = a + bx$$

For regression in the linear case, we use the following equations to determine values of a and b so that we might plot the *least squares regression line:*

$$\Sigma y = a \cdot N + b \cdot \Sigma x$$
$$\Sigma xy = a \cdot \Sigma x + b \cdot \Sigma x^2$$

These equations can be solved simultaneously. The value for a may be found and substituted to find b. The value b is usually called the *regression coefficient.* With a and b known, various values of x and y may be selected and used in the general equation for the line in order to plot the regression line. One interesting feature of the least squares regression line is that it passes through the means of both the X and Y variables. This can be used for a visual check of the plotted line.

In regression analysis, we are generally interested in how well the regression line summarizes the actual values of the dependent variable (Y). For each actual value of the independent variable (X), there are two corresponding values of y: y-observed and y-estimated from the regression line. The difference between these two is called the *residual* (or Z_i) value:

$$Z_i = (y\text{-observed}) - (y\text{-estimated})$$
$$Z_i = y - y'$$

The Standard Error of the Estimate

How well the regression line, or the computed values of a and b, fits the data can be estimated by calculating the **standard error of the estimate.** This is actually a standard deviation of the residual values discussed above. It is calculated as follows:

$$SEy' = \sqrt{\frac{\Sigma(y - y')^2}{N - 2}} \quad \text{or} \quad \sqrt{\frac{\Sigma Z^2}{N - 2}}$$

The standard error expresses the *spread* of the plotted observed points around the regression line, as well as how accurate X is in predicting Y. It may be recalled that for normally distributed data, certain percentages of the observations fall between plus and minus 1, 2, or 3 standard deviations. Likewise, the observed values of Y fall between plus and minus one, two, or three standard errors from the regression line. About 68 percent of the observations fall between ± 1 standard errors of the regression line, and about 95 percent between ± 2 standard errors. Observations beyond ± 2 are extremely rare—only 5 percent.

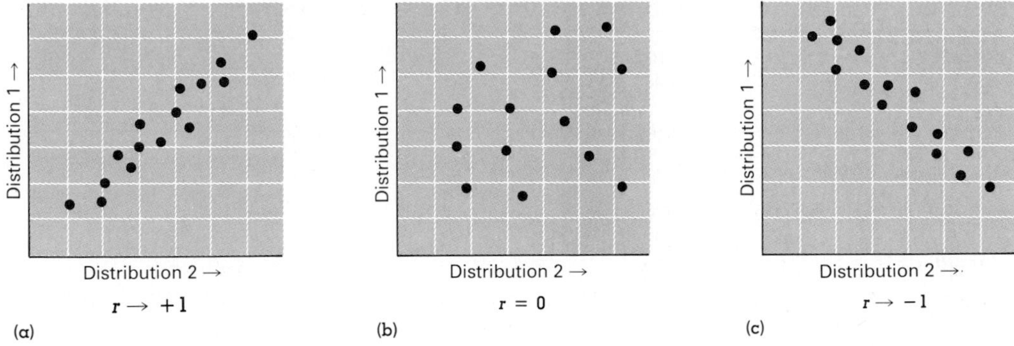

Figure 5.11. Scattergram patterns for different correlation (r) values.
In (a), the correlation approaches a strong positive relationship (the points would all fall on a line if $r = 1.0$). In (c), the relationship is strongly negative (inverse). No apparent linear correlation exists in (b), which yields an r value of 0.

Correlation

Statistical **correlation** is a method of gauging the *strength* of the mathematical association between variables. The statistic is called r and varies between +1 and −1. Typically, plotted points on scattergrams of different r values will appear as in Figure 5.11. The formula for r is usually written

$$r = \frac{\Sigma(y - \overline{Y})(x - \overline{X})}{\sqrt{\Sigma(y - \overline{Y})^2 \Sigma(x - \overline{X})^2}}$$

Examination of the scattergrams of Figure 5.11 illustrates that a perfect correlation can exist for either direct or inverse relationships. The correlation coefficient indicates how well X is performing as an independent variable relative to Y. The statistic r is dimensionless, not in the units of the original variables.

The value of r and its overall significance normally depend on the number of observations being compared. However, coefficients below about .25 are generally not indicative of any particular association between two variables. Between .25 and .50, there is a weak to moderate association, and r values between .50 and about .75 usually indicate a strong association. A coefficient of .75 or more suggests a very strong association between variables.

Mapping Residuals from Regression

Besides the numerical indices of association between two variables, correlation and regression analysis is important in *mapping residuals.* It is useful in identifying and locating those cases that are *not* well explained by the regression line (observed values beyond ±1 or ±2 standard errors of the estimate). The goal is to determine if their spatial patterns reveal other factors affecting their behavior in the data set.

Such analysis begins with calculation of the correlation coefficient, the regression line, and the standard error of the estimate. Mapping is then done if there is sufficient indication that it may yield further information. An example is provided in Tables 5.12 and 5.13, and in Figures 5.12, 5.13, and 5.14. A particular segment of the population (the independent variable, X) is thought to hold certain attitudes regarding health treatment for newborn infants: it is believed to increase infant mortality rates

Table 5.12 Residuals from Regression Table: Variables, Values, and Residuals

Map Region	Independent Variable (X)	Dependent Variable (Y)	Predicted Y (Y_e)	Y Residuals ($Y_o - Y_e$)
A	5	5	21.25	16.25
B	10	22	25.30	3.30
C	35	55	45.55	9.45
D	30	40	41.55	1.50
E	45	35	53.65	18.65
F	50	50	57.70	7.70
G	20	100	33.40	66.60
H	50	10	57.70	47.70
I	50	95	57.70	37.30
J	15	10	29.35	19.35
K	15	30	29.35	.65
L	25	25	37.45	12.45
M	25	15	37.45	22.45
N	25	55	37.45	17.55
O	15	20	29.35	9.35
P	35	30	45.55	15.55
Q	45	65	53.65	11.35
R	55	50	61.75	11.75
S	40	50	49.60	.40
T	55	75	61.75	13.25
U	35	75	45.55	29.45

Table 5.13 Formula Values for Data in Table 5.12

X Variable	Y Variable
$N = 21$	$N = 21$
$\overline{X} = 32.38$	$\overline{Y} = 43.42$
$\Sigma x = 680$	$\Sigma y = 912$
$\Sigma x^2 = 26{,}850$	$\Sigma y^2 = 54{,}634$
$\sigma_x = 15.1$	$\sigma_y = 26.75$
$\Sigma xy = 33{,}445$	

$\Sigma y = a \cdot N + b \cdot \Sigma x$

$\Sigma xy = a \cdot \Sigma x + b \cdot \Sigma x^2$

$b = .810$

$a = 17.2$

$\Sigma(Y_O - Y_E) = 11{,}857.012$

r (correlation coefficient) $= .4593$

S_E (standard error of estimate) $= 24.98$

(the dependent variable, Y). Health authorities realize that these thoughts are not shared by all members of the population group but are not sure where the deviations occur most frequently.

A simple linear regression is performed, a correlation coefficient determined, and the standard error of the estimate calculated. A scattergram and regression line are plotted, with standard error bands added on each side of the regression line. (See Figure 5.13.) Another map (Figure 5.14) is developed to show three groups of residuals: group 1 of those map regions within ± 1 standard error, group 2 of the regions between ± 1 and ± 2 standard errors, and group 3 of those regions greater than ± 3 standard error. The health authorities wanted especially to locate the regions in group 3, because they would be most unique. Only

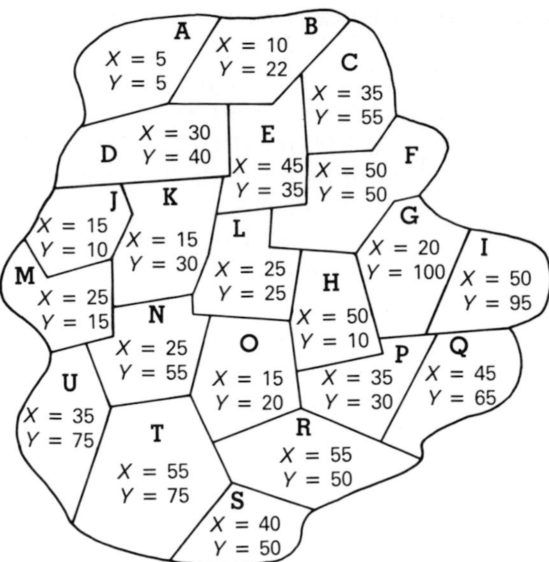

Figure 5.12. Data map used in mapping residuals from regression example.
It is extremely difficult to discern a relationship between X and Y by examination of a data map such as this, especially when the relationship is not very strong.

3, because they would be most unique. Only region G fit into the latter group. This hypothetical study would conclude with a visit of the health authorities to region G for a field investigation, in the hope of developing better health information methods.

Data Classification

Cartographers are called on to find ways to map large amounts of data. Recognizing the need for some sort of generalization and simplification, the cartographer is faced with a problem of data **classification.** Scientists classify objects or events for two main reasons:[13]

1. To reduce a large number of individuals (objects or events) to a smaller number of groups in order to facilitate description and illustration
2. To define phenomena—classes about which general statements can be made

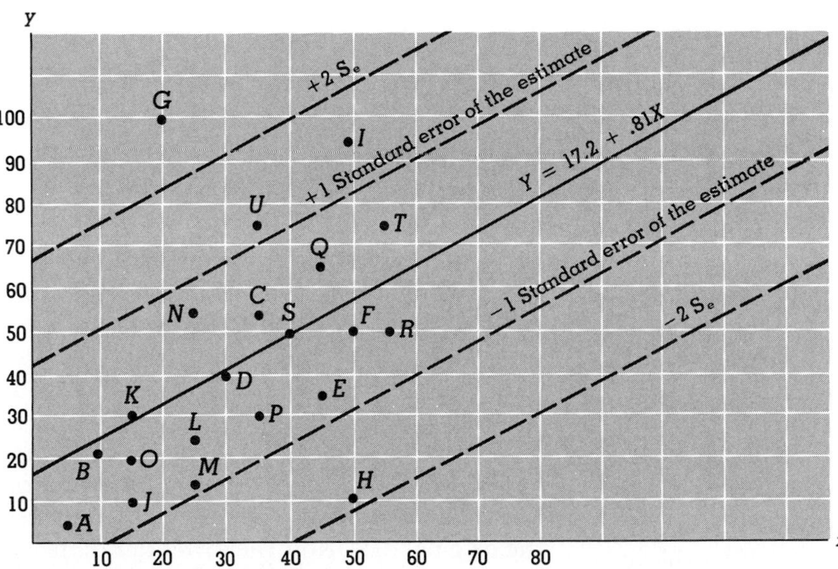

Figure 5.13. Scattergram, regression line, and standard error of the estimate plots for the example used in the text.

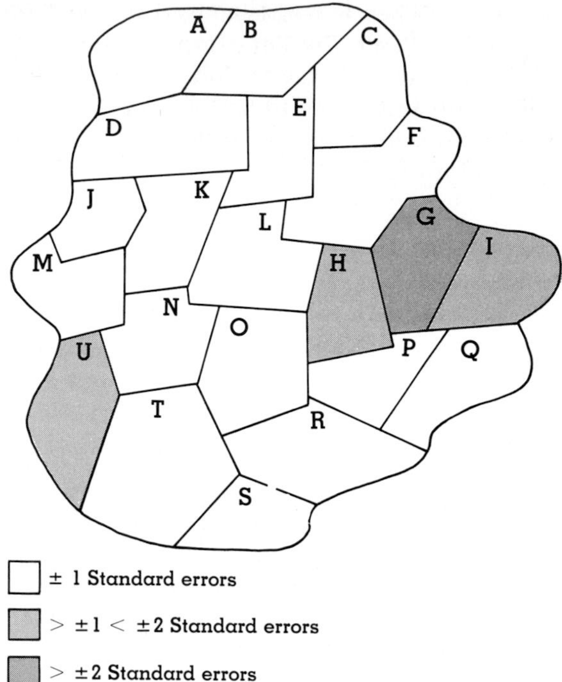

Figure 5.14. Map of residuals from regression.
The geographical areas having Y values considerably under- or overestimated relative to the regression line are mapped with identifiable area symbols. These sections of the study area need to be investigated more closely. Further data may be needed to determine why the dependent variables in these sections behave as they do. Identifying deviate areas is a major application of residual mapping.

All scientific disciplines, including geography, attempt classification to facilitate interpretation of the enormously complex world. Although classification may lead to loss of detail, it usually makes meaningful interpretation possible.

Clarification is a primary goal of classification:

> Every classification or "Taxonomy" is a simple but implicit theory. If by trial and error we are able to produce classifications and maps of spatial phenomena which suggest or expose important relationships, then our choices are good choices. Our classifications have served their purposes.[14]

We can learn at least two important ideas from this statement. First, classification is for a specific purpose: to show something not presented before. It is not a means in itself, but a method to assist the researcher in uncovering spatial relationships. Second, there are no rigidly defined rules for classification in all circumstances. Any number of methods may be selected—even by trial and error.

Most cartographic data classification is of the agglomerative type. Data values are placed into groups based on their numerical proximity to other values along a number array. Classification is necessary because of the impracticability of symbolizing and mapping each individual value.[15] The goal is to do the grouping in a way that reveals the spatial patterns that serve the thematic purpose of the mapping activity.

Classification is More

As an example of how classification leads to a loss of detail yet provides more interpretative power, consider the distribution of murder and non-negligent manslaughter in the United States, shown in Figure 5.15. Those states above the average rate (8 per 100,000 people) have been rendered as a group in Figure 5.15b. They are set apart by an important criterion: all are above average. The geographical pattern shows that the heavily populated states east of the Mississippi River, the Old South states plus Texas and Oklahoma, and the western states of California, Nevada, Arizona, New Mexico, and Alaska are part of this class.

A different geographical picture emerges in Figure 5.15c, after classifying the two groups even further by computing an average of the above-average states and mapping only those states above this new average. The new group represents a class having the extreme values of the original data. Now the lower tier of the Old South states (plus Texas) appears as a region in the southeast. Four other states also remain in this group: Alaska, California, Nevada, and New Mexico.

Figure 5.15. Classifying data to reveal more.
See the text for an explanation.

Unclassed mapped data present a geographical pattern that is not particularly revealing and has little interpretive power. As the data are grouped into similar classes having identical numerical characteristics, there is an increase in their ability to convey information about which generalizations can be made.

Classification is introduced in this chapter because it is an integral part of data preprocessing. A variety of methods of data classification are treated in the next chapter where choropleth mapping is discussed, but the methods can be applied to other mapping techniques as well.

Notes

1. Phillip Muehrcke, *Thematic Cartography* (Association of American Geographers, Resource Paper No. 19—Washington, D.C.: Association of American Geographers, 1972), p. 16.
2. Gouri K. Bhattacharyya and Richard Johnson, *Statistical Concepts and Methods* (New York: John Wiley, 1977), p. 1.
3. Linton C. Freeman, *Elementary Applied Statistics for Students in Behavioral Science* (New York: John Wiley, 1968), pp. 12–16.
4. Bhattacharyya and Johnson, *Statistical Concepts,* p. 30.
5. S. Gregory, *Statistical Methods and the Geographer* (London: Longman Group, 1978), p. 30.
6. Bhattacharyya and Johnson, *Statistical Concepts,* p. 36.
7. David Ebdon, *Statistics in Geography,* 2nd ed. (Oxford: Basil Blackwell, 1985), pp. 28–31.
8. *Ibid.*
9. David Unwin, *Introductory Spatial Analysis* (London: Methuen, 1981), pp. 189–90.
10. Peter J. Taylor, *Quantitative Methods in Geography* (Boston: Houghton Mifflin, 1977), p. 27.
11. *Ibid.,* pp. 27–30.
12. *Ibid.,* pp. 29–30.
13. R. J. Johnston, *Classification in Geography* (London: Institute of British Geographers, 1976), p. 4.
14. Ronald Abler, John S. Adams, and Peter Gould, *Spatial Organization: A Geographer's View of the World* (Englewood Cliffs, NJ: Prentice-Hall, 1971), p. 157.
15. This statement is generally accurate, although recent computer methods have allowed the mapping of unclassed data. This idea will be presented in more detail in the chapter dealing with choropleth maps.

Glossary

areal mean method of calculating the spatial balance of a set of data points

arithmetic mean customary one-number descriptor of an interval/ratio data set; defined simply as

$$\frac{\Sigma x}{N}$$

array in statistical terminology, an ordered arrangement of values; examples are ascending or descending arrays

classification scientific reduction of a large number of individual observations, events, or numbers into smaller groups to facilitate explanation

coefficient of areal correspondence method of comparing areal spatial distributions; defined as a ratio:

$$\frac{\text{area covered jointly by both phenomena}}{\text{total area covered by both phenomena}}$$

correlation method of showing the mathematical association between two or more variables

data preprocessing processing geographical data before mapping in order to reduce, enhance, retain key features, or show primary spatial patterns; preliminary activity of data symbolization

equation for a straight line $y = a + bx$; important in regression and correlation analysis

histogram a graphic way of presenting the frequency or relative frequency of occurrence of a variable

kurtosis numeric value indicating "peakiness" in a frequency distribution

location quotient a measure of geographical concentration; illustrates deviation from assumed proportional share

median that place in a ranked ordinal data set that neither exceeds nor is exceeded in rank by more than 50 percent of the observations; used to describe ordinal data

mode the class in a frequency distribution containing the highest relative frequency; used to describe nominal data

normal distribution frequency distribution represented by a bell-shaped curve; used as a basis for comparison in many statistical measures

percentage a proportion multiplied by 100 for ease of numerical comparison

percentile the place in a ranked data set that divides the number of observations into specified portions of all the observations

proportion a special ratio that expresses the relationship between the amount in one class and the total in all classes; *percentage* is a proportion developed by multiplying the decimal fraction by 100

quartile one method of describing dispersion in an ordinal data set; defined as the inner quartile range (i.e., the upper quartile minus the lower quartile) divided by two

ratio a fraction used to express the relationship between two variables

regression line drawn on a graph to depict the relationship between two variables; linear regression is a common form

relative frequency of the total frequency occurrence in a data set, the part belonging to a specific class

residuals from regression differences between observed y (or x) values and those estimated by the regression line

scattergram diagram containing a plot of data points, each of which has a value in two dimensions; a graphic way to illustrate mathematical correlation

skewness numeric value of deviation from the normal, or bell-shaped, frequency distribution

standard deviation the square root of the variance; used to describe dispersion around the arithmetic mean in an ordinal/ratio data set; defined as

$$\sqrt{\frac{\Sigma(x - \overline{X})^2}{N}}$$

standard distance measure for depicting dispersion around an areal mean

standard error of the estimate the standard deviation of residual values; illustrates the dispersion of residuals (differences) of y (or x) around the regression line; used to measure the appropriateness of the regression line in describing the relationship between two variables

statistics concepts and methods used to collect and interpret data for drawing conclusions

value an individual numerical observation of a variable

variable raw data used in a statistical analysis

variance the statistic used most commonly to describe dispersion around the arithmetic mean in an ordinal/ratio data set; defined as

$$\frac{\Sigma(x - \overline{X})^2}{N}$$

variation ratio used to define dispersion around the mode in a nominal data set; defined as

$$1 - \frac{f_{\text{modal}}}{N}$$

Readings for Further Understanding

Abler, Ronald; John S. Adams, and Peter Gould. *Spatial Organization: A Geographer's View of the World.* Englewood Cliffs, NJ: Prentice-Hall, 1971.

Barker, Gerald M. *Elementary Statistics for Geographers.* New York: Guilford Press, 1988.

Bhattacharyya, Gouri K., and Richard Johnson. *Statistical Concepts and Methods.* New York: John Wiley, 1977.

Blalock, H. M. *Social Statistics.* New York: McGraw-Hill, 1972.

Dixon, Wilfred J., and Frank J. Massey, Jr. *Introduction to Statistical Analysis.* New York: McGraw-Hill, 1957.

Ebdon, David. *Statistics in Geography,* 2nd ed. Oxford: Basil Blackwell, 1985.

Freeman, Linton C. *Elementary Applied Statistics for Students in Behavioral Science.* New York: John Wiley, 1968.

Gregory, S. *Statistical Methods and the Geographer.* London: Longman Group, 1978.

Hammond, Robert, and Patrick McCullough. *Quantitative Techniques in Geography.* Oxford: Clarendon Press, 1974.

Johnston, R. J. *Classification in Geography.* London: Institute of British Geographers, 1976.

King, Leslie J. *Statistical Analysis in Geography.* Englewood Cliffs, NJ: Prentice-Hall, 1969.

Muehrcke, Phillip. *Thematic Cartography.* Association of American Geographers, Resource Paper No. 19. Washington, D.C.: Association of American Geographers, 1972.

Neft, David S. *Statistical Analysis for Areal Distributions.* Monograph Series Number Two. Philadelphia: Regional Science Research Institute, 1966.

Sneath, Peter H. A., and Robert R. Sokal. *Numerical Taxonomy.* San Francisco: W. H. Freeman, 1973.

Taylor, Peter J. *Quantitative Methods in Geography.* Boston: Houghton Mifflin, 1977.

Unwin, David. *Introductory Spatial Analysis.* London: Methuen, 1981.

Part

II

Techniques of Quantitative Thematic Mapping

The chapters grouped into this part naturally cluster to form the quantitative thematic mapping techniques portion of this book. Each chapter presents a different kind of thematic technique, and each includes the rationale for its adoption. This part is introduced by an examination of the choropleth map, perhaps one of the most widely used forms today. Enumeration data are symbolized by area patterns or colors in this form of map. The common dot map is addressed in the next chapter with examples selected to emphasize the proper technique in its use. This kind of map is not encountered as often as in earlier times, except perhaps in agricultural atlases. Symbolizing quantities at points is presented in the next chapter where the technique of mapping by proportional symbols is introduced. Much of the research literature of the past thirty years is devoted to this map technique and the issues are presented here along with current design standards.

Chapter 9 introduces volume, or isarithmic mapping, a technique similar to the mapping of relief by contours of elevation, a form of map familiar to many. The thematic cartographer will deal with the isoplethic variety where mapped values are assumed to occur at points, or the isometric form where values do actually occur at points. Value-by-area mapping, or mapping by transforming familiar geographical space into some other space, yielding a cartogram, is dealt with in the next chapter. This form of mapping is more abstract than real, yet can yield interesting results and is often used to attract the reader's attention. It is a useful pedagogical device in geography. The design of flow maps concludes the chapters in this part. This century-and-a-half old technique is often used for mapping the movement of things or ideas, connections and interactions, spatial organization, and often find their way into economic presentations. Flow maps are often referred to as dynamic maps. They present very interesting design challenges for the cartographer.

Chapter

6

Mapping Enumeration Data: The Choropleth Map

Chapter Preview

Choropleth mapping is a common technique for representing enumeration data. For the most part, the rationale of the choropleth technique is easily understood by map readers. Major concerns of the cartographer are data classification, areal symbolization, and legend design. From a variety of classification methods, the designer is faced with selecting the one that best serves the purpose of the map and best depicts the spatial array of the data. Methods include constant and variable interval formats. One result of grouping data is the creation of different maps based on the classification scheme chosen by the designer. A new type of choropleth map is the unclassed variety, chosen when the map designer simply wants to give an unstructured picture of the spatial data. Activities in designing choropleth maps include area symbol selection, understanding reader behavior of gray tints, cross-screening techniques, and providing base materials. Adequate base information must be provided to enhance the information-carrying capacity of choropleth maps.

This chapter presents material necessary for the understanding of a class of quantitative thematic maps called **choropleth maps.** This form of statistical map was introduced early in the nineteenth century. It was used by the Bureau of the Census in several statistical atlases in the last half of that century and has been a favorite of professional geographers and cartographers ever since. Its name is derived from the Greek words *choros* (place), and *pleth* (value). Choropleth mapping has also been called *area* or *shaded* mapping.

An entire chapter is devoted to this form of mapping because of its widespread use and appeal, not only for professionals, but for the general public as well. Because of this, the student must look at the choropleth map in considerable detail to learn its advantages and the standards for its use in a variety of mapping situations.

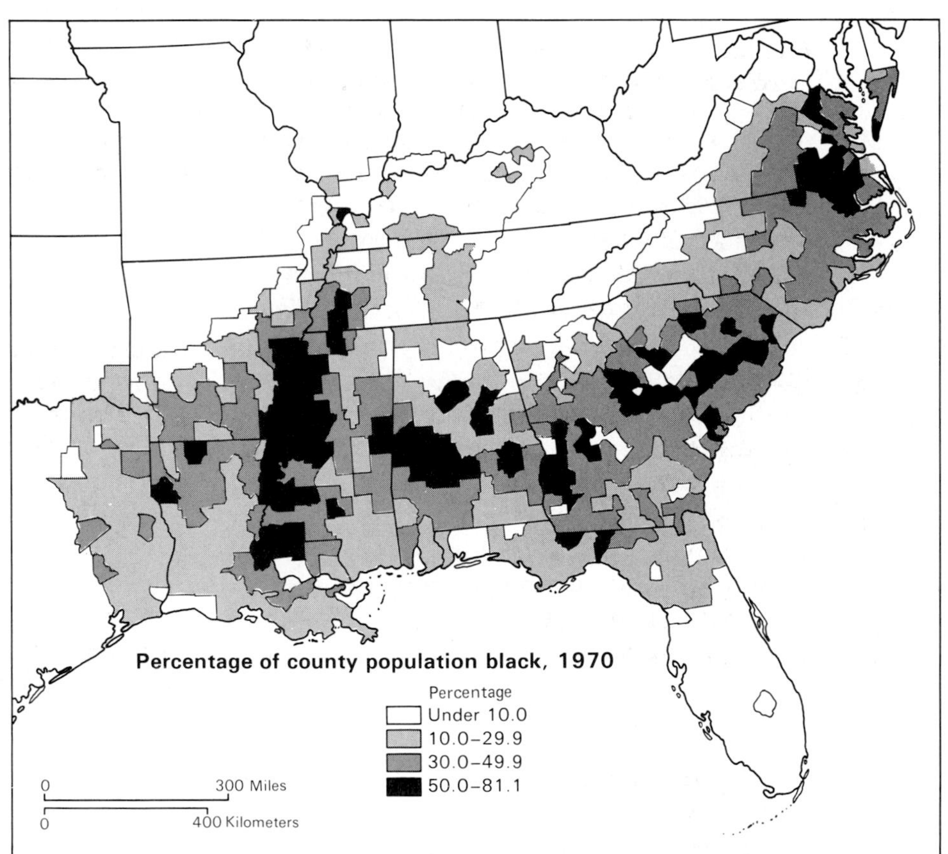

Figure 6.1. A typical choropleth map.
Each enumeration unit, in this case a county, has an areal symbol applied to it, depending on the class in which its data falls. Over the entire map, it is possible to determine spatial variation of the data.

Illustration from Stephen S. Birdsall and John Florin, Regional Landscapes of the United States and Canada, p. 200. Copyright John Wiley and Sons, Inc., 1981. Redrawn by permission.

Selecting the Choropleth Technique

What guides cartographers in selecting one mapping technique over another when approaching a given design task? When is a given technique not appropriate? The following section describes under what conditions the choropleth map should be chosen.

Mapping Rationale

The choropleth technique is defined by the International Cartographic Association as follows: "A method of cartographic representation which employs distinctive colour or shading applied to areas other than those bounded by isolines. These are usually statistical or administrative areas."[1] Because this form of mapping often uses data gathered and aggregated by administrative area, it is often called *enumeration* mapping. (See Figure 6.1.)

Choropleth mapping may be thought of as a three-dimensional histogram or stepped statistical surface. (See Figure 6.2.) A choropleth map is simply a planimetric representation of this three-dimensional **data model.** In the model, the height of each prism is proportional to the value it represents. The planimetric way of looking at the model incorporates areal symbolization to depict the heights of the prisms. In black-and-white mapping, the higher prisms are normally represented by darker areal

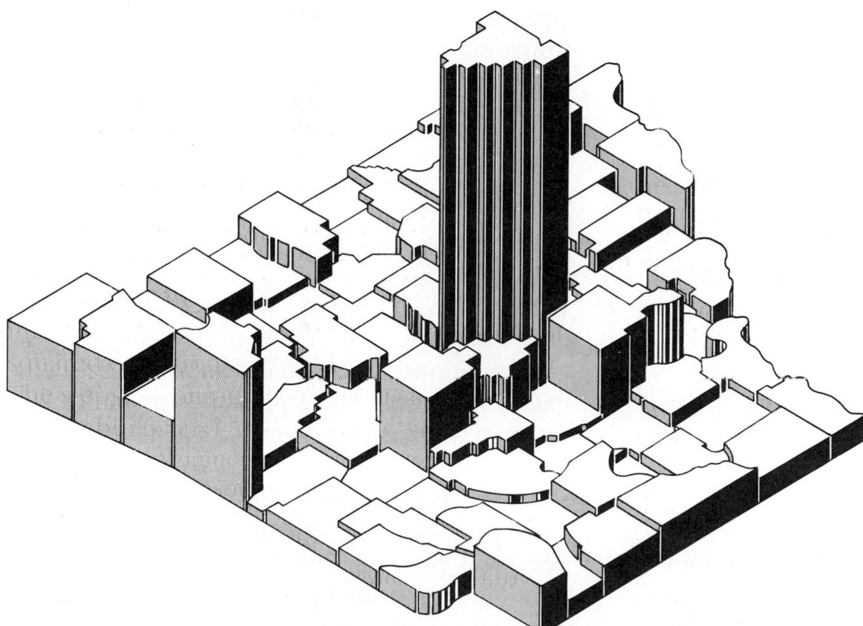

Figure 6.2. The data model concept in choropleth mapping.
In this conceptual model, each enumeration unit is a prism raised vertically in proportion to the value it represents.

U.S. Data Model

A unique and interesting pedagogic computer program is available for demonstrating the data model concept. A three-dimensional prism model of the United States, by states, is generated, and the user can specify viewing direction (NE, SE, SW, NW) and viewing elevation (20, 20, or 40 degrees). Title, the plot, a scaled vertical prism legend, and the direction and viewing elevation selected are presented on the screen. Eight data sets are provided: population per square mile (1983), percent population completing high school (1988), percent voting Republican in the 1980 presidential election, percent population receiving public aid (1983), percent farm population (1980), average acres per farm (1982), and per capita income (1983). An interesting classroom exercise would be for students to select one data set, examine the model carefully, and then prepare a choropleth map of it. The program was written by Richard E. Groop (1986) for an IBM environment, and the most recent version utilizes the EGA video mode. Additional information may be obtained by writing the Department of Geography at Michigan State University.

symbols; conversely, the lower prisms are represented by lighter areal symbols. It is helpful to think of this conceptual model when beginning a choropleth design task. (See box, "U.S. Data Model.")

Mapping techniques normally require the cartographer to collect data by statistical or administrative area—referred to as a **chorogram.** An areal symbolization scheme is then devised for these values, and the symbols are applied to those areas on the map whose data fall into the symbol classes. (See Figure 6.3.) Techniques developed through computer mapping can alter this general procedure; the newer techniques will be introduced later in the chapter.

There are three ways in which map readers use choropleth maps: to ascertain an actual value associated with a geographical area, to obtain a sense of the overall geographic pattern of the mapped variable with attention to individual values, and to compare one choropleth map's pattern to another's. It has been argued convincingly that the reader who wants only to find individual values should simply consult a table of values, and not bother with the map.

Using two or more choropleth maps to compare geographical distributions (usually to look for positive correlations) is an acceptable application of choropleth mapping, but this subject is not treated in detail here. Our presentation concentrates on methods useful in the production of an individual choropleth map whose purpose is to portray a single geographical theme.

Data Appropriateness

The choropleth technique should be selected only when the form of data is appropriate. Certain assumptions can often be made about the data. Typically, choropleth maps are constructed when discrete data occur or can be attributed to definite enumeration units—either statistical as used by the Bureau of the Census, or administrative political subdivisions. Geographical phenomena that are continuous in nature should not be mapped by the choropleth technique since their distributions are not controlled by political or administrative subdivisions. For example, to map average annual temperature by this method would not be ap-

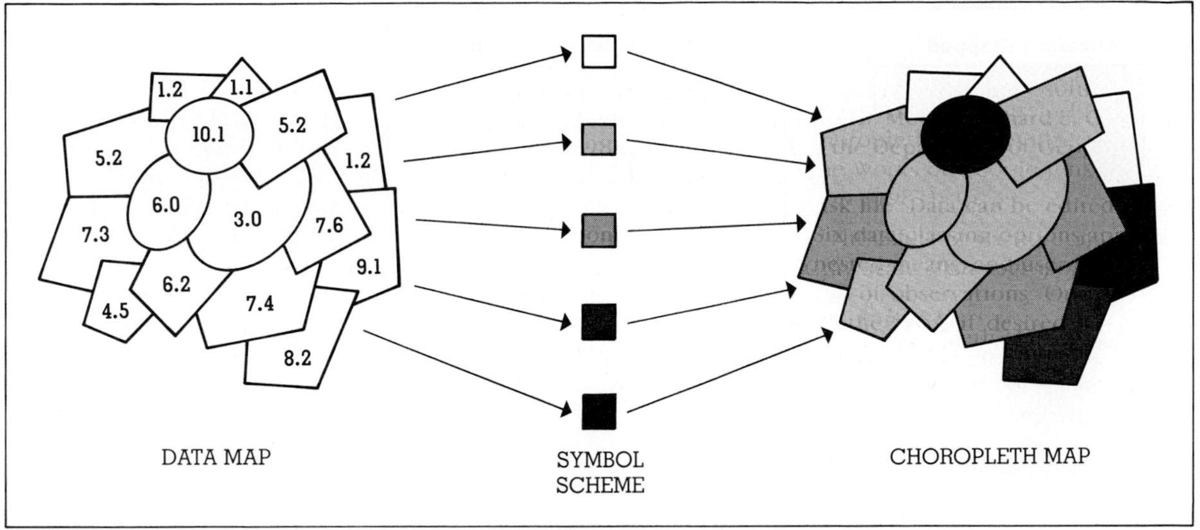

Figure 6.3. The choropleth technique.
Cartographic designers begin with a data model—a map of the enumeration units and all data values. Areal symbol classes are selected to represent range-graded numerical classes. The final map is then developed by applying appropriate areal symbols to the enumeration units, depending on each unit's data value.

propriate, but to map the number of hospital beds per 1000 people would be.

Enumeration data may be of two kinds: *totals* or *derived values* (rates or ratios). The number of people living in a census tract is an example of the former, and average annual income is an example of the latter. It is traditionally not acceptable to map total values when using the choropleth technique. This convention is based on sound reasoning: In most choropleth mapping situations, the enumeration units are unequal in area. The varying size of areas and their mapped values will alter the impression of the distribution. (See Figure 6.4a.) In addition, uniform distributions may be masked when enumeration total values are used. (See Figure 6.4b.) It has therefore become customary to use *ratios involving area* or *ratios independent of area.* Thus, data involving area must be areally standardized.

A familiar example of a ratio involving area is density, of which many kinds could be named. Population per square mile is often used, as is crop yield per acre. Ratios independent of area include per capita income, infant deaths per 100,000 live births, and so on. Proportions, including percentages, are also used.

Much of the enumeration data available to cartographers is in the form of aggregated areal data, such as average value of farm products sold, median family income, average annual income, and average persons per household unit. These are treated in the same manner as ratios or proportions in choropleth mapping.

The most important assumption made in choropleth mapping is that the value in the enumeration unit is uniformly spread throughout the unit. (See Figure 6.5.) The top of each prism in the conceptual model is horizontal and unchanging. Wherever one places

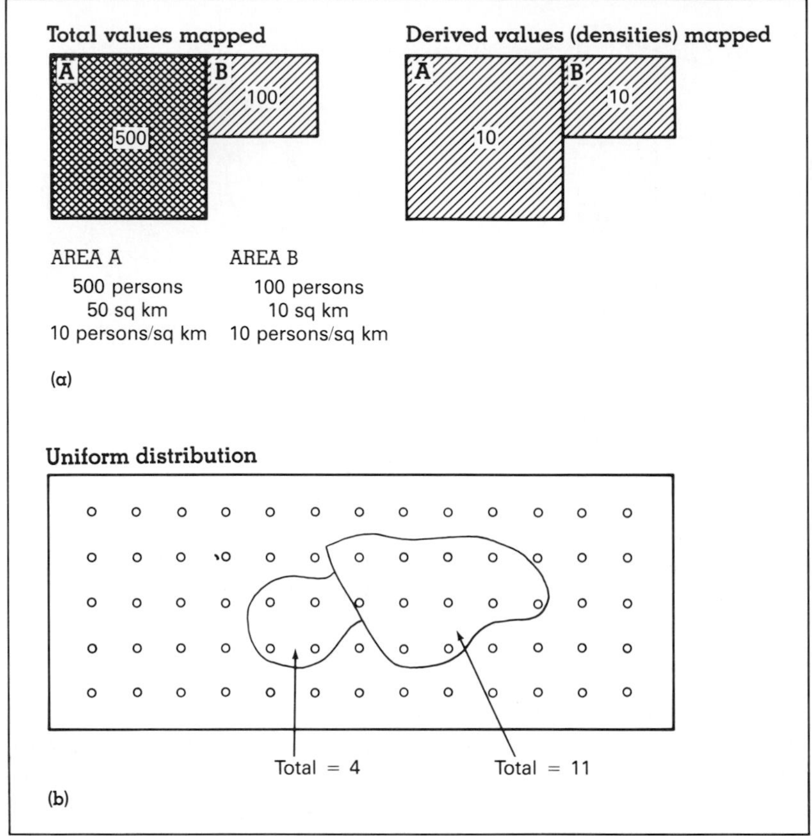

Figure 6.4. Total values should not be mapped by the choropleth method.
In (a), mapping data totals masks the even densities, because the areas are of unequal size. Uniform distributions, as in (b), will also be obscured when totals alone are mapped.

a pencil point in the area, one finds the amount of the variable chosen to represent the entire area. Thus, the choropleth technique is insensitive to changes of the variable that may occur at scales larger than the chosen enumeration unit. If the variable is changing within the enumeration unit, the change cannot be detected on the choropleth map.

In choropleth mapping, the boundaries of the chorograms have no numerical values associated with them. They function only to separate the enumeration areas and signify the geographical extent to which the enclosed area values apply. This is in contrast to the lines on the isopleth map which do have values, and which will be discussed in a later chapter.

When to Use the Choropleth Map

In summary, the choropleth technique is appropriate whenever the cartographer wishes to portray a geographical theme whose data are discrete and occur within well-defined enumeration units. If the data cannot be dealt with as ratios or proportions, they should not be portrayed by the choropleth technique. Nor should choropleth (or any other mapping technique) be used if the interest is to show

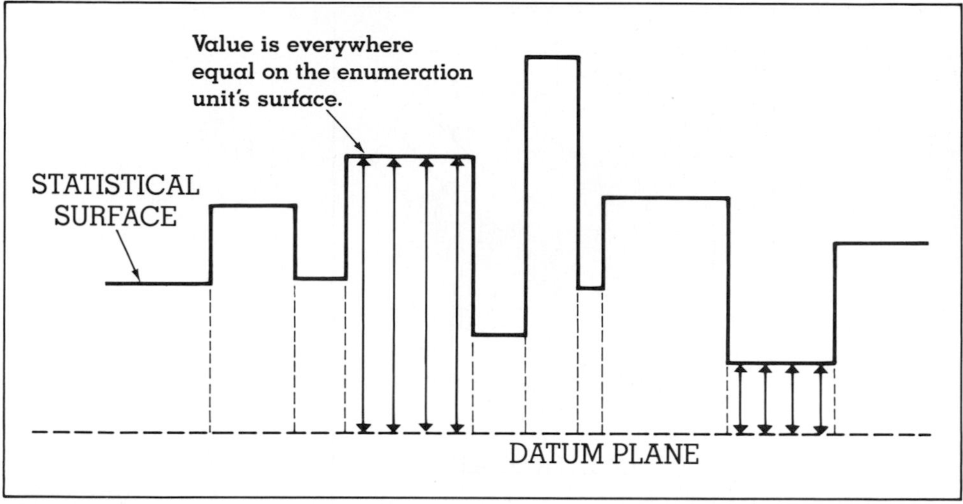

Figure 6.5. The stepped statistical surface.
The choropleth technique assumes a stepped statistical surface on which the value in each unit is constant. That is, the unit's surface is everywhere equidistant from the map datum plane. The view shown here is as if a vertical profile or a slice were taken out of the three-dimensional data model.

actual, precise values within enumeration units. Choropleth mapping is simple and should be used only when its assumptions are acceptable to the cartographer and to the eventual reader. The method should not be made overly complex.[2]

Preliminary Considerations in Choropleth Mapping

Important considerations in the design of a choropleth map include thorough examination of the geographical phenomenon and its elements, map scale, number and kind of areal units, data preprocessing, data classification, areal symbolization, and legend design.

Geographical Phenomenon

All map design begins with careful analysis of just what it is that is being mapped. A careful designer assembles facts that will help in understanding the mapping activity. The beginner may attempt to develop maps showing dairying without knowing anything about cows! In a mapping problem to illustrate the geographical aspects of retailing, what measures should be used? Dollar sales, payrolls, and number of employees might be appropriate. What industrial or trade indices are commonly accepted and used by analysts? What surrogate measures might be used? What other geographical variables accompany the one being mapped? How does this phenomenon behave spatially or aspatially, with or without other geographical phenomena?

Cartographic designers must equip themselves with as much knowledge about the map subject as possible. In many cases, consultants are brought in during the early phases of design. Regardless of the individual steps taken, premapping research is necessary for effective map design, just as a product designer must measure the proportions of the human hand before designing an electric drill.

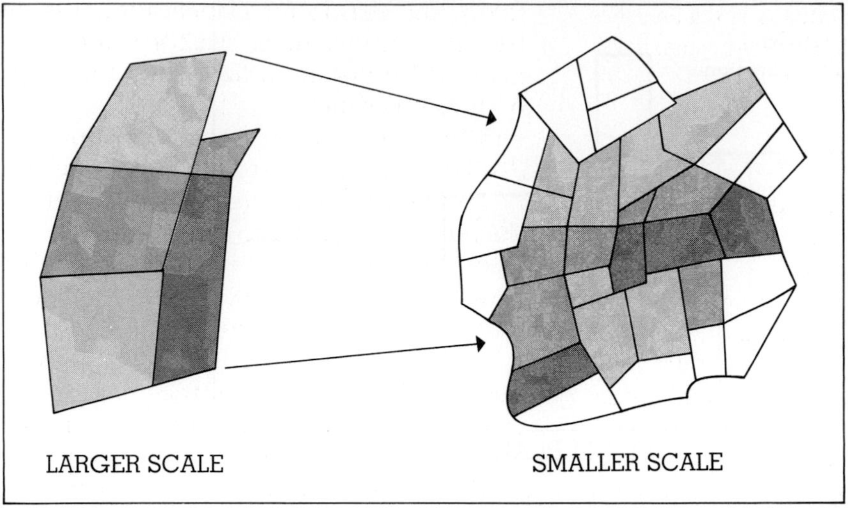

Figure 6.6. Choropleth detail and map scale.
At larger scales, distributional detail is lost, relative to the size of the enumeration units. As more units are used at smaller scales, more geographic resolution is gained, and a better overall picture of the spatial attributes of the distribution emerges. If mapping scale remains the same, the identical effect is achieved by using smaller enumeration units.

Map Scale

Map scale as it relates to choropleth design involves two considerations: necessity and available space.[3] The cartographer strives for balance between these two. Necessity dictates that the scale be sufficiently large to accommodate symbolization of the phenomenon—the areal units must be large enough for the reader to see and differentiate areal patterns. In most cases, however, the cartographer is forced to operate in a map space smaller than the ideal. The designer seeks the balance that best serves the purpose of the map.

Number and Kind of Areal Units

For practically all choropleth mapping, the larger the number of areal units used, the more details of the geographical distribution the map can show. Scale is significant in this respect. Spatial detail is added as the number of enumeration units is increased; conversely, spatial coarseness increases as the number of units is decreased. (See Figure 6.6.)

Symbolization also has a direct effect on determination of the number of areal units. As the number of units increases, their sizes decrease, making it more difficult to differentiate symbols.

The choice of how many areal units to use depends also on such variables as time, cost, map purpose, map size and scale, and symbolization. Each design task will have its own set of constraints.

The kind of areal unit is usually dictated by map purpose, level of acceptable generalization, data availability, and the scale considerations mentioned above. For much of choropleth mapping, the kind of areal unit is determined before actual mapping begins—most often dictated by availability of data. Cartographers seldom have the option of specifying areal units, which are usually those used by local, state, or federal census sources. This is not always undesirable, however; standardization leads to easier comprehension.

Data Preprocessing

Ideally, the designer would like to live in a world in which data are presented in mappable form (though this would also eliminate the need for the designer). This is not the case. Choropleth mapping requires that data be in ratio or rate form, most often necessitating the preprocessing of raw data according to the purpose of the map. Electronic calculators, microcomputers, or large computers can be used to ease the burden when dealing with large data sets.

Data preprocessing may require consultation with experts familiar with the purpose of the map, such as the map client. Data preprocessing is an integral part of the total design activity and deserves careful attention.

Data Classification Techniques

Just a few years ago, before the computer plotting of thematic maps, the usual steps in the choropleth technique included combining mapped values into groups or classes, symbolizing each group with a unique areal symbol, and applying the appropriate areal symbol to each enumeration unit according to the class in which its value fell. The entire mapping process was then completed. This will henceforth be referred to as the **conventional choropleth technique.**

Values are grouped into classes in order to simplify mapped patterns for the reader. The number of classes became somewhat standardized when it was learned that map readers could not easily distinguish between more than eleven areal-symbol gray tones.[4] In practice, no more than six classes are recommended,[5] and a minimum of four is also good practice. Actually, the reason for classification has been better management of symbol selection and map readability, not primarily for the purpose of grouping data.[6]

With modern computer plotters and video displays, it is possible to symbolize each value by its own unique areal line symbol, thus producing an *unclassed* choropleth map.[7] Such maps have not become widely used partly because of the need for expensive computer facilities. Also, not all cartographers approve of the result of not classifying the data: an unstructured or ungeneralized view of the mapped phenomenon. Differences between the conventional choropleth technique and the unclassed variety will be addressed below.

The Importance of Classification

The classification operation performed for choropleth mapping behaves pretty much like a group of stacked sieves. (See Figure 6.7.) Each sieve acts as a class boundary, and only values of certain sizes are allowed to pass into one of several classes. As we look at classification techniques, it is wise to recall from the previous chapter that in the activity of classification we wish the results to be meaningful and revealing. The sieve analogy is presented simply to remind you that the class boundaries, or sieves, may be anywhere, but must be chosen wisely.

Assigning values into groups on the choropleth map is a form of data classification that leads to simplification and generalization.[8] Details may be lost, but as with all kinds of generalization, the resulting structures allow more information to be transmitted. Classification allows the designer to structure the message of the thematic communication. How well this is done depends largely on the designer's ability to understand the geographic phenomenon. It is not possible to structure (and hence generalize) messages with the unclassed method, except in the sense that the cartographer has chosen what to map.

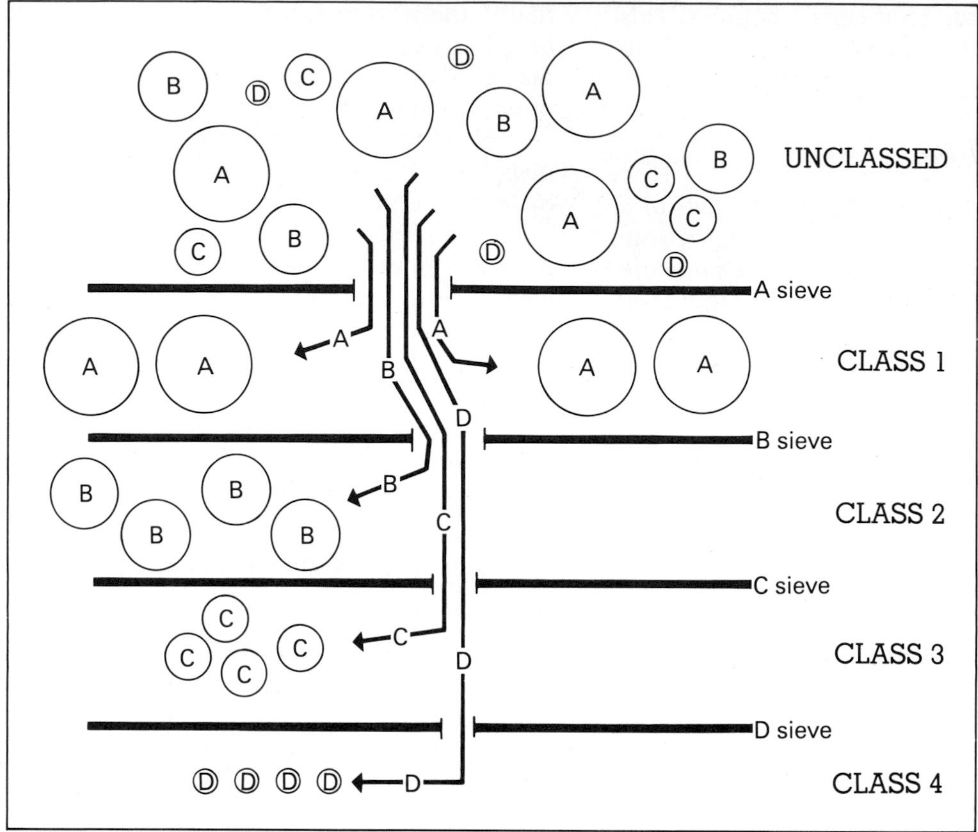

Figure 6.7. The sieve analogy in classification.
Each sieve functions as a screen allowing only balls of a certain size to drop through to the next level. Each sieve can be compared to taxonomic criteria established for the particular study. Spaces between the sieves become the classes, and the sieves are the class boundaries.

The purpose of the choropleth map dictates its form. If the map's main purpose is to simplify a complex geographical distribution for the purpose of imparting a particular message, theme, or concept, the conventional choropleth technique should be followed. On the other hand, if the goal is to provide an inventory for the reader in a form that the reader must simplify and generalize, then the unclassed form should be chosen.

Methods of Data Classification

Before discussion of individual methods of classifying data, it is necessary to examine what happens when values are grouped. In the three-dimensional data model, it is clear that grouping data values tends to smooth the model and reduce its irregularities. (See Figure 6.8.) This procedure, however, reduces the number of individual data values transmitted to the reader. The reader can only determine within a specified range of values what a particular enumeration unit's value is. Classification—determining class intervals and class boundaries—consists conceptually of passing planes through the imaginary model. The various methods of classification are simply ways of determining the vertical position of the planes. To assist the designer in selecting map classes, there are numerous methods which may be grouped into four types: exogenous, ar-

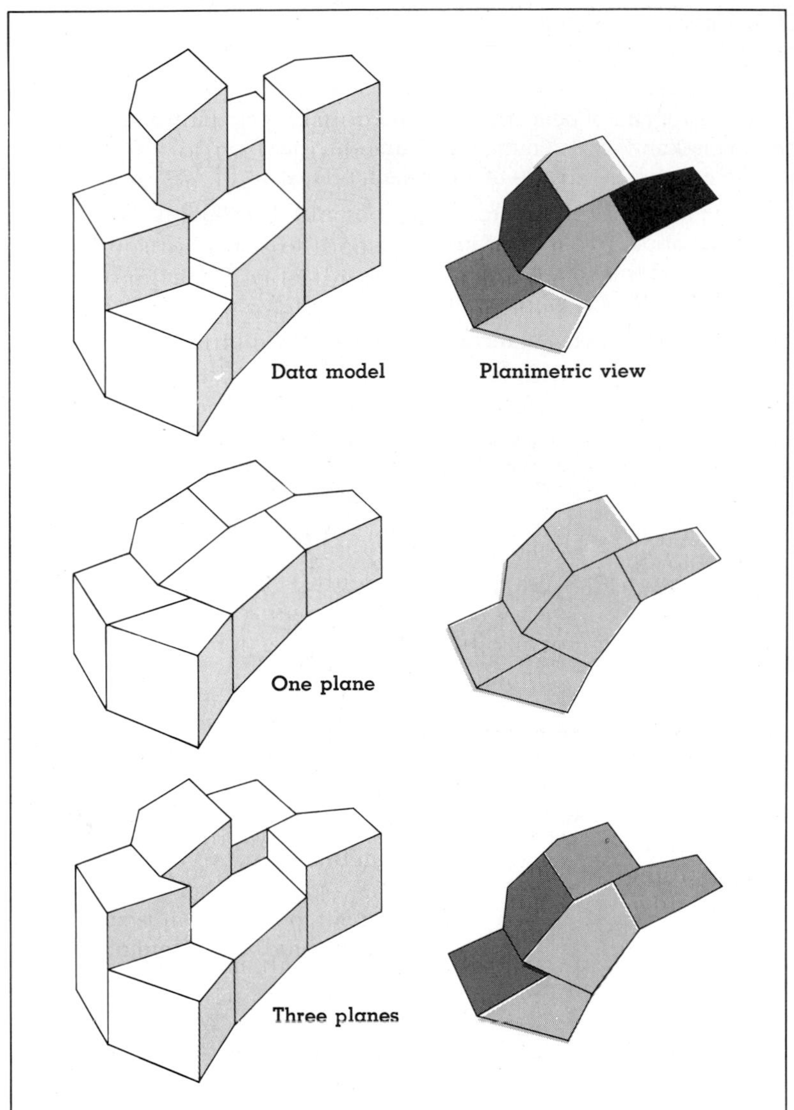

Figure 6.8. Planimetric maps.
Different planimetric maps result from passing one or more planes through the data model.

bitrary, idiographic, and serial.[9] **Exogenous data classification** includes schemes in which values not related to the data set are chosen to subdivide the values into groups. An income level selected to define poverty is one example. A certain disease incidence rate may be critical for the epidemiologist, and this becomes a class limit. **Arbitrary data classification** methods use regular, rounded numbers having no particular relevance to the distribution as class divisions, such as 10, 20, 30, 40, and so on. This system has usually been used for reasons of convenience. **Idiographic data classification** methods, long used by cartographers, are determined by particular events in the data set. The "natural breaks" type is often used; this is an example of the graphic form. (See Table 6.1.) Quartiles are also frequently

Table 6.1 Summary of Major Class Interval Methods

Method	Characteristics
Equal steps	Particularly useful when histogram of data array has a rectangular shape (rare in geographical phenomena) and when enumeration units are nearly equal-sized. In such cases, produces an orderly map. Example of a constant interval.
Standard deviation	Should be used only when the data array approximates a normal distribution. The classes formed yield information about frequencies in each class. Particularly useful when purpose is to show deviation from array mean. Understood by many readers. Usually limited to six classes. Example of a constant interval.
Arithmetic progression	A variable, systematic, mathematical class interval system. General form will vary depending on positive or negative form of the common difference. Used only when the shape of data array approximates the shape of an arithmetic progression.
Geometric progression	Same characteristics as arithmetic progressions. Very useful when frequency of data declines continuously with increasing magnitude—which commonly happens with geographical data.
Quantiles	Good method of assuring an equal number of observations in each class. Can be misleading if the enumeration units vary greatly in size, although this can be offset by areal weighting. Does not yield information about frequencies in each class.
Natural breaks	Good, graphic way of determining natural groups of similar values by searching for significant depressions in frequency distribution. Minor troughs can be misleading and may yield poorly defined class boundaries. Used in conjunction with other methods.
Optimal	An extension of the natural breaks method, but one that yields a quantitative index. The most appropriate optimal technique is one that forms classes that are internally homogeneous and at the same time retains heterogeneity among classes. As with the natural breaks method, class intervals are irregular and class boundaries may look unorganized. A clustering activity.

used. **Serial** methods include standard deviational units, equal intervals, and arithmetic and geometric progressions. As the material in Table 6.1 indicates, the nature of the data set helps in selecting a method.

Another way to view the classing activity is to consider the map user. By so doing it is possible to form two groups of classing objectives—statistical and geographical.[10] In the former, the user views the map to gain some statistical attribute of the data array, such as medians, averages, standard deviations, or other statistical measures. Specific, limited numbers of significant planes are passed through the data model. In the second group of users, the map reader views the map for the purpose of seeing how well the map, and its classification, has replicated the original data array. Several classing techniques achieve this objective, such as natural breaks, arithmetic and geometric progressions, and others. Here, reproduction of the data model is attempted with several well-selected planes. As with all other forms of thematic mapping, map purpose must guide the way.

There is no one best way of devising class intervals and class boundaries for choropleth maps. Simplicity is a major goal, whether the system is devised on purely visual or more rigidly formed mathematical grounds.[11] It is also worth considering that the class interval system should include the full range of the data, have no overlapping classes, and reflect some logical division of the data array in order to portray the purpose of the map.

Developing class intervals and class boundaries for conventional choropleth maps remains largely an experimental activity. Each geographical phenomenon being mapped presents a unique set of circumstances and nuances for grouping. Mapping purposes vary, as does the intended readership. The cartographer must look at each new choropleth map from a fresh perspective, without feeling constrained to follow a given method of classification.

Constant Intervals

A common way of expressing data classes is the *equal-step* method, which is analogous to passing planes through the three-dimensional data model so that the vertical distances between them are equal. Devising class boundaries this way encloses equal amounts of the range of the mapped quantity within each class interval. Calculate equal-step class intervals as follows:

1. Calculate the range of the data (R):

$$R = H - L$$

where H is the highest reported value and L the lowest reported value.

2. Obtain the common difference (CD):

$$CD = \frac{R}{\text{number of classes}}$$

3. Obtain the class limits by calculating:

$$\begin{aligned} L + 1 \cdot CD &= \text{first class limit} \\ L + 2 \cdot CD &= \text{second class limit} \\ L + (n - 1) \cdot CD &= \text{last class limit} \end{aligned}$$

where n is the number of class limits, which will be one less than the number of classes.

Maps that are classed by this method generally have intuitive appeal. Their legends tend to appear orderly. If the enumeration units are nearly equal in size and the numerical distribution rectangular, such maps appear neat and organized. Unfortunately, most histograms are not rectangular.

Standard Deviations

If the data set displays a normal frequency distribution, then class boundaries may be established by using its standard deviation value. Class intervals designated this way should be used only when it can be assumed that the reader understands them. Assessments are made about the number of values in each class. Class boundaries are compiled by computing the mean and standard deviation, then determining the boundaries by adding or subtracting the deviation from the mean. No more than six classes are usually needed to account for most of the values in a normal distribution. Special symbolization problems arise with this method, since class boundaries are arrayed around a central value instead of ascending from the lowest value, as is usually the case. (See Figure 6.9.) Nonetheless, this method yields constant class intervals because the standard deviation is unchanging.

Variable Intervals

It is more common, and usually more desirable, for cartographers to develop variable intervals in the class system. Generally, these methods are better for depicting the actual distribution of the mapped quantity. Variable intervals are classified as either *systematic* or *irregular* forms.

Arithmetic and Geometric Intervals

These mathematically defined interval systems produce class boundaries and intervening distances that change systematically. They should be used only when a graphic plot of the mapped values tends to replicate the mathe-

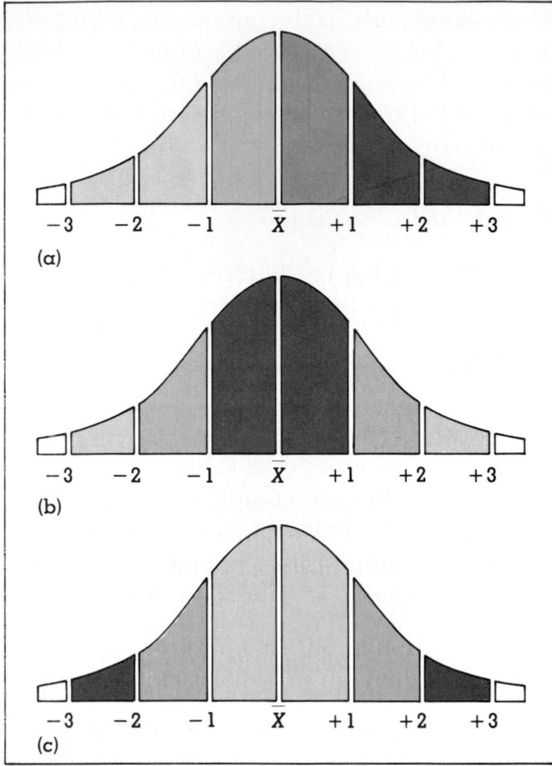

Figure 6.9. Alternative ways of symbolizing classes by standard deviations.
In (a), the classes range in visual importance from −3 to +3 in a continuum. Greater importance is given those values close to the mean in (b). In (c), greater importance is assigned to those values farthest from the mean. The purpose of the map will dictate the choice of a symbolization method. Because of the bi-directional nature of the standard deviation, however, there appears to be little intuitive appeal for method (a).

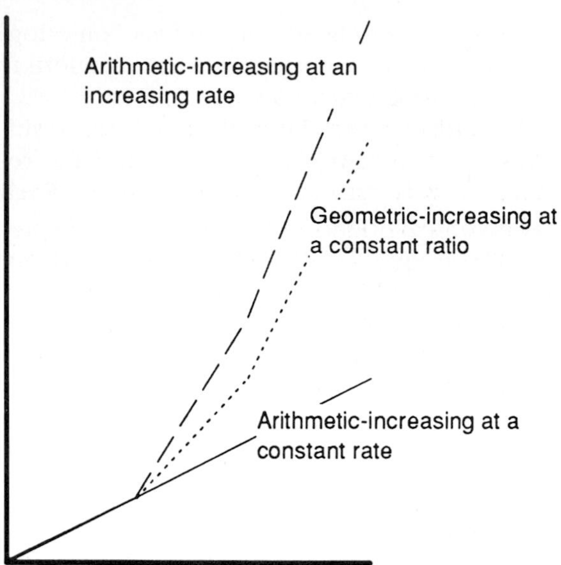

Figure 6.10. Common mathematical progressions.
A graphic plot (or array) of the data values on arithmetic paper is visually compared to common mathematical progression plots. This can assist in selecting an interval plan.

matical progressions. (See Figure 6.10.) From this plot, it is possible to see if an orderly mathematical function exists by comparing the shape of the curve to that of a typical arithmetic or geometric progression.

An *arithmetic progression* (identical in idea with the equal-step method described above) is defined as

$$a, a + d, a + 2d, a + 3d, \ldots a + (n - 1)d$$

where

- a = first term
- d = common difference
- n = number of terms
- l = last term = $a + (n - 1)d$

A geometric progression is defined as

$$a, ar, ar^2, ar^3, \ldots ar^{n-1}$$

where

- a = first term
- r = common ratio
- n = number of terms
- $l = ar^{n-1}$

Different expressions of d and r will yield a variety of progression curves.

Quantiles

Quantile class interval systems produce irregular variable intervals, such as quartiles (25 percent), quintiles (20 percent), deciles (10 percent), or any similar value. Developing class boundaries of quantiles assures an equal number of values in each class and minimizes the importance of the class boundaries. If the enumeration areas vary considerably in area, this method can be misleading, although the values can be weighted by area.

Class boundaries are produced in this way:

1. Array all values in ascending order
2. Solve for K, the number of values in each class:

$$K = \frac{\text{number of enumeration areas}}{\text{number of classes}}$$

3. Beginning with the lowest value, K values are included in the first class, K values in the next class, and so on. The class boundary is normally the mean value between the two adjacent values separating adjoining classes.

Groups Based on Similarities—Natural Breaks

Most classification tasks faced by cartographers involve *numerical classification,* in which numbers along an imaginary number line are grouped according to some criteria of similarity. The cartographer usually attempts to form number groups so that the numerical differences within groups are less than the differences between groups. Regardless of the method, most cartographers agree today that *any* method that strives to class the data by these criteria are superior to those that do not.

Visual inspection. One graphic method is the histogram. (See Figure 6.11.) Class boundaries are placed at breaks in the distribution. These breaks signal changes in the data and may be lower (smaller frequencies) or open

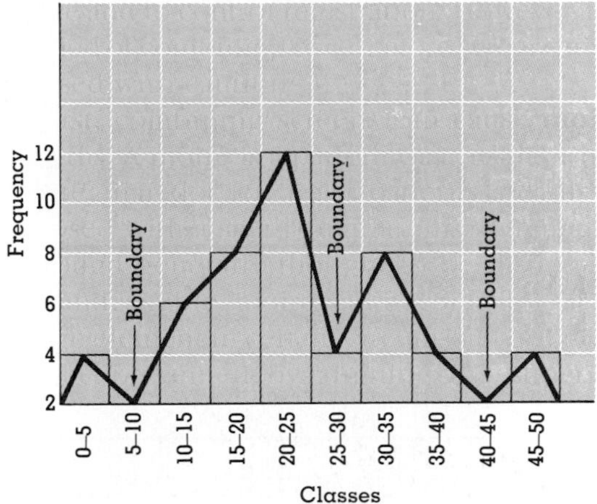

Figure 6.11. Histograms or frequency polygons are used for data classing. Boundaries separating classes are usually placed at locations having lower frequencies or breaks in the array.

classes in the histogram. Of course the number of classes selected for the activity will alter the locations chosen for the boundaries.

Another method calls for the construction of a **graphic array.** (See Figure 6.12.) A graphic array, a kind of rank-order graph, is begun by using arithmetic grid paper and graduating the horizontal axis so that it accommodates the number of observations in the variable array. The vertical axis is graduated to accommodate the range of data values. For this example, we are using the values displayed in Table 6.2. Producing the graphic array requires that the values in the data set are arranged (sorted) in ascending order.

Construction of the plot begins by placing a dot on the graph corresponding to each value's position in the array on the horizontal axis, and correctly placed according to its position on the vertical axis. After all dots are located on the graph, they may or may not be joined by a line. This completes the construction of the graphic array.

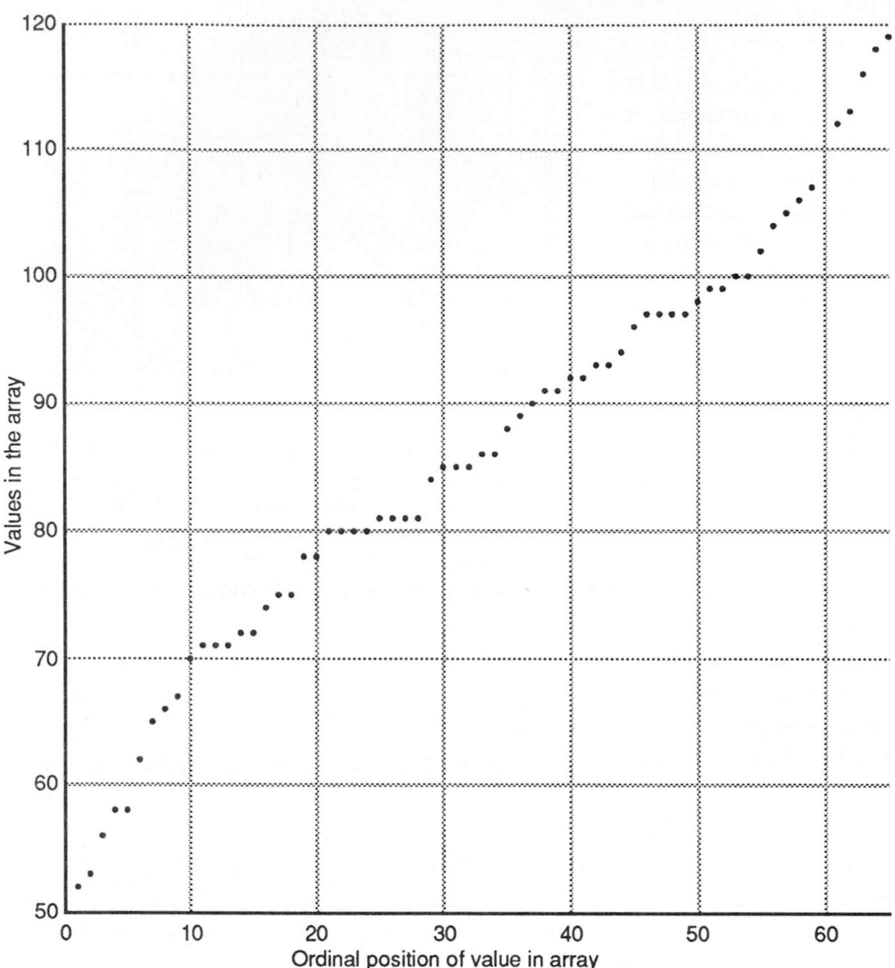

Figure 6.12. Graphic array.
Values in the data set are plotted in an array and class boundaries are identified at places where slopes change remarkably.

Data classing is accomplished by visual inspection of the dots or line. Class limits are selected where the slope of the line changes dramatically. These points tend to define the boundaries between clusters of similar values in the array, and helps achieve the goal in numerical classing as defined above. These places can also be seen from a tabular list but the graphic plot greatly assists in the endeavor. As before, any natural selection may be influenced by the number of classes imposed on the project. Choices must be made, however, within these constraints because an unlimited number of classes is possible.

Developing class boundaries by construction and visual inspection of either the histogram or graphic array is troublesome for very large data sets. Just where it becomes too much trouble depends on the project, and on whether there is a computer close by to assist the cartographer. Computer solutions eliminate the tedium of numerical processing, but do not make choices.

One-way analysis of variance. After graphing methods are used to determine classes, more sophisticated numerical means may be employed. One method used is based

Table 6.2 Data Table for Graphic Array

52	75	86	97
53	75	86	98
56	78	88	99
58	78	89	99
58	80	90	100
62	80	91	100
65	80	91	102
66	80	92	104
67	81	92	105
70	81	93	106
71	81	93	107
71	81	94	112
71	84	96	113
72	85	97	116
72	85	97	118
74	85	97	119

on what statisticians call **one-way analysis of variance.**

The fundamental idea of one-variable analysis of variance is to determine whether there is any significant difference between the means of several subsets of values obtained as samples from a larger population. This kind of test is often used to measure the significance of different treatments given to sample groups. It employs the variances of the subsets and of the whole sample.

The statistic used is denoted as F and is a ratio:

$$\frac{\textbf{Variance of all subgroup means (variance between groups)}}{\textbf{Average of the variances in each subgroup (variance within groups)}}$$

$$F = \frac{V_m}{V_c}$$

The likelihood of any F value occurring can be read directly from a statistical F-table. We need not determine significance in using this procedure for simply classifying numbers.

This test will determine if the average within-group variance is larger or smaller than the overall variance, so as to assure sufficient *similarity* within groups. These steps should be followed:

Table 6.3 Data Table for One-Way Analysis of Variance Problem

1.2	17.5	24.6	34.2
4.2	17.9	24.8	34.5
10.7	18.9	24.8	34.9
11.2	20.0	24.9	37.9
14.0	21.1	26.0	38.0
14.7	22.9	27.3	48.0
15.1	23.1	32.0	
16.2	23.9	32.1	
16.4	24.6	33.5	

1. Develop the numerical classes. The frequency histogram is a good place to begin.
2. Compute the means of each class.
3. Compute the variance of these means, designated V_m.
4. Compute the variance in each class.
5. Calculate the *average* of the class variance. This is called V_c.
6. Test the classing scheme by assuring that $V_c < V_m$.

Simply determining that $V_c < V_m$ may not be sufficient; there can be other class structures that yield $V_c < V_m$. The most appropriate classing scheme is the one that yields the greatest F value. There is no simple solution to finding the class scheme that yields the highest F. It can be found by *iteration*—by repeating the procedures outlined above for each new class scheme until the highest value is found. If a computer is available, the task is simplified; otherwise, it may be formidable.

A practical solution, if a computer is not available, is to develop the classes by the histogram method, or other criteria established by the map problem at hand, and determine F. Small changes can then be made at the class boundaries by including or excluding adjacent values only, and new Fs calculated. If a few such alterations make little difference in F, others are probably not needed. The final class determination is the one having the largest F.

Our sample problem deals with the data array shown in Tables 6.3 and 6.4.

Table 6.4 Classing by One-Way Analysis of Variance

First Classing Iteration to Determine F of Sample Data

Class 1	Class 2	Class 3	Class 4	Class 5
1.2	10.7	26.0	32.0	48
4.2	11.2	27.3	32.1	
	14.0		33.5	
	14.7		34.2	
	15.1		34.5	
	16.2		34.9	
	16.4		37.9	
	17.5		38.0	
	17.9			
	18.9			
	20.0			
	21.1			
	22.9			
	23.1			
	23.9			
	24.6			
	24.6			
	24.8			
	24.8			
	24.9			
$\overline{X}_1 = 2.7$	$\overline{X}_2 = 19.36$	$\overline{X}_3 = 26.65$	$\overline{X}_4 = 34.63$	$\overline{X}_5 = 48$
$\sigma^2 = 4.49$	$\sigma^2 = 22.56$	$\sigma^2 = .844$	$\sigma^2 = 5.24$	$\sigma^2 = 0$

$$F = \frac{V_m}{V_c} = \frac{285}{6.62} = 43.05$$

Second Classing Iteration to Determine F of Sample Data

Class 1	Class 2	Class 3	Class 4	Class 5
1.2	10.7	22.9	32.0	48
4.2	11.2	23.1	32.1	
	14.0	23.9	33.5	
	14.7	24.6	34.2	
	15.1	24.6	34.5	
	16.2	24.8	34.9	
	16.4	24.8	37.9	
	17.5	24.9	38.0	
	17.9	26.0		
	18.9	27.3		
	20.0			
	21.1			
$\overline{X}_1 = 2.7$	$\overline{X}_2 = 16.14$	$\overline{X}_3 = 24.69$	$\overline{X}_4 = 34.63$	$\overline{X}_5 = 48$
$\sigma^2 = 4.49$	$\sigma^2 = 10.30$	$\sigma^2 = 1.63$	$\sigma^2 = 5.24$	$\sigma^2 = 0$

$$F = \frac{V_m}{V_c} = \frac{299.29}{4.33} = 69.12$$

It illustrates the groups and statistics necessary to calculate F for the first and second iteration of our sample data set. The calculated F in the first iteration is 43.05. The second iteration yields an F of 69.12, clearly superior to the first class system. It appears likely that additional changes will not produce large differences. However, it should be pointed out that changes in the *number of classes* can affect F. Experimentation will determine the appropriate class number, if it is not dictated by other design considerations.

Jenks optimization. A method of classification that is analogous to a one-way analysis of variance has been introduced by George F. Jenks.[12] This procedure, now frequently called the **optimization method,** incorporates the logic most consistent with the purpose of data classification: forming groups that are internally homogeneous while assuring heterogeneity among classes. Computers are normally necessary to perform the calculations, except in the rare case of only a few values—where an electronic calculator is still indispensable. Whenever possible, this method must be explored.

The measure produced by this optimization technique is called the **Goodness of Variance Fit (GVF).**[13] The procedure is a minimizing one in which the smallest sum of squared deviations from class means is sought. The steps in computing the GVF are as follows (see Table 6.5):

1. Compute the mean ($\overline{X}$) of the entire data set and calculate the sum of the squared deviations of each observation (x_i) in the total array from this array mean:

   $$\Sigma\ (x_i - \overline{X})^2$$

   This will be called SDAM (squared deviations, array mean).

2. Develop class boundaries for the first iteration. Compute the class means ($\overline{Z}_c$s). Calculate the deviations of each x from its class mean ($x_i - \overline{Z}_c$), square these, and calculate the grand sum

   $$\Sigma\Sigma\ (x_i - \overline{Z}_c)^2$$

   This will be called SDCM (squared deviations, class means).

3. Compute the goodness of variance fit (GVF):

   $$\text{GVF} = \frac{\text{SDAM} - \text{SDCM}}{\text{SDAM}}$$

   The computed difference between SDAM and SDCM is the sum of squared deviations between classes.

4. Note the value of GVF for iteration one. The goal through the various iterations is to *maximize* the value of GVF.

5. Repeat the above procedures until the GVF cannot be maximized further.

The choropleth map that has a class for each value is considered to be the purest form. In this case, a class mean would be identical to the one value in the class, so the squared deviation would be zero. Across all classes, then, SDCM = 0.0, and GVF = 1.0, the maximum value of GVF. Of course, we do not have classes for each value in conventional choropleth mapping. Real solutions for GVF will be less than 1.0, and "best" solutions will only tend toward 1.0.

It should be apparent that applying an optimization technique requiring experimentation with class intervals is a timely process. However, a "one-pass" (no iteration) solution is available in a computer program available from Michigan State University. (See box, "Jenks Classing Program.")

In the previous chapter, the measures of kurtosis and skewness were defined as numerical ways of describing the structural characteristics of data distributions. In a recent study conducted to determine the relative effectiveness of the Jenks Optimization method, when compared to four other classing methods

Table 6.5 An Example of the Optimization Method of Developing Choropleth Class Limits

Array Values	Deviations from Array Mean	Squared Deviations from Array Mean	Solution 1 (special case with a class for each value)			Solution 2			Solution 3		
	$(x-\bar{X})$	$(x-\bar{X})^2$		$(x-\bar{Z}_c)^2$	$(x-\bar{Z}_c)^2$		$(x-\bar{Z}_c)$	$(x-\bar{Z}_c)^2$		$(x-\bar{Z}_c)$	$(x-\bar{Z}_c)^2$
2	−4.54	20.61	$Z_1 = 2$	0	0		−1.0	1.0	$Z_1 = 2.5$	.5	.25
3	−3.54	12.53	$Z_2 = 3$	0	0	$Z_1 = 3.0$	0	0		.5	.25
4	−2.54	6.45	$Z_3 = 4$	0	0		1.0	1.0		−1.0	1.0
5	−1.54	2.37	$Z_4 = 5$	0	0	$Z_2 = 5.5$	.5	.25	$Z_2 = 5.0$	0.0	0
6	−.54	.29	$Z_5 = 6$	0	0		.5	.25		1.0	1.0
7	.46	.21	$Z_6 = 7$	0	0	$Z_3 = 7.0$	0	0		−.33	.11
7	.46	.21	$Z_7 = 7$	0	0		0	0	$Z_3 = 7.33$	.33	.11
8	1.46	2.13	$Z_8 = 8$	0	0		1.5	2.25		.67	.45
9	2.46	6.05	$Z_9 = 9$	0	0	$Z_4 = 9.5$	−.5	.25		−1.0	1.0
10	3.46	11.97	$Z_{10} = 10$	0	0		.5	.25	$Z_4 = 10.0$	0.0	0
11	4.46	19.89	$Z_{11} = 11$	0	0		1.5	2.25		1.0	1.0

$\bar{X} = 6.54$

Solution 1: $\text{SDCM} = \Sigma\Sigma(x-\bar{Z}_c)^2 = 0$; Solution 2: $\text{SDCM} = \Sigma\Sigma(x-\bar{Z}_c)^2 = 7.5$; Solution 3: $\text{SDCM} = \Sigma\Sigma(x-\bar{Z}_c)^2 = 5.17$

$\text{SDAM} = \Sigma(x-\bar{X})^2 = 82.72$

$\text{GVF} = \dfrac{\text{SDAM} - \text{SDCM}}{\text{SDAM}} = 1.0$ $\quad \text{GVF} = \dfrac{\text{SDAM} - \text{SDCM}}{\text{SDAM}} = .909$ $\quad \text{GVF} = \dfrac{\text{SDAM} - \text{SDCM}}{\text{SDAM}} = .937$

(quartile, equal interval, standard deviation, and natural breaks), the following results were achieved:[14]

1. For the quartile method, GVF scores decreased with increasing skewness, and decreased similarly with increasing kurtosis.
2. For the equal interval method, GVF scores showed little correspondence with skewness, that is, GVF scores remained little changed with increasing skewness. GVF scores did not change greatly with increasing kurtosis, but did show to be somewhat better with flat distributions.
3. For the standard deviation method, GVF scores generally declined with increasing skewness, but not always. The relationship between kurtosis and GVF was similar.
4. For the natural breaks method, GVF varied greatly with skewness and kurtosis. "In general, distributions with high coefficients of skewness and kurtosis class the data more accurately than those which are close to normal."
5. For the Jenks Optimization method, high GVF scores result for distributions ranging from low to high skewness, and from low to high kurtosis. "This procedure produces consistently high GVF indicies irrespective of the degree of skewness and kurtosis because the method selects classes on the basis of the GVF criteria."
6. None of the methods proved wholly reliable because the GVF did vary, but the Jenks Optimization method proved best overall.

The implications for the designer are apparent. The structural characteristics of the data distribution to be classed must be considered before classing is begun. Skewness and kurtosis should be calculated, and GVF scores computed. Here is the point: It appears abundantly clear that any numerical clustering procedure is better used than none at all, at least

Jenks Classing Program

A computational, interactive microcomputer program is available that includes a "one-pass" solution to the Jenks Optimal method of class interval determination. Written by James B. Moore, Richard E. Groop, and George F. Jenks, version 3 (copyright 1983–1988), is available from the Department of Geography, Michigan State University.

This program allows the user to input data from the keyboard or a disk file. Data can be edited and presented on screen or printer, and may be saved on disk for later use. Six data classing options appear on the menu: Jenks method, quartiles, even steps, standard deviations, nested means, and user-defined. Any number of classes may be specified, but may not exceed the number of observations. Output includes a graphic array plot, with class boundaries appearing as boxes in the array, if desired. For the method specified, a table showing class number, number of observations in each class, limits of each class, mean of each class, and class variance is produced. Total variance (the sum of class variances), and low and high values of the data array are presented.

Classes produced by the Jenks method option are optimal—that is, for the number of classes specified, no better class interval solution is possible that yields a smaller total variance. Minimum within-class error and maximum distance between classes results. The algorithms used in this program are derived from W. D. Fisher and John A. Hartigan. GVF values are not produced by the program.

There are many computational advantages to this program. Results are produced quickly, and comparison between classing methods is easily done by inspecting the total variances. This program is a welcome companion for the quantitative thematic cartographer.

See Richard E. Groop, *JENKS: An Optimal Data Classification Program for Choropleth Mapping.* Technical Report 3. Department of Geography, Michigan State University, March, 1980.

when the geographical objective of classing is of primary concern.

Iteration and optimization methods are cumbersome if not impossible without the employment of computers. For small mapping tasks, they can be approached with only an electronic calculator, although this soon becomes troublesome. The most reasonable solution is to reduce beforehand the number of possible iterations by visual inspection of the histogram of the array of values, so that only a limited number will need to be processed.

Different Maps from the Same Data

One consequence of having a variety of methods of classing is that different maps can result from the same data. (See Figure 6.13.) Either an alteration in the number of classes or in the class limits can cause such change. The designer has the task of mapping the data in the manner that best represents the distribution of the geographical phenomenon. This requires thorough knowledge of the mapped phenomenon and awareness that the map reader's perception of choropleth map patterns is the result of several conditions:[15]

1. The map reader's experience with cartographic materials
2. The cartographic method used to portray the distribution
3. The complexity of the map pattern

Pattern complexity has been studied by cartographic researchers in recent years, and most agree that complexity should be reduced. Pattern complexity has been defined in these terms:

1. Number of regions (contiguous enumeration units within the same class)
2. Fragmentation index

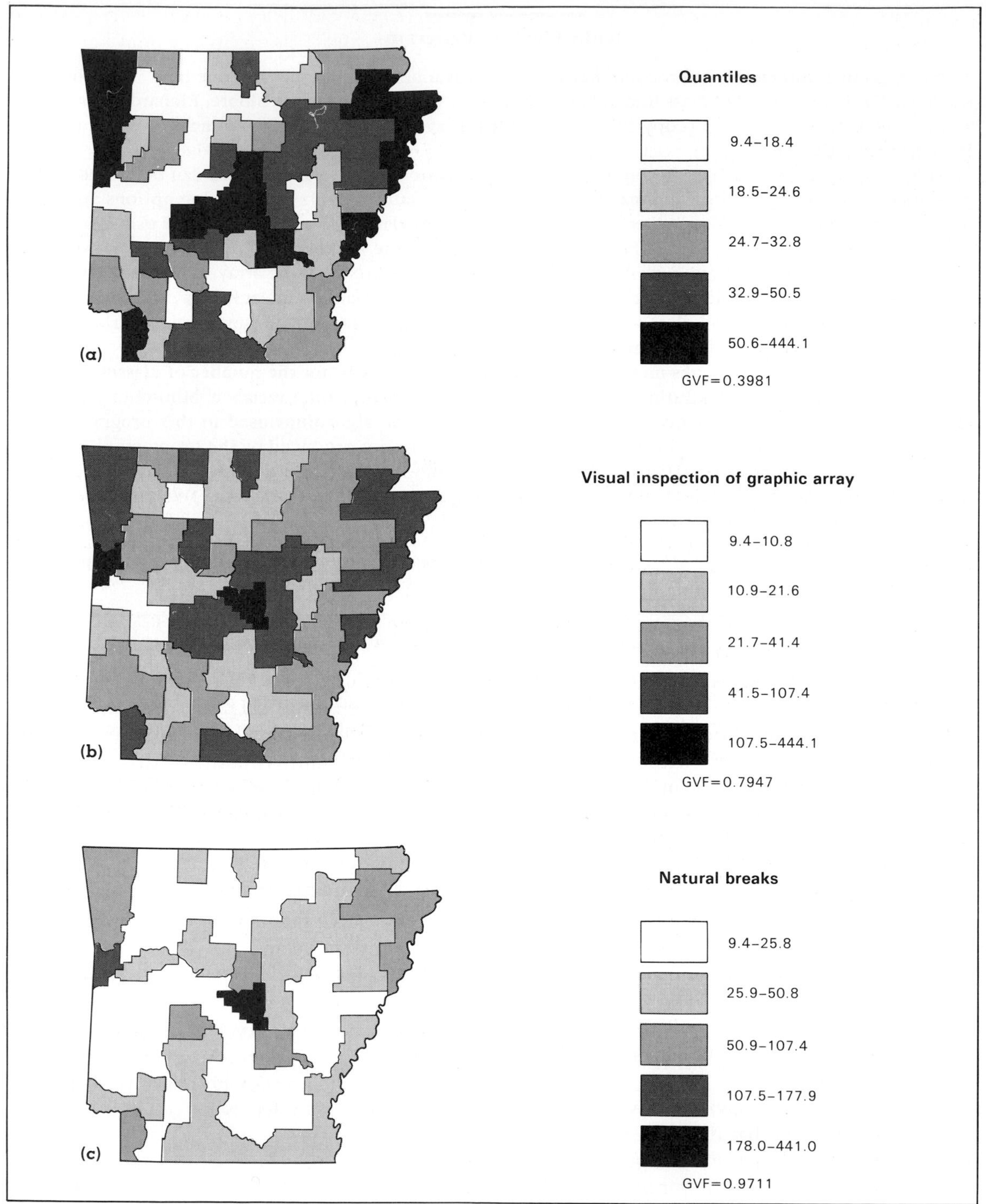

Figure 6.13. Different choropleth maps from the same data.
Different maps result because of the way the data have been classed. In this example, the goodness of variance fit (GVF) has been computed for each.

3. Aggregation index
4. Contrast index
5. Size disparity index

Space will not permit a detailed explanation of these indices. However, at least one researcher has discovered that map readers tend to judge choropleth maps having different classing schemes as similar in all pattern complexity indices. This has led him to conclude that only *one* index, perhaps number of regions, can be used to compare pattern complexity among different maps.[16]

Cartographic designers need to examine closely the visual pattern that results from a particular classing method. Although there are no exact indices that cover all cases, it is best to look for a pattern that results in a *simple* picture. The purpose of the map, the background and experience of the reader, and the mapping method must all be considered.

Unclassed Choropleth Maps

The **unclassed choropleth map,** made possible with the advanced technologies of computer plotting, CRT display, and facsimile can remove the classification burden from the designer.[17] (See Figure 6.14.) In such maps, however, the cartographer loses the ability to direct the message of the communication.[18] Map generalization, simplification, and interpretation is left to the reader. There is increasing evidence that readers can make reasonably good judgments of the mapped values by inspection of the legends that accompany the map bodies, and specifically, that the unclassed method can convey values more accurately to most map readers than classed maps having fewer than six classes.[19]

Cartographic designers must look to the purpose of their design activity and make individual judgments regarding the use of unclassed maps. They may be used in conjunction with ordinary classed maps to provide the reader with an unstructured inventory of the mapped area, or they might be used as a base for an overprinted classed map. Other possible uses (other than as a main map) may yet be discovered.

Legend Design, Areal Symbolization, and Base Map Design

Class interval specification is not the only activity that occupies the designer's time while preparing a choropleth map. Other design considerations include legend design, symbol selection and scaling, base map preparation, and attention to reproduction details.

Sources of Map Reading Error and the Need for Accurate Design Response

Reading and interpreting a single choropleth map is at best a difficult assignment. The design must be planned carefully to reduce the reading task and eliminate as many sources of error as possible. One of the chief sources of error results from **range-grading** of the map data—that is, creating classes. Effective classification and corresponding legend design will help eliminate this source of method-produced error. Another source of considerable error lies in the behavioral characteristics of readers in dealing with graded-tone series. Fortunately, studies have been done to help the designer minimize perceptual error. Yet another area that creates problems is the choice of areal symbols; some symbols are easier to read and less distracting to the eye. Finally, the provision of adequate base materials for the reader is an area that has not received much attention from cartographic researchers. Unfortunately, most choropleth maps do not have enough base information.

Range-Grading, Legend Treatment, and Effective Design

A significant amount of error in reading choropleth maps arises from range-grading the data array. In Figure 6.15a, the reader can be certain

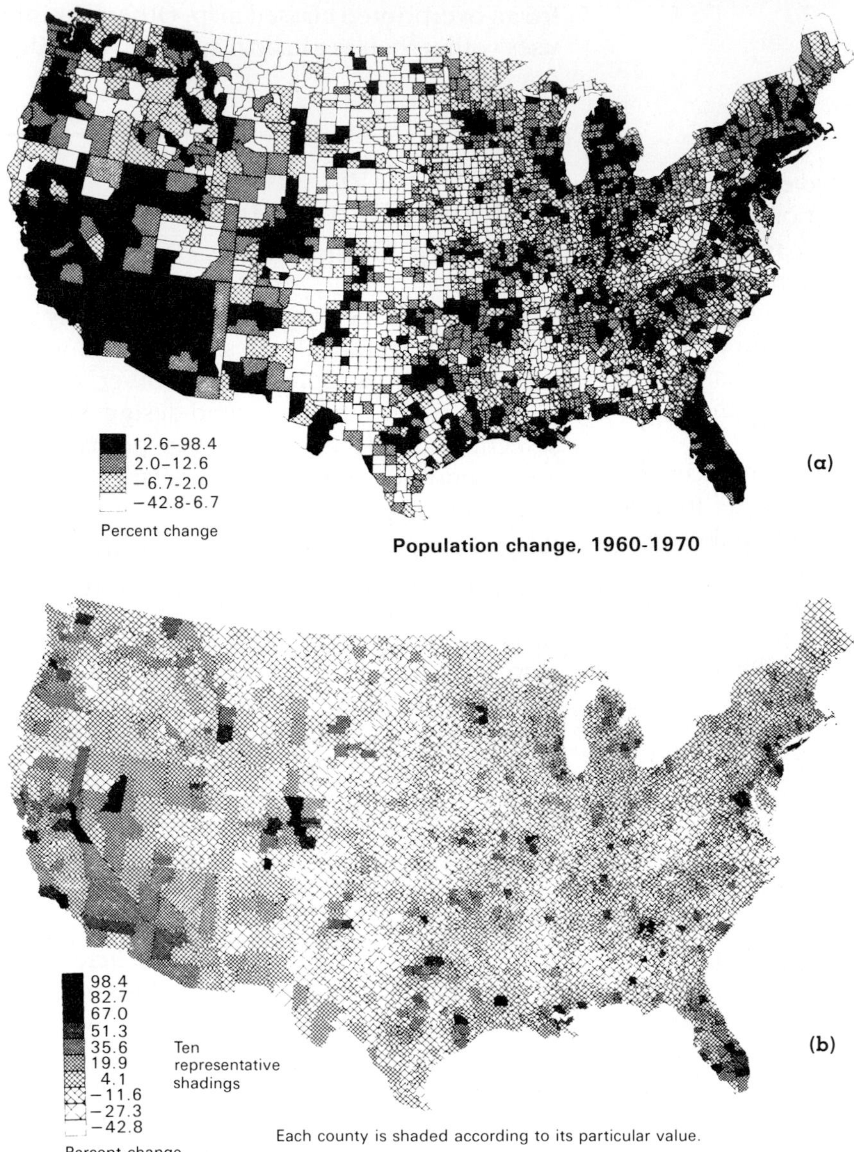

Figure 6.14. Classed and unclassed choropleth maps.
A four-class conventional map (a) and an unclassed map (b). The same data were used for both maps. Notice the different legends.

Used by permission from Michael P. Peterson, "An Evaluation of Unclassed Cross-Line Choropleth Mapping," *American Cartographer* 6 (1979): 22–23.

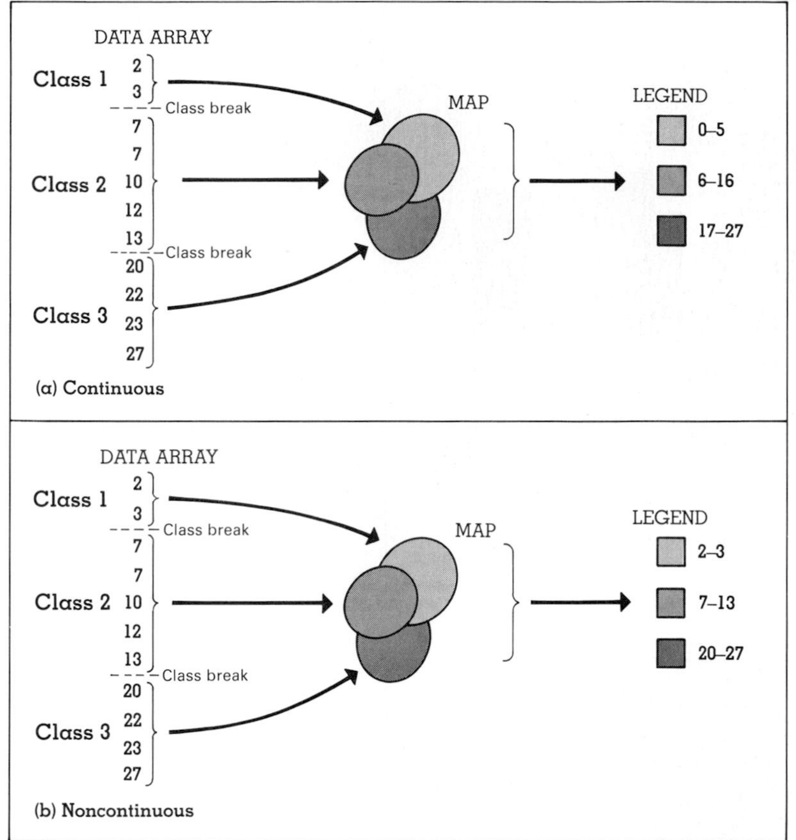

Figure 6.15. Continuous and noncontinuous legend and classing designs.
Map interpretation errors may result from the continuous design (a), because the reader is led to believe that data values may exist in a class. Less confusion results when the legend is scaled in a noncontinuous fashion (b), because only existing data values are used in forming the legend classes.

only that the value in the middle chorogram is somewhere between 6 and 16. Because this classification scheme does not have vacant classes, further error is introduced. For example, the reader might guess that a value of 15 occurs in class 2, but it does not because the value 15 is not in the data array.

Better alternatives exist for this condition. (See Figure 6.15b.) It would be more desirable to report the class breaks in the legend so that only the values actually existing in each class form the legend scores. The reader's estimate of the actual value is then narrowed, resulting in greater accuracy. There is no reason for widening the ranges of each class to serve continuity when map reading error will result.

Another improvement in legend design is always to include a frequency histogram of

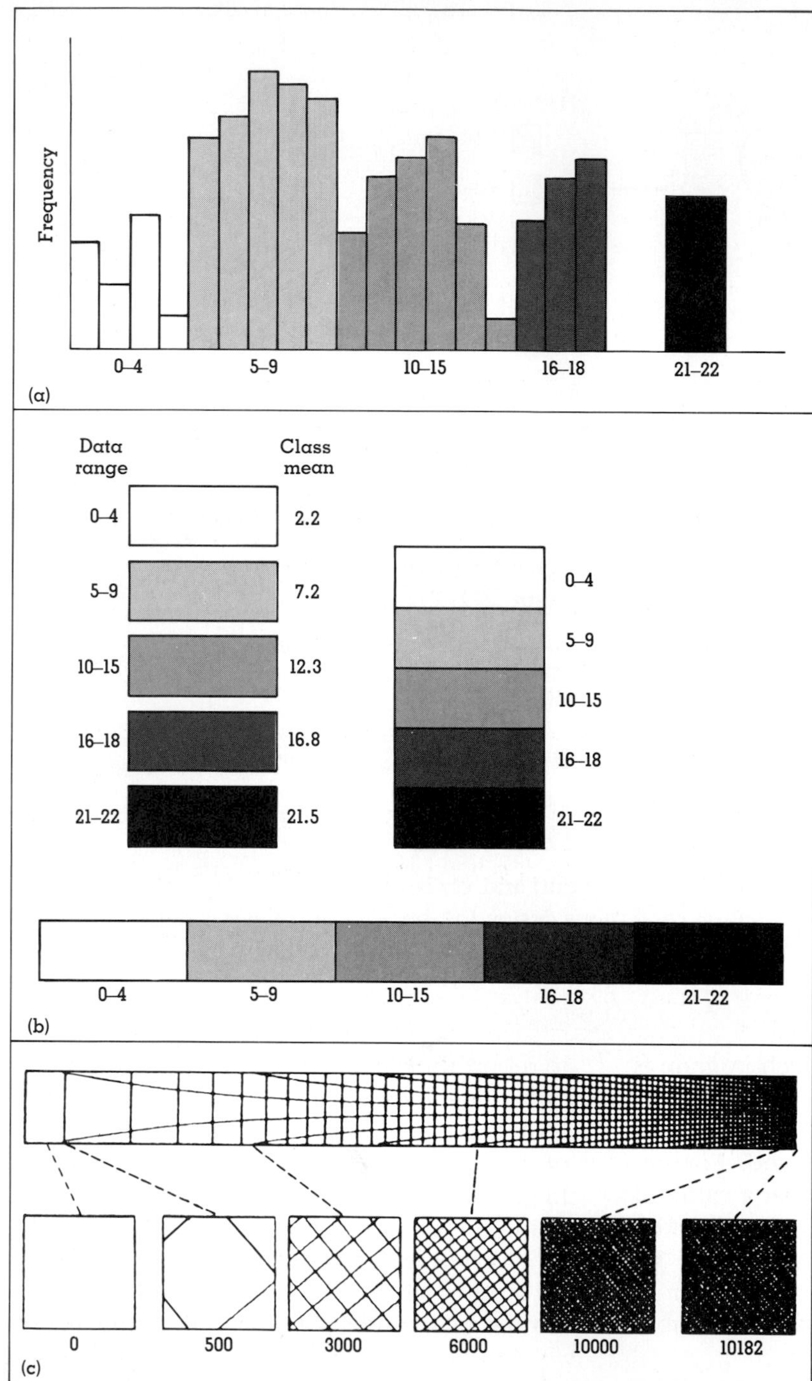

Figure 6.16. Various legend designs for choropleth maps.
Class symbols applied to a data histogram, as in (a), are desirable when possible. Different forms of conventional legends are shown in (b), with one case showing class means. A legend for an unclassed map is offered in (c).

Legend (c) is used by permission from Michael P. Peterson, "An Evaluation of Unclassed Cross-Line Choropleth Mapping," ***American Cartographer*** *6* (1979): 22–23.

the data array. Class symbolization and class boundaries may be included as part of the histogram. (See Figure 6.16a.) This gives the reader additional information about the aspatial qualities of the data array, and provides background on how the class intervals and boundaries have been selected. In light of the communication goals of thematic mapping, it is important to provide this element for the reader. The inclusion of class means, or other measures of central tendency for each class, is likewise desirable. These may be labeled alongside class boundaries. (See Figure 6.16b.)

Symbol Selection for Choropleth Maps

The selection of symbols for choropleth maps involves a philosophical thicket of questions. There are two opposing points of view on the issue of whether it is desirable for the map reader to make easy distinctions among the areal symbols, thus maintaining the integrity of each chorogram on the map. If so, the designer chooses symbols (1) whose *patterns* are different (dots, lines, hachures); (2) whose *texture* varies (small spacing between elements to large spacing of the elements in the same pattern); or (3) whose *orientation* changes (vertical, horizontal, or oblique to the map border); and (4) at the same time display a light-to-dark gradation series in keeping with the overall technique of the choropleth map. (See Figure 6.17a.)

On the other face of the coin is the argument that the differentiability of areal symbols is not as important as the spatial variation of tone *over the whole* map. This position concentrates on showing the overall geographical distribution, whereas the former is chiefly concerned with providing an areal table. Providing a smooth gradation of tone is easily accomplished, either by applying preprinted adhesive pattern sheets or by using mechanical screens during photolithography. (See Figure 6.17b.)

Dot screens cause little eye irritation during reading because they do not lead the eye to wander. Line screens, on the other hand, should be avoided unless they are screened during

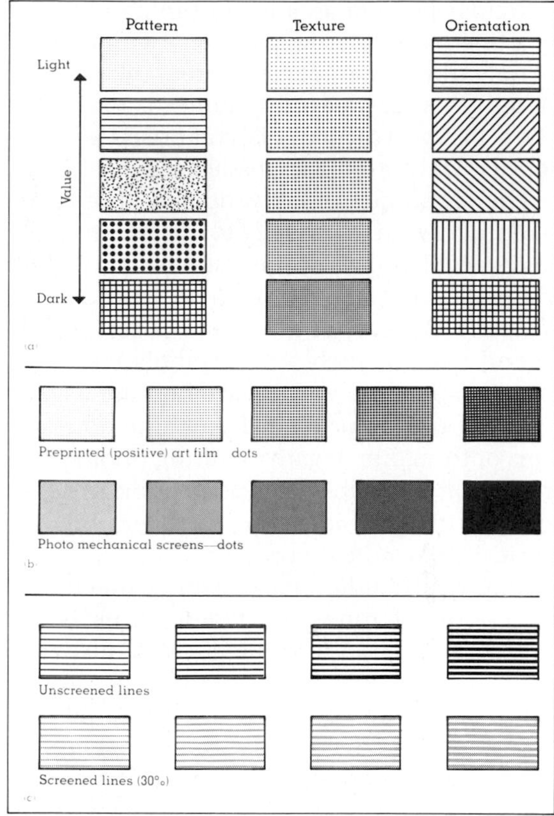

Figure 6.17. Areal symbols for black-and-white choropleth maps.
Pattern, texture, and orientation (a) are areal symbol features that can be used to differentiate units. In most cases, we strive to make the pattern's values range from light to dark. In (b), we see a comparison of photomechanical screens and preprinted media sheets. In (c), two ways of using lines for area patterns are shown. Screened lines are preferred because they are less harsh.

negative preparation. (See Figure 6.17c.) Black lines are harsh, cause severe irritation, and can mislead the eye to trace paths along their axes. Screening line patterns is particularly useful when considerable other base material is provided on the map.

Because there has been a rather steady increase in the production of choropleth maps by computer plotters incorporating line patterns for area symbols, increasing concern about scaling these patterns is noted. Research findings have disclosed that a gray tone of a particular percent area inked can be replaced by a line pattern (from a plotter) with approximately the same percent area inked.[20] Although this may be true, caution is expressed when using area symbols comprised of lines because of our remarks made earlier.

The technique of conventional choropleth mapping requires the cartographer to symbolize each enumeration unit according to where its value falls in the range-graded series of values. Because of the nature of scaling and legend design, such areal symbols take on a stepped appearance. Areal symbols are usually chosen from preprinted dot or line adhesive films or from film negatives during photo preparation of the maps. The perception of these shading tones is intended to reflect the numerical values they represent.

Empirical studies have demonstrated that perception of tones in graded series is not linear.[21] In a series of printed tones, most map readers do not perceive the tones as varying from light to dark in perfect correspondence with the physical stimulus (the percentage of area inked). Moreover, the perception of gray tones in a graded series is affected by whether the individual elements of the texture of the area symbol are perceptible. Two design avenues are thus possible, depending on which kind of areal symbol is used.

Perception of Areal Symbols—Perceptible Pattern

Most cartographical and psychological studies of the perception of value have utilized tonal areas made up of small dot patterns. If the pattern of dots has a texture of less than 75 lines of dots per inch, it is considered coarse, and the individual dots are usually perceived by map readers as having pattern. Above this level, most viewers see the tonal area as having value only, and the individual dots are no longer distinguishable. The percentage of area inked is normally used as a gauge in differentiating areas where the only value is perceived from areas having distinguishable pattern.

In the case of areal symbols that have perceptible pattern, the relationship between percentage of area inked and reader response is shown by the Williams **curve of the gray spectrum.** (See Figure 6.18.) This curve shows a nonlinear relationship between physical stimulus and response—it is easier to distinguish tonal differences at the lighter and darker ends of the spectrum than in the middle. Cartographic designers should select areal symbols based on this behavioral trait, following these steps:

1. After determining the number of classes, array their midpoint (or average) values from low to high. Consider the lowest value to be white (0 percent area inked), and the highest to be black (100 percent area inked).
2. Scale the values represented by the midpoints of the intermediate classes, in accordance with their relation to the 0 and the 100 already established.
3. For each class, enter the vertical axis of the Williams curve at the desired percentage, draw a horizontal line to the curve, and establish a perpendicular down to the horizontal axis. Read directly the percentage of area inked that must be chosen for the symbol.
4. Remembering that texture must be compatible for all symbols, assign each class an areal symbol that has the appropriate percentage of area inked.

Percentages of area inked are usually specified by manufacturers in product announcement brochures and catalogues for their preprinted screens. Numerous independent densitometer studies have shown, however, that these are usually nominal specifications. The actual percentages of area inked measured from the screen may vary by as much as 20 percent. Furthermore, these actual percentages may vary from sheet to sheet and from batch to batch. Care is necessary in choosing such materials.

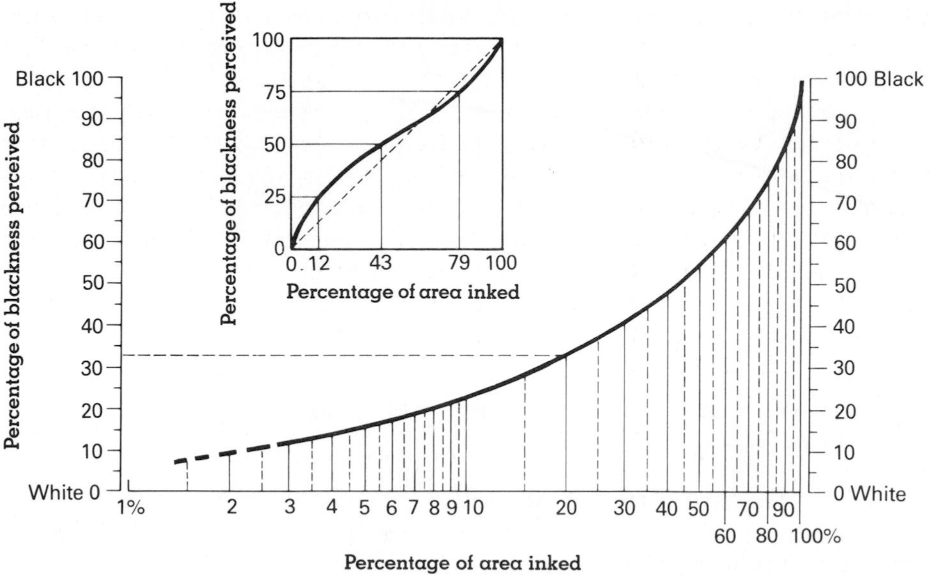

Figure 6.18. The curve of the gray spectrum, after Williams.
The inset curve is plotted on an arithmetic grid, and the curve on the main chart is plotted on a semi-log grid. Both charts are entered on the vertical axis with the desired percentage of blackness perceived. A horizontal line is drawn over to the curve; at this intersection, a vertical line is dropped to the horizontal axis. A dot screen tint having this percentage of area inked is then selected for the areal symbolization scheme. The example on the chart indicates that a 20-percent tint would be used if the reader is intended to perceive a 33-percent value.

If the designer chooses to use areal symbols containing perceptible pattern, such as preprinted "stick-on" film sheets, the Williams curve is a useful design aid. Today, however, mechanical film screens are widely used to produce area symbols. These call for different standards.

Perception of Areal Symbols—Indistinguishable Pattern

The perception of value of tonal areas having no recognizable pattern has been studied by cartographers and psychologists alike for several decades.[22] The results of several studies suggest that the equal value gray scale of A. E. O. Munsell is most accurate as to perception of tonal areas that have no distinguishable pattern. (See Figure 6.19.) The physical stimulus is the percentage of reflectance (the ratio of reflected light to incident light falling on the surface), and the response is perceived value. The relationship between perceived value and the reflectance can be approximated by this formula:[23]

$$V = 2.467R^{.33} - 1.636$$

where V is perceived value and R is reflectance. The value produced can be transformed into a more useful value-reflectance scale by multiplying the answer by 10.

Differences in gray tones at the darker end of the scale are more noticeable than equal differences at the lighter end. Therefore, the actual differences in tone must be made greater at the lighter end if appropriate value perception is to be obtained. A graph of the relationship can be used to determine appropriate amounts; little or no computation is needed.

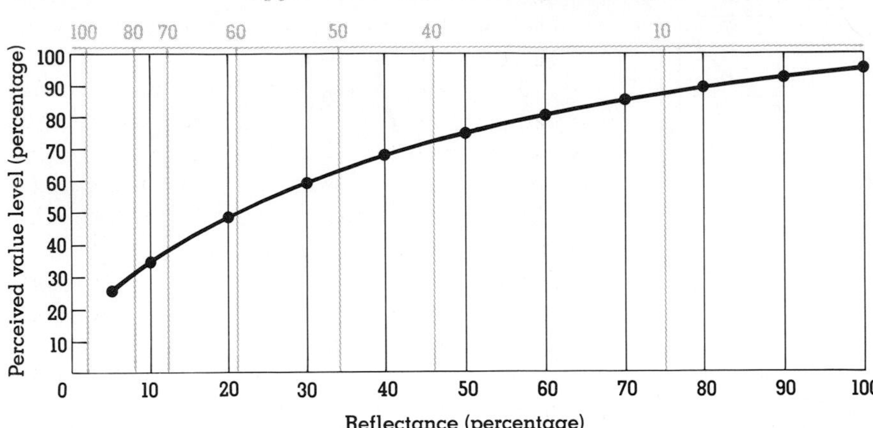

Figure 6.19. The equal value gray scale of A.E.O. Munsell.
This graph illustrates the relationship between percentages of reflectance of a tonal area and perceived value. Applied to the graph are approximate screen-tint values associated with various reflectances.
Source of screen tints: Carleton W. Cox, "The Effects of Background on the Equal Value Gray Scale," *Cartographica* 17 (1980): 53–71.

The procedure in selecting screen tints is similar to that for choosing preprinted patterns to produce an equal value gray scale. The following steps can be followed where the areal symbols are black and light gray at the extremes:

1. Determine the mean value of each class.
2. Scale the mean values by proportionality to a new perceived-value scale, from 0 to 100 percent.
3. For each value to be represented, determine the reflectance by computation or from a graph of the value-reflectance relationship. (See Figure 6.19.)
4. For each class, choose a screen tint whose reflectance approximates that obtained in step 3.

When map designers create choropleth maps on which map readers are faced with the task of distinguishing gray tones of finely textured screens between classes, they can be guided by the results of a recent study suggesting the following relationship:[24]

$$P = .398 \times W^{1.2}$$

where P is a succession of evenly spaced numbers from 0 to 100, and W are the gray tones in percent area white (which then must be subtracted from 100 to yield gray tones in percent area inked). Table 6.6 presents the calculated percent area inked for six equal contrast gray scales.

Texture of area patterns also influence the perception of gray tones. For example, in one study more than half of the subjects thought that a 133-line, 60 percent screen was darker than a 65-line, 70 percent screen.[25] It appears clear that texture must be carefully selected when choosing area patterns, and that equal value curves be applied only in cases where texture differences are limited.

Table 6.6 Percent Area Inked of Gray Tones for Six Equal Contrast Gray Scales

Five-Tone	Six-Tone	Seven-Tone
0	0	0
21.0	15.8	12.4
44.5	33.5	26.3
70.7	53.3	41.8
100	75.3	59.1
	100	78.4
		100

Eight-Tone	Nine-Tone	Ten-Tone
0	0	0
10.0	8.2	6.8
21.2	17.4	14.5
33.6	27.6	23.1
47.6	39.1	32.6
63.2	51.9	43.3
80.6	66.2	55.2
100	82.2	68.5
	100	83.4
		100

Source: Glen R. Williamson, "The Equal Contrast Gray Scale," *The American Cartographer* 9 (1982): 131–139, Table 3.

Perception of Areal Symbols—Cross-screening to Achieve Areal Tones

It has become quite common to achieve areal tones on choropleth maps by using the photolithographer's mechanical film screens. These are usually of better and more consistent quality than preprinted (positive art) adhesive films. In single screening, a film screen having a certain percentage of area inked is chosen, in accordance with the steps outlined above. Screens are usually available in 5-percent increments from 5 to 95 percent.

Cross-screening has also become a common practice. This method combines two or more mechanical screens to achieve the desired areal tone; it is followed in order to economize and to control registration. However, the method of combining two or more screens must be used with caution.

First, combining screens can cause **moiré** patterns. A moiré is a pattern that results when two regular dot (or line) patterns are superimposed with a small angular misalignment.[26] Although generally thought to be distracting and undesirable, moiré can become a positive design element in certain circumstances. Moiré can be prevented by aligning the superimposed screens with a 30° separation.

Second, and most importantly, dot-tint screens cannot simply be "added" together to achieve the desired percentage of area inked. For example, two 30-percent screens do not yield a tone of 60 percent area inked, but one of 51 percent. Furthermore, a 40-percent and a 20-percent screen do not yield the same 51 percent, but rather 52 percent. This must be understood if cross-screening is to be used effectively.

The relationship can be expressed with this formula:[27]

$$\mathrm{PAI}_{A+B} = \mathrm{PAI}_A + \mathrm{PAI}_B - \frac{(\mathrm{PAI}_A \cdot \mathrm{PAI}_B)}{100}$$

where PAI = percent area inked, and A and B are two different screens. The values in Table 6.7 have been computed from the above formula. When it is desirable to combine three screens, the following formula must be used:

$$\mathrm{PAI}_{A+B+C} = (\mathrm{PAI}_A + \mathrm{PAI}_B + \mathrm{PAI}_C) - \left[(\mathrm{PAI}_A \cdot \mathrm{PAI}_B) + (\mathrm{PAI}_B \cdot \mathrm{PAI}_C) + (\mathrm{PAI}_A \cdot \mathrm{PAI}_C)\right] \cdot 100 + \frac{(\mathrm{PAI}_A \cdot \mathrm{PAI}_B \cdot \mathrm{PAI}_C)}{10{,}000}$$

Combining more than three screens is not recommended, since tonal quality decreases.

Table 6.7 Percent Area Inked (PAI) Resulting from Overprinting Two Dot Screens

		Screen A								
		10	20	30	40	50	60	70	80	90
Screen B	10	19	28	37	46	55	64	73	82	91
	20		36	44	52	60	68	76	84	92
	30			51	58	65	72	79	86	93
	40				64	70	76	82	88	94
	50					75	80	85	90	95
	60						84	88	92	96
	70							91	94	97
	80								96	98
	90									99

Source: A. Jon Kimerling, *"Visual Value as a Function of Percent Area Inked for the Cross-Screening Technique,"* American Cartographer 6 (1979): 141–148, Table 2. These percentages are based on screen-dot sizes, not printed dot sizes.

Cross-screening must be approached with caution because of the likelihood of introducing unwanted moiré. It is good practice to discuss the details with the photolithographer early in the reproduction phase.

The formulas used above are for the determination of tint screen percentages prior to printing. At least one agency, the Department of Defense, has determined the relationship between the screen percentage and that which results after printing. (See Figure 6.20a.) As one would expect, the printed percentage is higher (approximately 7–10 percent). A gray scale has been developed to show the relationship between cross-screen percent area inked (after combining screens and using *printed* percentages) and visual value. (See Figure 6.20b.) This gray spectrum should be used whenever two or more film negative screens are used to develop tonal areas on quantitative maps.

Providing Adequate Base Information

One of the principal failures of choropleth map designers has been a reluctance to place adequate base information on the map. This is no doubt due to the difficulty of making the base map data compatible with often-used dark or black areal symbols at the high end of the symbol spectrum. It becomes virtually impossible to communicate the base information in a pleasing way. The designer is thus faced with a choice of alternatives: providing minimal base material or adding base information and changing the symbolization scheme to accommodate it.

As with most other decisions in thematic cartography, the map purpose and the experience and needs of the reader will dictate the choice. A great many choropleth maps are made for the general public—people who want to be able to tie the choropleth map pattern to other features that can provide locational cues. Unless the reader is especially familiar with the mapped area, much of the information-carrying potential of the choropleth map is lost.

Scale, coupled with map purpose, can act as a constraint in the choice of providing base information. A choropleth map of the United States that uses counties as enumeration areas has the capacity to show regional variations of the mapped phenomenon, if this is the main purpose. On the other hand, a marketing analyst working in an urban area might want plotting of political boundaries, interstate

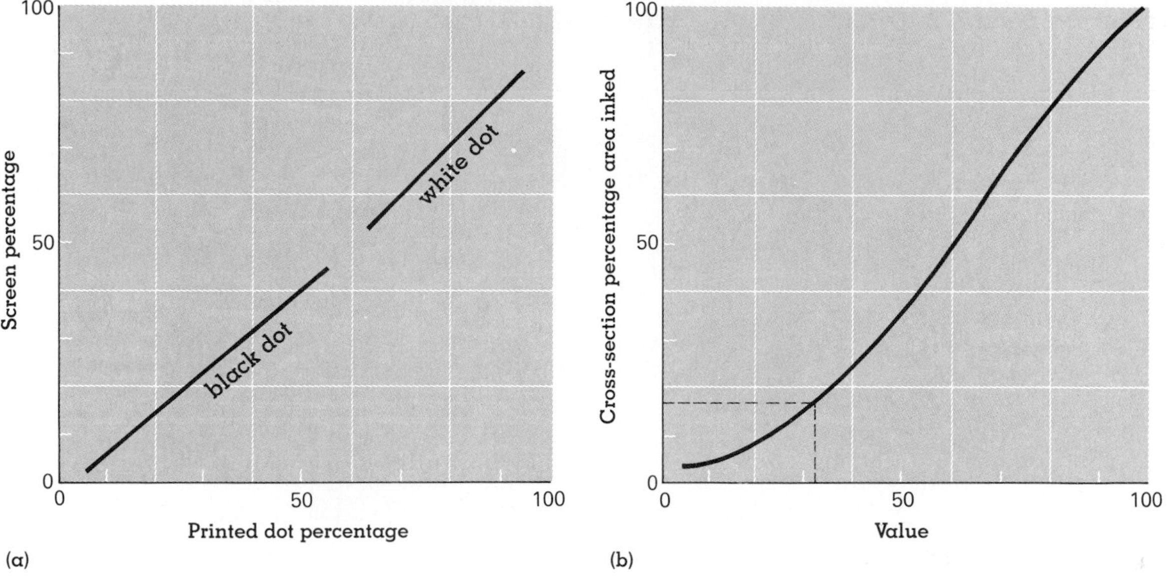

Figure 6.20. Percentage of area inked and perception relationships on areal patterns achieved by cross-screening.
In (a), the chart shows photomechanical screen percentages and resultant percentages of area inked, for various screens and for both black and white dots. These should be used when photomechanical screens are employed to develop areal tones. Perception of tonal values when cross-screening is used is shown in (b). Notice that this curve of the gray spectrum is a little different from that offered by Williams. Here, a 17.5-percent tint would be used when the reader is intended to perceive a 33-percent value. The chart in (b) should be used whenever the cross-screening method is used to develop tonal areas.

Source: A. Jon Kimerling, "Visual Value as a Function of Percent Area Inked for the Cross-Screening Technique," *American Cartographer* 6 (1979): 147.

highways, major thoroughfares, or the competition. The designer must make arbitrary decisions; the needs of users and map clients must be taken into consideration.

Designers faced with providing adequate base information need to choose areal symbols carefully and avoid the black and dark hues. In black-and-white mapping, line patterns can be screened so as to make the line pattern effect less objectionable and to gray the patterns to accommodate overprinting. (See Figure 6.21.) In color mapping, more care is needed in providing a color spectrum (probably more pastel in appearance) that will accommodate overprinting. Surprisingly, this is not all that difficult with a little experimentation.

One cartographer has addressed the importance of providing base information this way:

> The simplistic notion that base data are merely elements of graphic design, subject to the whims of individual map makers, has reinforced the notion that these data are separate, unrelated map components. The cartographic solution to the problem of base-data selection must be to evaluate the relative appropriateness of all data to map purpose; the same criterion should be used in selection of both base and thematic data—their power to contribute to map purpose.[28]

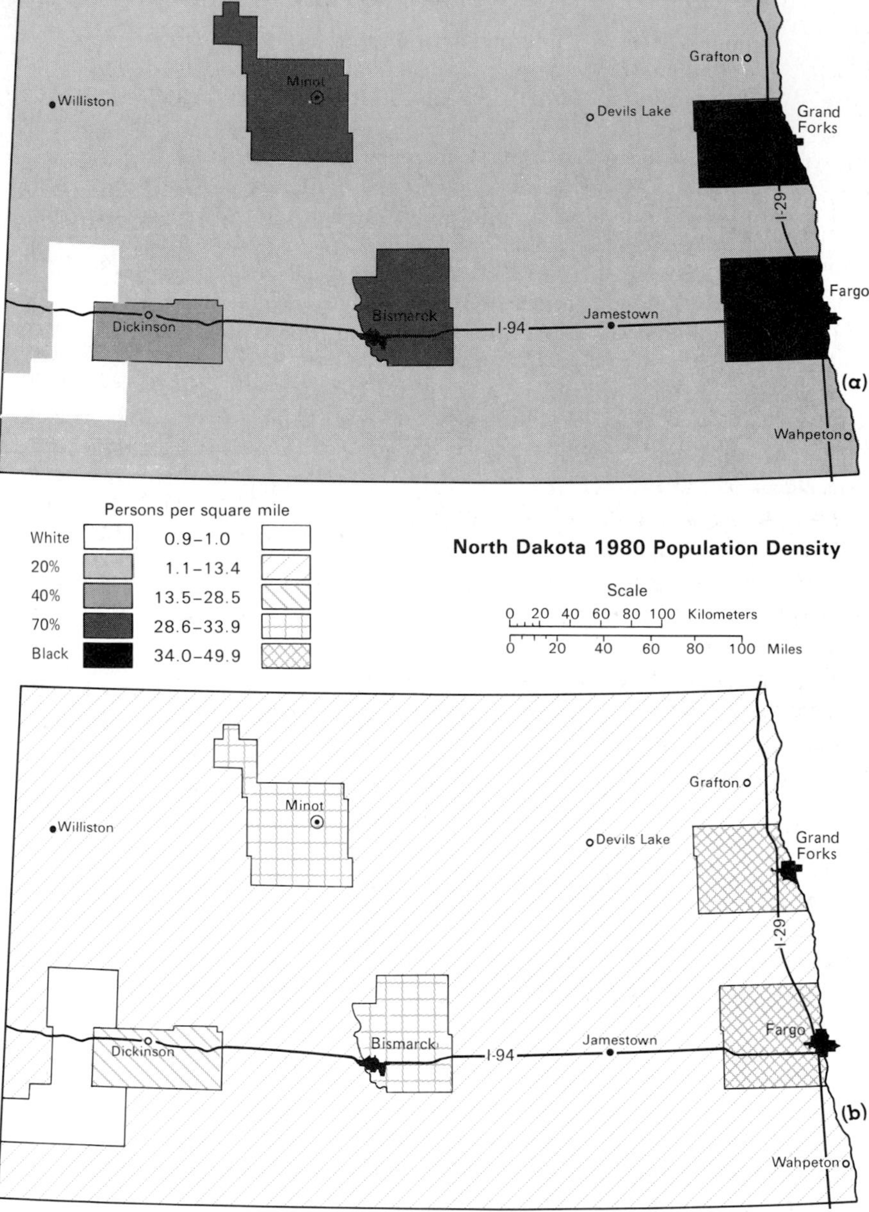

Figure 6.21. Alternative designs for choropleth symbolization and base map material. It is difficult to provide base material when dark hues (or black) are used in the symbol array (a). A solution is provided in (b) by screening the area symbols so that base material is more distinguishable. The alternative in (b) should only be used when it is compatible with the purpose of the map.

Notes

1. E. Meynen, ed., *Multilingual Dictionary of Technical Terms in Cartography* (International Cartographic Association, Commission II—Wiesbaden, West Germany: Franz Steiner Verlag, 1973), p. 123.
2. The discussion in this chapter does not include bivariate choropleth mapping. In this form of mapping, two data sets are portrayed. Each enumeration unit is symbolized according to which bivariate class (such as a combined class based on income and education levels) it falls. For more information on these types of maps, see Vincent P. Barabba, "The Graphic Presentation of Statistics," in *The Census Bureau: A Numerator and Denominator for Measuring Change.* Technical Paper 37. U.S. Bureau of the Census. (Washington: USGPO, 1975), pp. 89–103; Lawrence W. Carstensen, Jr., "Bivariate Choropleth Mapping: The Effects of Axis Scaling," *American Cartographer* 13 (1986): 27–42; and Stephen Lavin and J. Clark Archer, "Computer-Produced Unclassed Bivariate Choropleth Maps," *American Cartographer* 11 (1984): 49–57.
3. Jean-Claude Muller, "Mathematical and Statistical Comparisons in Choropleth Mapping," unpublished Ph.D. dissertation (University of Kansas, Department of Geography, 1974), p. 21.
4. George F. Jenks and Duane S. Knos, "The Use of Shading Patterns in Graded Series," *Annals* (Association of American Geographers) 51 (1961): 316–34.
5. Mark S. Monmonier, *Maps, Distortion, and Meaning* (Resource Paper No. 75-4—Washington, D.C.: Association of American Geographers, 1977), p. 27.
6. Kurt E. Brassel and Jack J. Utano, "Design Strategies for Continuous-Tone Area Mapping," *American Cartographer* 6 (1979): 39–50.
7. Waldo R. Tobler, "Choropleth Maps Without Class Intervals," *Geographical Analysis* 5 (1973): 262–65.
8. George F. Jenks and Michael R. C. Coulson, "Class Intervals for Statistical Maps," *International Yearbook of Cartography* 3 (1963): 119–34.
9. Ian S. Evans, "The Selection of Class Intervals," *Transactions* (Institute of British Geographers) 1977: 98–124.
10. Michael R. C. Coulson, "In the Matter of Class Intervals for Choropleth Maps: With Particular Reference to the Work of George Jenks," *Cartographica* 24 (1987): 16–39.
11. Mark S. Monmonier, "Continguity—Biased Class—Interval Selection: A Method for Simplifying Patterns on Statistical Maps," *Geographical Review* 62 (1972): 203–28.
12. The Jenks Optimal method of classing has its roots in a technique published as George F. Jenks and Fred C. Caspall, "Error on Choropleth Maps: Definition, Measurement, Reduction," *Annals* (Association of American Geographers) 61 (1971): 217–44. A publication appeared as George F. Jenks, *Optimal Data Classification for Choropleth Maps,* Occasional Paper No. 2. Department of Geography, University of Kansas, May 1977, in which Jenks identified the Fisher algorithm as the best solution for a grouping technique (Walter D. Fisher, "On Grouping for Maximum Homogeneity," *American Statistical Association Journal* 53 (1958): 789–98), and operational programs of the same in John A. Hartigan, *Clustering Algorithms* (New York: John Wiley, 1975). The GVF first appeared in Arthur Robinson, Randall Sale, and Joel Morrison, *Elements of Cartography.* 4th ed. (New York: John Wiley, 1978), pp. 178–79, and was attributed to personal communication with Jenks. Jenks himself never published any papers on the GVF or any more on the actual operational techniques used by Fisher. Richard E. Groop, while at the University of Kansas in the late 1970's, worked on the algorithm with Jenks and converted the program from mainframe Fortran to Tectronics Basic, and later published a report as Richard E. Groop, *JENKS: An Optimal Data Classifications Program for Choropleth Mapping.* Technical

Report 3. Department of Geography, Michigan State University, March 1980. The procedures of the algorithm were not published in the report. The latest version of the Jenks Optimal method was developed by James B. Moore, Richard E. Groop, and George F. Jenks for microcomputer applications and is available from Michigan State University.
13. Michael R. C. Coulson, "Class Intervals," pp. 16–39.
14. Richard M. Smith, "Comparing Traditional Methods for Selecting Class Intervals on Choropleth Maps," *The Professional Geographer* 38 (1986): 62–67.
15. Mark S. Monmonier, "Measures of Pattern Complexity for Choropleth Maps," *American Cartographer* 1 (1974): 159–69.
16. Kang-tsung Chang, "Visual Aspects of Class Intervals in Choropleth Mapping," *Cartographic Journal* 15 (1978): 42–48.
17. A recent version of unclassed map is the facsimile-produced one, made with modern weather-satellite imagery processing equipment. A report of this method can be found in Jean-Claude Muller and John L. Honsaker, "Choropleth Map Production by Facsimile," *Cartographic Journal* 15 (1978): 14–19.
18. This point of view has been stressed by Michael W. Dobson, "Choropleth Maps without Class Intervals? A Comment," *Geographical Analysis* 5 (1973): 358–60. Others have voiced nearly identical positions.
19. Michael P. Peterson, "An Evaluation of Unclassed Cross-Line Choropleth Mapping," *American Cartographer* 6 (1979): 21–37.
20. Terry A. Slocum and Robert B. McMaster, "Gray Tone Versus Line Plotter Area Symbols: A Matching Experiment," *The American Cartographer* 13 (1986): 151–64.
21. Jenks and Knos, "The Use of Shading Patterns," pp. 316–34.
22. A. Jon Kimerling, "A Cartographic Study of Equal Value Gray Scales for Use with Screened Gray Areas," *American Cartographer* 2 (1975): 119–27; Carleton W. Cox, "The Effects of Background on the Equal Value Gray Scale," *Cartographica* 17 (1980): 53–71.
23. Cox, "The Effects of Background on the Equal Value Gray Scale," pp. 53–71.
24. Glen R. Williamson, "The Equal Contrast Gray Scale," *The American Cartographer* 9 (1982): 131–39.
25. Richard M. Smith, "Influence of Texture on Perception of Gray Tone Map Symbols," *The American Cartographer* 14 (1987): 43–47.
26. A. Jon Kimerling, "Cartographic Guidelines For the Use of Moiré Patterns Produced by Dot-Tint Screens," *Canadian Cartographer* 16 (1979): 159–67.
27. A. Jon Kimerling, "Visual Values as a Function of Percent Area Inked for the Cross-Screening Technique," *American Cartographer* 6 (1979): 141–48.
28. Dennis E. Fitzsimons, "Base Data on Thematic Maps," *The American Cartographer* 12 (1985): 57–61.

Glossary

arbitrary data classification for convenience, simple, rounded numbers are chosen as class boundaries

chorogram the enumeration unit used in choropleth mapping

choropleth map a form of statistical mapping used to portray discrete data by enumeration units; areal symbols are applied to enumeration units according to the values in each unit and the symbols chosen to represent them

conventional choropleth technique form of choropleth map in which the data array is classed into groups and the groups symbolized on the map with areal symbols

cross-screening method of overprinting photomechanical dot-tint screens used by photolithographers to achieve desired areal tones; procedure is not additive

curve of the gray spectrum human visual response pattern to the perception of gray tones in series; the Williams curve is considered the best replication of the pattern of map readers in responding to traditional dot areal patterns

data model conceptual three-dimensional model in which prisms are erected vertically from each chorogram in direct proportion to its mapped value

exogenous data classification approach to data classification in which certain critical values not associated with the data set are selected as class boundaries

goodness of variance fit (GVF) index devised by Jenks; an example of the optimization method in data classification

graphic array ordered graphic plot of data values to assist in class boundary determination

idiographic data classification particular events in the data array are selected as class boundaries; these events may be troughs (depressions) in the data histogram

moiré pattern that develops when two dot screens are overprinted with a small angular misalignment

optimization method classing method in which within class error is minimized and between class differences are maximized

pattern complexity measure of complexity of repeating elements in a visual image; used in studying choropleth maps

range-grading result of the data classification activity often used in choropleth map symbolization

serial data classification mathematical functions best fitted to the data array are chosen to determine class boundaries

unclassed choropleth map the data array is not classed into groups; the data are symbolized by continuously changing proportional area symbol

Readings for Further Understanding

Barabba, Vincent P. "The Graphic Presentation of Statistics." In *The Census Bureau: A Numerator and Denominator for Measuring Change.* Technical Paper 37. U.S. Bureau of the Census. Washington, D.C.: USGPO, 1975.

Brassel, Kurt E., and Jack J. Utane. "Design Strategies for Continuous-Tone Areal Mapping." *American Cartographer* 6 (1979): 39–50.

Castner, Henry W., and Arthur H. Robinson. *Dot Area Symbols in Cartography: The Influence of Pattern on Their Perception.* American Congress on Surveying and Mapping, Cartography Division, Technical Monograph No. CA-4. Washington, D.C.: American Congress on Surveying and Mapping, 1969.

Chang, Kang-tsung. "Visual Aspects of Class Intervals in Choropleth Mapping." *Cartographic Journal* 16 (1978): 42–48.

Coulson, Michael R. C. "In the Matter of Class Intervals for Choropleth Maps: With Particular Reference to the Works of George Jenks." *Cartographica* 24 (1987): 16–39.

Cox, Carlton W. "The Effect of Background on the Equal Value Gray Scale." *Cartographica* 17 (1980): 53–71.

Dobson, Michael W. "Choropleth Maps Without Class Intervals? A Comment." *Geographical Analysis* 5 (1973): 358–60.

Evans, Ian S. "The Selection of Class Intervals." *Transactions* (Institute of British Geographers, 1977): 98–124.

Fisher, Walter D. "On Grouping for Maximum Homogeneity." *American Statistical Association Journal* 53 (1958): 789–98.

Groop, Richard E. *JENKS: An Optimal Data Classification Program for Choropleth Mapping.* Technical Report 3. Department of Geography, Michigan State University, 1980.

Hartigan, John A. *Clustering Algorithms.* New York: John Wiley, 1975.

Jenks, George F. "Generalization in Statistical Mapping." *Annals* (Association of American Geographers) 53 (1963): 15–26.

———. "Contemporary Statistical Maps—Evidence of Spatial and Graphic Ignorance." *American Cartographer* 3 (1976): 11–19.

———. *Optimal Data Classification for Choropleth Maps.* Occasional Paper No. 2. Department of Geography, University of Kansas, 1977.

Jenks, George F., and Fred G. Caspall. "Error on Choropleth Maps: Definition, Measurement, Reduction." *Annals* (Association of American Geographers) 61 (1971): 217–44.

Jenks, George F., and Michael R. C. Coulson. "Class Intervals for Statistical Maps." *International Yearbook of Cartography* 3 (1963): 119–34.

Jenks, George F., and Duane S. Knos. "The Use of Shading Patterns in Graded Series." *Annals* (Association of American Geographers) 51 (1961): 316–34.

Kimerling, A. Jon. "A Cartographic Study of Equal Value Gray Scales for Use with Screened Gray Areas." *American Cartographer* 2 (1975): 119–27.

———. "Cartographic Guidelines for the Use of Moiré Patterns Produced by Dot Tint Screens." *Canadian Cartographer* 16 (1979): 159–67.

———. "The Comparison of Equal-Value Gray Scales." *The American Cartographer* 12 (1985): 132–42.

———. "Visual Value as a Function of Percent Area Inked for the Cross-Screening Technique." *American Cartographer* 6 (1979): 141–48.

Lavin, Stephen, and J. Clark Archer. "Computer-Produced Unclassed Bivariate Choropleth Maps." *The American Cartographer* 11 (1984): 49–57.

Lloyd, Robert, and Theodore Steinke. "Visual and Statistical Comparison of Choropleth Maps." *Annals* (Association of American Geographers) 67 (1977): 429–36.

Mackay, R. R. "An Analysis of Isopleth and Choropleth Class Intervals." *Economic Geography* 31 (1955): 71–81.

Meynen, E., ed. *Multilingual Dictionary of Technical Terms in Cartography.* International Cartographic Association, Commission II. Wiesbaden, West Germany: Franz Steiner Verlag, 1973.

Monmonier, Mark S. "Contiguity—Biased Class-Interval Selection: A Method for Simplifying Patterns on Statistical Maps." *Geographical Review* 62 (1972): 203–28.

———. "Flat Laxity, Optimization, and Rounding in the Selection of Class Intervals." *Cartographica* 19 (1982): 16–26.

———. *Maps, Distortion, and Meaning.* Association of American Geographers, Resource Paper No. 75–4. Washington, D.C.: Association of American Geographers, 1977.

———. "Measures of Pattern Complexity for Choropleth Maps." *American Cartographer* 1 (1974): 159–69.

Muller, Jean-Claude. "Mathematical and Statistical Comparisons in Choropleth Mapping." Unpublished Ph.D. Dissertation, University of Kansas, Department of Geography, 1974.

Muller, Jean-Claude, and John L. Honsaker. "Choropleth Map Production by Facsimile." *Cartographic Journal* 15 (1978): 14–19.

Olson, Judy M. "Autocorrelation and Visual Map Complexity." *Annals* (Association of American Geographers) 65 (1975): 189–204.

Peterson, Michael P. "An Evaluation of Unclassed Cross-Line Choropleth Mapping. *American Cartographer* 6 (1979): 21–37.

Robinson, Arthur H.; Randall Sale, and Joel Morrison. *Elements of Cartography,* 4th ed. New York: John Wiley, 1978.

Scripter, Morton W. "Nested Means Map Classes for Statistical Maps." *Annals* (Association of American Geographers) 60 (1970): 385–93.

Slocum, Terry A., and Robert B. McMaster. "Gray Tone Versus Line Plotter Area Symbols: A Matching Experiment." *The American Cartographer* 13 (1986): 151–64.

———. "Influence of Texture on Perception of Gray Tone Map Symbols." *The American Cartographer* 14 (1987): 43–47.

Smith, Richard M. "Comparing Traditional Methods for Selecting Class Intervals on Choropleth Maps." *The Professional Geographer* 38 (1986): 62–67.

Tobler, Waldo R. "Choropleth Maps Without Class Intervals." *Geographical Analysis* 5 (1973): 262–65.

Tufte, Edward R. *The Visual Display of Quantitative Information.* Cheshire, CT: Graphics Press, 1983.

Williams, Robert L. "The Misuse of Area in Mapping Census-Type Numbers." *Historical Methods Newsletter* 9 (1976): 213–16.

Williamson, Glen R. "The Equal Contrast Gray Scale." *The American Cartographer* 9 (1982): 131–39.

Chapter 7

Mapping Point Phenomena: The Common Dot Map

Chapter Preview

Mapping discrete geographical phenomena can be accomplished by dot mapping, also called areal frequency mapping. Its main purpose is to communicate variation in spatial density. In use for over 100 years, the method is popular due to its simple mapping rationale. Although the simplest form is one-to-one mapping, where one dot represents one item, it is more common for one symbol to represent multiple items (one-to-many mapping). Significant design decisions involve the placement of dots and the selection of dot value and size. Legend design is also important, especially because recent research indicates that map readers do not perceive dot numerousness or density in a linear relationship. Representative densities must therefore be shown in the legend. It is now possible to generate dot maps by computer—a process that is very cost effective, at least for agencies producing many dot maps. The Bureau of the Census has produced computer-generated dot maps for some of its agricultural map products. Dot mapping is now possible using software written for microcomputers, although the methods used are far from adequate.

From the previous chapter's focus on choropleth mapping, a form of area mapping, we now proceed to treat *point* mapping, notably dot-distribution maps. These are also quantitative maps; information pertaining to density is gained by visual inspection of the spatially arrayed symbols to arrive at relative magnitudes. Precise numerical determination of density is subordinate in this form of mapping. Although any point symbol can be used, it has become customary to use small dots—thus the name dot mapping.

Mapping Technique

Dot-distribution maps were introduced as early as 1863,[1] so it is not surprising that their use and acceptance is widespread. In the simplest case, the technique involves the selection of an appropriate point symbol to represent each discrete element of a geographically distributed phenomenon. The symbol form does not change, but its number changes from place to place, in proportion to the number of objects being represented. (See Figure 7.1.) Precise location of each symbol is critical to the method. As in other forms of thematic mapping, a variety of types have evolved over the years.

A Classification of Dot Maps

Professor Richard E. Dahlberg has found that dot maps can be classified according to the way the dots are used:[2]

1. Dot equals one. In this case, there is a one-to-one correspondence between object and symbol. There may or may not be precise control over dot location.
2. Dot equals more than one. In the many varieties of this group, one symbol refers to more than one object.
   a. Dots are located so that the original data could be recovered by counting the dots.
   b. Dots are positioned at the centers of gravity of the objects they represent.
   c. Dots are used as shading symbols, either in a uniform pattern resembling a dot screen or across a statistical unit to represent smooth transition.
   d. Dots represent a part of stratified data, such as rural population, with urban populations represented by proportional point symbolization.
   e. Dots that represent a large part of the total data are rather large; dots may be arranged for counting rather than for the purpose of illustrating precise location.
   f. Two distributions are shown on the same map, each represented by dot symbols.
   g. Dots represent a distribution mapped along with another feature, to show functional relationships.

Advantages and Disadvantages of Dot Mapping

The number of newly drawn dot maps has declined during the last two decades, probably due to their high cost of production. Dot maps are very time consuming to construct and require considerable research. Although machine production is now possible, it has not yet led to a proliferation of new dot maps. This may occur when the advantages and disadvantages of the method are better understood.

The advantages of dot mapping include the following:

1. The rationale of mapping is easily understood by the map reader.
2. It is a fairly effective way of illustrating spatial density, especially suitable for showing geographical phenomena that are discontinuous (consisting of discrete elements).
3. Original data can be recovered from the map if the map has been designed for that purpose.

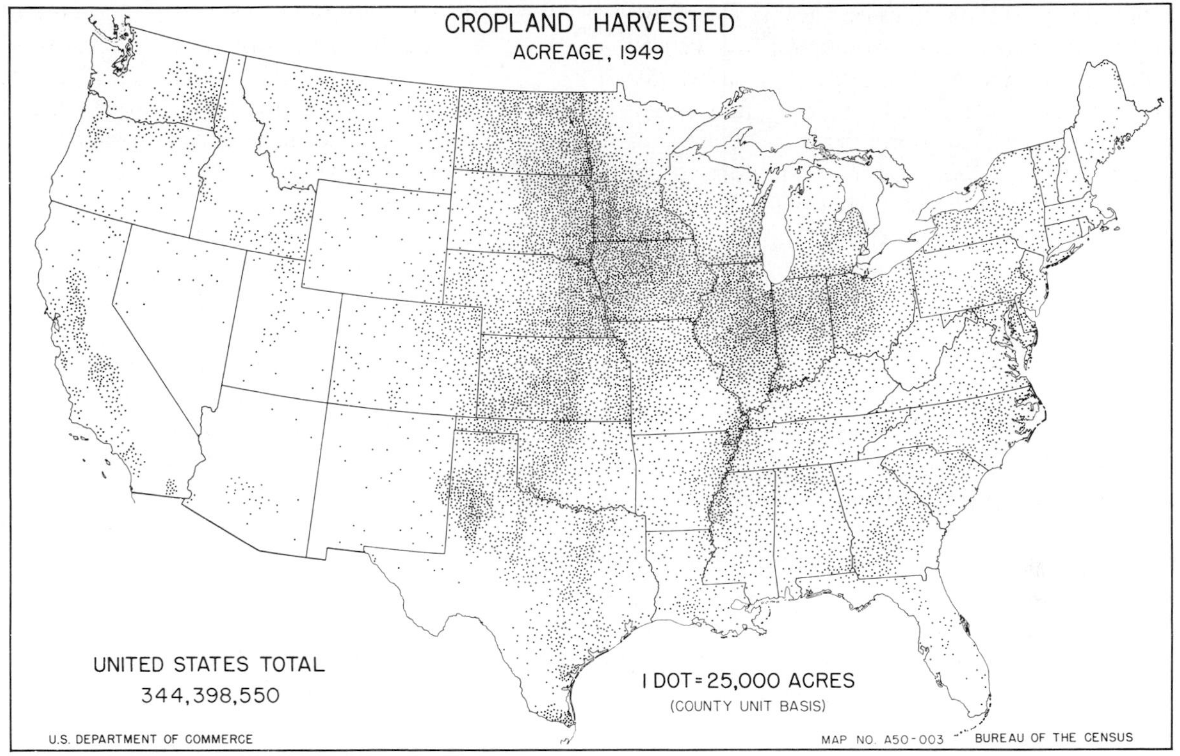

Figure 7.1. A common dot map.
The elements of the common map include point symbols, political boundaries, and a legend that includes dot value and if possible, representative densities taken from the map. The map reader gets the idea that the amount of the item varies from place to place.

Source: United States Bureau of the Census. *Portfolio of U.S. Census Maps,* 1950 (Washington, D.C.: USGPO, 1953), p. 16.

4. More than one data set may be illustrated on the same map. This is recommended only if there is a distributional or functional relationship between the sets.
5. Final execution, though time consuming, is rather straightforward and presents few production or reproduction problems.

Possible disadvantages would include these:

1. Map interpretation is not one-to-one; perception of relative density is not linear.
2. The method is relatively time consuming and therefore costly. Considerable research into factors controlling distribution is necessary, and provision must be made for the acquisition of ancillary resource materials.
3. When the map has been designed for maximum portrayal of relative spatial density, it is practically impossible for the reader to recover original data values.

Map designers must weigh these advantages and disadvantages carefully before deciding to apply dot mapping to a given project. As in all design choices, the map purpose, readership, and costs are central concerns in making the decision.

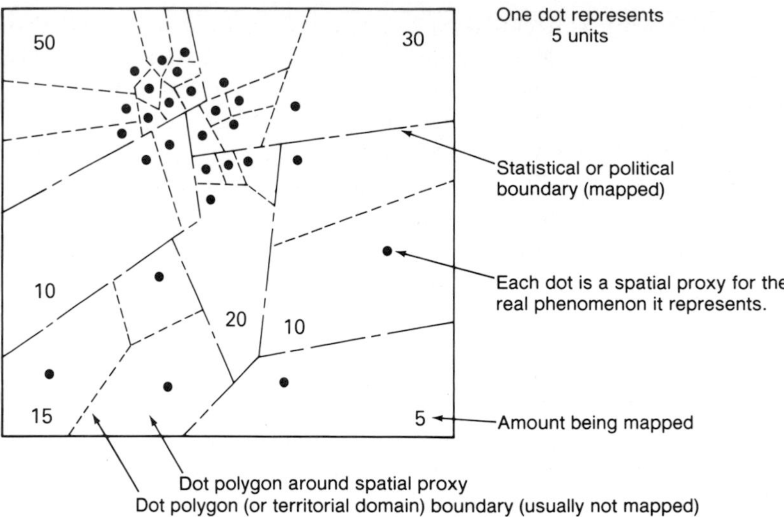

Figure 7.2. The conceptual elements of a dot map.
Each dot functions as a spatial proxy for the item it represents. Around each spatial proxy is its polygon or territorial domain. Political or statistical units are normally included to provide the reader with spatial cues, but the boundaries of territorial domains are not. These domains help the map maker in conceptualizing and producing the map.

The Mapping Activity

The ideal dot map would be at a scale that allowed for a one-to-one mapping, that is, one symbol for each mapped element. This requires an amount of informational and geographical detail not normally at the disposal of the cartographic designer. It is possible when individualized field studies have been carried out, but most map designers are not so fortunate. They must usually construct the dot map with data gathered in enumeration form, which requires a different approach.

In most dot mapping circumstances, a many-to-one situation exists: Each dot represents more than one mapped element. In this case, each dot can be thought of as a **spatial proxy** because it represents at a point some quantity that actually occupies geographical space.[3] (See Figure 7.2.) These geographical spaces are called **dot polygons** or *territorial domains.* They do not normally appear on the final map but are assumed to exist around each dot, making up the total enumeration area.

It is essential in dot mapping that associated resource materials showing functionally related variables be gathered together prior to mapping, so that greater precision in dot placement can be achieved. Each mapping task has its own peculiar set of related materials. Examples are physical relief diagrams; soil maps; maps showing lakes, ponds, rivers, and streams; land-use maps; national forest or park maps; and military reservation maps. The related distribution maps tend to operate as filters and serve to control the placement of the dots. (See Figure 7.3.)

For example, referring to county-level road maps at a scale of about one mile to the inch, where individual houses can be shown, is useful in guiding the placement of symbols on a population dot map. Other helpful resource maps in population dot mapping include forest and park land maps, military reservation maps, and physical relief maps. Maps that show the location of towns are frequently useful, as are resource maps that show where population is large and where it is limited.

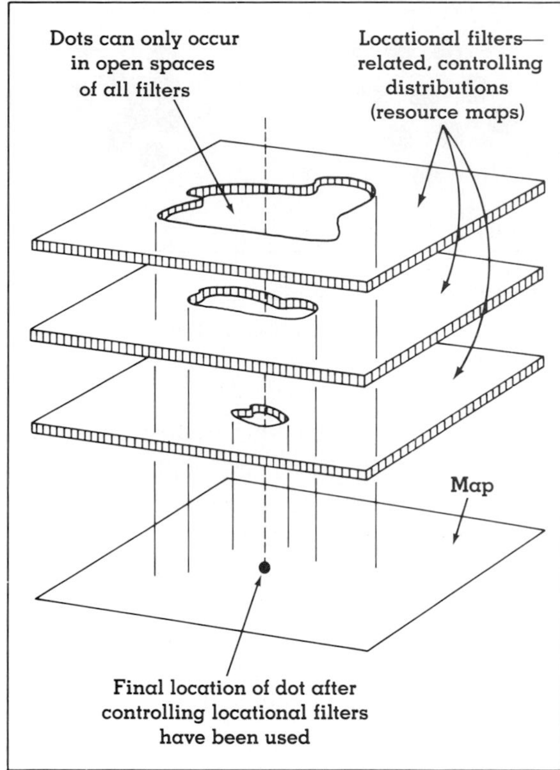

Figure 7.3. Dot placement.
Locational filters (related distributions) determine the location of dots. Each dot's territorial domain can only exist in that part of the map which has not been influenced by other distributions.

Once all relevant information is at hand, the designer can begin constructing the map. This involves decisions regarding scale, dot value, and dot size. These are closely interrelated in the construction of the map; a decision regarding one affects the others. In manual production of dot maps, work generally begins in pencil on work sheets. In this phase, the preliminary decisions are tested, largely by trial and error.

Size of Enumeration Unit

As mentioned above, dot mapping usually involves the presentation of geographical quantities that have been collected by statistical area, usually political or administrative units. Under most conditions, the smaller the statistical unit in relation to the overall size of the map, the greater will be the accuracy of the final dot distribution.[4] (See Figure 7.4.) Smaller statistical units mean a smaller territorial domain for each dot, reducing the chance of locational error. Large-scale maps (small earth areas) require quite small statistical units. In fact, it is often found that common statistical units used for enumeration data (block, tract, or county-sized units) are too large for dot mapping at large scales. Professional cartographers normally use intermediate to small scales for dot mapping.

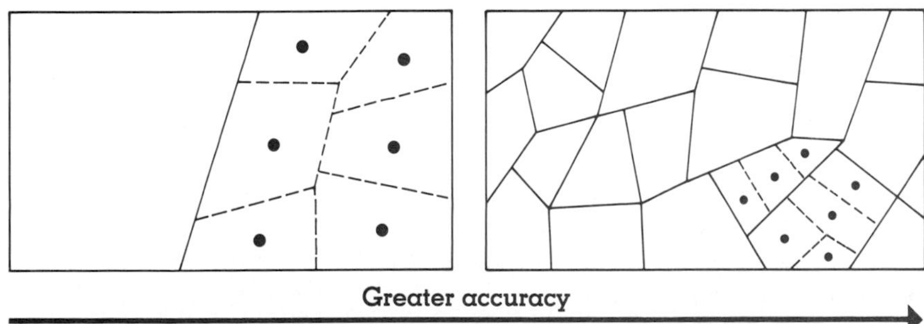

Figure 7.4. Accuracy.
Greater accuracy can be achieved by the use of statistical units that are smaller in relation to the entire mapped area. Statistical units provide locational control to a great degree. The more numerous (hence smaller) these units are, relative to the mapped area, the greater will be the degree of location accuracy produced. The territorial domains of the dots decrease in size with smaller enumeration units; greater locational accuracy is thus possible.

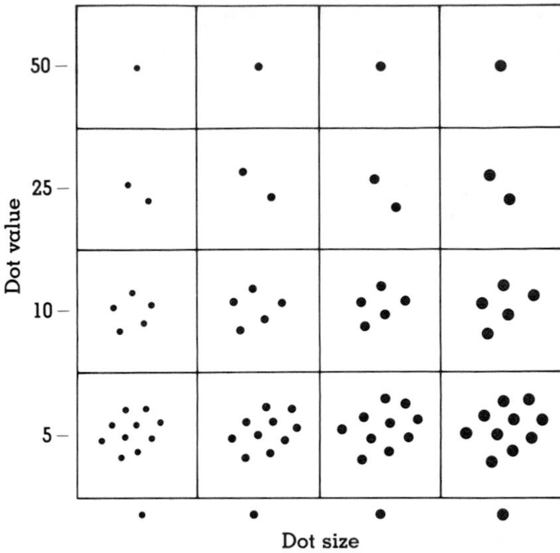

Figure 7.5. The effect of changing dot value and size.
The visual effects of changing dot value and size can be dramatic. The extremes of emptiness and crowdedness are easily seen.

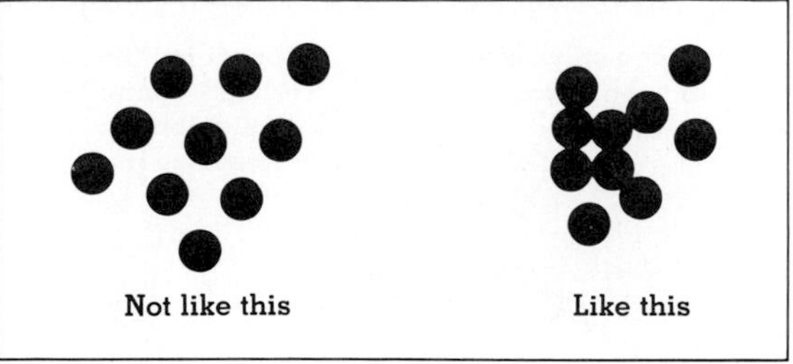

Figure 7.6. Dot coalescence.
Dots should begin to coalesce in the densest part of the map.

Dot Value and Size

Closely related to map scale is the determination of **dot value**—the numerical value represented by each dot. (See Figure 7.5.) Of course, one-to-one mapping involves no choice, since each symbol represents only one item. In many-to-one dot mapping, extreme care must be exercised in selecting dot value and size, since they affect the map reader's impression of accuracy. Small dots with small values cause a dot map to appear more accurate than it may actually be. The reverse is also true: large dots with large values appear crude and give the impression of unprofessional compilation.

It is difficult to list definite rules for selection of dot values and size, but here are some general guidelines:

1. Choose a dot value that results in two or three dots being placed in the statistical area that has the least of the mapped quantity.
2. Choose a dot value and size such that the dots just begin to **coalesce** in the statistical area that has the highest density of the mapped value. (See Figure 7.6.)
3. It is preferable to select a dot value that is easily understood. For example, 5, 500, and 1000 are better than 8, 49, or 941.
4. Select dot value and size to harmonize with the map scale, so that the total impression of the map is neither too accurate nor too general. This will require experimentation.

Dot scaling begins with an examination of the range of the data. The smallest value should be divided first by two, then by three. For example, if the lowest value in the array is 500 units, this divided by two will yield a dot value of 250; if divided by three, 166. Any dot value greater than 500 would eliminate a dot in this lowest enumeration area and should not be used.

Suppose a value of 250 were selected for this example. This preliminary dot value should be divided into the number of units present in the enumeration area that contains the highest value. The result is the total number of dots that must be placed in the highest enumeration area; it may be a large number. At this time, the designer begins to consider dot size. What size dots, when the correct number of them are placed in the highest enumeration area, will result in coalescence there?

During these trial-and-error tests, the designer should develop some working sketches of representative enumeration areas, at final map scale, and have several dot sizes at hand with which to experiment. The enumeration

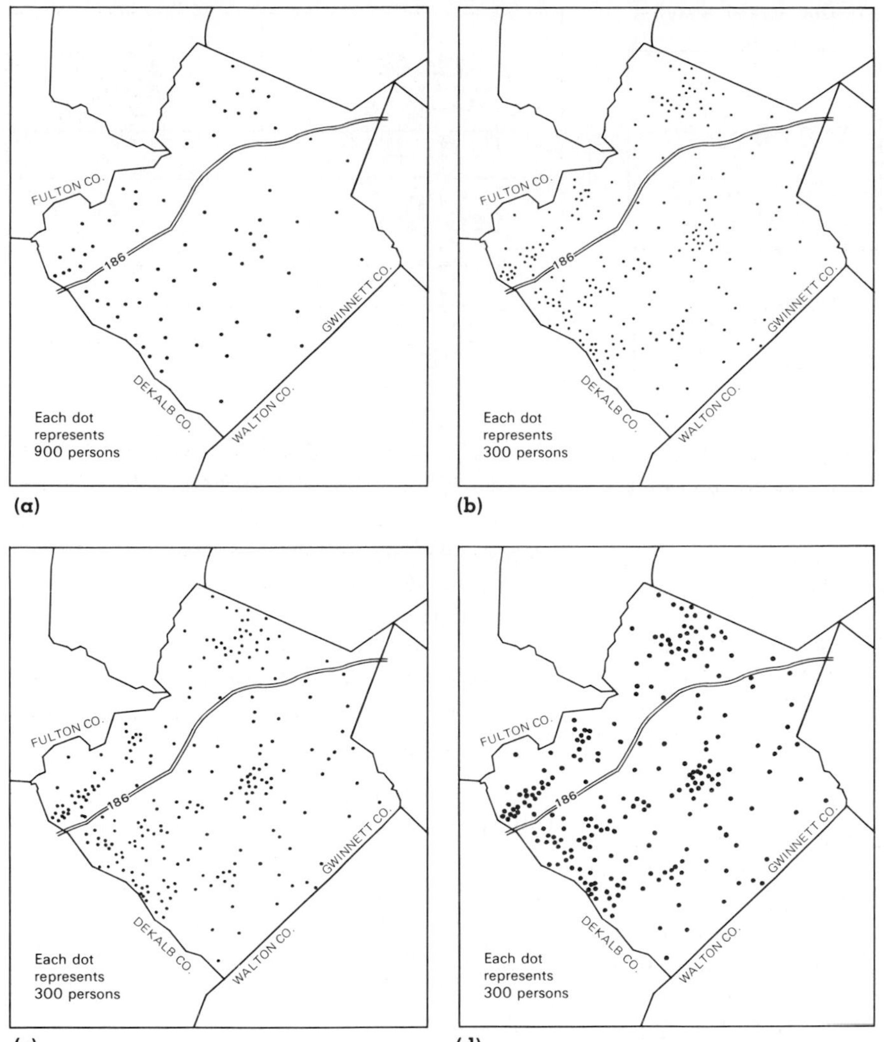

Figure 7.7. Dot value and size.
Changes in dot value and size will produce very different maps. The dot value in (a) is too large, leading to a map that appears too empty. In (b), the dot size is too small, again leading to a map that looks too empty. Dot size in (c) is still too small, although larger than in (b). Dot value in (d) remains at 300 persons, but dot size is now larger giving a better overall impression for the map.

area containing the largest number of dots is likely to be the one with the highest density. It is useful to know beforehand which enumeration units contain the highest and lowest densities of the mapped quantity. Experimental designs can then be tested on these extreme cases.

In summary, the designer first adopts a unit value for each dot, divides this value into the data total in each enumeration area to obtain the number of dots to be placed there, selects a dot size, and proceeds to place the dots. Tabular listing helps in organizing this material.

Several different dot maps can result from identical data because the selection of dot value and size is subjective. (See Figure 7.7.) There is no quantitative index to tell the designer which is best. Knowledge of the distribution

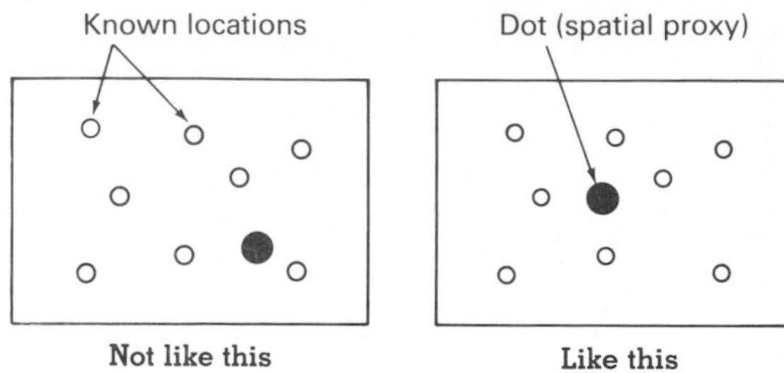

Figure 7.8. The proper placement of dots. Dots should be placed at the approximate center of gravity of the elements they represent.

plus an intuitive sense of what looks right are the standards against which any dot map must be judged.

Dot Placement

In most circumstances, the general rule is to locate the dots as close to the real distribution as possible, using the **center of gravity principle.** (See Figure 7.8.) Since each dot is a spatial proxy, it must be located so as to best represent all its referents. This requires considerable knowledge of the real distribution, and relevant resource materials are of invaluable assistance in this respect.

Geographical phenomena rarely occur uniformly over space, so a geometric arrangement of the dots within an enumeration area is not recommended. (See Figure 7.9a.) Only under unusual circumstances, or when the real distribution dictates it, should such a pattern be used. It is also good practice not to allow the political boundaries of the enumeration areas to exercise too much control over the location of the dots. (See Figure 7.9b.)

Legend Design

Every dot distribution map should have a legend containing two components: a statement indicating the unit value of the dot and a set of at least three squares (or other areas) that

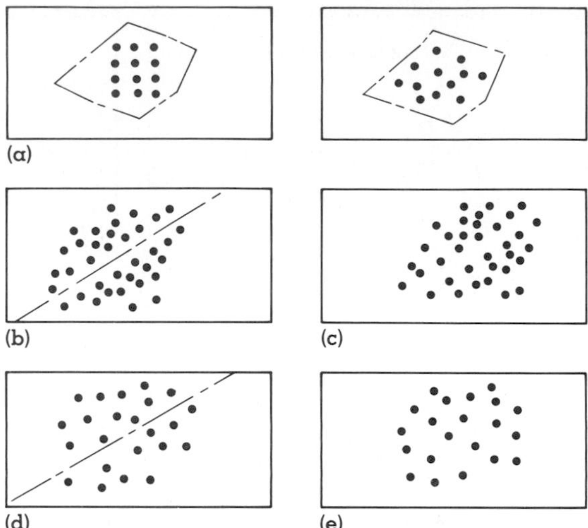

Figure 7.9. Acceptable and unacceptable methods of dot placement. In (a), the irregular pattern on the right is superior to the regular one on the left. Enumeration unit boundaries should not exercise too much control over the location of the dots; when the boundary in (b) is removed, its effect is still felt (c). The dots in (d) have been located in such a way that, if the boundary is removed, its control will not be noticed (e).

illustrate three different densities taken from the map. (See Figure 7.10.) It is good practice for the legend to indicate that a dot *represents* a value but does not equal it. The three squares may be arranged in any fashion that is unambiguous yet harmonious with the remainder of the map, but the legend must be designed for clarity.

Dot Map Production

There are no unique problems in dot map production. For many years, cartographers expressed concern over the production of uniform dots. With the development of preprinted dry-transfer dot sheets, or preprinted self-adhesive dot sheets, this problem has been all but eliminated. (See Figure 7.11.) Several

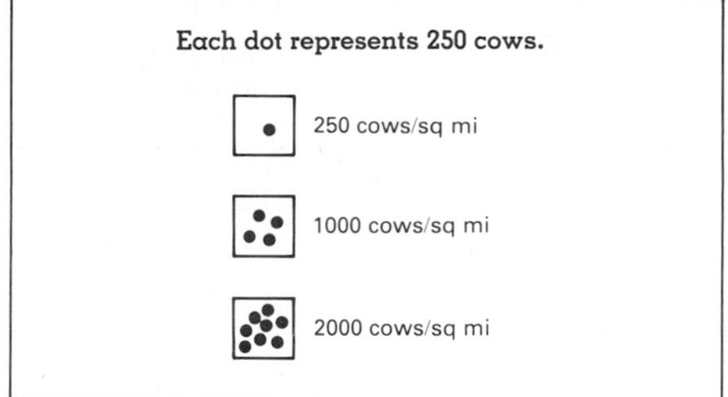

Figure 7.10. Designing dot legends.
A typical dot map legend includes a notation of dot value and should include representative densities across the range of data values. The size of the density boxes and the densities represented are determined in accordance with the map scale.

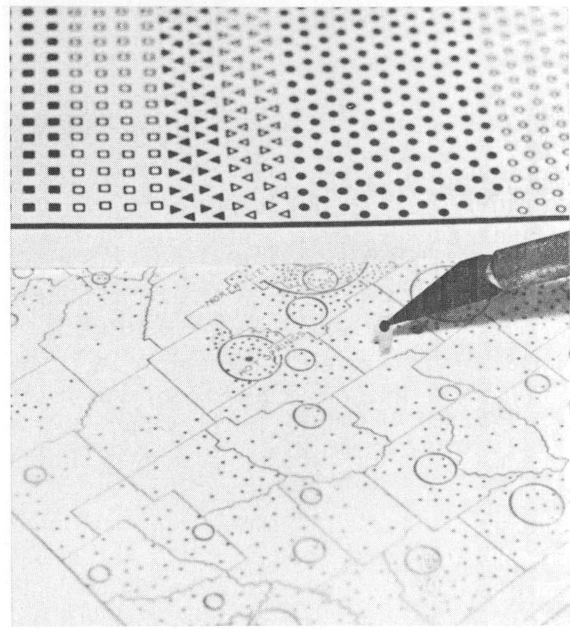

Figure 7.11. The production of uniform dots.
In the manual production of dot maps, uniform dot sizes can be obtained by using preprinted, adhesive-backed dots.

different manufacturers supply dots of different sizes. If these products are not used, uniform dots can be rendered by the careful use of technical pens. In some cases, the size of the map may prevent the use of preprinted dots; technical pens are then a suitable alternative.

Considerable attention must be given to any eventual reduction of the artwork, because dots may begin to appear grouped in areas not planned for coalescence. The dots should be placed so that they will appear as planned after reduction. The use of a reducing glass or the reduction of a copy can test for these situations prior to costly printing. Another way is to calculate precisely the reduced size and develop art at that size for those areas of the map that may present problems. For highly complex dot mapping, this may be worth the effort and expense. It is better to spend small amounts to guarantee satisfactory results than to go to press and make costly mistakes that cannot be rectified easily or cheaply.

Visual Impressions of Dot Maps

It was suggested earlier that reader perception of dot densities is not a one-to-one or linear relationship. This section treats this important concept in somewhat more detail.

Whatever dot value and dot size are used it should be borne in mind that an area full of dots on the map will carry a psychological impression that on the ground the same area is full of whatever is being mapped, and in the same way, too sparse a sprinkling of dots conveys an unmistakable impression of emptiness.

Source: G. C. Dickinson, *Statistical Mapping and the Presentation of Statistics* (London: Edward Arnold, 1973), p. 49.

Questions of Numerousness and Density

The reader of a dot map is asked to judge relative densities and to make certain assessments about the spatial distributions of these densities. In most dot mapping, we do not expect the reader to recover the original data—an impossibility if there is coalescence. The dot map is used to present data at the ordinal level in the sense that all the reader must do is to judge that there is more of the item in one place and less of it in another. **Apparent density** is the subjective reaction of the map reader to the physical stimulus to the actual density of dots per unit area.[5]

Numerousness Is the subjective reaction to the physical number of objects in the visual field, without actually counting the objects during perception. Dot maps are usually selected as a way of illustrating the spatial distribution of discrete objects; the reader is not normally expected to judge the actual number. Cartographers select this mapping technique to convey the spatial variableness of density and nothing more.

Recent cartographic research has demonstrated conclusively that most map readers underestimate the number of dots on a dot map. By extension, apparent densities are likewise guessed on the low side.[6] The evidence suggests that, when a person perceives an area as having 10 more dots than another with 1 dot, the actual ratio is probably closer to 15 or 16 to 1. This holds positive lessons for the cartographic designer. The following design ideas are suggested by recent research:[7]

1. Include a legend with examples of low, middle, and high densities—*visual anchors.* Provide at least a high-end anchor, preferably full-range anchors, to counteract the map reader's tendency to underestimate relative densities.
2. When recovery of original data is important, do not use coalescing dots; noncoalescing dots improve the estimation of dot number.

Rescaling Dot Maps

There has been some recent investigation by cartographers into ways of **rescaling** dot maps to allow more accurate reading. New dot values are assigned to compensate for the reader's tendency to underestimate. One method calls for each dot to represent a *range of values,* not simply a unit value.[8] The method is very time consuming; it appears to be particularly applicable to computer-generated maps.

Computer-Generated Dot Maps

Producing dot maps by automated methods has attracted the attention of cartographers during the last two decades. One program, developed by Wendell K. Beckwith under the supervision of Professor Joel Morrison, was used for the production of many dot maps for the Census of Agriculture Graphic Summary of 1969.[9] The program utilized county data, areas of each county, and centroids (geographic coordinates) for each county in the United States. For each map, the program computed the number of dots for each county (after a dot value was selected), and the dots in each county were positioned for greater accuracy by using the 16 land-use categories as mapped in the *National Atlas of the United States of America.* The land-use classification information was converted into nearly half a million cells to provide controls over the placement of dots. The current version of Beckwith's program retains most of the conceptual elements of the original, but has

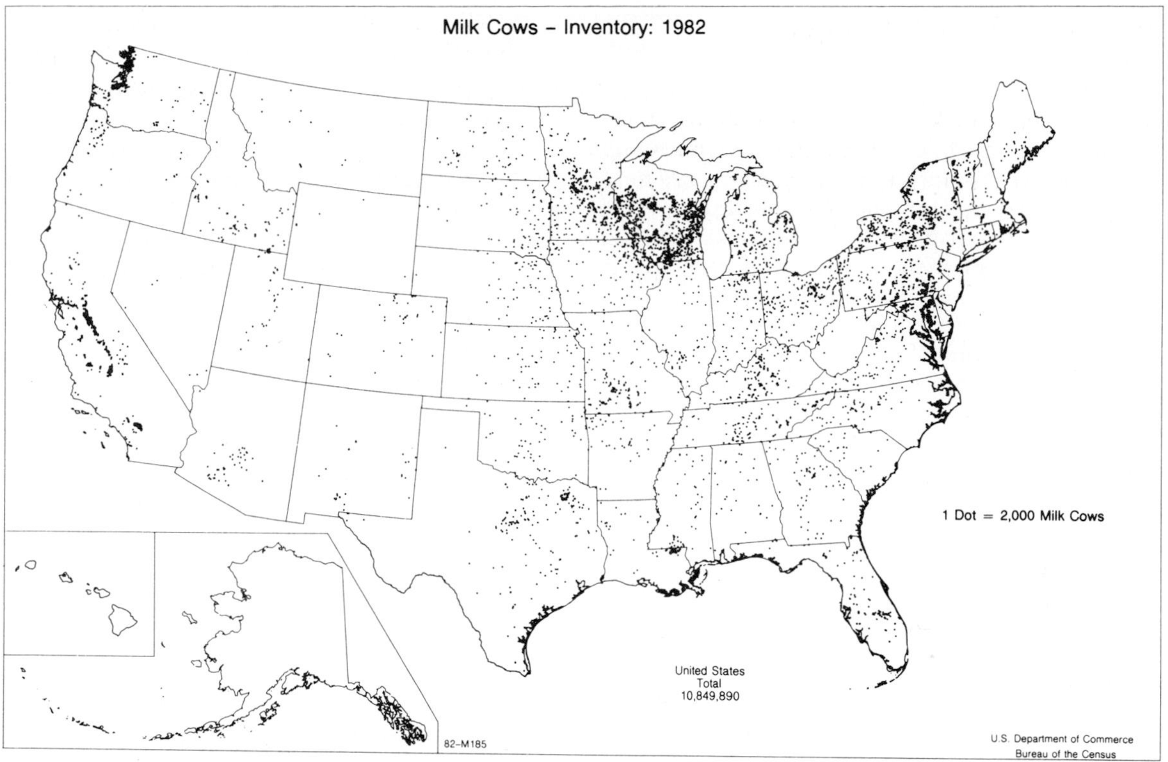

Figure 7.12. Dot map produced by computer and machine.
The computer output is a plot tape that drives a COM (computer output on microfilm) unit, which plots the map directly on publication-size 310mm microfilm.

Source: United States Bureau of the Census, *1982 Census of Agriculture,* vol. 5, Special Reports, p. 1, Graphic Summary (Washington, D.C.: USGPO, 1985), pt. 115.

added algorithms that take into account the impact of political boundaries on the location of the dots. The Bureau of the Census completed the publication of the Graphic Summary to accompany both the 1978 and 1982 Census of Agriculture, each including 200 computer-generated dot maps. (See Figure 7.12.) The estimated production cost of each map in 1978 was $75, including human labor, management, and computer time costs. These maps are completely machine-produced, even their lettering and other peripheral information. The version of the program used in 1969 did not produce these elements automatically.[10]

Dot maps developed by software programs written for microcomputers are now possible, but they do not have the design flexibility as do those designed for large mainframe or minicomputers. For example, programs written for smaller computers do not accommodate locational filters.

Microcomputer Dot Mapping

Choropleth mapping programs for microcomputers are more common than those that produce dot maps. However, the productions of such maps are possible, although they come with considerable constraint. Two mapping programs, *Atlas AMP* and *Atlas*Graphics* (both trademarks of Strategic Location Planning, Inc., San Jose, California) are in common use today. That they include the dot mapping option at all is a wonder, and certainly the company deserves credit for doing so.

Accurate dot mapping requires considerable knowledge about locational controls over dot placement. This requirement unfortunately cannot be met with low-cost microcomputer mapping software. The solution is usually to provide for the random placement of dots within an enumeration district, which is the way both these programs go about their business of dot mapping. As a consequence, without warning, for example, you may find dots representing people clustered in the Mojave Desert or the Carson Sink! The illustration below for example shows four different dot maps all produced from the same data.

Random location is a way of achieving a "quick and dirty" dot map, but the map maker and map user must beware.

Much to the credit of such software producers as Strategic Location Planning, however, is their allowing the designer to have control over two important dot mapping variables: dot value and dot size. This illustration shows two maps produced with the same data but with different dot sizes and values.

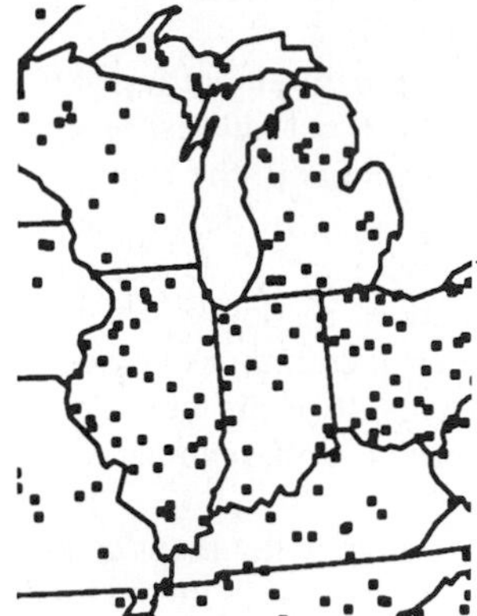

Thus, the cartographer has some control over the final appearance of the map. Until more sophistication is possible with these programs, the designer is cautioned to provide a statement on the map that the dots have been located randomly.

Notes

1. R. P. Hargreaves, "The First Use of the Dot Technique in Cartography," *Professional Geographer* 13 (1961): 37–39.
2. Richard E. Dahlberg, "Towards the Improvement of the Dot Map," *International Yearbook of Cartography* 7 (1967): 157–67.
3. *Ibid.*
4. T. W. Birch, *Maps: Topographical and Statistical* (Oxford: Oxford University Press, 1964), p. 154.
5. Judy M. Olson, "Rescaling Dot Maps for Pattern Enhancement," *International Yearbook of Cartography* 17 (1977): 125–37.
6. Robert W. Provin, "The Perception of Numerousness on Dot Maps," *American Cartographer* 4 (1977): 111–25.
7. *Ibid.*
8. Olson, "Rescaling Dot Maps," pp. 125–37.
9. United States Bureau of the Census, *Census of Agriculture, 1969,* vol. V, Special Reports, pt. 15, Graphic Summary (Washington, D.C.: USGPO, 1973), p. 14.
10. The information on the dot mapping program at the Bureau of the Census was generously shared with the author by Mr. Frederick R. Broome, Chief, Computer Mapping Staff, Geography Division, United States Bureau of the Census.

Glossary

apparent density the subjective response to the physical stimulus of the actual density of the objects in a visual field; density is underestimated in reading dot maps.

dot mapping method of producing a map whose purpose is to communicate the spatial variability of density of discrete geographical data; also called *areal frequency mapping*

dot polygon the area represented by each dot on a dot map; also referred to as the dot's *territorial domain*

dot value the numerical value represented by each dot in dot mapping; sometimes referred to as *unit value*

center of gravity principle placement of the dot in its territorial domain in the way that best represents all the individual elements in the domain

coalescence merging of the dots in dense areas; the visual integrity of individual dots is lost

numerousness the subjective response to the physical stimulus of the actual number of objects in a visual field; not the result of the conscious activity of counting; number is underestimated in reading dot maps

rescaling method used to enhance dot maps by assigning new dot values to compensate for the reader's underestimation of density

spatial proxy the dot's function of representing geographical space

Readings for Further Understanding

Birch, T. W. *Maps: Topographical and Statistical.* Oxford: Oxford University Press, 1964.

Dahlberg, Richard E. "Towards the Improvement of the Dot Map." *International Yearbook of Cartography* 7 (1967): 157–67.

Dickinson, G. C. *Statistical Mapping and the Presentation of Statistics.* London: Edward Arnold, 1973.

Geer, Sten de. "A Map of the Distribution of Population in Sweden: Method of Preparation and General Results." *Geographical Review* 12 (1922): 72–83.

Hargreaves, R. P. "The First Use of the Dot Technique in Cartography." *Professional Geographer* 13 (1961): 37–39.

MacKay, R. Ross. "Dotting the Dot Map: An Analysis of Dot Size, Number, and Visual Tone Density." *Surveying and Mapping* 9 (1949): 3–10.

Olson, Judy M. "Rescaling Dot Maps for Pattern Enhancement." *International Yearbook of Cartography* 17 (1977): 125–37.

Provin, Robert W. "The Perception of Numerousness on Dot Maps." *American Cartographer* 4 (1977): 111–25.

Smith, Guy-Harold. "A Population Map of Ohio for 1920." *Geographical Review* 18 (1928): 422–27.

United States Bureau of the Census. *Census of Agriculture,* 1969. Vol. 5, Special Reports, pt. 15, Graphic Summary. Washington, D.C.: USGPO, 1973.

———. *Census of Agriculture,* 1978. Vol 5, Special Reports, pt. 1, Graphic Summary. Washington, D.C.: USGPO, 1982.

———. *Census of Agriculture,* 1982. Vol. 5, Special Reports, pt. 1, Graphic Summary. Washington, D.C.: USGPO, 1985. Dot maps on pp. 25, 29, 55, 60, 81, 102, 104, 115 (milkcows), 128 (hogs and pigs sold), 140 (corn), and 153 (soybeans), are relatively good examples produced by automated methods. Students are encouraged to examine these to amplify work in this chapter.

Chapter

8

From Point to Point: The Proportional Symbol Map

Chapter Preview

Proportional or graduated quantitative point symbols have long been used in thematic mapping. The circle is one of the most popular forms, probably because of its compact size and ease of construction; others include the square and the triangle. Cartographic and psychological research into how map readers estimate the size of symbols indicates that the physical elements of area and volume are underestimated during perception. This has led to apparent scaling methods. Recent research has concluded that underestimation can be controlled by careful legend design even though absolute scaling is employed, so that apparent scaling methods may no longer need to be used. A method that eliminates the need of any true direct scaling is to present symbols in range-graded series, using point symbols of differentiable size, as in ordinal scaling. Regardless of the method of scaling, graduated point symbols must be rendered graphically so that they appear as dominant, unambiguous figures in perception. Proportional symbol maps can also be generated by computers, but the designer must still understand the basic concepts of this kind of mapping.

This chapter deals with one of the most flexible and popular thematic map techniques ever devised: *proportional point symbol mapping,* called *graduated* or *variable* point symbol mapping. The conceptual basis for this technique is easily understood by most map readers. Its popularity among cartographers is likely to continue for some time, but, like other processes in cartographic design, this method must be clearly understood if effective designs are to be created. Although much has been learned about this form of mapping in the last thirty years (notably in the area of symbol scaling), there is no single generally accepted design approach.

Conceptual Basis for Proportional Point Symbol Mapping

Proportional point symbol mapping is based on a fundamentally simple idea. The cartographer selects a symbol form (circle, square, or triangle) and varies its size from place to place, in proportion to the quantities it represents. Map readers can form a picture of the quantitative distribution by examining the pattern of differently sized symbols. (See Figure 8.1.) This simple approach can be spoiled by overloading the symbol form with too much information or by selecting scaling methods inappropriate to perceptual principles. The basic concepts of this form of quantitative thematic mapping are nonetheless easily grasped by most map readers. It is no doubt due to this fact that the method has reached its present level of popularity.

When to Select the Method

There are two commonly accepted instances when graduated point mapping is selected by the cartographer: when data occur at points and when they are aggregated at points within areas. (See Figure 8.2.) Of course, occurrence at points is to be interpreted relative to map scale. Practically any magnitude can be symbolized this way, including totals, ratios, and proportions. Density, however, is normally symbolized by the choropleth technique. Whenever

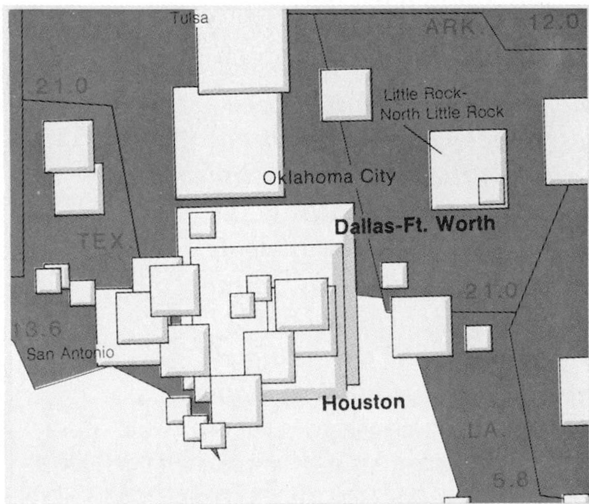

Figure 8.1. A typical proportional symbol map.
In this case squares have been used at points to represent map quantities.

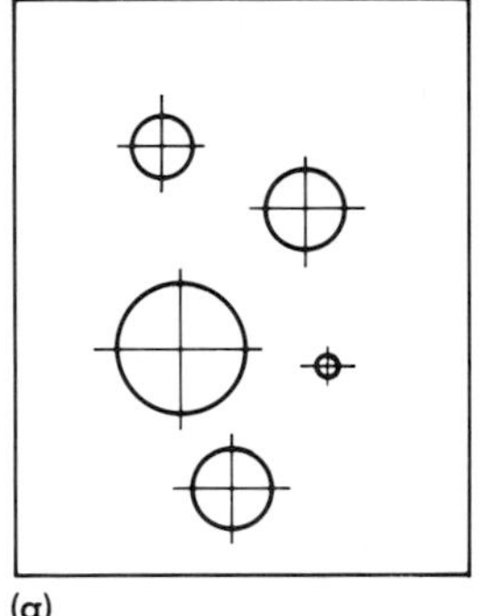

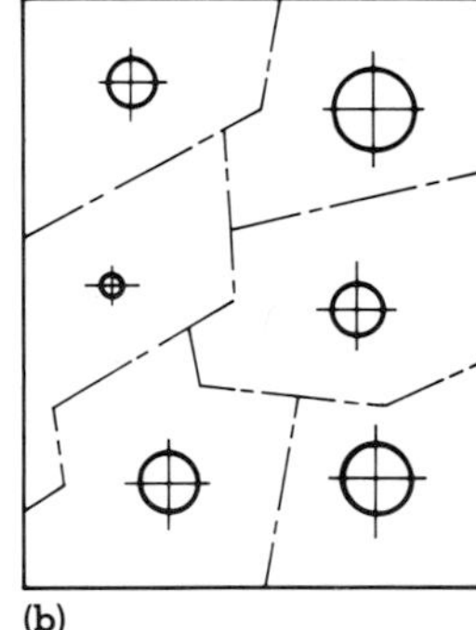

Figure 8.2. Two kinds of location and the use of proportional symbols.
In (a), circles are placed at the exact points where quantities being mapped are located (cities, towns, stores, etc.). Areal data assumed to be aggregated at points, as in (b), may also be represented by proportional point symbols. In each case, the symbols are used to show different quantities at distinct locations.

the goal of the map is to show relative magnitudes of phenomena at specific locations, the graduated symbol form of mapping is appropriate.

The kinds of data that can be mapped are varied: total population, percentage of population, black population, value added by manufacturing, retail sales, employment, any agricultural commodity, tonnage shipped at ports, and so on. Historically, proportional symbol maps have been extensively used to portray economic data, but magnitudes from the physical and cultural worlds also may be illustrated.

A Brief History of Proportional Point Symbols

The earliest history of proportional point symbols on maps is really a history of the use of the **graduated circle;** other symbol forms were not used until later. In 1801, William Playfair used graduated circles to depict statistical, noncartographic, data.[1] It is most interesting that Playfair used circle **areas** rather than scaling them to diameters or circumferences. We persist in using this scaling method today. It was apparently not until 1837 that circles were employed on maps. Henry Drury Harness used them to map city populations.[2] In 1851, they were used by August Peterman in the mapping of city populations in the British Isles.[3] Professor Arthur Robinson states that they were used by Minard, in France, by 1859.[4] Minard's circles portrayed port tonnages.

From these beginnings, the use of proportional symbols on thematic maps gradually increased. Graduated symbols were adopted slowly in the United States. It was not until the early part of the twentieth century that professional geographers and cartographers began to use them to any degree. They appeared first in professional journals. The graduated circle continues to be the most often used form of proportional symbol, although other choices are possible.

Many cartographers use bar graphs, line graphs, or other linear graphs as proportional symbols. The history of this form of point symbolization is unclear. The practice is discouraged because the spatial distribution of phenomena becomes muddled. It is virtually impossible to see relative magnitudes at points when a bar, for example, may stretch nearly half-way across the map. Proportional point symbol mapping requires compact geometrical symbols at points.

A Variety of Choices

Circles, squares, and triangles are the most common forms of proportional symbols, with the circle the dominant form. In each case, *area* is the geometric characteristic that is customarily scaled to geographical magnitudes. (See

In the case of graduated circle maps. . . . Some might argue that the immediate visual impression is of vital importance, others might suggest that any serious map user will be inclined to attempt some form of quantitative comparison between symbols. Although such stances may be considered contradictory, in essence they both possess a quantitative basis. However, they tend to differ on three counts: firstly, the former takes a more holistic viewpoint, while the latter is more concerned with specific detail; secondly, the former relies less heavily than the latter upon the existence of a suitably informative legend; thirdly, the former aims at the nominal and ordinal levels of scaling, while the latter ranges between the ordinal and ratio levels.

Source: T. L. C. Griffin, "Group and Individual Variations in Judgment and Their Relevance to the Scaling of Graduated Circles," *Cartographica* 22 (1985): 21–37.

Figure 8.3.) Circles have no doubt become more common because of several advantages they have over the other symbols:

1. Their geometric form is compact and far easier to construct than other forms.
2. Circle scaling is less difficult than with other geometrical forms (with the exception of the square), because the square root of the radius can be used.
3. Circles are more *visually stable* than other symbol forms and thus cause little eye irritation.

Squares are increasingly used but are more difficult to render than circles. Because they have straight sides, they must be oriented to the map border, to other symbols, or to structural elements of the map. This requires greater production time and cost and provides opportunities for production error. The square is also considered less visually stable than the circle. It is not without advantages, however—these will be discussed later in the context of scaling. Triangles have disadvantages similar to those of squares; they are used even less.

It is not uncommon for two proportional forms to illustrate two distributions on the same map. This is to be discouraged because it introduces too much complexity to the map and can undermine the communicative effort. A better solution would be to develop two separate maps.

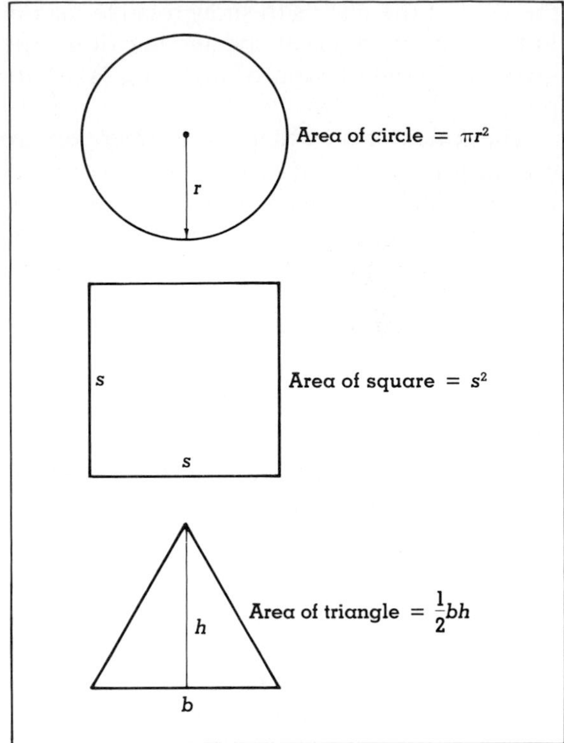

Figure 8.3. Three two-dimensional symbols (circle, square, triangle) and their area formulas.

Three-dimensional Symbols

Cartographers and geographers have experimented with point symbols of three-dimensional appearance, including spheres, cubes, and other geometrical volumes. (See Figure 8.4.) The use of **three-dimensional point symbols** can result in very pleasing, plastic, and visually attractive maps. Their chief difficulty lies in the inability of most map readers to gauge their scaled values correctly. Potential solutions to these shortcomings will be treated more fully below.

In addition to the visual effects, another important advantage of three-dimensional symbols lies in their scaling. With proportional circles, areas are scaled to the square root of the data (Table 8.1), but three-dimensional symbols, including the sphere, ordinarily have their volumes scaled to the cube root of the data. The net result is that, in three-dimensional mapping, the necessary range of symbol sizes is reduced. Greater data ranges can thus be handled on the map when spherical symbols are used. Less crowding of symbols also results, because the areas covered by symbols are reduced.

The difficulty is with reader perceptual scaling. Most map readers cannot correctly assess *scale*—quantitative three-dimensional symbols. We respond visually to the areas covered on the map by the three-dimensional-looking symbols and do not "see" relative volumes.

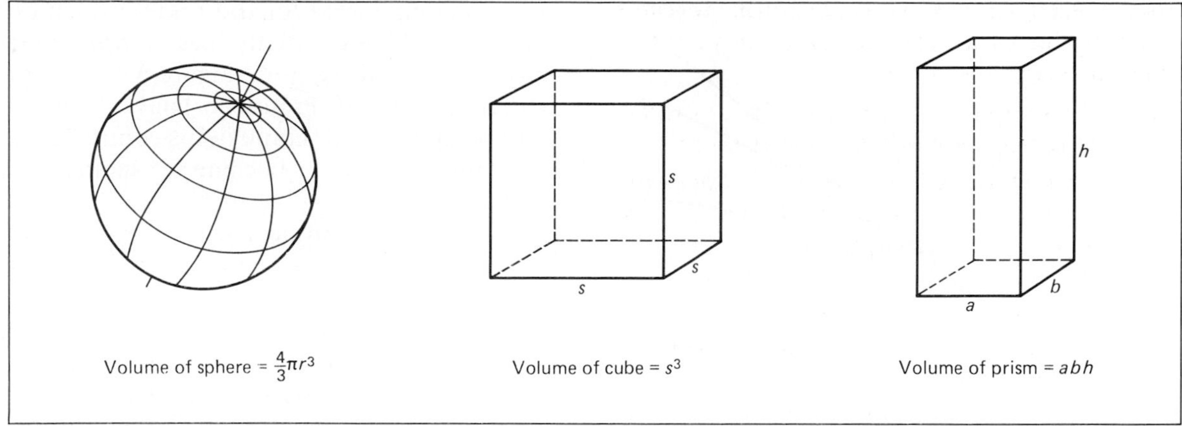

Figure 8.4. Three three-dimensional symbols (sphere, cube, and rectangular prism) and their volume formulas.

Table 8.1 Differences in Mapped Areas of Circles and Spheres

Symbol	Formula	Quantity Represented		Radii		Ratio of Large Radius to Small Radius	Area on Map Covered by Small Symbol (sq units)	Area Covered by Large Symbol (sq units)	Ratio of Area of Large Symbol to Area of Small Symbol
		Small	Large	Small Symbol	Large Symbol				
Circle	πR^2	100	1000	.1	.3162	3.162	.0628	.6278	9.99
Sphere	$\frac{4}{3}\pi R^3$	100	1000	.1	.2154	2.154	.0628	.2913	4.6

Proportional Symbol Scaling

We may now more formally introduce several ideas relating to how map readers perceive quantitative symbols. For several decades, psychologists, cartographers, and graphical statisticians have researched the mechanisms whereby map readers perceptually scale quantitative symbols. The researchers study response patterns to stimuli—the symbols and their characteristics such as length, area, and volume. The research attempts to document carefully how stimulus and response interact, usually by way of mathematical description. These studies are usually referred to as *psychophysical investigations.*

Psychophysical Examination of Quantitative Thematic Map Symbols

The general psychophysical relationship between visual stimulus and response can be described by this formula.[5]

$$R = K \bullet S^n$$

where *S* is stimulus, *R* is response, *n* is an exponent describing the mathematical function

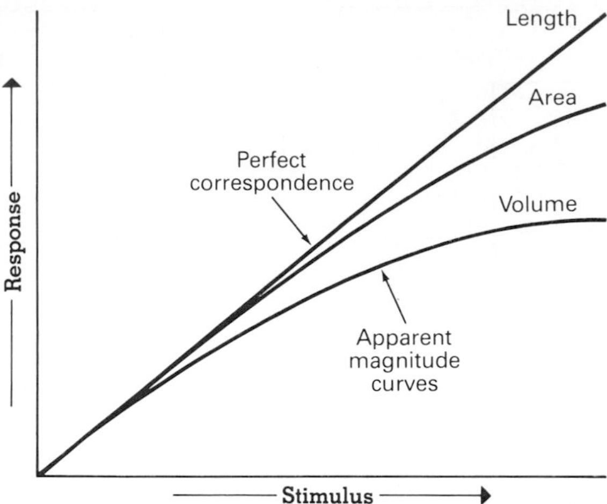

Figure 8.5. Apparent magnitude curves for length, area, and volume.
In the visual psychophysical world, the area and volume of geometric figures are underestimated. Length is usually correctly perceived. Underestimation means that for every unit of physical stimulus perceived less than unity is reported in response.

between *S* and *R*, and *K* is a constant of proportionality defined for a particular investigation. This relationship is sometimes referred to as the **psychophysical power law** (because $R = S$ raised to a power, n).

The results of the research efforts suggest overall that most people do not (or cannot) respond to the geometric properties of quantitative symbols in a linear fashion. (See Figure 8.5.) A linear response pattern means that if 10 units of stimulus are viewed, 10 units of response will be measured, and this one-to-one relationship will hold throughout the range of all stimuli. It has been found, overall, that length is correctly perceived, but area and volume are not. The geometrical property of area is increasingly underestimated as higher magnitudes of stimulus are judged. Volume is underestimated to a greater degree than area.

These findings have had considerable impact on map designers. The initial results, especially in the case of proportional circles, caused quick adoption of methods to adjust symbol sizes to compensate for the underestimation. This led to perceptually based **apparent-magnitude scaling.** (See Figure 8.6.) The pioneering work of Professor James Flannery must be cited in this context; his contributions to scaling adjustments became the standard for nearly 30 years.[6]

Recent psychophysical studies have had a different impact on designers. They suggest that apparent-magnitude scaling might be abandoned in favor of **absolute scaling,** but with more careful attention given to legend design.[7] Absolute scaling is simply the direct proportional scaling of magnitudes to the symbol's area.

For example, if we have two values, 6400 and 1600, that are to be symbolized and scaled by proportional circles, we must first set up a proportion, remembering that the area of a circle is πR^2 (R = radius of the circle), thus:

$$\frac{\pi R_1^2}{\pi R_2^2} = \frac{\text{data value 1}}{\text{data value 2}}$$

This reduces to

$$\frac{R_1^2}{R_2^2} = \frac{\text{data value 1}}{\text{data value 2}}$$

and then, taking the square root of both sides of the equation, we get:

$$\frac{R_1}{R_2} = \frac{\sqrt{\text{data value 1}}}{\sqrt{\text{data value 2}}}$$

for this example, we next have

$$\frac{R_1}{R_2} = \frac{\sqrt{6400}}{\sqrt{1600}}$$

and then,

$$\frac{R_1}{R_2} = \frac{80}{40} = \frac{2}{1}$$

or the radius of value 1 is twice that of value 2.

In practice a radius is selected to represent the smallest data value that yields a circle that looks good on the map, then a radius is calculated for the largest data value to determine if the circle size is appropriate, and if so, all intermediate circle sizes for the remaining

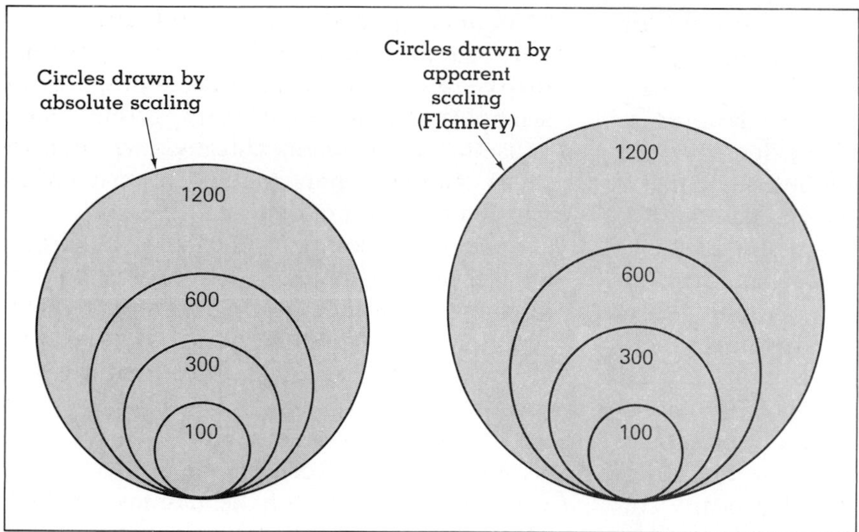

Figure 8.6. Absolute and apparent scaling of proportional circles.
Because the areas of circles are not perceived linearly, Flannery has introduced a correction factor, causing circles to appear larger as their values increase. This produces a range of circle sizes scaled by apparent magnitude.

values are calculated. If not, the process is begun again and a final set of circle sizes are computed and drawn.

The Proportional Circle

Scaling for circles has received more attention from psychologists and cartographers than scaling for any other form of symbol, because of the circle's popularity as a proportional symbol. When Flannery made clear that circle sizes need to be adjusted to compensate for underestimation, a standard was adopted based on his findings.

The psychophysical relationship, expressed as $R = K \bullet S^n$, is usually transformed into logarithmic form:

$$\log R = \log a + b \bullet \log S$$

This is an equation for a straight line where a is the constant K (and the intercept on the R-axis), b is equivalent to the slope of the line (and the n of the original power equation), and S and R are stimulus and response as before. By regression methods, a line will then be fitted to the experimental data, and a value for n is produced.

In the original equation $R = S^n$ (omitting K for simplicity), if a perfect relationship exists between R and S, then the exponent n would be 1.0. Flannery's experimental data yielded a value of $n = .8747$. For example, if a circle of area value 2 is judged experimentally, it would be seen as having an area of

$$2^{.8747} = 1.833$$

Flannery arrived at his *adjustment factor* by transposing the equations. In logarithmic terms, the square root of number N is (log n)(.5). To compensate for underestimation, the symbols are enlarged somewhat by using .5716 instead of .5. This value is tied directly to the value of .8747 for n.

The procedure to follow to scale circles this way includes the following steps:

1. Obtain logarithms of data values.
2. Multiply these log-values by .5716 (or rounded to .57).
3. Determine the antilogarithms of each.

4. Scale all values to the new set of antilogarithms in ordinary fashion.

In the psychophysical power law, n expresses how the response can be described relative to the stimulus. We might assume that n is a constant for circle judgment tests, but it is not. Recent research indicates that n fluctuates because of these factors: testing method, instructions, standard (or anchor) stimulus used, and the range of stimuli used in the experiments.[8] These n values range from .58 to 1.20. It is now evident that no one correction factor developed for an apparent magnitude scale is the answer. This is especially true in light of parallel findings which show that nearly correct responses can be achieved with carefully designed legends.

Available studies also indicate that apparent-magnitude scaling is not that much more efficient than absolute scaling if the range of required circle size judgments is ten or less.[9] One final caution, too, is worthy of note. Some map readers simply cannot make proportional symbol judgments well even if apparent-scaling is used. Map designers will need to weigh all these factors when making decisions regarding this form of quantitative map.

It must be pointed out, finally, as already shown in Figure 8.6, that selecting the apparent-magnitude scaling method requires much greater map space, and this is often a critical factor in design as space is usually limited.

Perception of circles among circles. Most cartographic research on the judgment of circle size has measured rather abstractly how people perceive circle size in nonmap settings. Moreover, and more troubling, the influences of neighboring circles on size judgment have been largely overlooked. However, at least one

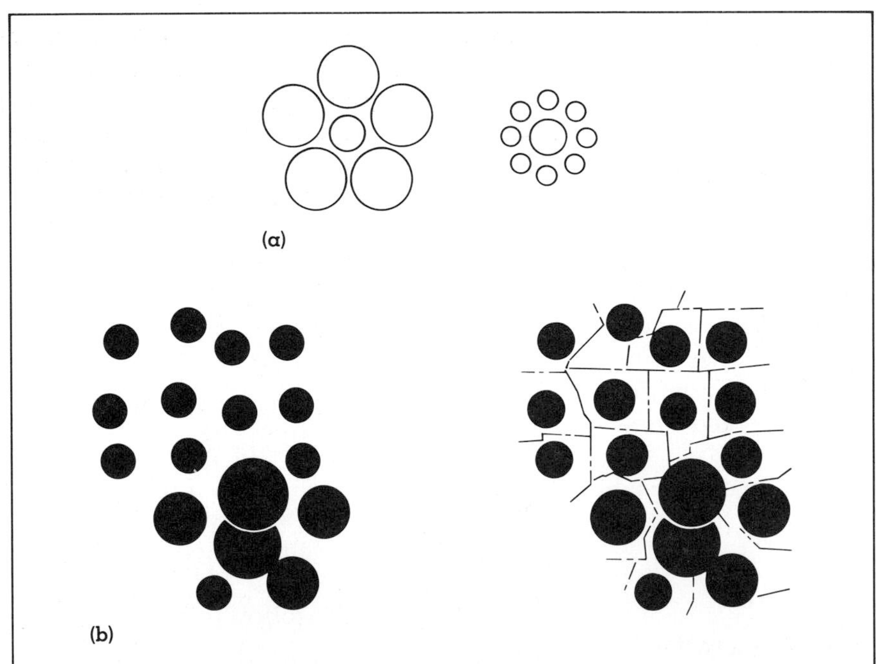

Figure 8.7. The effect of neighboring symbols on the judgment of circle size.
In (a), the inner circles in the right and left drawings are of the same size. Because of the contrast effects produced by neighbors, the inner circle on the right appears larger than its counterpart on the left. In (b), the provision of internal boundaries reduces the contrast effect of neighboring symbols. Unfortunately, cartographic designers have little control over variables of this kind.

recent study has investigated this phenomenon and has produced these interesting results.[10]

1. When a circle is seen among circles that are smaller, it is seen on the average about 13 percent larger than it would ordinarily be seen if judged among circles that are larger. (See Figure 8.7a.)
2. Circles surrounded by circles of the same size were judged both high and low relative to the surrounding circles.
3. The effect of surrounding circles can be reduced if internal borders are used on the map. (See Figure 8.7b.)

Thus, circle judgment is affected by the size of neighboring circles on the map. At the present time, no generally applicable solution is available to control these undesirable effects. If repeated design alternatives show that this problem cannot be solved, perhaps a different form of map should be constructed. Designers need at least to know that this problem can arise, so that solutions can be sought.

One very important aspect of proportional symbol maps is their ability to show spatial numerousness. (See Figure 8.8.) This feature is extremely effective when symbol sizes are carefully selected for the base map being used. Although overlapping symbols make it difficult to estimate individual symbol sizes, a sense of visual cohesiveness results, and this may lead to memorability, certainly a goal in cartographic communication.[11] This dilemma, to show overall pattern vs. specific symbol quantity, usually can be dealt with by careful examination of map purpose. If not, one solution is to provide an areal table along side the map, and in the long run, develop special scaling methods that take into account symbol coverage.[12]

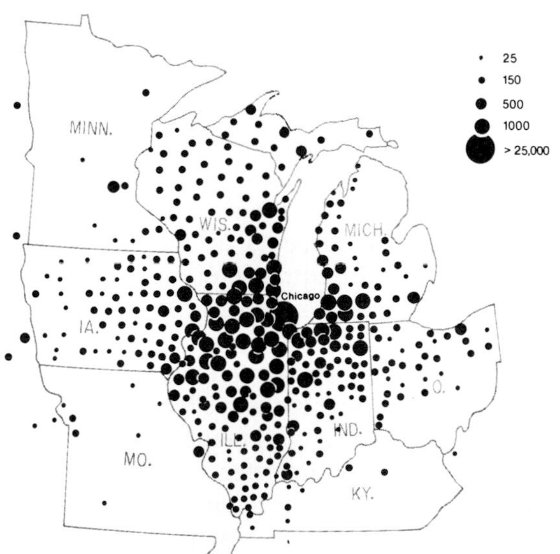

Figure 8.8. Proportional symbol maps show spatial numerousness.
In this illustration the spatial variableness of newspaper circulation (the Chicago Tribune) is clearly shown by the proportional symbol mapping method.
Source: Reprinted courtesy of Duxbury Press.

The Square Symbol

Graduated squares are used in quantitative point symbol mapping in much the same fashion as circles. The areas of the squares are scaled to data amounts yielding simple proportional relationships between the square roots of data and the sides of the squares:

$$\frac{S_1^2}{S_2^2} = \frac{\text{data value 1}}{\text{data value 2}}$$

$$\frac{S_1}{S_2} = \frac{\sqrt{\text{data value 1}}}{\sqrt{\text{data value 2}}}$$

Squares have never received quite the attention by researchers that circles have been given, probably because they are not as commonly used. As mentioned earlier, they are more difficult to handle in production. Visually, they introduce a rectangularity to the map's design which is often difficult to coordinate with other map elements. (See Figure 8.9.) Nonetheless, they can show geographical distribution—clustering, dispersion, or regularity—as well as circles.

One important consideration for the selection of squares is that their proportional areas are nearly perfectly perceived. In one study, the

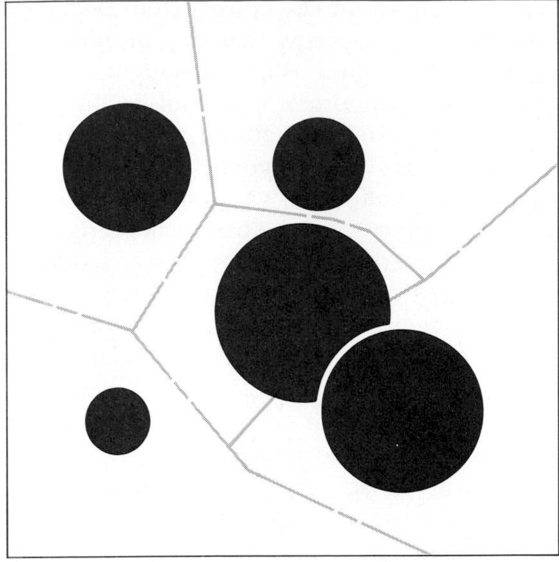

(a)

(b)

Figure 8.9. Comparison of the appearance of proportional circles and squares.
Some cartographic designers dislike using squares because they introduce "rectangularity" on the map. Circles produce symbols that are compact and visually stable; they need not be reoriented to other map elements during production. However, squares are perceived more accurately than circles.

exponent value *n* in the psychophysical power law was experimentally determined to be .93.[13] Other psychological and cartographic studies suggest that the areas of squares are more accurately visually judged than circle areas. This does not mean that circles should be abandoned in favor of squares. With appropriately designed legends, both symbols can communicate well. The selection should be made in concert with other design elements and mapping goals.

Range-grading: A Probable Solution

Apparent-magnitude scaling methods (the introduction of corrective factors into absolute scales) are too coarse for universal adoption. They are based on the "average" map reader, who does not exist outside of mathematical tables. Moreover, the landmark corrective factor of Flannery, although quite important in the early assessment of symbol scaling, is not applicable in all cases because it was developed from a single psychophysical study with one experimentally derived exponent. Recent investigations have demonstrated that the exponent itself varies depending on test conditions.

Legend designs have also been examined closely. This avenue of research is based on the theory of **adaptation level.** This theory holds that perceived judgments are based on an adaptation level (a neutral reference point) and that the perceptual anchors from which judgments are made affect this point.[14] Two **anchor effects** have been shown to exist: *contrast* and *assimilation.* The contrast effect is displacement of perceptual judgment away from the anchor; by assimilation, perceptual judgments are displaced toward the anchor. The size of the anchor has an effect on how perceptual judgments of the symbols on the map are made. In the case of proportional circles and squares, judgments are displaced toward the anchor, whether it is a large or small anchor. This is evidence that assimilation is taking place.

One study has produced important findings regarding anchor effects in proportional circles and squares. There are important design implications.[15]

1. When graduated circles are scaled according to the apparent value (after Flannery), there is greater underestimation of symbol sizes when a small legend value is used, less overestimation when a large symbol is used, only slight overestimation of circle sizes when *three* legend circles are used, and fairly accurate estimates when middle-sized legend circles are used. The variation in responses is less when three legend symbols are used.
2. For squares, small- and middle-sized legend squares lead to underestimations. (The squares are scaled to the square root of the data, with no corrective factor.) A large legend square yields overestimations, and three legend squares (small, middle, and large) lead to better estimates overall—principally at the higher end of the scale. Cox, author of the study, suggests that perhaps only a middle-sized and a large square symbol are needed.
3. *Apparent scaling does not compensate for underestimation.* The use of several differently sized legend symbols, scaled conventionally (absolute scaling), may be sufficient for accurate judgment of circle and square size.

It has become abundantly clear that certain proportional point symbol formats are not good design solutions:

1. Never provide apparent scaling for circles unless several circle sizes are used throughout the data range in the legend.
2. Do not provide absolute scaling for circles when only one legend (anchor) circle is used.
3. Do not scale any symbols to their volumes (cube roots of the data).

Traditional proportional point symbol mapping called for each value in the data to be represented by an absolutely scaled symbol (**square root scaling** for two-dimensional symbols and **cube root scaling** for three-dimensional symbols). As experimental studies showed that perception of these symbols was not linear, apparent scaling was introduced. Subsequent research has shown that apparent scaling may not be the total answer either. It appears that range-grading proportional symbols may be a better choice. Certainly, classifying data to show spatial distribution is an appropriate design approach.[16]

Range-grading is achieved in a manner similar to the development of classes for choropleth maps. The data array is divided into groups; each group is represented by a proportional symbol that is clearly distinguishable from other symbols in the series. In this scaling method, symbol size discrimination is the design goal, rather than magnitude estimation. Range-grading of proportional symbols is not a new alternative to symbol scaling; it was first suggested by Professor Meihoefer in the late 1960s.[17] His studies led him to devise a set of ten circle sizes that were easily and consistently discriminated and were applicable to most small-scale thematic maps. (See Table 8.2 and Figure 8.10.)

Table 8.2 Acceptable Range-Graded Circle Sizes for Small-Scale Thematic Maps

Radius of Circle (mm)	Area of Circle (sq mm)	Difference (mm)
1.27	5.0	
1.99	12.5	7.5
2.82	25.0	12.5
3.99	50.0	25.0
5.64	100.0	50.0
7.98	200.0	100.0
9.77	300.0	100.0
11.28	400.0	100.0
13.82	600.0	200.0
15.96	800.0	200.0

Source: Hans Joachim Meihoefer, "The Utility of the Circle as an Effective Cartographic Symbol," *Canadian Cartographer* 6 (1969): 105–117.

Figure 8.10. Representative circle sizes useful in a range-graded series for small-scale thematic maps.
Each circle is easily differentiated from its neighbor—the fundamental criterion in a range-graded series. No more than five classes are advisable for this form of scaling. When using this chart, any adjacent set of circles may be selected from the array.

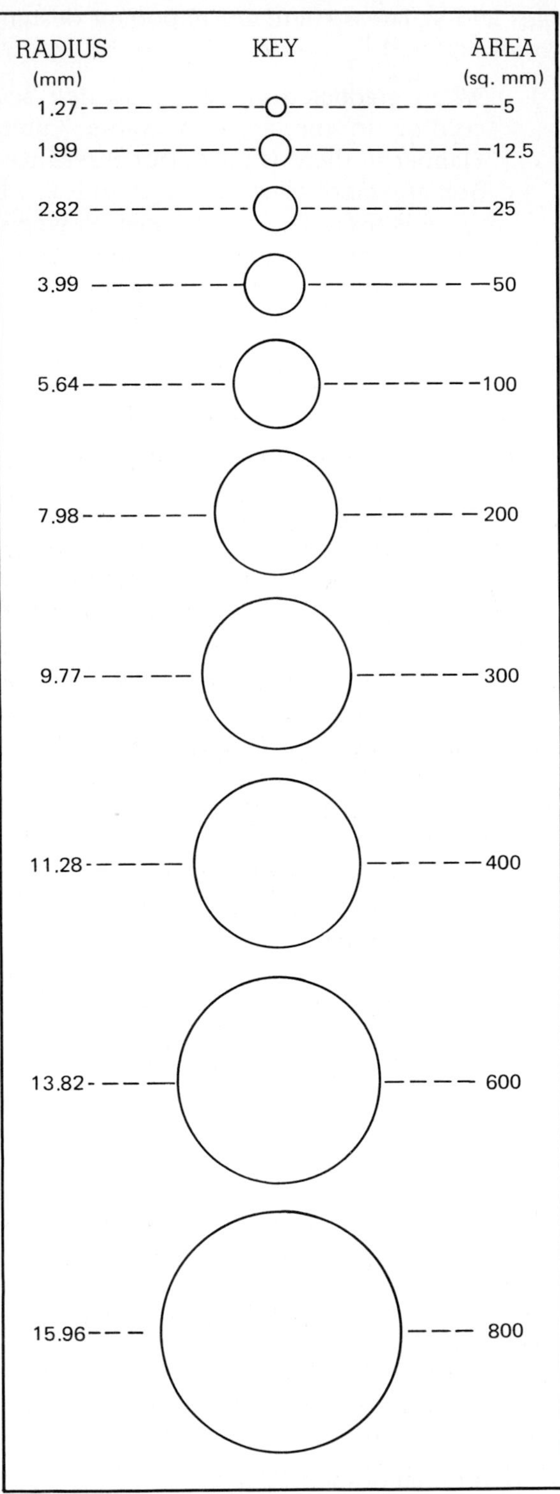

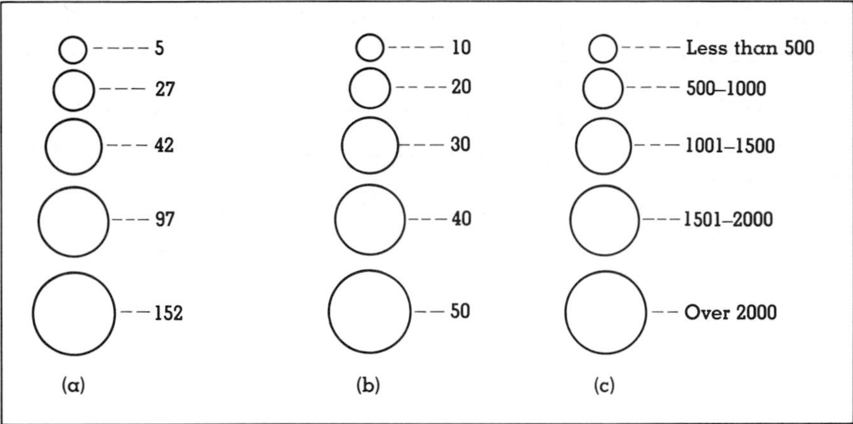

Figure 8.11. Different ways to illustrate range-graded series by legend circles.
In (a), a representative number is presented, perhaps an average of each homogeneous class. No rational relationship between numbers is apparent. In (b), an attempt to provide preferred numbers and even intervals is shown. Finally, in (c), data class boundaries of a continuous distribution are shown adjacent to each class circle. No one method is preferred. The goal is to make the legend readable and understandable and the range-graded series apparent. It may be necessary to explain the classing scheme in a legend note.

Range-grading of proportional point symbols raises the issue of class interval determination, just as with choropleth class intervals. The most applicable method appears to be the optimization approach. Just as determining choropleth class intervals, the goal is to devise classes that have the greatest degree of internal homogeneity and a high degree of heterogeneity between classes.[18] Although Meihoefer used ten circles in his research, we recommend that perhaps only four or five classes be used. The legend may appear in a variety of ways, but it must be clear that the circles are range-graded and do not represent individual values. (See Figure 8.11.)

Graphic Design Considerations for Proportional Point Symbol Maps

Several issues relating to the design of proportional point symbols have been presented above, focusing on the conceptual basis of their use and on scaling methods. Other design concerns deal more with their graphical treatment, discussed in this concluding section.

The Nature of the Data

In a quantitative mapping task, cartographers must deal with various attributes of the statistical data set before them and slect a mapping method that best represents the particular aspect of the data. The data attributes have been described as density, clustering or dispersion, extent, orientation, and shape (of the statistical distribution). Proportional point symbol mapping should not be selected when data variation is small or when the nature of the classing method renders a map of homogenous appearance. (See Figure 8.12.) The result will be a monotonous map that looks dull and undiscriminating (although it may be technically accurate). The intuitive appeal of the design can be the criterion of a map's appropriateness in such cases.

Overloaded Proportional Point Symbol

Maps and their quantitative symbols are unique mechanisms for the communication of spatial concepts. All too often, however, cartographic designers fall into the trap of pushing the

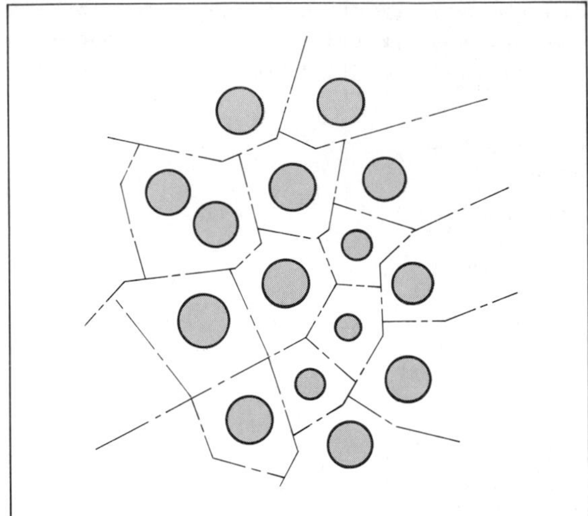

Figure 8.12. Inappropriate use of proportional symbols.
Maps in which unvarying geographical data are symbolized by proportional symbols are monotonous and reveal little. Designers should not use this mapping method with distributions of this kind.

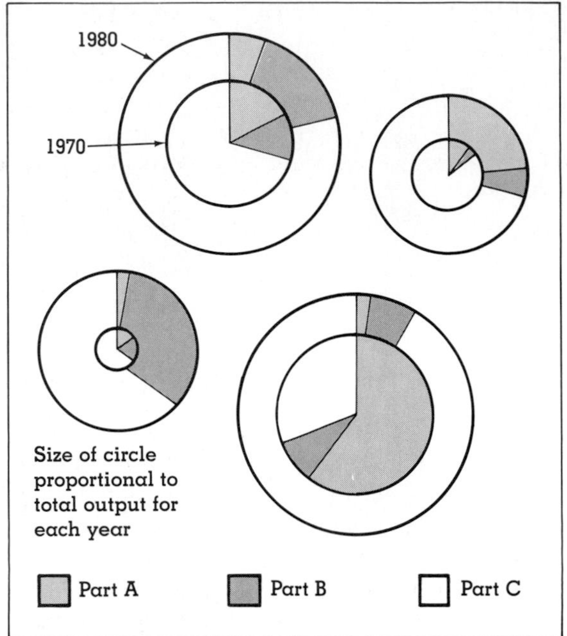

Figure 8.13. An example of an overloaded proportional point symbol, using a segmented circle (showing proportional parts) for two different years.
The map reader is expected to make an easy and quick assessment of the change in the geographical distribution of each part, at several locations. The map designer should seek simple solutions rather than use symbols so unwisely.

system beyond its capacities. Because it is possible to cram a lot of information into one symbol, the designer often tries for too much of a good thing. **Symbol overload** results when it is no longer easy for the reader to make judgments about the quantitative nature or spatial pattern of the distribution. Unfortunately, overload is reached at a point much earlier than most designers realize.

Probably more than any other symbol, the proportional circle has been misused by unthinking designers. Since it can be sized, segmented, colored, sectored, or whatever, why not do all of these on one map? (See Figure 8.13.) Unfortunately, the map readers can no longer see the distribution or its intent clearly and easily. The cartographer should not have troubled him- or herself or the reader. In at least two maps in the *National Atlas of the United States of America,* a proportional circle carries three different data sets.[19] These maps lose their thematic goals and exhibit complex graphic displays that almost totally destroy their communicative efforts.

A good design approach is to limit the number of variables symbolized by proportional point symbols to one, possibly two, but never three or more. When two or more variables are being represented, keep only one variable at the interval/ratio level and the other at the nominal or ordinal level. Do not use two at the interval/ratio levels, since this leads quickly to excessive complexity and symbol overload.

Graphic Treatment of Proportional Symbols

Consideration already given to the scaling of circles and squares applies equally to black or gray-tone circles (and probably squares). Gray-tone treatments do not require special considerations. Cartographers find several advantages in using gray proportional symbols in black-and-white mapping.[20]

1. The amount of information can be expanded by showing two distributions, one with black symbols and one with gray.
2. The use of gray-tone symbols allows the cartographer greater freedom for expression in design.
3. Gray-tone symbols allow for the better ordering of information on the map. Symbols can be rendered in tones of gray in proportion to the values they represent.
4. Some symbols can be deemphasized (gray) relative to others (black).

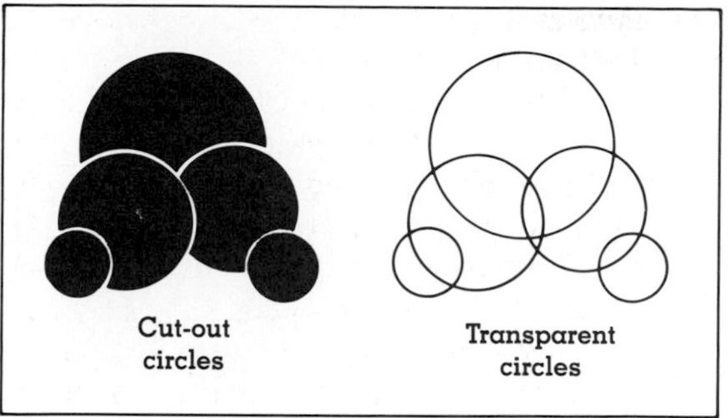

Figure 8.14. Cut-out and transparent circles.
Readers tend to make errors of perceptual judgment with cut-out symbols, in proportion to the amount of the circle obscured by the overlap. Transparent circles, on the other hand, are perceived about as well as circles that do not overlap. The designer must weigh these results against other design elements and the purpose of the map. Cut-out circles add a three-dimensional (plastic) quality to the map, whereas a map containing many transparent circles looks flat and uninteresting.

According to one current investigation of the effect of color on the perception of graduated circles, it is apparent that different hues do not affect the estimation of circle size differences.[21]

Open and "cut-out" circle treatments have been carefully examined by cartographic researchers. (See Figure 8.14.) Groop and Cole, for example, made these discoveries:[22]

1. Transparent circles (rendered so that edges of both are visible even after overlap) are more accurately perceived than "cut-out" circles and are seen similarly to circles having no overlap.
2. With "cut-out" circles (in which the overlap obscures part of one circle), there is a strong relationship between the amount of overlap and estimation error.

Because of the complex nature of many proportional symbol maps, the effects of overlapping, tightly clustered circles on readers' estimations of circle sizes are not known with certainty. In some cases, the reader is asked to estimate the magnitudes of individual symbols, in others, range-graded series. The designer must be aware that the reader's judgment is influenced by the graphic treatment of the symbols.

In closing, it is important that proportional point symbols be made to appear as strong figures in perception. The graduated point symbol should be clear and unambiguous in meaning, its edges sharp and not easily confused with background material, and its surface character easily differentiated from other surfaces and textures on the map. Its scaling should be precise, not confusing for the reader. The symbol

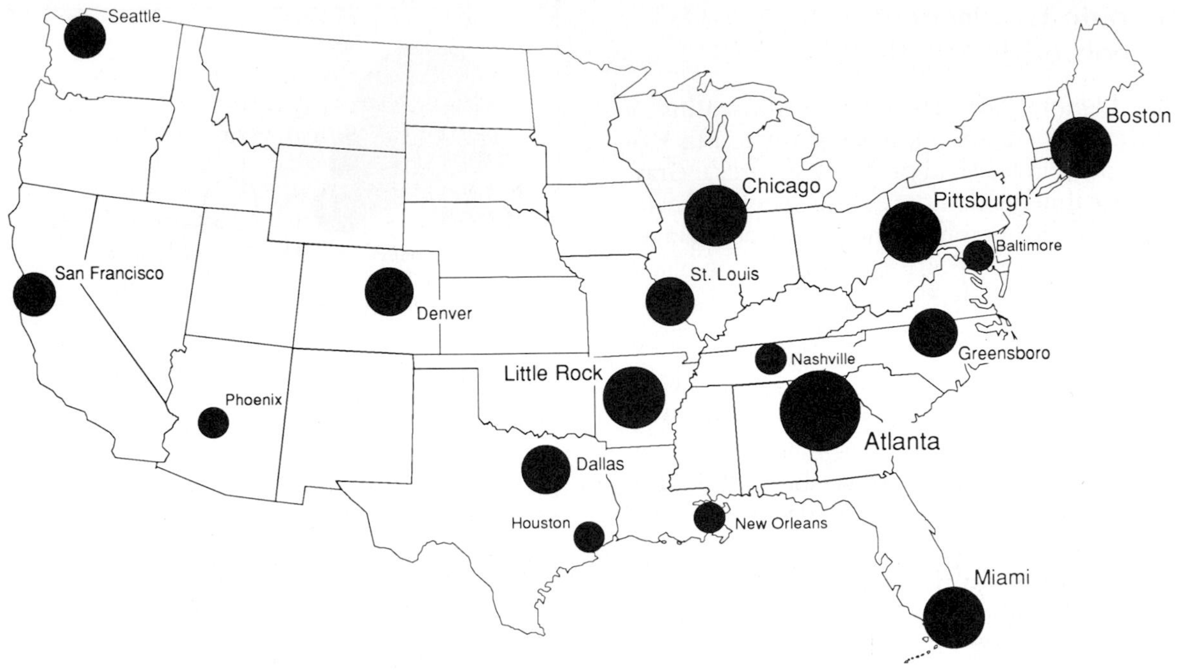

Figure 8.15. An example of a computer-generated proportional symbol map.
This map was produced by *Designer* software (a trademark of Micrografx, Inc.), and printed on a *QMS* laser-jet, 300-dpi printer.

should be made to stand out from its surroundings, both graphically and intellectually. If these simple guidelines are followed, the final map will perform well in the communication of spatial concepts.

Computer Maps

It is possible to generate computer maps using proportional point symbols. (See Figure 8.15.) Design considerations remain the same, however, with respect to computer maps. Questions of scaling (absolute vs. apparent), range-grading, and graphic appearance of the circles must be confronted by the designer; the development of computer programs to produce these maps has not reduced the burden. In fact, the ease with which these maps may be produced on color CRT devices, for example, should spur further research into the psychophysical nature of quantitative point symbols and color. Absolute scaling is much less time consuming in computer-generated maps. This may make this option more appealing but also cause a dilemma, given the present trend to present circles in range-graded fashion. Designers sensitive to the needs of clients and map readers alike must make the difficult design decisions, whether or not the new technology is employed.

Notes

1. James J. Flannery, "The Graduated Circle: A Description, Analysis, and Evaluation of a Quantitative Map Symbol," unpublished Ph.D. dissertation, Department of Geography, University of Wisconsin, 1956, p. 7.

2. Arthur H. Robinson, "The 1837 Maps of Henry Drury Harness," *Geographical Journal* 121 (1955): 440–50.
3. Flannery, "The Graduated Circle," pp. 8–9.
4. Arthur H. Robinson, "The Thematic Maps of Charles Joseph Minard," *Imago Mundi* 21 (1967): 95–108; and Arthur H. Robinson, *Early Thematic Mapping in the History of Cartography* (Chicago: The University of Chicago Press, 1982) pp. 207–208.
5. Kang-tsung Chang, "Circle Size Judgment and Map Design," *American Cartographer* 7 (1980): 155–62.
6. Flannery, "The Graduated Circle." Professor Williams of Yale University was also working on scaling methods as Flannery was doing his research. Flannery referred to his work in his dissertation.
7. Chang, "Circle Size Judgment," pp. 155–62.
8. Among the studies that have direct bearing on this subject are: Chang, "The Graduated Circle"; Kang-tsung Chang, "Visual Estimation of Graduated Circles," *Canadian Cartographer* 14 (1977): 130–38; Carleton W. Cox, "Anchor Effects and the Estimation of Graduated Circles and Squares," *American Cartographer* 3 (1976): 65–74; and James J. Flannery, "The Effectiveness of Some Common Graduated Point Symbols in the Presentation of Quantitative Data," *Canadian Cartographer* 8 (1971): 96–109.
9. T. L. C. Griffin, "Groups and Individual Variations in Judgment and Their Relevance to the Scaling of Graduated Circles," *Cartographica* 22(1985): 21–37.
10. Patricia P. Gilmartin, "Influences of Map Context on Circle Perception," *Annals* (Association of American Geographers) 71 (1981): 253–58.
11. Michael P. Peterson, "Evaluating a Map's Image," *American Cartographer* 12(1985):41–55; see also Terry A. Slocum, "A Cluster Analysis Model for Predicting Visual Clusters," *The Cartographic Journal* 21(1984):103–11.
12. Peterson, "Evaluating a Map's Image," pp. 41–55.
13. Paul V. Crawford, "The Perception of Graduated Squares as Cartographic Symbols," *Cartographic Journal* 10 (1973): 85–88.
14. Cox, "Anchor Effects," pp. 65–74.
15. *Ibid.*
16. George F. Jenks, "Contemporary Statistical Maps—Evidence of Spatial and Graphic Ignorance," *American Cartographer* 3(1976):11–19.
17. Hans Joachim Meihoefer, "The Utility of the Circle as an Effective Cartographic Symbol," *Canadian Cartographer* 6 (1969): 105–17.
18. Michael W. Dobson, "Refining Legend Values for Proportional Circle Maps," *Canadian Cartographer* 11 (1974): 45–53.
19. United States Geological Survey, *The National Atlas of the United States of America* (Washington, D.C.: USGPO, 1970), pp. 223, 240.
20. Paul V. Crawford, "Perception of Gray-Tone Symbols," *Annals* (Association of American Geographers) 61 (1971): 721–35.
21. Richard E. Lindenberg, "The Effect of Color on Quantitative Map Symbol Estimation," unpublished Ph.D. dissertation, Department of Geography, University of Kansas, 1986, p. 121.
22. Richard E. Groop and Daniel Cole, "Overlapping Graduated Circles: Magnitude Estimation and Method of Portrayal," *Canadian Cartographer* 15 (1978): 114–22.

Glossary

absolute scaling directly proportional scaling of area or volume of symbols to data values

adaptation level in psychological theory, a neutral reference point on which perceived judgments are based; affected by the perceptual anchors from which judgments are made

anchor effects the size of the visual anchor affects the estimate of magnitude of an unknown symbol; the contrast effect causes estimation to be away from the anchor, and the assimilation effect causes judgments to be toward the anchor

apparent-magnitude scaling scaling of proportional symbols that incorporates correction factors to compensate for the normal underestimation of a symbol's area or volume

cube root scaling when three-dimensional symbols are used, they are scaled to the cube root of the data values

graduated circle a popular form of proportional point symbol; compact and easy to construct

proportional point symbol mapping type of quantitative thematic map in which point data are represented by a symbol whose size varies with the data values; areal data assumed to be aggregated at points may also be represented by proportional point symbols

psychophysical power law the mathematical expression that describes the relationship between stimulus and response; general form is

$$R = K \bullet S^n$$

where R is response, S is stimulus, n is exponent describing the relationship between R and S, and K is a constant

range-grading a symbol represents a range of data values; differently sized symbols are chosen for differentiability for each of several ranges in the series of values being mapped

square root scaling when two-dimensional symbols are used, they are scaled to the square root of the data values

symbol overload too much information in a symbol, so that readers have difficulty making assessments about the quantitative nature of the data

three-dimensional point symbol any point symbol made to appear three-dimensional, e.g., a sphere or cube

Readings for Further Understanding

Alexander, John W. *Economic Geography.* Englewood Cliffs, NJ: Prentice-Hall, 1963. Students are encouraged to examine this book for its many good examples of proportional symbol maps, especially those on pp. 300, 303, 335, and 405.

Carstensen, Lawrence W. "A Comparison of Single Mathematical Approaches to the Placement of Spot Symbols." *Cartographica* 24(1987):46–63.

Chang, Kang-tsung. "Circle Size Judgement and Map Design." *American Cartographer* 7 (1980): 155–62.

———. "Visual Estimation of Graduated Circles." *Canadian Cartographer* 14 (1977): 130–38.

Clarke, John I. "Statistical Map Reading." *Geography* 44 (1959): 96–104.

Cox, Carleton W. "Anchor Effects and the Estimation of Graduated Circles and Squares." *American Cartographer* 3 (1976): 65–74.

Crawford, Paul V. "Perception of Graduated Squares as Cartographic Symbols." *Cartographic Journal* 10 (1973): 85–88.

Dobson, Michael W. "Benchmarking the Perceptual Mechanism for Map-Reading Tasks." *Cartographica* 17 (1980): 88–100.

———. "Refining Legend Values for Proportional Circle Maps." *Canadian Cartographer* 11 (1974): 45–53.

Dodd, Donald B. *Historical Atlas of Alabama.* University, AL: University of Alabama Press, 1974.

Ekman, G., and K. Junge. "Psychophysical Relations in the Perception of Length, Area, and Volume." *Scandinavian Journal of Psychology* 2 (1961): 1–10.

Ekman G.; R. Lindman, and W. William-Olson. "A Psychophysical Study of Cartographic Symbols." *Perceptual and Motor Skills* 13 (1961): 335–68.

Flannery, James J. "The Effectiveness of Some Common Graduated Point Symbols in the Presentation of Quantitative Data." *Canadian Cartographer* 8 (1971): 96–109.

———. "The Graduated Circle: A Description, Analysis, and Evaluation of a Quantitative Map Symbol." Unpublished Ph.D. Dissertation, Department of Geography, University of Wisconsin, 1956.

Gilmartin, Patricia P. "Influences of Map Context on Circle Perception." *Annals* (Association of American Geographers) 71 (1981): 253–58.

Griffin, T. L. C. "Group and Individual Variation in Judgment and Their Relevance to the Scaling of Graduated Circles." *Cartographica* 22 (1985): 21–37.

Groop, Richard E., and Daniel Cole. "Overlapping Graduated Circles: Magnitude Estimation and Method of Portrayal." *Canadian Cartographer* 15 (1978): 114–22.

Guelke, Leonard. "Perception, Meaning and Cartographic Design." *Canadian Cartographer* 16 (1979): 61–69.

Jenks, George F. "The Evaluation and Prediction of Visual Clustering in Maps Symbolized with Proportional Circles." In *Display and Analysis of Spatial Data.* J. C. Davis and M. J. McCullogh, eds. New York: John Wiley, 1975, pp. 311–27.

———. "Contemporary Statistical Maps—Evidence of Spatial and Graphic Ignorance." *American Cartographer* 3 (1976): 11–19.

Keates, J. S. *Understanding Maps.* New York: Halsted Press, 1982.

Lindenberg, Richard E. "The Effect of Color on Quantitative Map Symbol Estimation." Unpublished Ph.D. Dissertation, Department of Geography, University of Kansas, 1986.

Meihoefer, Hans Joachim. "The Utility of the Circle as an Effective Cartographic Symbol." *Canadian Cartographer* 6 (1969): 105–17.

———. "The Visual Perception of the Circle in Thematic Maps—Experimental Results." *Canadian Cartographer* 10 (1973): 68–84.

Peterson, Michael P. "Evaluating a Map's Image." *American Cartographer* 12 (1985): 41–55.

Robinson, Arthur H. "The 1837 Maps of Henry Drury Harness." *Geographical Journal* 121 (1955): 440–50.

———. "The Thematic Maps of Charles Joseph Minard." *Imago Mundi* 21 (1967): 95–108.

———. *Early Thematic Mapping in the History of Cartography.* Chicago: The University of Chicago Press, 1982.

Robinson, Arthur H.; Randall Sale; and Joel Morrison. *Elements of Cartography,* 4th ed. New York: John Wiley, 1978.

Slocum, Terry A. "A Cluster Analysis Model for Predicting Visual Clusters." *The Cartographic Journal* 21 (1984): 103–11.

Tufte, Edward R. *The Visual Display of Quantitative Information.* Cheshire, CT: Graphics Press, 1983.

United States Geological Survey. The *National Atlas of the United States of America.* Washington, D.C.: USGPO, 1970.

Chapter

9

Capturing Geographical Volumes: The Isarithmic Map

Chapter Preview

Isarithmic mapping involves mapping a real or conceptual three-dimensional geographical volume with quantitative line symbols. This form of quantitative thematic map dates back to the mid-sixteenth century, when isobaths were first charted. Today, two forms are recognized: isometric *and* isoplethic. *Each involves the planimetric mapping of the traces of the intersections of horizontal planes with the three-dimensional surface. The isarithm is placed by "threading" it through a series of control points at which magnitudes are assumed to exist. Control points are locations where measurements are taken or places chosen to represent unit areas. All isarithmic maps contain error, as do other quantitative thematic maps, but the designer can learn to recognize potential sources of error and reduce their effects on the overall map. In the total map design, isarithmic lines should be placed at the top of the visual/intellectual hierarchy and made to appear as figures in perception. Legends should be clear and unambiguous and should specify isoline units. There are no particular manual production problems; machine-produced isometric and isoplethic maps have become quite common.*

Six distinct kinds of quantitative thematic maps are treated in this book: choropleth maps, dot maps, proportional symbol maps, isarithmic maps, value-by-area cartograms, and flow maps. Of these, the isarithmic map may be the most difficult conceptually. Isarithmic mapping requires fresh thinking. This form of mapping has enjoyed greater popularity than it now enjoys, but it is still widely used in both professional and popular publications. Isarithmic mapping is an important part of the cartographer's repertoire and should therefore be mastered. It has particular potential in the growing realm of automated mapping.

The Nature of Isarithmic Mapping

The basic concepts, diversity, and history of isarithmic mapping are first introduced, so that the full range of the activity and its product, the isarithmic map, will be better understood.

Fundamental Concepts

An **isarithmic map** is a planimetric graphic representation of a three-dimensional volume. (See Figure 9.1.) The graphic image or map that results from **isoline mapping** is a system of *quantitative line symbols* that attempt to portray the undulating surface of the three-dimensional volume. Regardless of the items being mapped or any complexities associated with the map content, a third dimension must exist or be assumed to exist if a mapping technique is to be called isarithmic.[1] Whether produced by hand or by machine, the map attempts to present a three-dimensional model of reality.

The model may be a scaled-down version of a real three-dimensional volume, or it may be an abstract mental construct representing some varying geographical distribution. For example, the lithosphere is a real geographical volume whose top forms what is called the *topographic surface.* It is possible to make a scaled-down model of the volume, or a portion

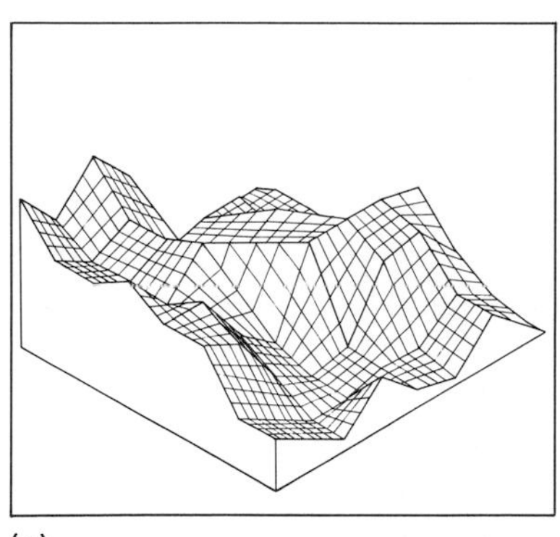

(a)

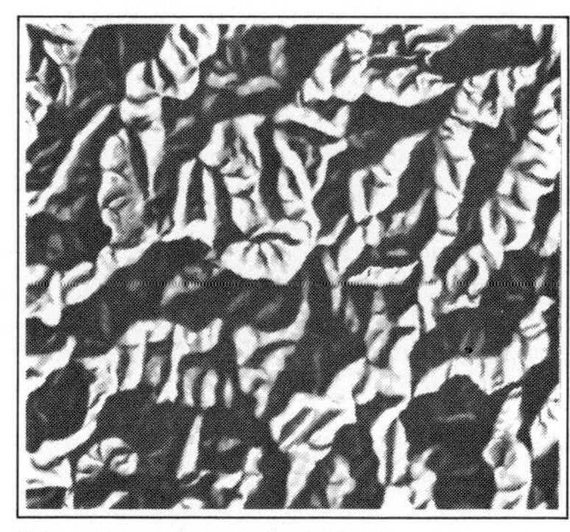

(b)

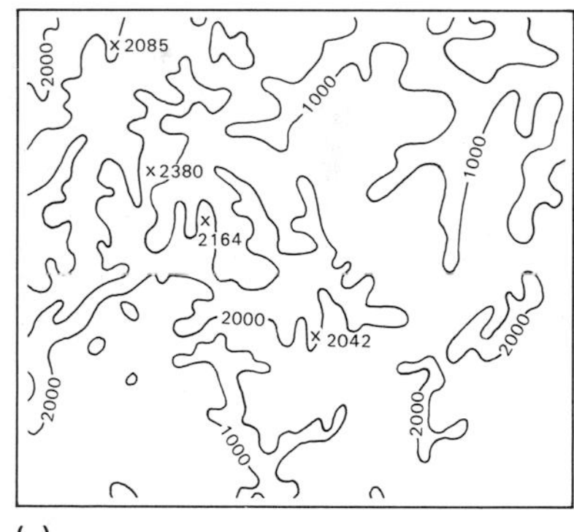

(c)

Figure 9.1. Isarithmic map.
The isarithmic map is a planimetric representation of a three-dimensional volume. The model (a) is voluminous and contains a continuous surface. One way to represent the model is by vertical shaded relief (b), although this method lacks commensurability. The isarithmic map (c) is planimetric, shows the form of the model's surface, and is also commensurable.

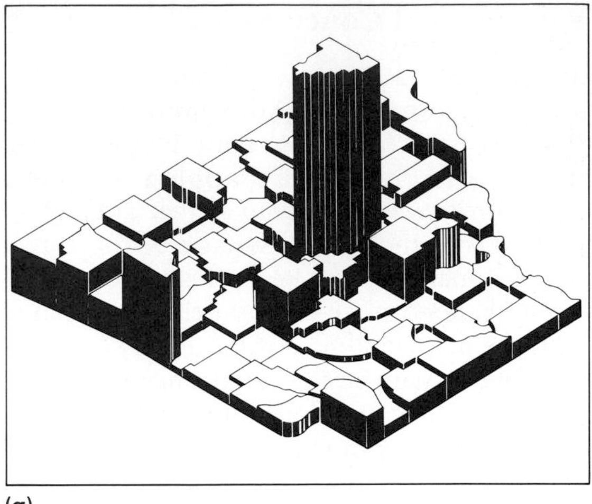
(a)

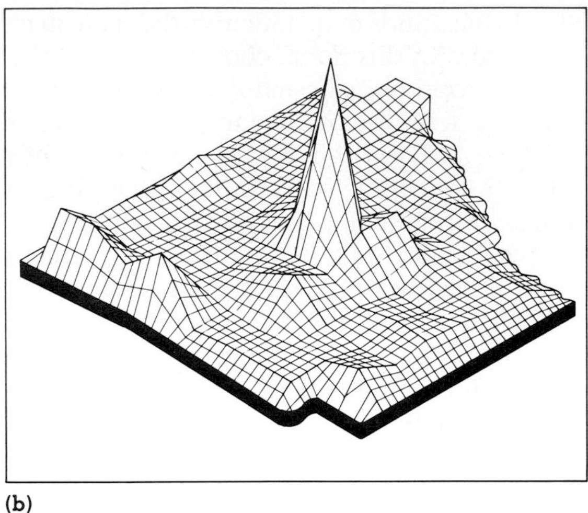
(b)

Figure 9.2. Isarithmic mapping.
Isarithmic mapping involves the representation of a continuous surface. If the phenomenon being mapped is a stepped surface, as in (a), it should not be mapped by the isarithmic technique. A continuous surface is appropriate, as in (b).

of it. An attempt to map the surface, using quantitative line symbols, results in an isarithmic map. On the other hand, the mind can form a construct of a geographic quantity that has volume—say, population density, which is high in some areas and low in others. This abstract mental picture can also be mapped isarithmically. Either real or abstract models can provide the content for isarithmic maps, but the third dimension must be present or assumed.

Another requirement of the isarithmic technique is that the volume's surface be continuous in nature, rather than stepped.[2] (See Figure 9.2.) If the mapped volume's surface contains discrete magnitudes (or can be thought to have discreteness), isarithmic mapping is inappropriate. On the other hand, if the surface of the volume has an underlying continuous nature, representing something that exists everywhere in some amount, the volume and its surface can be treated isarithmically. Geographical phenomena such as the locations of factories are discrete—values do not occur between points. Temperature, however, exists everywhere, both at and between observation points. Some geographical phenomena, such as population density, can be assumed to exist everywhere and thus can be mapped isarithmically.

It may be useful to compare the common frequency histogram and the frequency polygon. Chapter 5 presented the histogram, which graphically portrays the occurrence of statistical data using bars with horizontal tops. An extension of the histogram is the **frequency polygon,** which is used to represent statistical data that are continuous in nature. It is developed by joining the centers of each bar of the histogram with straight lines, forming a polygon. The model represented by the isarithmic map, whether real or abstract, can be likened to a three-dimensional frequency polygon.[3] (See Figure 9.3.)

Isarithmic Forms and Terminology

Isarithmic maps can have two distinct forms: the **isometric map** and the **isoplethic map.** The construction of each is achieved in similar

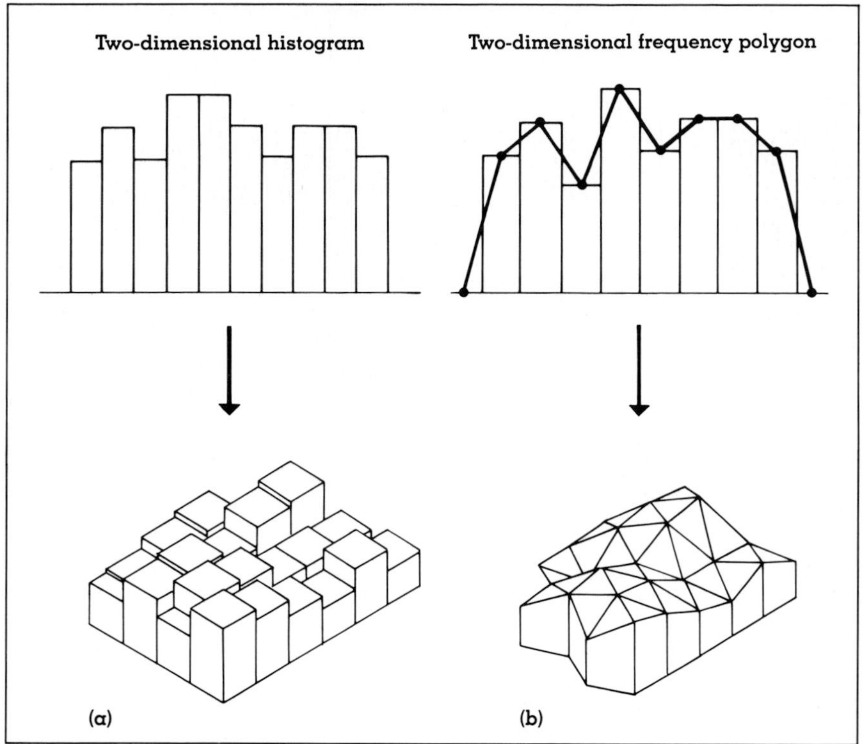

Figure 9.3. Surfaces.
The normal two-dimensional frequency diagram or histogram in (a) produces a three-dimensional stepped surface. Isarithmic mapping can be more appropriately compared to a three-dimensional continuous surface that is developed from a two-dimensional frequency polygon, as in (b).

fashion, but the nature of the data from which they are generated is quite distinct. Isometric maps are generated from data that occur or can be considered to occur at points; isoplethic maps result from mapping data that occur over geographical areas.

A distinct organization of data types and isarithmic forms has developed.[4] (See Figure 9.4.) Data that can occur at points, to be mapped isometrically, can be divided into *actual* and *derived* values. Actual data values include such things as temperature, precipitation, and elevation. Such values are generally obtained by recording instruments or by other means of point sampling in the field. Derived values are subdivided into two groups. One group includes such statistical measures as means and measures of dispersion; the other deals with such magnitudes as proportions and ratios.

Isopleth maps are generated from data that occur over geographical areas called **unit areas.** The values that can be represented include ratios that directly or indirectly involve area, for example, population density or crop yield per acre. (See Table 9.1.) Professional cartographers do not illustrate absolute values isoplethically. If data are being used that represent areas, they must first be converted into ratios or proportions that involve the areal magnitude. For example, if population totals by census tract are being mapped, these should be changed into density values before mapping.

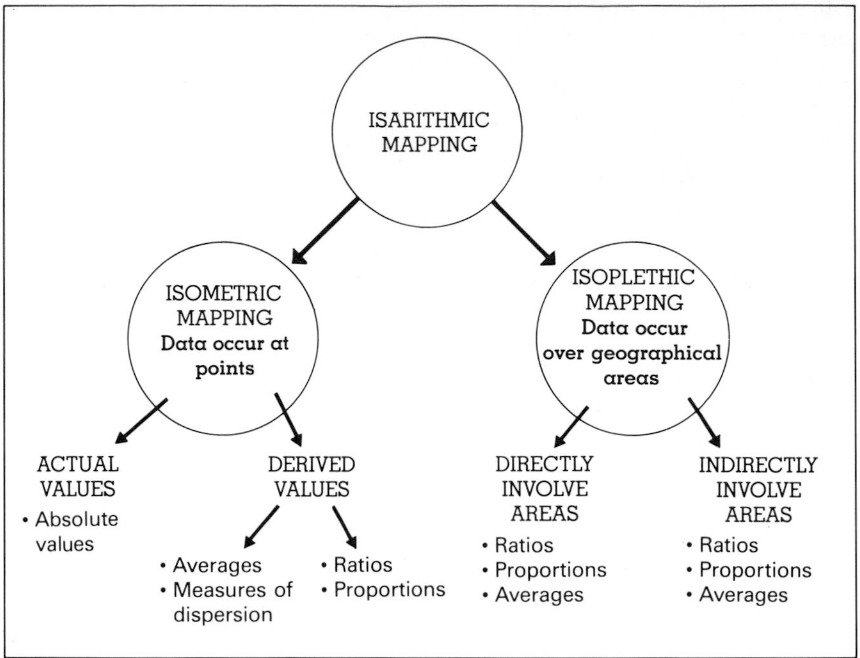

Figure 9.4. Data types and isarithmic form.
Notice that absolute values are never mapped isoplethically.

Isopleth mapping is more difficult conceptually than isometric mapping, especially for the beginner. It is not easy to conceive of an areal magnitude existing along a line. How can people per square mile exist in such a fashion? It is perhaps best not to think of these values in this way, but rather to envisage the line as a *surface element* below which is forward the volume of population density. The following discussion of construction methods may also assist in understanding the conceptual basis behind this form of mapping.

The Basis of Isarithmic Construction

The mechanical methods of isarithmic construction are rather straightforward but require a grasp of the basis behind the technique. Isarithmic mapping depicts the surface of a volume by quantitative line symbols. In both isometric and isoplethic mapping, construction of the lines begins by imagining a series of pins erected at the **data points** so that they extend vertically in proportion to the magnitude they represent at each point. (See Figure 9.5.) This is the vertical scale. The *tops* of all the pins will form the surface of the new volume. The line symbols are the planimetric traces of the intersections of hypothetical planes with this three-dimensional undulating surface. The lines represent the surface, not the volume.

The vertical positions of the hypothetical planes are selected relative to a **datum** adopted for the particular map. The range of data values represented at the points dictates the value of the datum (usually zero). Each plane has an assumed value associated with it depending on its placement relative to the vertical scale (the same scale used when erecting the pins proportional to the magnitudes they represented).

Table 9.1 Sample Ratios Useful in Isopleth Mapping

Type	Dividend and Divisor	Examples of Ratios
Percentages, proportions	Same units, or with dividend a portion of the divisor	$\frac{\text{20 acres wheat}}{\text{200 acres cultivated land}}$ = 10% of cultivated land in wheat
Density	Different units, with the divisor the total area of the statistical division	$\frac{\text{3500 people}}{\text{100 square miles}}$ = population density of 35 people per square mile
General ratios	Different units	$\frac{\text{3500 people}}{\text{70 square miles of cultivated land}}$ = population density of 50 people per square mile of cultivated land
		$\frac{\text{14,000 acres in farms}}{\text{100 farms}}$ = average size of farm of 140 acres
		$\frac{\text{260 bushels of oats}}{\text{10 acres in oats}}$ = yield of 26 bushels of oats per acre

Source: Adapted from J. Ross MacKay, "Some Problems and Techniques in Isopleth Mapping," *Economic Geography* 27 (1951): 1–9.

The magnitude or value of the isarithmic lines represents their vertical distance from the datum. Because the planes are constructed parallel to the datum, each isarithm will maintain an *unchanging magnitude* or vertical distance from the datum. Actually, these isarithms are traces of locations on the surface that are equally distant from the datum.

The total effect of these traces or isarithms on the surface is to show the varying amounts of the volume beneath the surface. The interpretation of the **pattern of isarithms** is the critical element in reading these maps. Pattern elements are *magnitude, spacing,* and *orientation.*

A Brief History of Isarithmic Mapping

Isarithmic mapping in each of its two forms has had a fairly long history, at least in relation to thematic mapping in general. **Isobaths** (isometric lines showing depth of the ocean floor) were first used as far back as 1584.[5] This early use was no doubt the result of the urgent need for reliable map information for commercial and military navigation.

In 1777, the **isohypse** line was proposed by Meusnier as a way of depicting surface features. An actual map using the isohypse was made by du Carla-Dupain-Triel in 1782. The isohypse is an isometric line; isopleths were not used until somewhat later.

Isogones, lines showing equal magnetic declination, were first used around 1630 by Borri, an Italian Jesuit.[6] They were used on a thematic map by Edmond Halley in 1701. The renowned naturalist-scientist-geographer Alexander von Humboldt mapped equal temperatures using isometric lines called **isotherms.**

Professor Arthur Robinson, in tracing the genealogy of the isopleth, places its inception

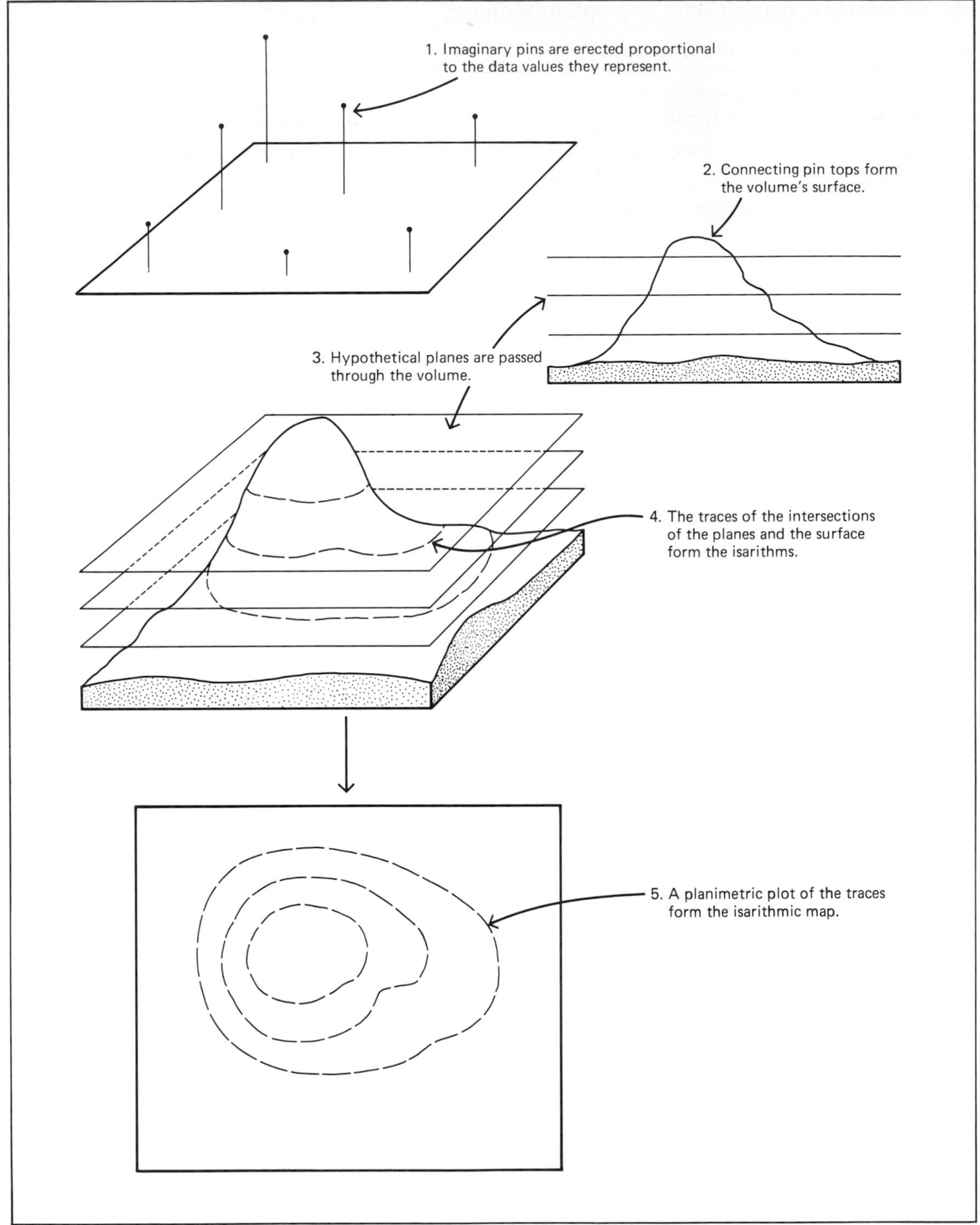

Figure 9.5. The conceptual development of an isarithmic map.

with Léon Lalanne, a Frenchman, in 1845.[7] Although based on the use of isometric lines, such as isobaths, isohypses, and isotherms, this marked a clear departure because point data representing area were part of the method. Lalanne's description of the method is still accurate today.

From these beginnings, the isometric and isoplethic techniques gained wide usage. Isopleth mapping came later and is clearly based on isometric examples. Isopleth mapping was adopted by American professional geographers late in the nineteenth century and earned a place of prominence during the early decades of this century.[8] Its use peaked just prior to World War II. During these early decades, the isopleth technique was most used in agricultural mapping, showing intensity of crop production and rural population density.[9] The number of differently named isolines has reached sizable proportions. (See Table 9.2.)

Isoplethic mapping has declined among professional geographers during the last few decades. Not only has it been used less and less, but scholarly research concerning the isopleth has also declined. Several reasons for this decline have been given, the most notable of which is uncertainty of its scientific nature, in

Table 9.2 List of Isoline Names

Isobath	depth below a datum (e.g., mean sea level)
Isogonic line	magnetic declination
Isocline	magnetic dip (inclination) or angle of slope
Isohypse (contour)	elevation above a datum (e.g., mean sea level)
Isodynamic line	intensity of the magnetic field
Isotherm	temperature (usually average)
Isobar	atmospheric pressure (usually average)
Isohyet	precipitation
Isobront	occurrence of thunderstorms
Isanther	time of flowering of plants
Isoceph	cranial indices
Isochalaz	frequency of hail storms
Isogene	density of a genus
Isospecie	density of a species
Isodyn	economic attraction
Isohydrodynam	potential water power
Isostalak	intensity of plankton precipitation
Isovapor	vapor content in the air
Isodynam	traffic tension
Isophot	intensity of light on a surface
Isoneph	degree of cloudiness
Isochrone	travel time from a given point
Isophene	date of beginning of a plant species entering a certain phenological phase
Isopectic	time of ice formation
Isotac	time of thawing
Isobase	vertical earth movement
Isohemeric line	minimum time of (freight) transportation
Isohel	average duration of sunshine in a specified time
Isodopane	cost of travel time

Source: These names were selected from a larger list found in Norman J. W. Thrower, *Maps and Man* (New York: Prentice-Hall, 1972), Appendix B.

the context of today's greater precision in mapping.[10] However, any thematic map is a form of generalization, so the isoplethic method is surely as conceptually correct as others.

The use of isometric maps has not declined appreciably. The conceptual basis on which they are formed has not received as much criticism as in the case of isopleths. The isometric map (and to a lesser degree the isoplethic map) was one of the first kinds to be done by automated methods.

When to Select the Isarithmic Method

Isarithmic mapping should be selected only if the advantages of its use contribute to achieving the goals of the mapping task. Certain additional requirements must be met before adopting this method:

1. The mapped data must be in the form of a geographical volume, or must be assumed to be voluminous, and must have a surface that bounds the volume.
2. It must be feasible to consider the mapped phenomena continuous in nature; discrete phenomena cannot be mapped isarithmically.
3. The cartographer must fully understand the distribution being mapped. A casual acquaintance with the phenomena will not suffice to develop a sensitive and accurate isarithmic solution to the mapped data.Extensive background research is generally required.

Various advantages to the isarithmic technique must be weighed in the selection process:

1. Isarithmic mapping shows the *total* form of a spatially varying phenomenon.
2. The method is commensurable (although less so at small scales) and graphic at the same time.
3. It is flexible and can easily be adapted to a variety of levels of generalization or degrees of precision.
4. The technique is easily rendered by automated methods.

The cartographic designer chooses the isarithmic technique on the basis of these advantages, as weighed against those of other methods. Of course, such matters as data availability, base map availability, and scale influence the ultimate selection. Beyond these considerations, other constraints are imposed by the method. Erwin Raisz, a noted cartographer and strong influence in the discipline during his life, once said, "Making true and expressive isopleth maps is something of an art and requires the best geographical knowledge of the region."[11] Another observer has expressed the same concern:

> The drawing of an isopleth map is not a simple and straightforward procedure, although it is often treated as such. It is frequently subject to greater individual variations in judgment and skill than is usual with isometric maps. This arises from several different factors inherent in each type of map. In drawing an isometric map, the cartographer does not choose the positions of the numbers which he is plotting and he is often guided in his placement of lines by a knowledge of the distribution which he is mapping. Thus the draftsman of a contour map is aided by his "topographic sense." In isopleth mapping, the cartographer deals with ratios for areas and not points, and he cannot always rely upon any "sense" to assist him in the placement of his lines.[12]

Isarithmic Practices

Like other forms of quantitative thematic mapping, the isarithmic variety contains elements and design strategies that are unique to the method. These are discussed in this section, along with error sources in isarithmic mapping.

Elements of Isarithmic Mapping

The cartographer must master all elements of the isarithmic process, since they directly in-

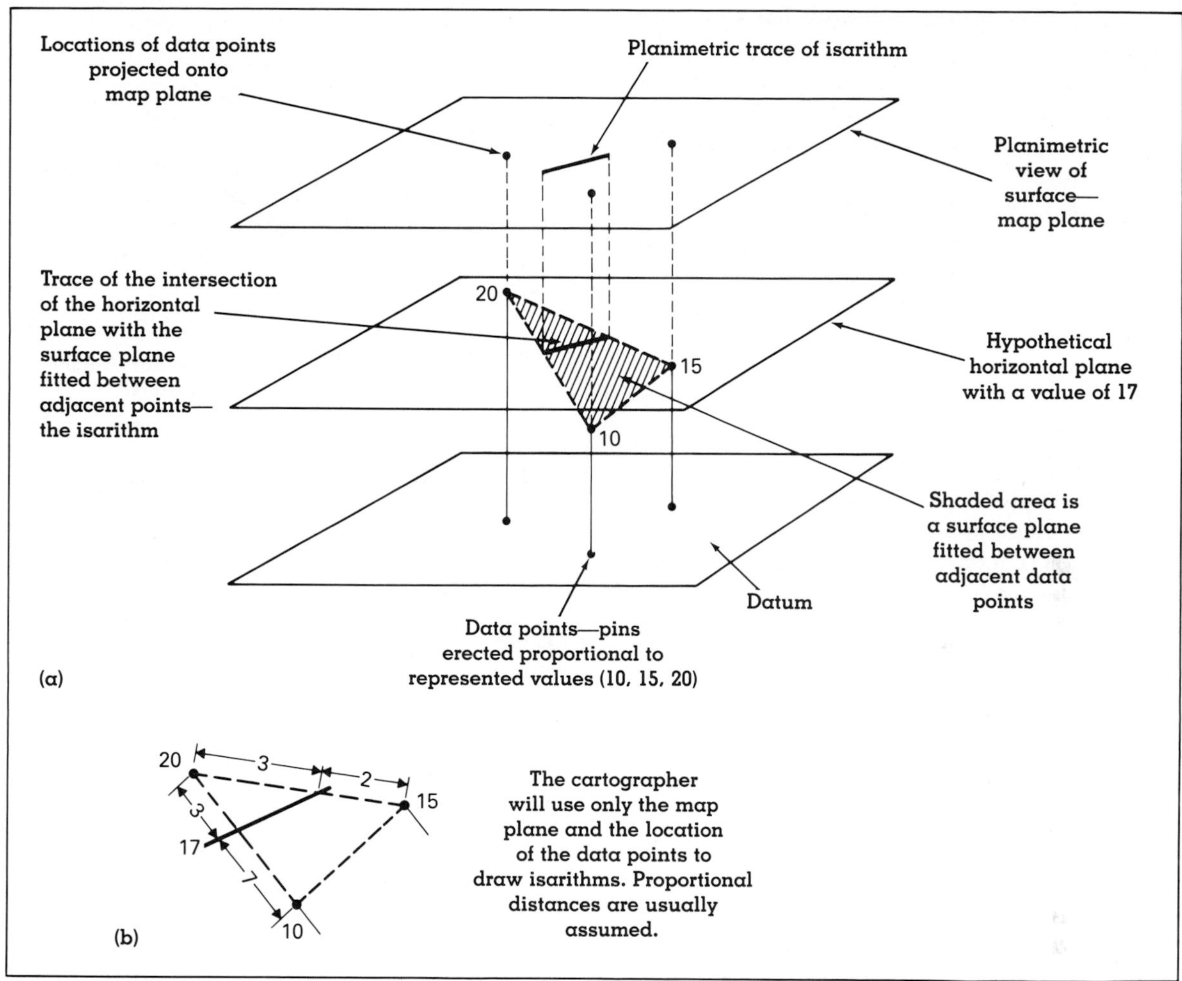

Figure 9.6. Isarithms.
Methods of projective geometry provide the basis for the planimetric location of isarithms.

fluence the quality of the finished map. The elements take on different significance in each mapping activity.

Placing the Isarithms—The General Case

In the construction of an isarithmic map, it is not usually necessary to erect pins or draw horizontal planes. Building a pin model is sometimes helpful in understanding a difficult, complex surface. The pin-and-plane model is, however, important conceptually. It should be thoroughly understood before beginning the planimetric placement of the isarithms with respect to the array of data points.

Methods of projective geometry form the basis of isarithm placement. (See Figure 9.6a.) The method assumes that some generalized surface plane is interposed between adjacent data points. Where a hypothetical horizontal plane intersects this new surface plane, a trace is formed. The orthogonal projection of this trace is made to the map plane, intersecting the surface proportionally between the data points.

In actuality the cartographer only deals with the map plane, the array of data points, and their values. (See Figure 9.6b.) The trace of the isarithm of a certain magnitude is ordinarily placed by assuming linear distances between the data points.

First, all values of the horizontal planes are chosen (see class interval selection, below). The cartographer then completes the rough map by "threading" the isolines through the array of data points. In the next step, these sharp, straight-line segments are *smoothed*—generalized—to give the appearance of a continuously varying, undulating surface. (See Figure 9.7.)

This procedure describes how virtually all isarithmic maps are made. The most notable exception is placement of the isoline between the data points—a matter of *interpolation,* addressed more fully below. Although the above procedure is generally followed, certain choices among the elements will affect the accuracy and appearance of the final map.

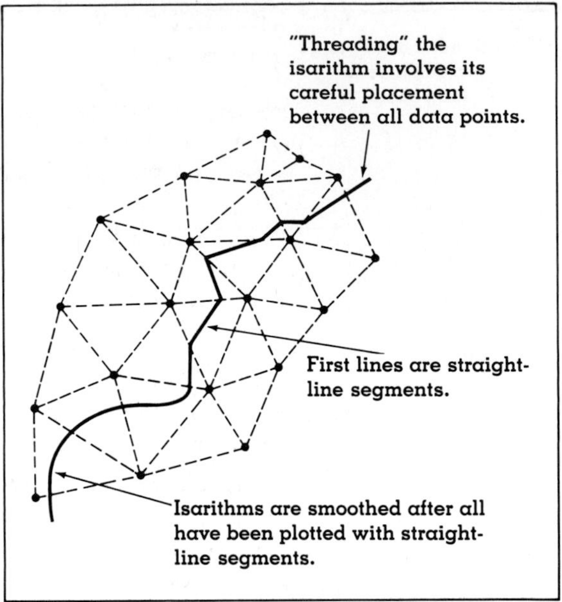

Figure 9.7. Plotting isarithms.
Isarithms are first drawn as straight-line segments and then smoothed prior to final rendering.

Locating Data Points

Data points, also called **control points,** have their locations specified by grid notation. Exact position can be determined either by an *x-y* rectangular coordinate reference or by geographic coordinates. Specification of location differs between isometric mapping and isoplethic mapping.

In isometric mapping, the positions of control points can usually be specified exactly, because the positions of recording instruments are well known. Problems usually relate to the uneven spacing of these points, since the lack of uniform spacing leads to varying precision in interpolation. In cases where the cartographer does have control over the pattern of spacing, the selection of a point pattern that yields a regular *triangular* net is most desirable.

Isopleth data point location is a more difficult procedure. A data point is usually selected for each enumeration district of the study and a value assigned which represents an average magnitude for each area. The magnitude is some ratio or proportion that directly or indirectly involves area. The location of the data point within the area is a matter of some concern, because it will affect the accuracy and appearance of the whole map.

The actual location of the distribution affects the accurate position of the data point. Since the data point is selected to represent the distribution in the entire unit area, its location should reflect the spatial attributes of the distribution. If the unit area is regularly shaped and the distribution within it is known to be uniformly and evenly spaced, the geographical center of the unit area may be chosen for the location of the data point. The exact measure-

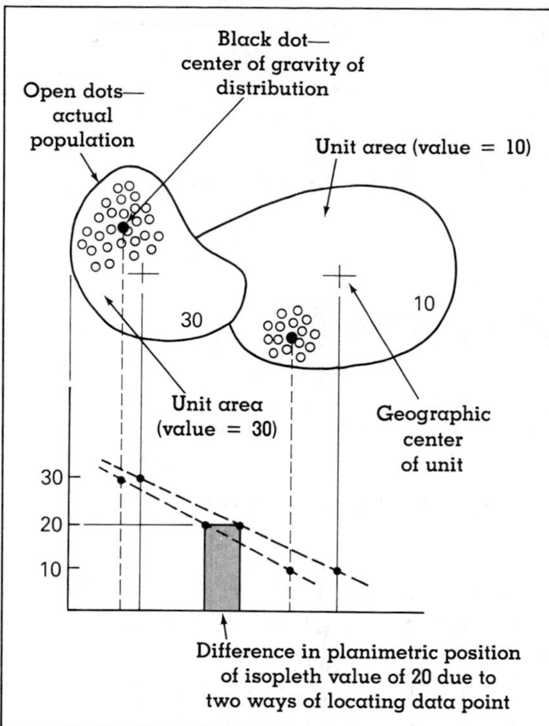

Figure 9.8. Data point placement affects the final maps.
Planimetric displacement may result from alternate ways of locating data points on isopleth maps. Over an entire map, the discrepancy can be considerable.

ment of the center is not required; a visual approximation will suffice.

If the distribution is spatially skewed and clustered within the unit area, the location of the data point should reflect this pattern. The control point should be placed in the center of gravity of the actual distribution. When the geographical phenomenon is spatially skewed, the two distinct data point location schemes can bring about considerable difference in the placement of an isoline. (See Figure 9.8.) When applied over an entire map, the discrepancy can be quite extensive; completely different maps will result. Most cartographic scholars agree that a data point's location should be adjusted in the direction of the distribution. This requires a thorough understanding of the distribution being mapped.

Unit Areas of Isopleth Mapping

Unit areas in most isopleth mapping are civil enumeration districts or political area units: states, counties, minor civil divisions, and census tracts. In any given study, the cartographer must deal with several questions regarding the size, number (for a specified scale, number is a function of size), and shape of the unit areas.

Size and number are inseparable, and there is a functional relationship between them and overall map precision. Whenever there is a large number of unit areas at a small mapping scale, the need for precision in locating isopleths drops off. Conversely, when a small number of units is used at a large mapping scale, the need for precision increases. Unfortunately, there are no quantitative guidelines as to how many isopleths are required for a given level of precision at various map scales. Intuition, understanding of the distribution, and the purpose of the map determine the decision.

When unit areas are of the enumeration variety, it is good practice to use only those that are similarly shaped and sized. (See Figure 9.9.) Irregularity of size in particular can lead to uneven precision over the map and should be avoided. Considerations of control point location make it inadvisable to use unit areas of highly differing shape. Especially for highly clustered distributions, the uneven spacing of control points can lead to a map containing varying levels of precision.

In rare cases, the cartographer can specify the shape and number of the unit areas prior

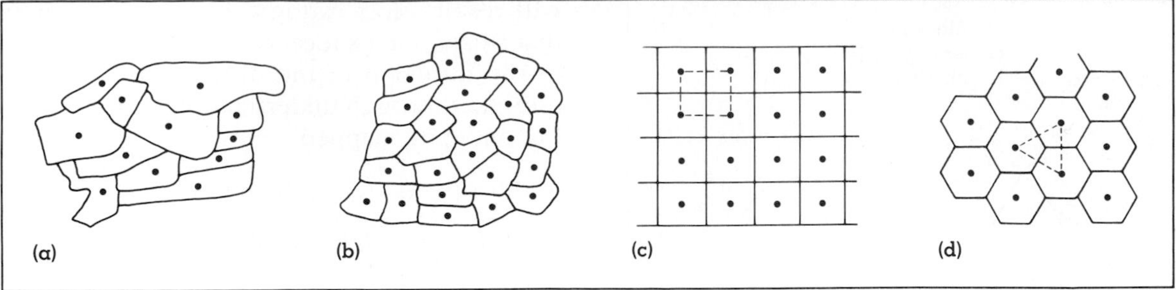

Figure 9.9. Different unit area patterns.
Irregularly sized and shaped enumeration areas used as unit areas, as in (a), are less desirable than uniformly sized and shaped ones, as in (b). When unit area shapes can be designated by the cartographer, square area, as in (c), should be avoided, since the pattern of control points leads to interpolation problems. A better solution is the hexagon-shaped unit area in (d), because the resultant data point pattern is triangular.

to mapping. This generally occurs only in experimental designs in research. The best overall solution in such cases is to adopt hexagonal pattern, for two reasons. First, this geometrical shape is space-filling. Second, connecting the centers of hexagons produces a triangular net, which is desirable in interpolation for isopleth location. (See Figure 9.9.)

Interpolation Methods

Probably the key activity in isarithmic mapping is **interpolation** for isoline placement—a procedure for the careful positioning of the isolines in relation to the values of the data points. A number of options exist; with the introduction of automated mapping, these options are proliferating rapidly. For the sake of organization, this important subject is divided into manual and automated methods.

Manual. Through projective geometry, the trace of the intersection of hypothetical plane with a surface segment results in a line that falls proportionally between the data points. This is true, but the case described in Figure 9.6 was a linear fit. In *linear interpolation* between data points on the map plane, the cartographer positions an isarithm with a certain value between adjacent control points at proportionate distances from each. (See Figure 9.10a.) The linear fit assumes that the gradient of change in magnitude between adjacent data points is even and regular. Equal change in the unit over equal horizontal distances produces a uniform gradient.

Adoption of this method assumes that the distribution being mapped changes in linear fashion, with a uniform gradient between data points. This may not be the case, however; many geographical phenomena do not behave this way. An example is population density in urban areas, which displays a more complicated pattern of change over distance.[13] Adopting a non-uniform gradient in preference to a linear one can result in considerable local displacement of the isoline. (See Figure 9.10b.) However, because of the underlying complexity of interpolating isoline position using a non-uniform gradient between all data points, and because the exact pattern of change from place to place is often unknown prior to mapping, most manual methods of interpolation assume uniform or linear change between data points.

This practice can be defended. Because linear interpolation is used throughout the whole map, the total form of the mapped distribution is not badly represented. Since most

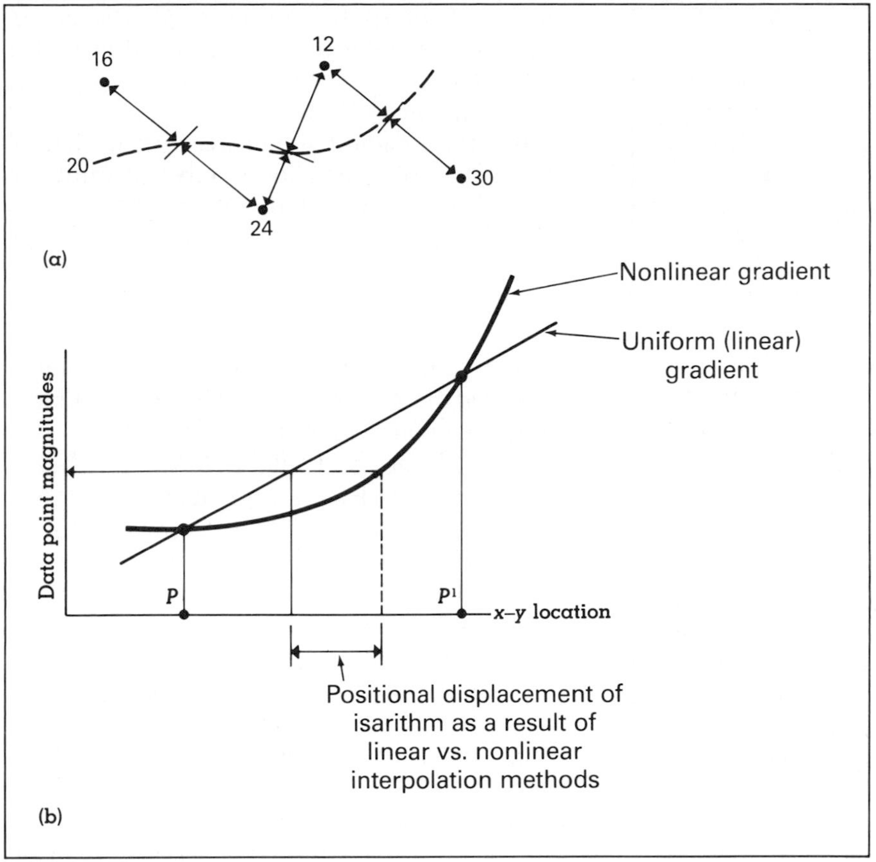

Figure 9.10. The interpolation model and the planimetric location of the isarithm. Linear interpolation is standard in isarithm placement, as in (a). Planimetric displacement can occur due to the gradient selected for interpolation between adjacent data points, as in (b).

gradients of change are made up of both convex and concave portions along the slope, the linear or uniform gradient is a kind of compromise. Any error that results from its use is likely to appear randomly throughout the map.[14] Error resulting from the linear approach can be decreased by intensifying the density of data points, thus reducing their spacing and making the linear fit more precise between adjacent data points.

Another interpolation problem deals with choice of alternatives. This often results where the data points are arranged in rectangular fashion and two opposite data values are higher and two values are lower than the isoline being interpolated.[15] (See Figure 9.11a.) The only reasonable solution is to average each pair, then average these as a new value for the center of the data points. Triangular patterns (often difficult to find when enumeration unit areas are used) never result in this troublesome choice of alternatives and should be selected whenever possible.

Automated. The introduction of computer mapping and digital plotting has made rapid and iterative calculations possible, and isarithmic mapping has become much easier. The conceptual basis remains the same, but automation makes it possible to vary interpolation schemes in accordance with the mapping task at hand.

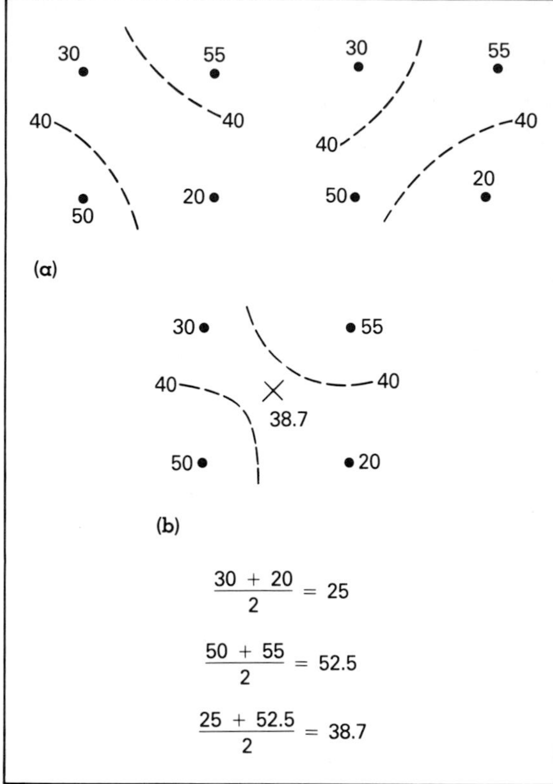

Figure 9.11. Choice of alternatives in interpolation.
When two opposite data points have higher values than two other opposite points, arranged in a square pattern, two interpolations are possible, as in (a). Averaging opposite pairs to determine an intermediate value, as in (b), is one solution to this dilemma.

Computer-generated isarithmic mapping usually proceeds in two steps. **Primary interpolation** involves the calculation of data point values (z-values) at all locations on a fine-mesh grid matrix for the mapped area. This interpolation stage may include any one of a number of mathematical models: inverse-distance weighted, planar, quadratic, or cubic.[16] The next or **secondary interpolation** stage positions the isarithms with respect to the grid matrix of interpolated points. This stage locates the lines by linear interpolation and is generally accurate to within the tolerances of the plotter's capabilities.[17]

In typical manual methods of interpolation, the cartographer considers only the values of adjacent data points to determine the position of an isoline or a point to be interpolated. Computer interpolation during the primary stage is likely to embrace several points at a time in this activity. The inverse-distance weighting model is popular; it is incorporated in one of the most widely used computer mapping programs, SYMAP.[18] The basis of this method is that points closer to the one being interpolated will have a greater effect on its value than points farther away. Any number of adjacent points may be used. A general formulation is[19]

$$Z_p = \frac{\sum_{i=1}^{n} \frac{Z_i}{d_i}}{\sum_{i=1}^{n} \frac{1}{d_i}}$$

where z is the height (value) of the ith point, d_i is the distance from the point to the point being interpolated (Z_p), and n is the number of points used in the interpolation. (See box "Computer-Generated Isarithmic Maps.")

The Selection of Isarithmic Intervals

It is impossible to map all isarithms because of space and time constraints. The cartographer is faced with the task of choosing the most appropriate ones and a suitable number of them. An **isarithmic interval** must be selected for the map.

The only reasonable and logical approach is to select a uniform interval. Isoline values of 2, 4, 6, 8, and 10 are appropriate, but 2, 5, 9, 14, and 20 are not. To put this another way, the hypothetical horizontal planes that pass through

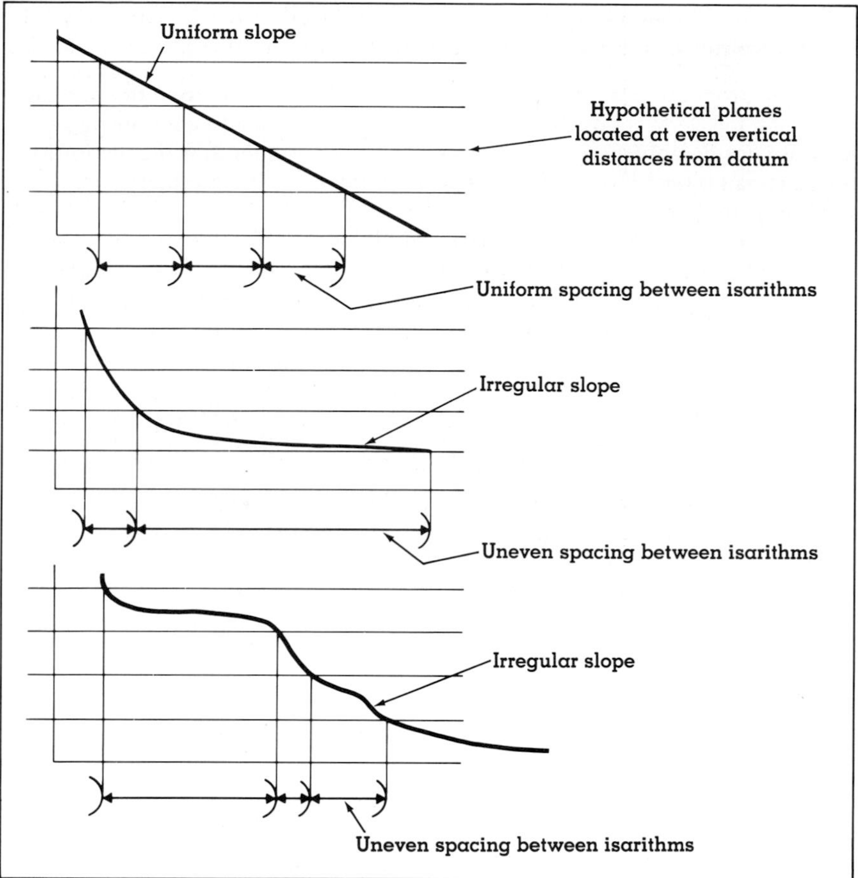

Figure 9.12. Isarithm spacing.
The spacing of planimetrically viewed isarithms determines the gradient of change of the surface. Isarithms spaced close together indicate rapid change (steep slope), and those spaced farther apart suggest a relatively gradual rate of change. Experience in reading isarithmic maps is required.

the three-dimensional surface should be *uniformly spaced vertically.* Only through this approach can the isolines show the total or integrated form of the three-dimensional surface. (See Figure 9.12.) This results in a particularly knotty problem at times, since most geographic distributions are skewed. In mapping population density, for example, extremely high values at urban centers lead to very crowded isopleths around cities. At best, the designer can choose map scale and isopleth interval carefully to avoid crowding; the result will be a compromise.

Another aspect of interval selection is the determination of the lowest isoline value. Because of the uniform interval, the lowest isoline greatly influences where subsequent isolines fall in relation to the whole map. Will the chosen interval show the distribution best? No universal rules apply here. As in so many

Computer-Generated Isarithmic Maps

Micro-computer driven isoline map production is making rapid advances and offers to lessen the burden of manual production of these map forms. It is possible with only modest cost to add contouring programs to the workstation, and this considerably increases the options for the designer. It is the flexibility contained in such programs that expands the view for the cartographer and is providing them an incentive to do more of this kind of geographic mapping.

The SURFER program (a trademark of Golden Software, Inc., of Golden, Colorado) produces isarithmic maps and three-dimensional plots from irregularly spaced data points (each containing *x*-, *y*-, and *z*-values). In the latest version of SURFER, as many as 10,000 data triplets may be imported. In the GRID portion of SURFER, a grid is superimposed over the array of data points and the *z*-values at the intersections of the grid are interpolated. The user can select either the inverse distance interpolation for the gridding method, or the Kriging (regional) solution, which takes more time. In the inverse distance option, any weighting power (the default is 2) greater than one and equal to or less than 10 may be chosen. Search area parameters may also be set. Gridding time is considerably reduced by having a math coprocessor installed in the hardware.

TOPO is the part of SURFER that produces contour maps and these lines may be either isoplethic or isometric, depending on the data used. The method of producing the isolines, of course, applies to either case. TOPO is interactive and the user may select such features as contour interval, contour limits, frequency of labeled lines, smoothed or unsmoothed lines, dashed lines, hachure marks, line label sizes, tics and labels at study area boundaries, and other variables. Original data points may be posted or not. Map size can be varied for screen viewing or for output device, and a title may be added.

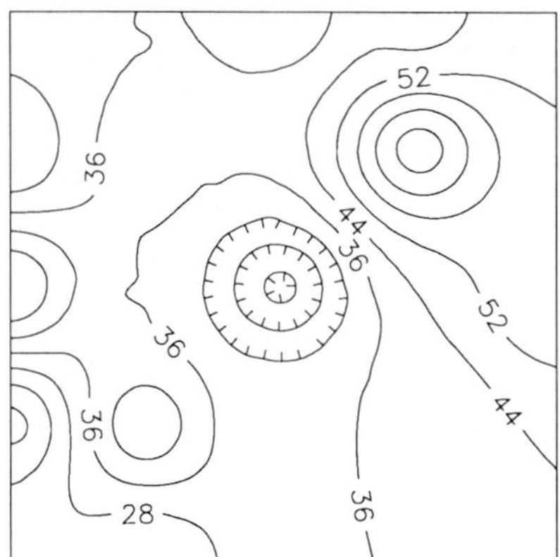

Gridded point values produced in GRID may also be used as input to SURF which develops three-dimensional plots of the gridded surface. The viewing azimuth (with respect to the original grid) as well as the viewing angle (angular height above the horizon) may be selected by the user. Stacked contours or fish net views are possible. Some of the menu options are perspective or orthographic projection, contour interval, vertical scaling and horizontal sizing, posted data points, top or bottom views of the surface (or both), and others. A title and directional legend may be added. A geographic boundary, or other line phenomena (such as streams and roads) may be added with a special file.

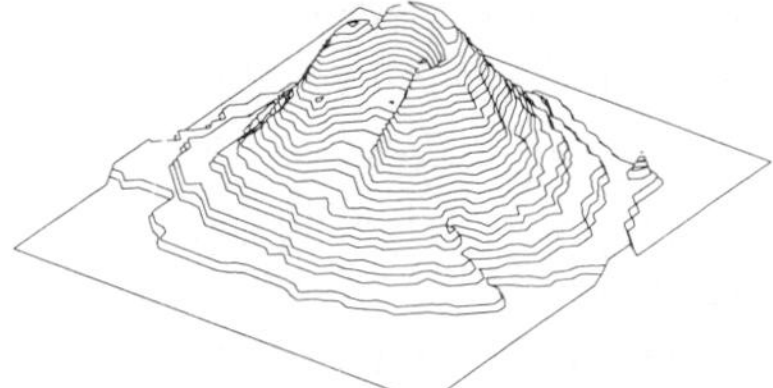

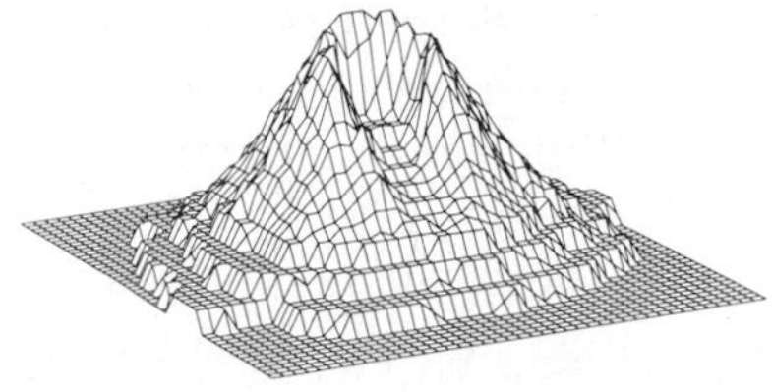

Plotted outputs from SURFER-TOPO are useful for creating isarithmic maps either as final or draft-quality versions. The cartographic designer, for example, might wish to use the TOPO plot simply as a fair-drawing and enhance it for final reproduction art. In this way, the laborious task of manual isoline interpolation is given over to the computer, which is efficient. The three-dimensional plots from SURF may be used to amplify isoline maps (e.g., to assist the map reader in understanding the form of the isarithmic surface), and especially, they are useful in teaching the principles of contour map reading.*

SURFER operates in an IBM environment that requires only 320K of RAM, a DOS of 2.0 or higher, and at least one disk drive. Output is to monitor, plotter, or printer. Drivers are supplied with the program. Laser printers are supported if they contain a HPGL (Hewlett-Packard graphics language) language interface. A math coprocessor, although not required, is highly desirable. Data input is by keyboard (the data editor supplied is easy to use), ASCII file, or from LOTUS files (a trademark of the Lotus Development Corporation). The sample maps used here are produced by a 300 dpi QMS-810 laserjet printer (QMS is a trademark of Quality MicroSystems, Inc.). Maps may be produced in panel sections, that when attached together, can yield a map up to 32 inches (80 cm) square.

Although cartographers now have rich resources such as SURFER to assist them in thematic map production, principles must be understood. Computer programs such as SURFER will not make decisions regarding the appropriateness of the method for the data at hand, or in helping the cartographer focus on the map purpose. These computer programs are not geographers. Map planning, data selection and appropriateness, and generalization decisions are aspects of design that remain with the cartographer, as well they should.

*See, for example, William R. Buckler, "Computer Generated 3-D Maps: Models for Learning Contour Map Reading," *Journal of Geography* 87 (1988): 49–58.

other areas of thematic cartography, experimentation, geographical knowledge, study, and experience are necessary.

A graphic procedure can assist in interval selection. The **graphic array** method mentioned in a previous chapter, plots the array of data point values in ascending order, along a horizontal axis to discover sharp changes in values in the data set. Isoline values are most revealing if they are selected to coincide with abrupt changes in slope on the graphic array. All places should be considered, and an interval chosen that best accommodates the places of rapid change and provides enough detail to show the form of the surface. An interval that provides too many isolines generally results in a map that looks "too accurate." A wide interval appears too generalized and reveals little. This is especially true in cases where a large number of isolines have been interpolated from a small number of control points.

In most cases, a wise choice of interval requires detailed knowledge of the distribution being mapped. The cartographic designer should therefore do ancillary studies, plot rough dot maps of the distribution, or perform other research to learn the attributes of the distribution so that the isarithmic interval produces an accurate picture. The final map represents only one of many possible solutions; it should therefore be based on all available information.

Sources of Error in Isarithmic Mapping

Every thematic map contains some sort of error. Error can come from a variety of sources, four of which have been documented: CRM, method-produced, production, and map reading and analysis.[20] **CRM errors** result from collection, recording, and the manipulation of machines or techniques, and range from faulty recording instruments to the use of inappropriate statistical measures. **Method-produced errors** result from the cartographic technique. In isarithmic mapping, anything from the assumption to the placement of the isolines can conceivably lead to errors. **Production errors** are those caused by the person or machine rendering the map: imprecise manual drafting, other graphic mistakes, incorrect registration, printing shortcomings, and the like. **Map reading and analysis errors** are generally caused by the psychological limitations of the map reader. These range from psychophysical aspects to cognitive considerations.

Isarithmic mapping is subject to all these sources of error. The types and sources of error differ somewhat between isometric and isoplethic mapping, although some error sources are mutual to both kinds. Of course, the designer wishes to reduce error as much as possible, but it is not feasible to eliminate all map error.

The chief CRM sources of error for isoplethic mapping relate to the quality of the data, which may be affected by observational bias or sampling errors. Observational error can be caused by either humans or machines. Faulty instruments or poor instrument-reading practices can lead to erroneous z-values. Bias or persistent errors may be the result of recording instruments that produce consistently high or low values. Sampling error is of considerable concern. Since the observations (z-values) used represent only one set from a much larger universe, the manner of selection (sampling) can become a source of error. In isoplethic mapping, the arrangement of the unit areas is usually not within the control of the cartographer; they are predetermined enumeration districts. The designer should therefore alert the map reader to the pattern of control points used. If the pattern of unit areas can be governed by the cartographer, one which yields a triangular pattern of control points is desirable because this will result in fewer interpolation errors.[21]

Method-produced errors for isarithmic maps, those caused by technique, differ between isoplethic maps and isometric maps. (See Table 9.3.) Sources of error can be found in the quantity of the data, the form of the data, and the

Table 9.3 Sources of Method-Produced Error in Isarithmic Mapping

Error Source	Map Type	
	Isometric	**Isoplethic**
Quantity of data	Number of data points	Number of data points
Form of data	Locational scatter of points	Size of unit area Shape of unit area Method of point assignment
Interpolation model	Mathematical formulation	Mathematical formulation

Source: Adapted from Joel L. Morrison, *Method-Produced Error in Isarithmic Mapping* (Technical Monograph No. CA-5—Washington, D.C.: American Congress on Surveying and Mapping, 1971), pp. 11–13.

interpolation model. In general terms, the quantity of data for both isometric and isoplethic maps is determined simply by the number of control points. Most cartographers agree that the greater the number, the less the potential for error. No definite number can be recommended for every mapping task, but 25 appears to be too few, and for most mappings a number between 49 and 100 is appropriate.[22]

No one mathematical interpolation model can be specified for all mapping cases. Certainly the model, whether linear or some other, will affect the number of errors. It is felt, at least for isoplethic mapping, that the linear interpolation model yields only *random* errors and therefore can be used with a certain degree of assurance.[23]

For isometric mapping, less error results from an areal sampling method that distributes control points in a stratified, systematic unaligned, scattered fashion, which is more uniform in spacing than random. In many instances, the distribution of data points is not controlled by the cartographer. The *x* and *y* positions are often determined by previously established networks of recording instruments.

The size and shape of the unit area and the method of assigning control points all affect the degree of accuracy of isopleth maps. "Map accuracy" may be a misleading term with such maps; there is no real measurable surface with which to compare the representation. Only general comments can be made about possible errors caused by the selection of the elements. If cartographers have control over their selection (and they most often do not), the preferable shape is hexagonal, for two reasons. First, a system of adjacent hexagons can completely cover a geographical area. Second, the centers of the hexagons, when connected by straight lines, form a triangular pattern, which results in fewer alternative choices during interpolation.

Most cartographers agree that the distribution should control the placement of control points. Less error is introduced if this plan is followed. The size of unit areas can affect accuracy and thereby reduce error, but specification of size is interconnected with map scale and the level of generalization planned for the map. At a given level, the greater the size of the unit area, the more potential for error.

The relationships of size, shape, and number of control points are critical to isopleth mapping and to potential error. At best, however, our knowledge is incomplete:

> At this stage of our knowledge, it is safe to conclude that the quality of isopleth mapping is greatly affected by the complexity of the original distribution, the variation in the pattern of unit areas (sizes and shapes), and the number of control points used. It is still not possible, however, to define precisely the manner in which these variables affect the fidelity of a map. For example, one cannot prepare a comprehensive summary which includes a complete classification of surfaces, unit area characteristics, and hexagonal patterns on the one hand and the probable measures of the quality of the isopleth maps on the other. The major problem seems to

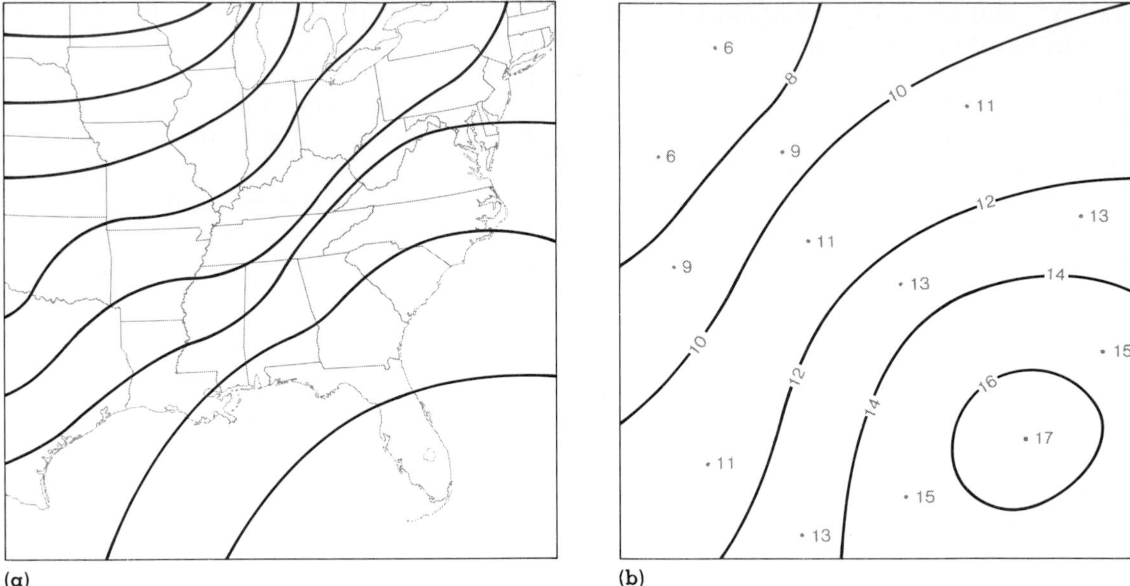

Figure 9.13. Isarithmic lines.
Isarithmic lines should be treated to make them appear dominant in map reading. In (a), the base material has been screened to reduce its contrast, thereby visually strengthening the isarithms printed in solid color. Data point locations and values may also be screened, as in (b), so that they will not interfere with the isarithms' function of showing total form.

be that the essential spatial characteristics of quantitative distributions and unit area patterns have not yet been numerically defined, and further understanding is needed concerning the effects of sample point location and generalization of areal data on the precision of the isopleth map.[24]

Production errors in isarithmic mapping are caused by either manual or computer means during the preparation of the final map. Errors caused by registration problems, misaligned media, or imprecise drafting techniques can all lead to inaccurate final renderings. In computer mapping, the establishment of specifications that exceed the machine's capabilities cause final production errors.

An isarithmic map that is perfect in every technical sense—has as little error as possible—is still subject to error caused by the inability of the reader to interpret it accurately. Isarithmic maps are perhaps the most difficult to understand of the ones discussed in this text, because the logic behind the method is the most abstract. Map reading and analysis error can be reduced by training and experience.

Preparing the Finished Isarithmic Map

Several design considerations related to the appearance of the completed isarithmic map must be learned. In addition, several production methods and media are especially useful in isarithmic map production.

Design Elements

At least three design elements are unusual in isarithmic maps: lines, labeling, and legend designs.

Making Isolines Appear as Figures

It is good design to make all isolines on the map appear dominant, as figures in perception. Thus they should be placed highest in the visual hierarchy for the map. The easiest way to accomplish this is to render the lines as solid color and screen much or all of the remaining

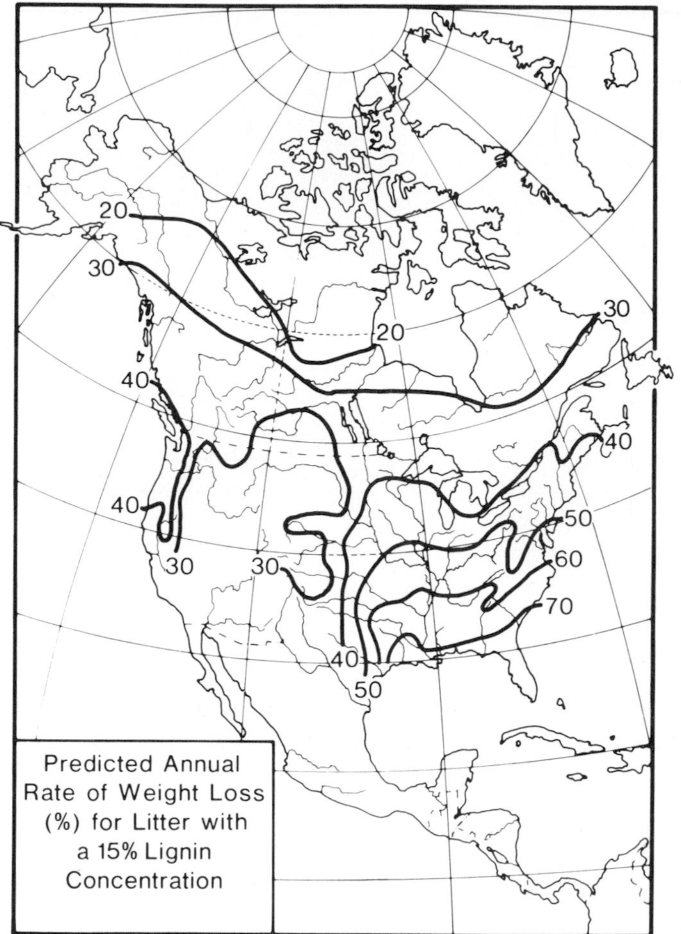

Figure 9.14. Isolines as figures.
Contrast of line weight is used to make isolines appear as figures on the map.
Reprinted by permission of the *Annals* (Association of American Geographers) 74 (1984): 557, fig. 3, V. Meentemeyer.

map information. (See Figure 9.13a.) In this way, the total form of the distribution becomes immediately recognizable, and the less important but necessary base map material does not interfere. In some circumstances, it may be necessary to provide control point locations, and even their values, on the map. The points and values may then be screened to reduce their contrast and thereby lessen their figural strength. (See Figure 9.13b.)

If screening is not a design option, the importance of all isolines may be heightened by making the isolines noticeably heavier than other line work on the map. (See Figure 9.14.) As in other forms of thematic mapping, contrast should be developed to emphasize the visual and intellectual hierarchy.

Isoline Labels

Isolines should be labeled periodically to make reading easy. Do not overlabel; this reduces the statistical surface shown by the lines. (See Figure 9.15a.) Labels should be easily readable, but not so large as to dominate the map. (See Figure 9.15b.) Beginning students often make an easily avoidable mistake: placing labels upside down. (See Figure 9.15c.)

Figure 9.15. Design alternatives for labeling isarithms.

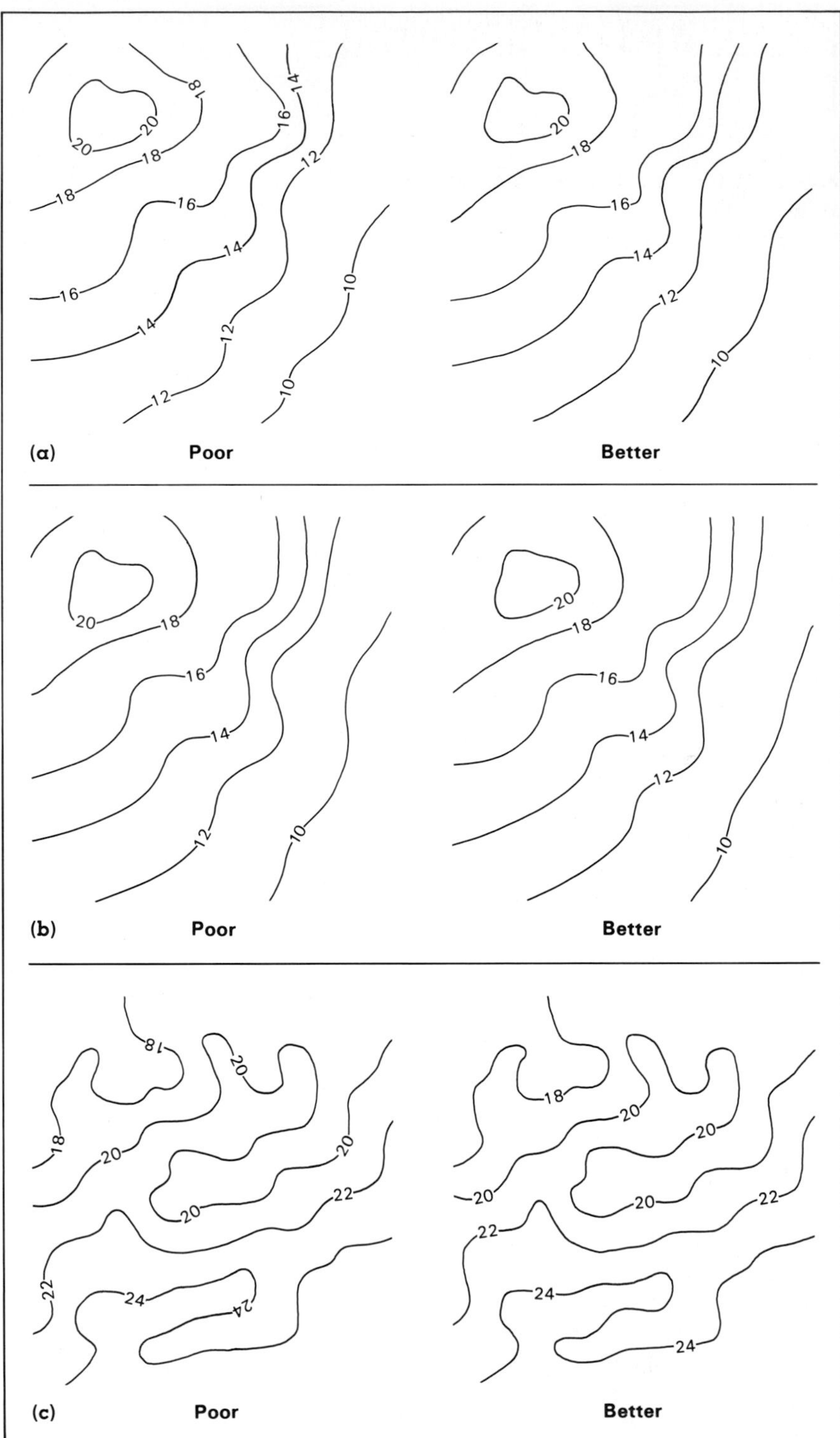

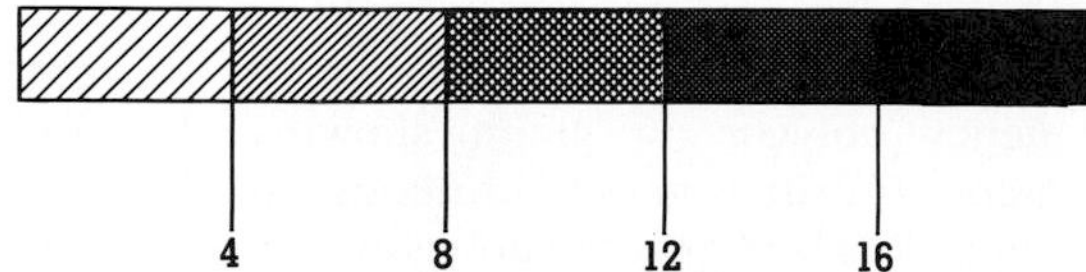

Figure 9.16. Legend design for isarithmic maps.
If areas between isolines are covered with a pattern or tint, these should be incorporated in the legend; the isarithmic interval also can be shown.

Legend Design

Legends should be designed for clarity. The legend material most often contains only verbal statements. These must describe at least 1) the units of the isolines (e.g., frost-free days, days of sunshine, average annual precipitation in inches, persons per square mile) and 2) the isoline interval. In the rare instances when shading is applied to areas between isolines, the tints, the isoline interval, and the representative values can be combined in one legend. (See Figure 9.16.) The practice of applying shading between isolines is not recommended because it gives the impression of a stepped surface, which is very misleading.

Production Methods

Production of isoline maps does not involve particularly difficult or unique circumstances. Manual rendering methods include inking, scribing, and the use of flexible border tapes. When inking, it is advisable to employ either rigid plastic (French) curves or flexible curves to reduce undesirable irregularities in the line. An alternative to inking in positive art preparation is the use of crepe border tape, which is available in a variety of widths, easy to apply, and effective. Care must be taken, however, that fuzzy edges are eliminated when using these tapes. Flexible-head engravers are used in scribing and generally produce high-quality lines.

Machine production of isolines is usually accomplished with pen plotters, which apply ink to paper, or with automated scribe plotters. The latter can use regular scribe blades to etch the scribe coat. New versions employ laser beams to do the same. Of course, special computer programs are necessary for the automated production of isoline maps.

Notes

1. Mei-Ling Hsu and Arthur H. Robinson, *The Fidelity of Isopleth Maps* (Minneapolis: University of Minnesota Press, 1970), p. 4.
2. *Ibid.;* and George F. Jenks, "Generalization in Statistical Mapping," *Annals* (Association of American Geographers) 53 (1963): 15–26. It may be pointed out that there are other ways of mapping continuous data although not in the line category. See for example, Richard E. Groop and Paul Smith, "A Dot Matrix Method of Portraying Continuous Statistical Surfaces," *American Cartographer* 9(1982): 123–30, and Stephen Lavin, "Mapping Continuous Geographical Distributions Using Dot-Density Shading," *American Cartographer* 13(1986): 140–50.
3. Calvin F. Schmid and Earle H. MacCannell, "Basic Problems, Techniques, and Theory of Isopleth Mapping," *Journal of the American Statistical Association* 50(1955): 220–39.
4. Arthur Robinson, Randall Sale, and Joel Morrison, *Elements of Cartography,* 4th ed. (New York: John Wiley, 1978), pp. 224–25.
5. The principal source for this discussion of the history of the isarithmic method is Arthur H. Robinson, "The Genealogy of the Isopleth," *Cartographic Journal 8* (1971): 49–53; see also Arthur H. Robinson, *Early Thematic Mapping in the History of Cartography* (Chicago: The University of Chicago Press, 1982).
6. *Ibid.*
7. *Ibid.*
8. Philip W. Porter, "Putting the Isopleth in its Place," *Proceedings of the Minnesota Academy of Science* 25–26 (1957–58): 372–84.

9. Wellington D. Jones, "Ratios and Isopleth Maps in Regional Investigation of Agricultural Land Occupance," *Annals* (Association of American Geographers) 20 (1930): 117–95.
10. Arthur H. Robinson, "The Cartographic Representation of the Statistical Surface," *International Yearbook of Cartography* 1 (1961): 53–61.
11. Erwin Raisz, *Principles of Cartography* (New York: McGraw-Hill, 1962), p. 201.
12. J. Ross MacKay, "Some Problems and Techniques in Isopleth Mapping," *Economic Geography* 27 (1951): 1–9.
13. Truman A. Hartshorn, *Interpreting the City: An Urban Geography* (New York: John Wiley, 1980), pp. 216–17.
14. Hsu and Robinson, *The Fidelity of Isopleth Maps,* p. 11.
15. J. Ross MacKay, "The Alternative Choice in Isopleth Interpolation," *The Professional Geographer* 5 (1953): 2–4.
16. Joel L. Morrison, "Observed Statistical Trends in Various Interpolation Algorithms Useful for First Stage Interpolation," *Canadian Cartographer* 11 (1974): 142–59.
17. Joel L. Morrison, *Method-Produced Error in Isarithmic Mapping* (Technical Monograph No. CA-5—Washington, D.C.: American Congress on Surveying and Mapping, 1971), pp. 15–16.
18. David Unwin, *Introductory Spatial Analysis* (New York:Methuen, 1981), p. 172.
19. *Ibid.*
20. Morrison, *Method-Produced Error,* pp. 3–4.
21. Hsu and Robinson, *The Fidelity of Isopleth Maps,* p. 8.
22. Morrison, *Method-Produced Error,* p. 61.
23. Hsu and Robinson, *The Fideltiy of Isopleth Maps,* p. 11.
24. *Ibid.,* p. 73.

Glossary

control points data points

CRM errors result from collection, recording, or the manipulation of machines and techniques

data points points at which magnitudes occur, or are assumed to occur, and from which isarithmic maps are constructed; also called *control points;* each point's magnitude is referred to as its z-value

datum the bottom-most horizontal base adopted for an isarithmic map; usually has a value of zero

frequency polygon a graph showing the frequency distribution of continuous data; a three-dimensional frequency polygon can be compared with geographical volume

graphic array a graph containing the plot of an array of data point values in ascending order; useful in selecting isoline intervals

interpolation procedure for the careful positioning of isolines between adjacent data points

isarithmic interval the vertical distance between the hypothetical horizontal planes passing through the three-dimensional geographic model; selection of the interval determines the degree of generalization and detail of the map

isarithmic map a planimetric graphic representation of a three-dimensional volume

isobaths isometric lines showing the depth of the ocean

isogones isometric lines showing equal magnetic declination

isohypse isometric line showing the elevation of land surfaces above sea level

isoline mapping isarithmic mapping

isometric map one form of isarithmic map; made from data that occur at points

isoplethic map one form of isarithmic map; constructed from geographical data that occur over area

isotherms isometric lines showing equal temperature

map reading and analysis errors caused by the inability of the reader to interpret map accurately

method-produced errors result from the cartographic technique

pattern of isarithms arrangement of isarithms, especially their magnitude, spacing, and orientation, to reveal the surface configuration of the geographical volume

primary interpolation first step in machine interpolation of data points; involves the mathematical interpolation of all points on a fine-mesh grid adopted for the map

production errors result from either manual or machine rendering of the final map

secondary interpolation second step in machine interpolation of data points; involves the location of isolines with respect to values at points previously determined through primary interpolation; usually a linear model

unit areas geographical areas containing the statistical data from which isoplethic maps are generated

Readings for Further Understanding

Barnes, James A. "Central Areas and Control Points in Isopleth Mapping." *American Cartographer* 5 (1978): 65–69.

Blumenstock, David I. "The Reliability Factor in the Drawing of Isarithms." *Annals* (Association of American Geographers) 43 (1953): 289–304.

Buckler, William R. "Computer Generated 3-D Maps; Models for Learning Contour Map Reading." *Journal of Geography* 87 (1988): 49–58.

Groop, Richard E., and Paul Smith. "A Dot Matrix Method of Portraying Continuous Statistical Surfaces." *American Cartographer* 9 (1982): 123–30.

Hartshorn, Truman A. *Interpreting the City: An Urban Geography.* New York: John Wiley, 1980.

Hsu, Mei-Ling. "The Isopleth Surface in Relation to the System of Data Derivation." *International Yearbook of Cartography* 8 (1968): 75–87.

Hsu, Mei-Ling, and Arthur H. Robinson. *The Fidelity of Isopleth Maps.* Minneapolis: University of Minnesota Press, 1970.

Jenks, George F. "Generalization in Statistical Mapping." *Annals* (Association of American Geographers) 53 (1963): 15–26.

Jones, Wellington D. "Ratios and Isopleth Maps in Regional Investigation of Agricultural Land Occupance." *Annals* (Association of American Geographers) 20 (1930): 177–95.

Lavin, Stephen. "Mapping Continuous Geographical Distributions Using Dot-Density Shading." *American Cartographer* 13 (1986): 140–50.

MacKay, J. Ross. "The Alternate Choice in Isopleth Interpolation." *Professional Geographer* 5 (1953): 2–4.

———. "Isopleth Class Intervals: A Consideration in Their Selection." *Canadian Cartographer* 7 (1963): 42–45.

———. "Some Problems and Techniques in Isopleth Mapping." *Economic Geography* 27 (1951): 1–9.

Monmonier, Mark S. *Computer-Assisted Cartography: Principles and Prospects.* Englewood Cliffs, NJ: Prentice-Hall, 1982.

Morrison, Joel L. *Method-Produced Error in Isarithmic Mapping.* Technical Monograph No. CA-5. Washington, D.C.: American Congress on Surveying and Mapping, 1971.

———. "Observed Statistical Trends in Various Interpolation Algorithms Useful for First Stage Interpolation." *Canadian Cartographer* 11 (1974): 142–59.

Porter, Phillip W. "Putting the Isopleth in Its Place." *Proceedings of the Minnesota Academy of Science* 25–26 (1957–1958): 372–84.

Raisz, Erwin. *Principles of Cartography.* New York: McGraw-Hill, 1962.

Robinson, Arthur H. "The Cartographic Representation of the Statistical Surface." *International Yearbook of Cartography* 1 (1961): 53–61.

———. "The Genealogy of the Isopleth." *Cartographic Journal* 8 (1971): 49–53.

———. *Early Thematic Mapping in the History of Cartography.* Chicago: The University of Chicago Press, 1982.

Robinson, Arthur; Randall Sale, and Joel Morrison. *Elements of Cartography,* 4th ed. New York: John Wiley, 1978.

Schmid, Calvin F., and Earle H. MacCannell. "Basic Problems, Techniques, and Theory of Isopleth Mapping." *Journal of the American Statistical Association* 50 (1955): 220–39.

Thrower, Norman J. W. *Maps and Man.* New York: Prentice-Hall, 1972.

Unwin, David. *Introductory Spatial Analysis.* New York: Methuen, 1981.

Chapter

10

Value-by-Area Mapping

Chapter Preview

Erwin Raisz called cartograms diagrammatic maps. Today they may be called cartograms, value-by-area maps, or spatial transformations. Whatever their name, cartograms are unique representations of geographical space. Examined more closely, the value-by-area mapping technique encodes the mapped data in a simple and efficient manner with no data generalization or loss of detail. Two forms, contiguous and noncontiguous, have become popular. Mapping requirements include the preservation of shape, orientation, contiguity, and data that have suitable variation. Successful communication depends on how well the map reader recognizes the shapes of the internal enumeration units, the accuracy of estimating these areas, and effective legend design. Complex forms include the two-variable map. Cartogram construction may be by manual or computer means. In either method, a careful examination of the logic behind the use of the cartogram must first be undertaken.

We are accustomed to looking at maps on which the political or enumeration units (e.g., states, counties, or census tracts) have been drawn proportional to their geographic size. Thus, for example, Texas appears larger than Rhode Island, Colorado larger than Massachusetts, and so on. The areas on the map are proportional to the geographical areas of the political units. (Only on non-equal-area projections are these relationships violated.) It is quite possible, however, to prepare maps on which the areas of the political units have been drawn so that they are proportional to some space other than the geographical. For example, the areas on the map that represent states can be constructed proportional to their population, aggregate income, or retail sales volume, rather than their geographical size. Maps on which these different presentations appear have been called *cartograms, value-by-area maps,* and *spatial transformations.*

This chapter introduces this unique form of map. In these abstractions from geographical reality, ordinary geographical area, orientation, and contiguity relationships are lost. The reader is forced to look at a twisted and distorted image that only vaguely resembles the geographic map. Yet cartograms are being used more and more by professional cartographers and others in the communications industry. Their eventual success as a communication device rests on the ability of the map reader to restructure them back into a recognizable form. Regardless of these complexities, cartograms are popular. Their appeal no doubt results from their attention-getting attributes.

The Value-by-Area Cartograms Defined

All **value-by-area maps,** or *cartograms,* are drawn so that the areas of the internal enumeration units are proportional to the data they represent. (See Figures 10.1 and 10.2.) This method of encoding geographical data is unique in thematic mapping. In other thematic forms, data are mapped by selecting a symbol (area shading or proportional symbol, for example) and placed in or on enumeration units. In the cartogram, the actual enumeration unit and its size carry the information.

Value-by-area cartograms can be used to map a variety of kinds of data. Raw or derived data, at ratio or interval scales, census data or specially-gathered data can be mapped in a cartogram. Because of the method of encoding, there is no data generalization. No data are lost through classing and consequent simplification. In terms of data encoding, the value-by-area cartogram is perhaps one of the purest forms of quantitative map, since no categorization is necessary during its preparation. Unfortunately, data retrieval is fraught with complexity, and readers may experience confusion because the base map has been highly generalized.

Brief History of the Method

As with so many other techniques in thematic mapping, it is difficult to pinpoint the beginning of the use of value-by-area maps. They ap-

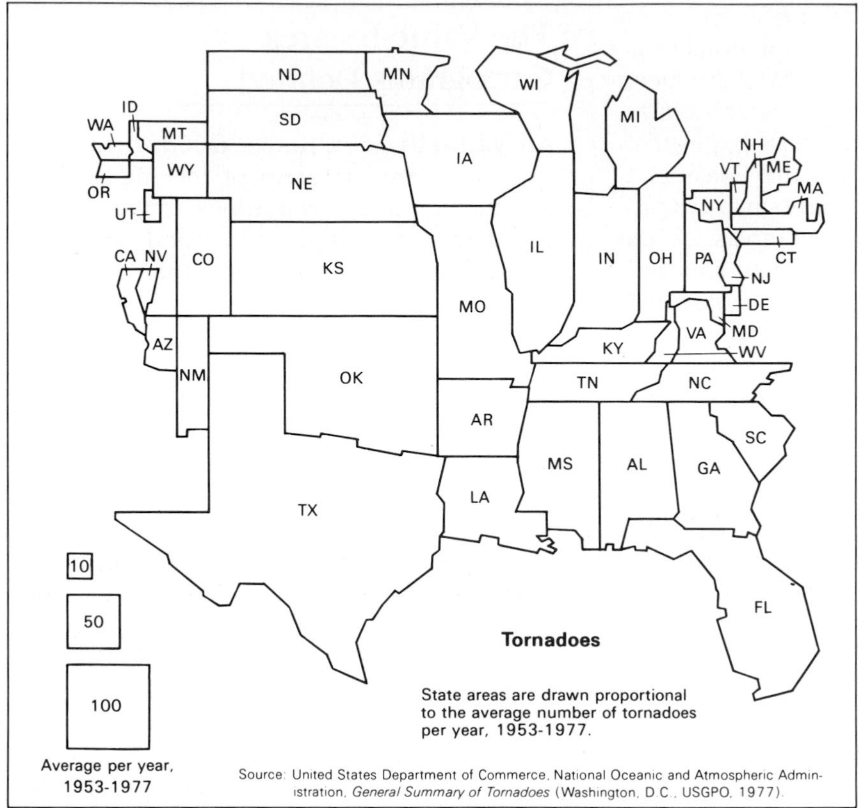

Figure 10.1. Typical value-by-area cartogram.

Source of data: United States Department of Commerce, National Oceanic and Atmospheric Administration, General Summary of Tornadoes (*Washington, D.C.:* USGPO, 1977).

parently were a practical solution to a graphic problem. The idea of the cartogram has been traced to both France and Germany in the late nineteenth and early twentieth centuries respectively.[1] Erwin Raisz was certainly among the first American cartographers to employ the idea; he wrote on the subject 50 years ago.[2] Cartogram construction techniques were treated by Raisz through several editions of his textbook on cartography.[3] In 1963, Waldo Tobler discussed their theoretical underpinnings, most notably their projection system, and concluded that they are maps based on unknown projections.[4] Cartograms have been used in texts and in the classroom to illustrate geographical concepts; their role in communication situations has recently been investigated.[5]

Since their introduction, cartograms have been used in atlases and general reference books to illustrate geographical facts and concepts,[6] but no book has been devoted entirely to these interesting maps.

Two Basic Forms Emerge

Two basic forms of value-by-area cartogram have emerged: contiguous and noncontiguous. (See Figure 10.3.) Each has its own set of advantages and disadvantages, which the designer must weigh in the context of the map's purpose.

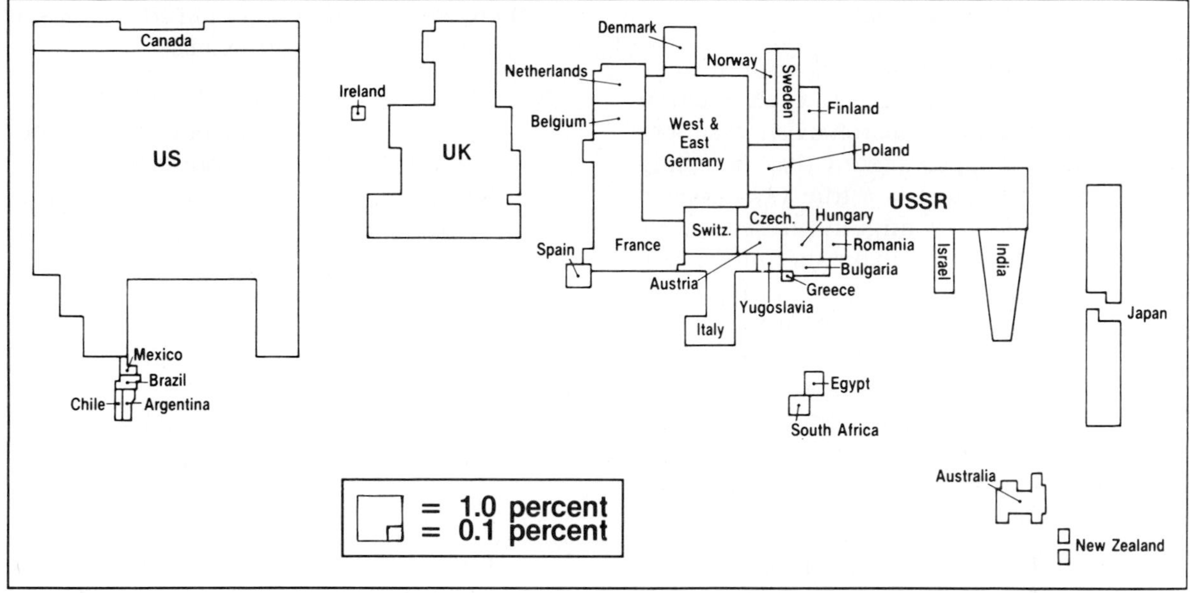

Figure 10.2. Contribution of countries to world scientific authorship.

Source: Anthony R. deSouza, "Scientific Authorship and Technological Potential" (editorial), *Journal of Geography* (July/August, 1985):138. Reprinted by permission of the National Council for Geographic Education.

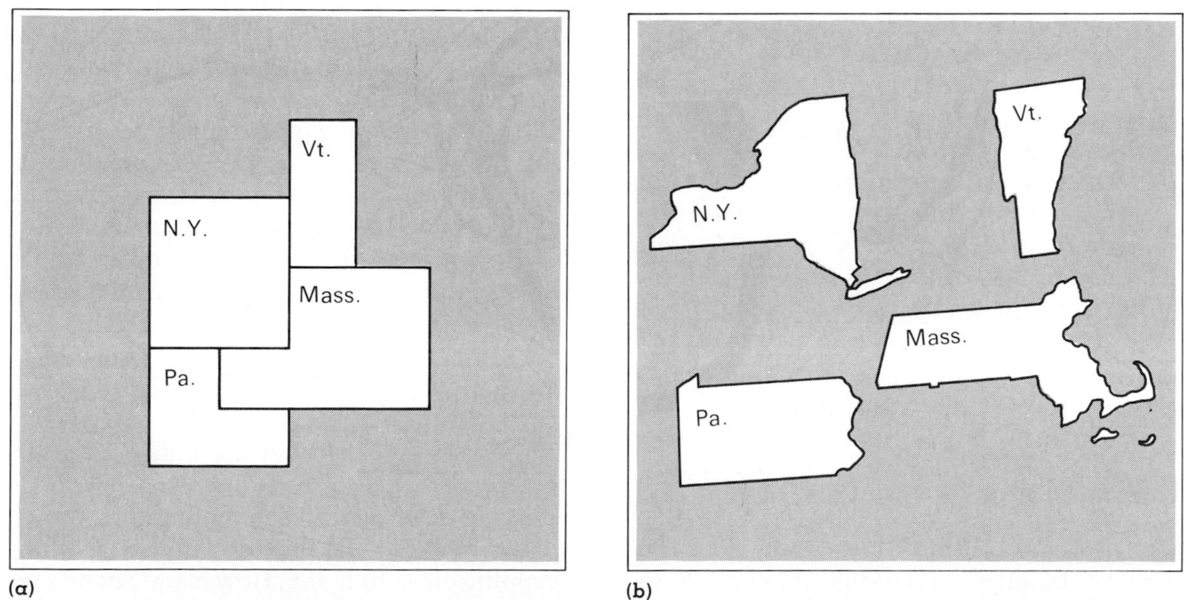

Figure 10.3. Contiguous and noncontiguous cartograms.

Continguous cartograms like (a) are compact and boundary relations are attempted. In noncontiguous cartograms, such as (b), enumeration units are separated and positioned to maintain relatively accurate geographic location.

Contiguous Cartograms

In **contiguous cartograms,** the internal enumeration units are adjacent to each other. Although no definitive research exists to support this position, it appears likely that the contiguous form best suggests a true (i.e., conventional) map. With contiguity preserved, the reader can more easily make the inference to continuous geographical space, even though the relationships on the map may be erroneous. Making the cartogram contiguous, however, can make the map more complex to produce and interpret, for both manual and computer solutions.

Several advantages may be listed for the contiguous form:

1. Boundary and orientation relationships can be maintained, strengthening the link between the cartogram and true geographical space.
2. The reader need not mentally supply missing areas to complete the total form or outline of the map.
3. The shape of the total study area is more easily preserved.

Disadvantages of this form include these:

1. Distortion of boundary and orientation relationships can be so great that the link with true geographical space becomes remote and may confuse the reader.
2. The shapes of the internal enumeration units may be so distorted as to make recognition almost impossible.

Noncontiguous Cartograms

The **noncontiguous cartogram** does not preserve boundary relations among the internal enumeration units. The enumeration units are placed in more or less correct locations relative to their neighbors, with gaps between them. Such cartograms cannot convey continuous geographical space and require the reader to infer the contiguity feature.

There are nonetheless certain advantages in using noncontiguous cartograms:

1. They are easy to scale and construct.
2. True geographical shapes of the enumeration units can be preserved.
3. Areas lacking mapped quantities (gaps) can be used to compare with the mapped units for quick visual assessment of the total distribution.[7]

On the negative side, these disadvantages may be listed:

1. Noncontiguous cartograms do not convey the continuous nature of geographical space.
2. They do not possess an overall compact form, and it is difficult to maintain the shape of the entire study area.

Mapping Requirements

Communication with cartograms is difficult at best, because it requires the reader be familiar with the geographic relations of the mapped space: the total form of the study area as well as the shapes of the internal enumeration units. This task may not be too difficult for students in the United States when the mapped area is their homeland and the internal units are states, but how many students in this country are familiar with the shapes of the Canadian provinces or those of the African nations? Likewise, are European students that knowledgeable of the shapes of the Canadian provinces or the states of the United States? On the other hand, by the very fact that they are unfamiliar with the mapped areas, map readers may pay more attention to the map than they otherwise would.

The situation can even be complex when mapping close to home. How many Tennessee residents know or could recognize the shapes of the counties in Tennessee? Georgia has 159 counties, Texas more than 200. Fortunately, most professional cartographers realize the futility of mapping little-known places with cartograms.

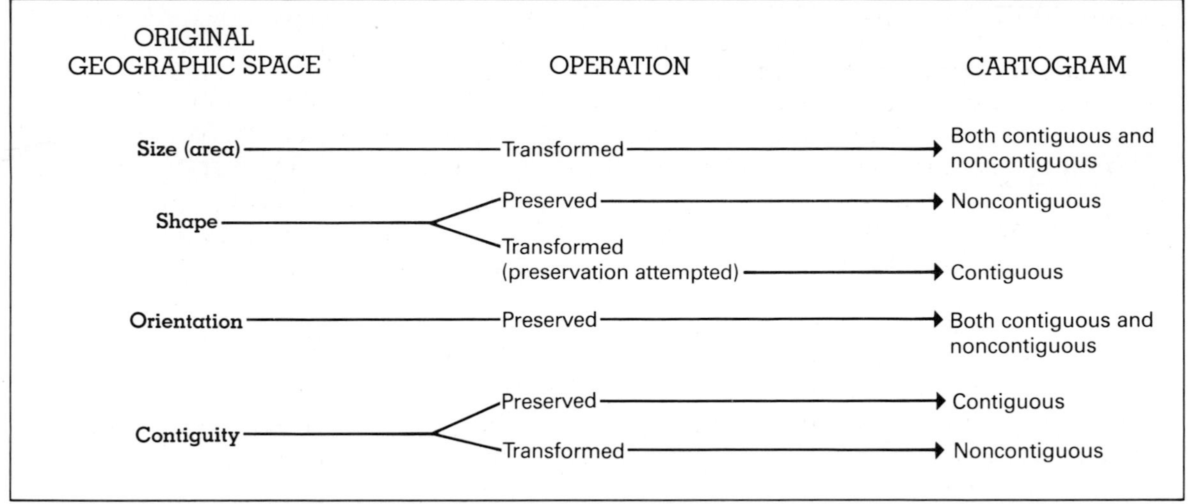

Figure 10.4. Cartographic operations in value-by-area mapping.

Cartograms can present a unique view of geographical space. Raisz said many years ago that cartograms "may serve to right common misconceptions held by even well-informed people."[8] Harris and McDowell have suggested that the value-by-area map is a good way to teach about geographical distributions.[9] Tentative evidence indicates that map readers can obtain information from value-by-area maps as effectively as from more conventional forms. For this to happen, however, certain qualities of the true geographical base map must be preserved during transformation. The first of these is the **shape quality.** Preservation of the general shape of the enumeration units is so crucial to communication that the cartogram form should not be used unless some approximation of true shape can be achieved.

Conventional thematic maps are developed by placing graphic symbols on a geographic base map. Regardless of the form of the thematic presentation, the symbols are tied to the geographical unit with which the data are associated. Thus, for example, graduated symbols are placed at the centers of the states. Value-by-area maps, however, are unique in that the thematic symbolization also forms the base map. In a way, the enumeration units are their own graduated symbols, in addition to carrying the information of the conventional base map. On an original geographic base map of, for example, the United States, each state contains four kinds of information—size, shape, orientation, and contiguity. (See Figure 10.4.) In value-by-area mapping, only size is transformed; the other elements are preserved as nearly as possible. Contiguity is somewhat special and may not be as important as the others in map reading.

Individual unit shapes on the cartogram must be similar to their geographical shapes. It is through shape that the reader identifies areas on the cartogram. Shape is a bridge that allows the reader to perceive the transformation of the original. (See Figure 10.5.) If the reader cannot recognize shape, confusion results and comprehension is difficult, if not lost altogether. The designer's problem is deciding how far it is possible to go along a continuum between shape preservation and shape transformation before the enumeration unit becomes unrecognizable to the majority of readers.

Geographical **orientation** is another important element in value-by-area mapping. Orientation is the internal arrangement of the enumeration units within the transformed space. Because the reader must be familiar with

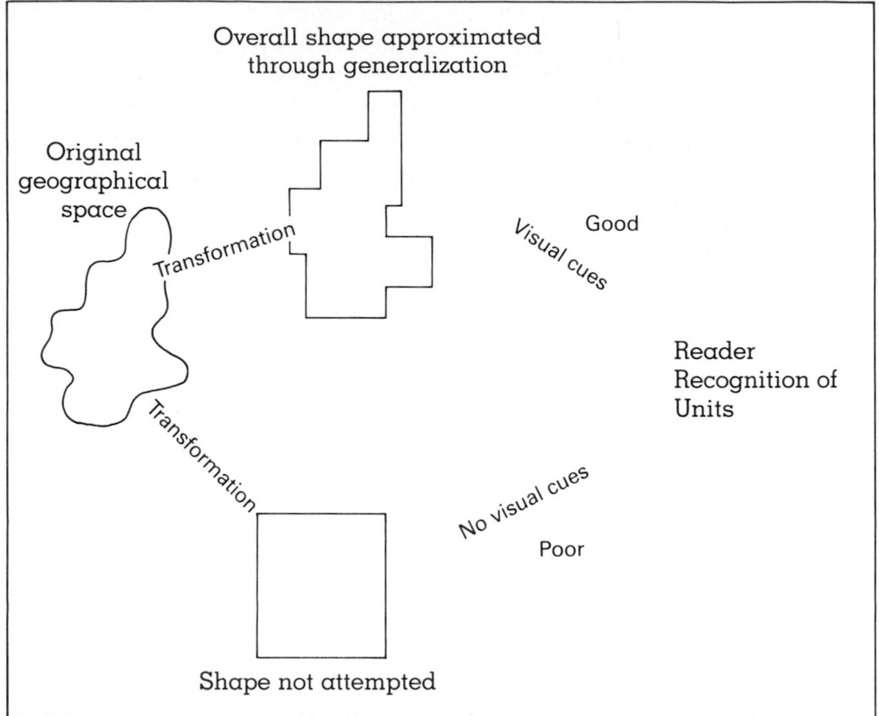

Figure 10.5. The importance of shape in cartogram design.
Shape preservation provides necessary visual cues for efficient reader recognition of original spatial units.

the geographical map of the study area to interpret a cartogram properly, the cartographer must strive to maintain recognizable orientation. When distortion of internal order occurs, communication surely suffers. How frustrating it would be to see Michigan below Texas!

Contiguity as an element in cartogram development relates, of course, only to the contiguous form. When producing this kind, it is desirable to maintain as closely as possible the original boundary arrangement from the geographical base. Of the elements mentioned thus far—shape, order, and contiguity—it appears that contiguity is the least important in terms of communication. It is likely that map readers do not use understanding of geographical boundary arrangements in reading cartograms. How many for us, for example, know how much of Arkansas is adjacent to Texas? On noncontiguous varieties, of course, contiguity per se cannot be preserved. It is possible, however, to maintain loose contiguity by proper positioning of the units, although gaps remain between the units.

Of the qualities mentioned (shape, order, and contiguity), shape is by far the most important. Use the value-by-area cartogram technique only where the reader is familiar with the shapes of the internal enumeration units. Do not overestimate the ability of the reader in this regard. Well-designed legends can be helpful, as discussed later in this chapter.

Data Limitations

Although value-by-area maps present numerous possibilities for the communication of thematic data, they are not without their limitations. Within the three principal ways of symbolizing data for thematic maps—point, line, and area—cartograms fall most comfortably into the category of area. Area is the element

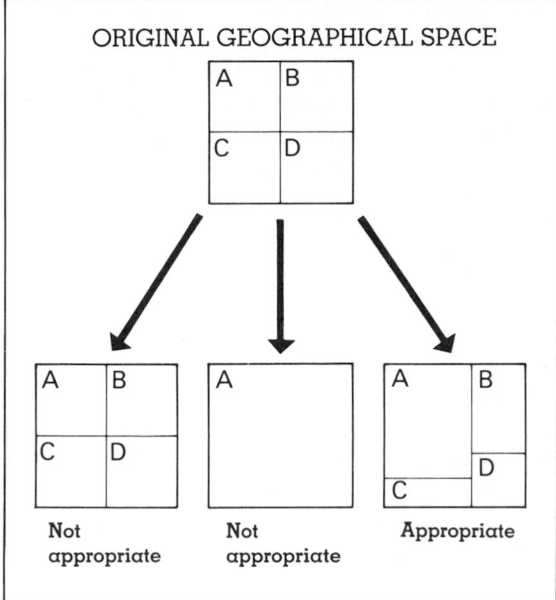

Figure 10.6. Data limitations and value-by-area mapping.
If the original data lead to spatial transformation that is unchanged from the original (as on the left), the value-by-area technique is inappropriate. Also inappropriate would be those cases resulting in only one enumeration unit remaining after transformation (as in the center). Most suitable would be those instances when original data are transformed into new spatial arrangements dramatically different from the original (as on the right).

that must vary within the cartogram, so there are obvious limits outside of which one should not attempt this kind of representation. The limits are dictated by the data and their variability. It would be fruitless to map data that are exactly proportional to the areas of the enumeration units of the geographical base. (See Figure 10.6.) The cartogram would then replicate the original. At the other extreme, there could be a single enumeration unit having the same area as the entire "transformed" space, in which case no internal variation would be shown. No cartogram (or any other map) would be needed. Within these general limits, there exists a range of possibilities.

The chief goal of the cartogram is to illustrate a thematic distribution in dramatic fashion, which requires that the data be compatible with the map's overall purpose. The data set should be compared to the enumeration units on the geographical base. If reversal is evident (large states having small numerical value or vice-versa), the cartogram is likely to be worthy of execution. Two measures, the linear regression and rank-order correlation indices, provide a degree of quantitative support. Unfortunately, these methods fall short in that they only provide overall indices of association; they do not indicate variation or agreement between data pairs within the total set. A residual from regression analysis proves useful, but arbitrary limits must still be selected. There are more informal ways of determining appropriateness that are easily workable.

Whatever procedure is chosen to determine the appropriateness of a data set for cartogram construction, such a determination should always be made before such a map is begun. Those not familiar with such maps often launch into a construction, only to find the results rather disappointing. If the map does not illustrate the distribution in a visually dramatic way, it is best abandoned.

Communicating with Cartograms

Success in transmitting information by the value-by-area technique is not guaranteed. There are at least three problem areas: shape recognition, estimation of area magnitude, and the stored images of the map reader. The designer should be familiar with the influences of each on the communication task.

Recognizing Shapes

It is by the shape of objects around us that we recognize them. We often identify three-dimensional objects by their silhouettes, and we can label objects drawn on a piece of paper by the shapes of their outlines. This holds true for recognition of outlines on maps. For example, South America can be seen as distinct

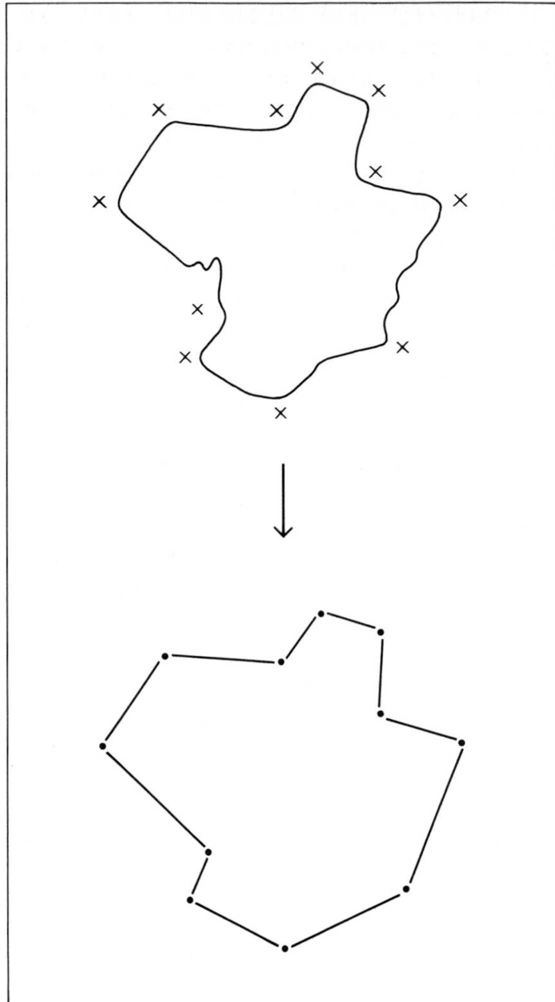

Figure 10.7. Straight-line generalization of the original shape.
Important shape cues are concentrated at points of major change in direction along the outline. These points should be retained in transformation and may be connected by straight lines to develop a reasonable approximation of the original shape.

from the other continents. The shape qualities of objects that make them more recognizable are simplicity, angularity, and regularity.[10] Simple geometric forms such as squares, circles, and triangles are easily identified. Shapes to which we can attach meaning are also easy to identify.

In the production of value-by-area maps, the cartographer ordinarily attempts to preserve the shapes of the enumeration units. How this is done is crucial to the effectiveness of the map. Many of the elements that identify the shape of the original should be carried over to the new generalized shape on the cartogram. The places along an outline where direction changes rapidly appear to be those that carry the most information about the form's shape.[11] Therefore, such points on the outline should be preserved in making the new map. These points can be joined by straight lines without doing harm to the generalization or to the reader's ability to recognize the shape. (See Figure 10.7.)

Estimating Areas

Since each enumeration unit in a cartogram is scaled directly to the data it represents, no loss of information has occurred through classification or simplification. If any error results, it is to be found somewhere else in the communication process—most likely in the reader's inability to judge area accurately. The psychophysical estimation of area magnitudes is influenced by the shapes of the representative areas used in the map legend.

Research suggests that for effective communication of area magnitudes, the shapes of the enumeration units should be irregular polygons (not amorphous shapes) and that at

Maps prepared using these transformations, however, from many points of view are more realistic than the conventional maps used by geographers.

Source: Waldo R. Tobler, *Map Transformations of Geographic Space* (unpublished Ph.D. dissertation, Department of Geography, University of Washington, 1961), pp. 162–63.

least one square legend symbol should be used at the lower end of the data range.[12] It is best to provide three squares in the legend, one at the low, one at the middle, and one at the high end of the data range. Of course, the overall communication effort may fail because the distortions of reality brought about by the method interfere with the flow of information.

A Communication Model

It has been stressed thus far that communicating geographic information with cartograms is difficult unless certain rules are followed. First, shape recognition clues along the outline of enumeration units must be maintained. Second, if the cartographer cannot assume that the reader knows the true geographical relationships of the mapped area, a geographical inset map must be included. Third, the cartographer should provide a well-designed legend that includes a representative area at the low end of the value range.

These three design elements are placed in a generalized model of value-by-area cartogram communication in Figure 10.8.[13] In this view, design strategies should accommodate the map-reading abilities of the reader. In Step 1, all the graphic components are organized into a meaningful hierarchical organization so that the map's purpose is clear.

Accurate shapes of the enumeration units are provided in Step 2 by retaining those outline clues that carry the most information—the places where the outline changes direction rapidly.

In the United States, people have been exposed from early childhood to maps of the country in classroom wall maps, road maps, television, and advertising. Recently, satellite photographs have added to the already clear mental images of the country's shape in the minds of the populace. How well these images are formed varies from individual to individual. Some people have well-formed images not only of the shape of the United States but also of the individual states; others have difficulty in choosing the correct outline from several incorrect ones. Successful cartogram communication may well rest on the accuracy of the reader's image of geographical space. Without a correct image, the reader cannot make the necessary match between cartogram space and geographical space. Confusion results if this connection is not made quickly.

In Step 3, the readers search through stored images of geographical areas in an attempt to match what they see with a stored image. Because the reader's stored images may be inaccurate, the designer should include a geographic map of the cartogram area in an inset map.

The map reader in Step 4 estimates the magnitudes of the enumeration units by comparing them with those presented in the legend. Effective legend design makes this task easier. Anchor stimuli in the legend should be squares, including at least one at the low end of the value range.

In Step 5, written elements, such as labels and explanatory notes, are included to assist the map reader in identifying parts of the map that may be unfamiliar at first. Finally, the designer should be willing to restructure the message to make the communication process better (Step 6). Since the cartographer may not know what the reader thinks, because they are usually separated in time and space, the first five of the cartographer's tasks become even more important.

Advantages and Disadvantages

Unfortunately, cartograms have not been studied in enough detail to reveal exactly what impresses map readers about them or exactly how they are read. Preference testing research has discovered that cartograms do communicate spatial information, are innovative and interesting, display remarkable style, and present a generalized picture of reality. Value-by-area maps are often stimulating, provoke considerable thought, and show geographical distributions in a way that stresses important aspects. On the other hand, they are viewed as difficult to read, incomplete, unusual, and different from reader's preconceptions of geographical space.

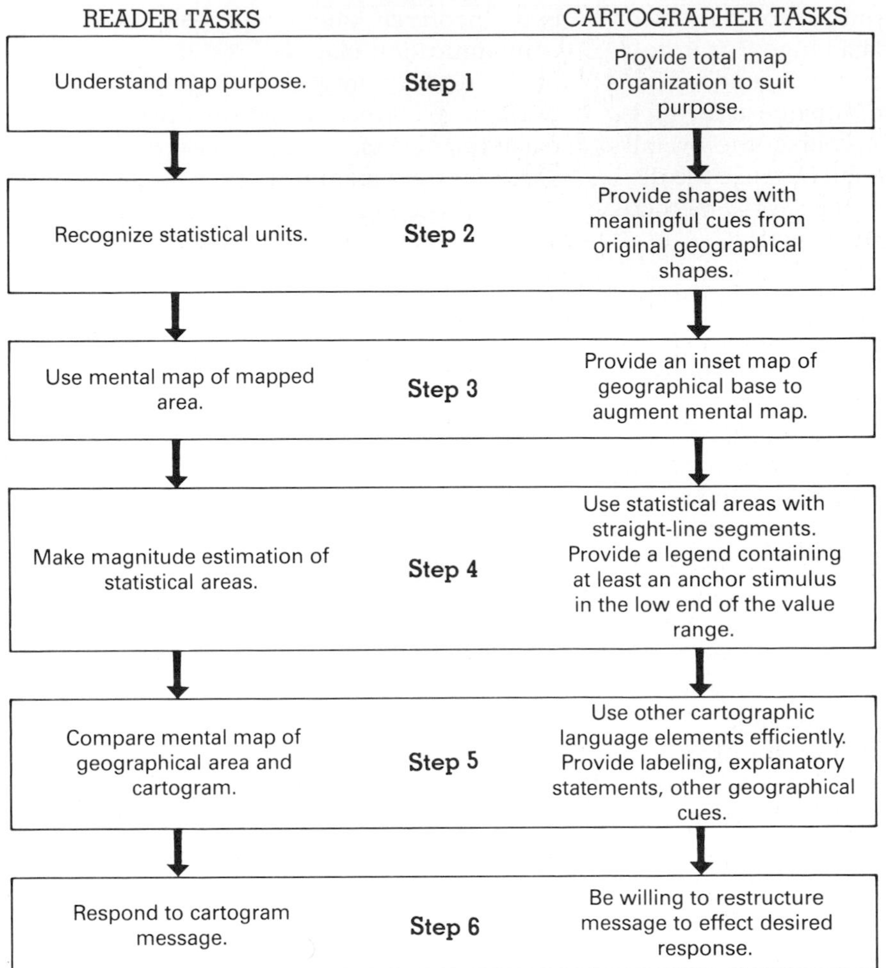

Figure 10.8. Cartographer and reader tasks in a generalized value-by-area cartogram communication model.
Many of the steps are likely to occur simultaneously, not sequentially—especially steps 2 through 5.
Source: Borden D. Dent, "Communication Aspects of Value-by-Area Cartograms," *American Cartographer* 2 (1975): 154–68.

Probably the most serious drawback is that no established methodology leads to consistent results. No two people devise identical cartograms of the same area. (This may be considered a strength rather than a drawback.) For the untrained map reader, the new configurations can cause visual confusion, detracting from the purpose of the map rather than adding to it.

The advantages of this thematic mapping technique include these:[14]

1. To shock the reader with unexpected spatial pecularities.
2. To develop clarity in a map that might otherwise be cluttered with unnecessary detail.
3. To show distributions that would, if mapped by conventional means, be obscured by wide variation in the sizes of the enumeration areas.

These disadvantages may be listed:

1. Some map readers may feel repugnance at the "inaccurate" base map that results from the study.

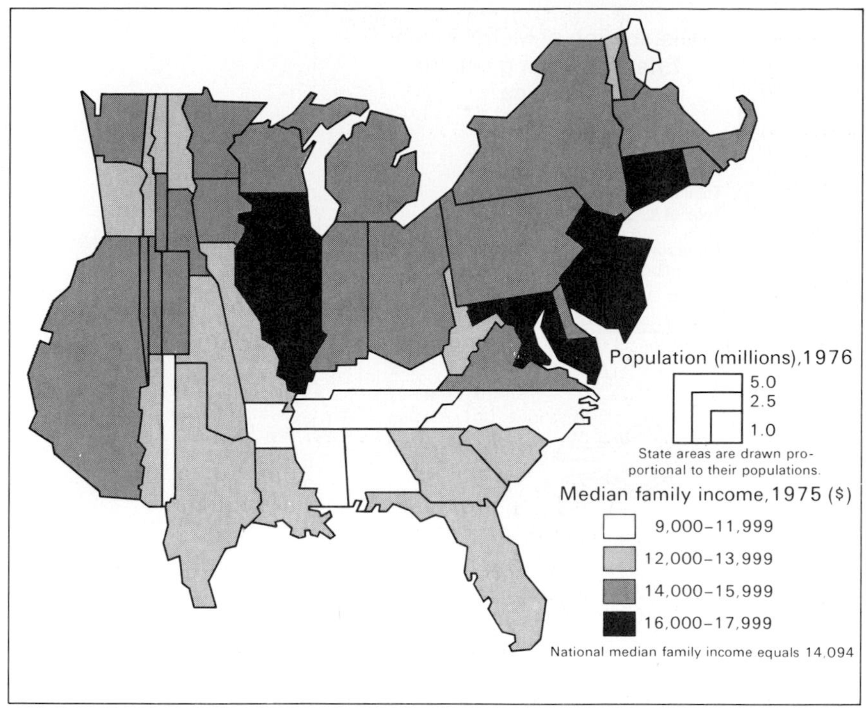

Figure 10.9. A two-variable value-by-area cartogram.
It is often interesting to map a second, related, variable on a cartogram base, thereby providing the reader with a different picture of two associated phenomena. In this cartogram, it is apparent that large numbers of people with high median family incomes are located in the upper Middle West, New England, the northern Atlantic coastal states, and California.
Population cartogram base designed by Karen McHaney. Used by permission.

2. Map readers may be confused by the logic of the method unless its properties are clearly identified.
3. Specific locations may be difficult to identify because of shape distortion of the enumeration areas.

Two-Variable Cartograms

The discussion thus far has concerned only the use of a single data set (variable), but it is possible to illustrate two or more data sets on a single cartogram. For example, on a cartogram of the United States in which the states are represented proportional to their populations, the cartographer can render individual states by gray tones, as on a choropleth map. The state areas may be represented as belonging to classes in another distribution. (See Figure 10.9.) This appears to be a very compatible representation of two distributions, as both relate to area. A choropleth map presupposes an even distribution throughout each enumeration unit, as does a cartogram. This form of **two-variable value-by-area cartogram** has been used successfully in mapping the spatial variation of socioeconomic data in Australian cities.[15]

Other second variables can be accommodated on cartograms by graduated point symbol schemes. The second distribution can be represented by placing a graduated symbol within each enumeration unit of the cartogram. The reader must make the visual-intellectual comparison between the size of the enumeration unit and the size of the scaled symbol. This may be difficult for some readers at first. Although little research has been done on either method,

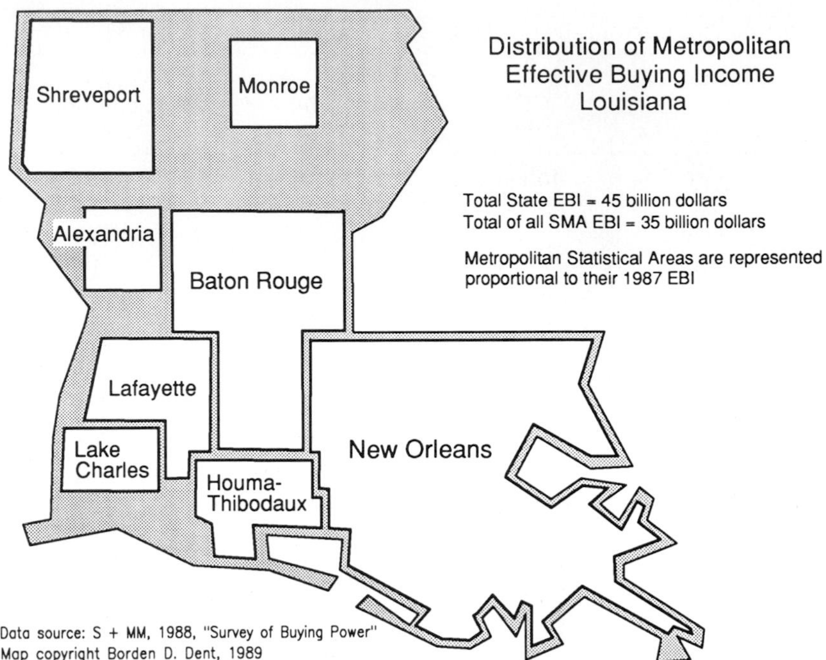

Figure 10.10. Cartogram to show geographical proportion.
In this presentation SMAs are drawn proportional to their buying power and are shown relative to the total buying power of the state. Shapes of the SMAs are not as important in this form of cartogram, although relative location is.

it would seem likely that they should be used only where there is a high degree of mathematical association between the two data sets. They certainly deserve further inquiry.

Another use related to two-variable mapping is to show how much of a total area is occupied by internal geographical divisions. (See Figure 10.10.) In this instance, the reader is asked to compare area proportions, and shape preservation is not often of central concern. The sizes of the internal areas are drawn proportional to the data being mapped.

Cartogram Construction

There are two ways of producing value-by-area cartograms: manually and by computer technology. At present, more maps are probably generated by manual methods.

Manual Methods

Manual techniques for the construction of value-by-area maps are quite simple. Suppose a cartographer wishes to construct a cartogram of total United States population. First, the total population is recorded for each state. The cartographer must then decide what the total area for the transformation is to be, and what proportion of the total population is represented by each state. Then the area for each state is computed on the basis of its share. (See Table 10.1.) Drafting can then begin. The cartographer must draft each state, preserving the shapes of the states while making their areas conform to the values computed. Of course, exact shapes are not preserved in contiguous cartograms.

To facilitate the drafting of the states, it is convenient to begin by computing what some small areal division represents in terms of pop-

Table 10.1 **Data Sheet for a Population Cartogram of the United States**

State	1980 Population	Number of Counting Units	State	1980 Population	Number of Counting Units
Alabama	3,890 006	60	Montana	786,690	14
Alaska	400,481	6	Nebraska	1,570,006	25
Arizona	2,717,866	42	Nevada	799,184	14
Arkansas	2,285,513	35	New Hampshire	920,610	14
California	23,668,562	350	New Jersey	7,364,158	116
Colorado	2,888,834	46	New Mexico	1,299,968	21
Connecticut	3,107,576	49	New York	17,557,288	277
Delaware	595,225	9	North Carolina	5,874,429	91
Florida	9,739,992	154	North Dakota	652,695	11
Georgia	5,464,265	88	Ohio	10,797,419	168
Hawaii	965,935	15	Oklahoma	3,025,266	49
Idaho	943,935	15	Oregon	2,632,663	42
Illinois	11,418,461	175	Pennsylvania	11,866,728	186
Indiana	5,490,179	84	Rhode Island	947,154	14
Iowa	2,913,387	46	South Carolina	3,119,208	49
Kansas	2,363,208	35	South Dakota	690,178	11
Kentucky	3,661,433	57	Tennessee	4,590,750	74
Louisiana	4,203,972	67	Texas	14,228,383	242
Maine	1,124,660	18	Utah	1,461,037	25
Maryland	4,216,446	67	Vermont	511,456	11
Massachusetts	5,737,037	91	Virginia	5,346,279	84
Michigan	9,258,344	147	Washington	4,130,163	67
Minnesota	4,077,148	63	West Virginia	1,949,644	32
Mississippi	2,520,638	39	Wisconsin	4,705,335	74
Missouri	4,917,444	77	Wyoming	470,816	7

Total population (excluding District of Columbia and Puerto Rico) = 222,670,654. Total map area adopted in cartogram = 35 sq in. Counting unit size adopted for project = .01 sq in. Total number of counting units = 3500. For each state, a ratio of the state's population to the national population was determined. The ratio was applied to the 3500 total counting units to compute the number of units assigned to the state. For computation in this table, population figures were rounded to the nearest thousand.

ulation. For example, the population of every .01 square inch can be calculated by dividing the total population into the total number of square inches determined for the cartogram (this unit size is selected simply because of convenience of obtaining this grid paper). By dividing the population determined for each .01 square inch unit into the state's total population, the number of these .01 **counting units** can be ascertained. The cartographer need only arrange these small counting units until the shape of the state is approximated. (See Figure 10.11.) After the shape is achieved, the cartographer may wish to check the accuracy of the state's area by a quick planimeter measurement. Digital readout planimeters are available for such uses.

Each state's shape is adjusted and fitted to adjacent states until the cartogram is completed. The shape of the entire study area must be roughly preserved throughout. This is not difficult but is time consuming and often frustrating. It is wise to construct the larger enumeration units first, then the smaller ones. If odd shapes result, the noncontiguous cartogram may be selected.

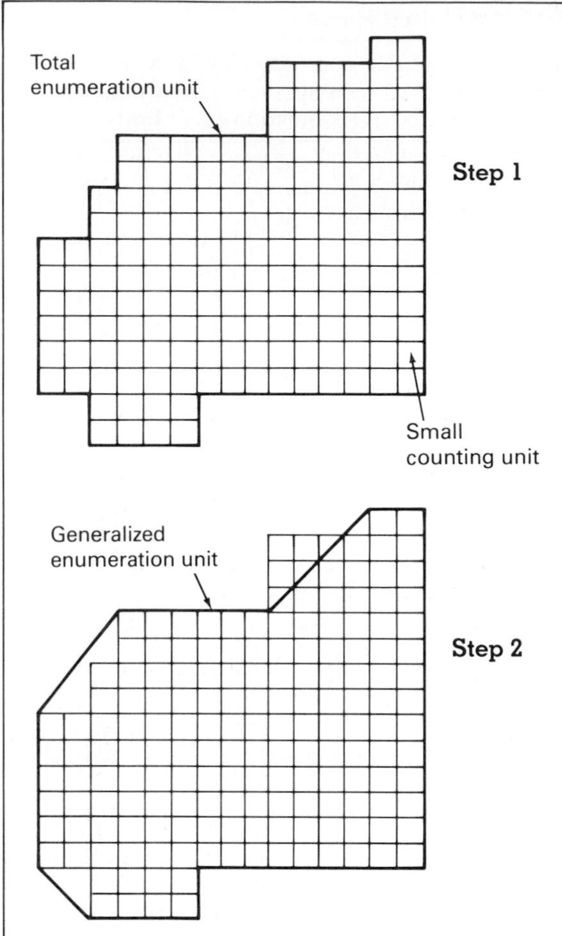

Figure 10.11. Constructing the cartogram. Small counting units are used to "build" the size and shape of the enumeration units (e.g., countries, states, counties) in step 1. Step 2 involves smoothing to the approximate final shape.

Constructing a noncontiguous cartogram involves a slightly different procedure after computations are made. A conventional (generalized, if desired) base map is drawn. By using an optical reducer-enlarger, the states can be reproduced at their proportionate sizes relative to one that has the same size as on the true base map.[16] After the individual state areas are determined and rough shapes are formed, the cartographer positions the state outlines on a draft map to form the shape of the total study area. Relative geographical position of each state is sought. The newly sized states may be positioned in accordance with the centers of the states on a conventional map. Of course, the advantage of the noncontiguous form is the preservation of individual state shapes.

Computer Solutions

There are computer programs available for the generation of contiguous spatial transformations, notably one by Tobler.[17] The chief drawback of these programs is their inability to preserve shapes accurately, since the goal is to achieve contiguity. They also reduce flexibility in design. For the noncontiguous type, the size of polygons can be scaled in the CALFORM program.[18] This has direct application in production and allows the cartographer to omit the use of an optical reducer-enlarger.

At least one author suggests that computer solutions may not be desirable because "the novelty of an automated approach may lead to intemperate haste in its utilization, whereby both the merits and weaknesses of topological transformation may be submerged in the deluge of products."[19] As in other computer applications in cartography, the machine can greatly reduce time and drudgery of production but it must not replace or interfere with the designer's choices.

Notes

1. John M. Hunter and Johnathan C. Young, "A Technique for the Construction of Quantitative Cartograms by Physical Accretion Models," *Professional Geographer* 20 (1968): 402–406.
2. Erwin Raisz, "The Rectangular Statistical Cartogram," *Geographical Review* 24 (1934): 292–96.
3. Erwin Raisz, *General Cartography* (New York: McGraw-Hill, 1948), pp. 257–58; and *Principles of Cartography* (New York: McGraw-Hill, 1962), pp. 215–21.

4. Waldo R. Tobler, "Geographic Area and Map Projections," *Geographical Review* 53 (1963): 59–78; see also Waldo R. Tobler, *Map Transformations of Geographic Space* (unpublished Ph.D. dissertation, Department of Geography, University of Washington, 1961), p. 146.
5. Borden D. Dent, "Communication Aspects of Value-by-Area Cartograms," *American Cartographer* 2 (1975): 154–68.
6. There are numerous examples of such atlases. The following are particularly interesting: Tony Loftas, ed., *Atlas of the Earth* (London: Mitchell Beazley, 1972); Rezine Van Chi-Bonnardel, *The Atlas of Africa* (New York: Free Press, 1973); and Michael Kidron and Ronald Segal, *The State of the World Atlas* (New York: Simon and Schuster, 1981).
7. Judy M. Olson, "Noncontiguous Area Cartograms," *Professional Geographer* 28 (1976): 371–80.
8. Raisz, "The Rectangular Statistical Cartogram," pp. 292–96.
9. Chauncey Harris and George B. McDowell, "Distorted Maps. A Teaching Device," *Journal of Geography* 54 (1955): 286–89.
10. Borden D. Dent, "A Note on the Importance of Shape in Cartogram Communication," *Journal of Geography* 71 (1972): 393–401.
11. *Ibid.*
12. Borden D. Dent, "Communication Aspects," pp. 154–68.
13. *Ibid.*
14. T. L. C. Griffin, "Cartographic Transformation of the Thematic Map Base," *Cartography* 11 (1980): 163–74.
15. *Ibid.*
16. Olson, "Noncontiguous Area Cartograms," pp. 371–80.
17. Waldo R. Tobler, "A Continuous Transformation Useful for Districting," *Annals* (New York Academy of Sciences) 219 (1973): 215–20.
18. CALFORM is a computer plotting program that develops shaded, conformant maps. It was developed by, and is obtainable from, Harvard University, Laboratory for Computer Graphics and Spatial Analysis.
19. Griffin, "Cartographic Transformation," pp. 163–74.

Glossary

cartogram name applied to a variety of representations; used synonymously with value-by-area map or spatial transformation

contiguous cartogram a value-by-area map in which the internal divisions are drawn so that they join with their neighbors

counting unit small spatial unit used in the manual preparation of value-by-area maps

noncontiguous cartogram a value-by-area map in which the internal divisions are drawn so that their boundaries do not join their neighbors; internal units appear to float in mapped space

orientation the internal arrangement of the enumeration unit within the total transformed region; cartogram communication relies heavily on the map reader's knowledge of the geography of the study area

shape quality a bridge allowing the reader to perceive the new value-by-area transformation of the original geographic base map; shape recognition is critical—without it, confusion results and communication fails

two-variable value-by-area cartogram a value-by-area map on which a second, related variable is mapped using area shading (chorograms) or graduated symbols

value-by-area map name applied to the form of map in which the areas of the internal enumeration units are scaled to the data they represent

Readings for Further Understanding

Burrill, Meredith. "Quickie Cartograms." *Professional Geographer* 7 (1955): 6–7.

Cole, John P., and Cuchlaine A. M. King. *Quantitative Geography*. London: John Wiley, 1968.

Cuff, David J.; John W. Pauling; and Edward T. Blair. "Nested Value-by-area Cartograms by Symbolizing Land Use and Other Proportions." *Cartographica* 21 (1984): 1–8.

Dent, Borden D. "A Note on the Importance of Shape in Cartogram Communication." *Journal of Geography* 71 (1972): 393–401.

———. "Communication Aspects of Value-by-Area Cartograms." *American Cartographer* 2 (1975): 154–68.

Eastman, J. R.; W. Nelson; and G. Shields. "Production Considerations in Isodensity Mapping." *Cartographica* 18 (1981): 24–30.

Getis, Arthur. "The Determination of the Location of Retail Activities with the Use of a Map Transformation." *Economic Geography* 39 (1963): 1–22.

Griffin, T. L. C. "Cartographic Transformation of the Thematic Map Base." *Cartography* 11 (1980): 163–74.

———. "Recognition of Areal Units on Topological Cartograms." *American Cartographer* 10 (1983): 17–28.

Haro, A. S. "Area Cartogram of the SMSA Population of the United States." *Annals* (Association of American Geographers) (1968): 452–60.

Harris, Chauncy. "The Market as a Factor in the Localization of Industry in the United States." *Annals* (Association of American Geographers) 44 (1954): 315–48.

Harris, Chauncy, and George B. McDowell. "Distorted Maps, A Teaching Device." *Journal of Geography* 54 (1955): 286–89.

Hunter, John M., and Melinda S. Meade. "Population Models in the High School." *Journal of Geography* 70 (1971): 95–104.

Hunter, John M., and Johnathan C. Young. "A Technique for the Construction of Quantitative Cartograms by Physical Accretion Models." *Professional Geographer* 20 (1968): 402–406.

Kelly, J. "Constructing an Area-value Cartogram for New Zealand's Population." *New Zealand Cartographic Journal* 17 (1987): 3–10.

Kidron, Michael, and Ronald Segal. *The State of the World Atlas.* New York: Simon and Schuster, 1981.

Loftas, Tony, ed. *Atlas of the Earth.* London: Mitchell Beazley, 1972.

Monmonier, Mark S. *Maps, Distortion and Meaning.* Resource Paper No. 75–4, Association of American Geographers. Washington, D.C.: Association of American Geographers, 1977.

———. "Nonlinear Reprojection to Reduce the Congestion of Symbols on Thematic Maps." *Canadian Cartographer* 14 (1977): 35–47.

Olson, Judy M. "Noncontiguous Area Cartograms." *Professional Geographer* 28 (1976): 371–80.

Raisz, Erwin. *General Cartography.* New York: McGraw-Hill, 1948.

———. *Principles of Cartography.* New York: McGraw-Hill, 1962.

———. "The Rectangular Statistical Cartogram." *Geographical Review* 24 (1934): 292–96.

Rowley, Gwyn. "Landslide by Cartogram." *Geographical Magazine* 45 (1973): 344.

———. "The World: Upside Down, Inside Out," *The Economist* (December 22, 1984): 19–24.

Tobler, Waldo R. "A Continuous Transformation Useful for Districting," *Annals* (New York Academy of Sciences) 219 (1973): 215–20.

———. "Geographical Area and Map Projections." *Geographical Review* 53 (1963): 59–78.

———. *Map Transformations of Geographic Space.* Unpublished Ph.D. Dissertation, Department of Geography, University of Washington, 1961.

Tufte, Edward R. *The Visual Display of Quantitative Information.* Cheshire, CT: Graphics Press, 1983.

Van Chi-Bonnardel, Rezine. *The Atlas of Africa.* New York: Free Press, 1973.

Chapter

11

Dynamic Representation: The Design of Flow Maps

Chapter Preview

Maps that show linear movement between places are called flow maps, and because of this quality, are sometimes referred to as dynamic maps. Symbols on quantitative flow maps are lines, usually with arrows to show direction, that vary in width. Some flow lines are uniform in thickness and these are often referred to as desire lines. Flow mapping began with those done by Henry Drury Harness in 1837. Little flow mapping was done by government agencies in the United States in the 19th century, but with the impetus supplied by the study of economic geography, flow mapping became common in geography textbooks in the early decades of the 20th century. Flow maps used in textbooks are divided into three classes: radial, network, and distributive. Particular attention must be paid to total map organization and figure-ground when designing flow maps because of their complex graphical structures. The selection of the projection for the flow map, line scaling and symbolization, and legend design are the chief design elements in flow mapping. Unique solutions, including computer graphics, should be explored in the design stages of producing a flow map.

The Purpose of Flow Mapping

Maps showing linear movement between places are commonly called **flow maps,** or sometimes **dynamic maps.** Flow line symbolization is used when the cartographer wants to show what kind of (qualitative) or how much of (quantitative) movement there is between two places. For the quantitative variety, the widths of the flow lines, or bands, connecting the places are drawn proportional to the quantity of movement represented. Any time the cartographer wishes to show movement between places, and has the data to support this theme, a flow map is appropriate. Purpose may

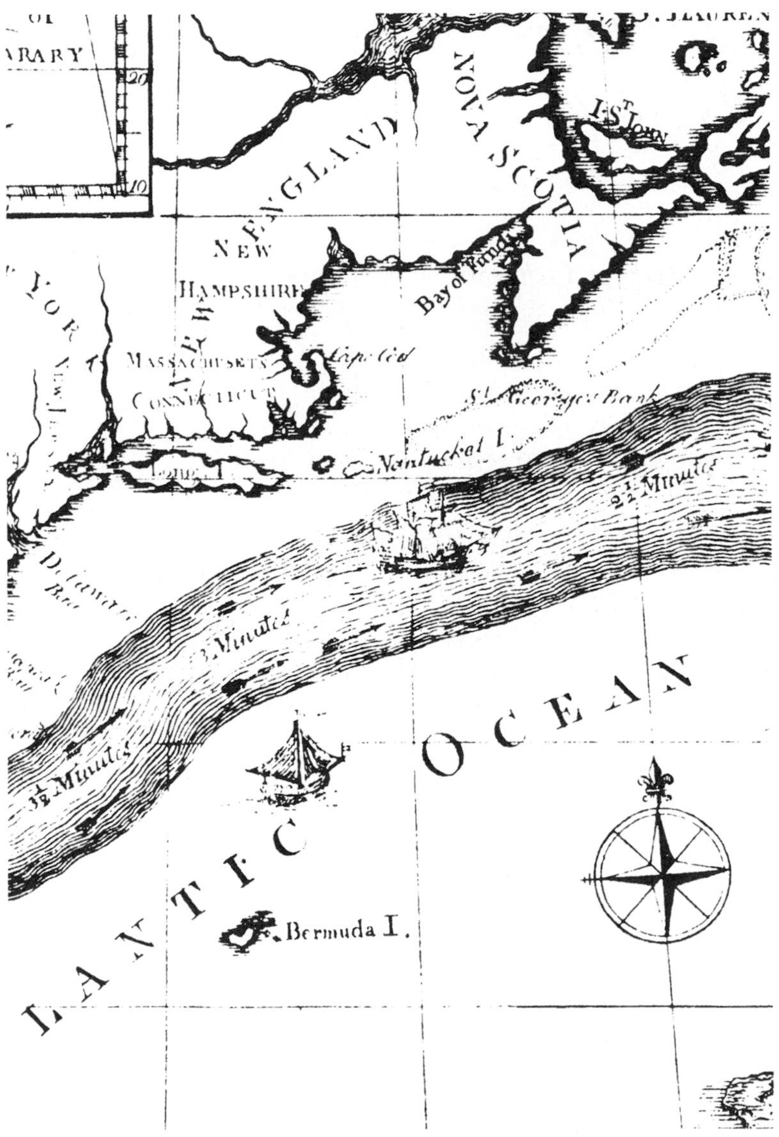

Figure 11.1. Chart of the Gulf Stream, Commissioned by Benjamin Franklin, 1786.
An early non-quantitative flow map.

be to show movement of actual items, or movement of ideas.

In most instances, except for the desire line case, the cartographer attempts to show the actual route taken by the movement, although this may be difficult because of map scale and the level of generalization selected for the map. Actually, because of the nature of the symbolization process, and inherent complexity of these maps, all flow line maps become highly generalized. In the last edition of his book, Raisz places them into a class of maps called cartograms.[1] However, not to be confused with value-by-area cartograms, flow maps are more correctly called linear cartograms.

Flow line symbols may be used on maps to show nonquantitative movement, as mentioned above. The lines are unscaled and generally have arrowheads to indicate direction of movement. Lines are usually shown of uniform thickness. Maps showing shipping routes, airline service maps, ocean currents, migration flows, and other similar presentations are examples. Early non-quantitative flow maps date back to the late 18th century. (See Figure 11.1.) It is possible to map different types of flow data on one map by using different line characters.

Quantitative Flow Maps

Lines on quantitative flow maps are scaled such that their widths are proportional to the amounts they represent. (See Figure 11.2.) Weight, volume, value (dollar), and amount or frequencies are often the units used on quantitative flow maps. Data may be in nominal, ordinal, or interval levels. Absolute or derived values may be used. Direction may or may not

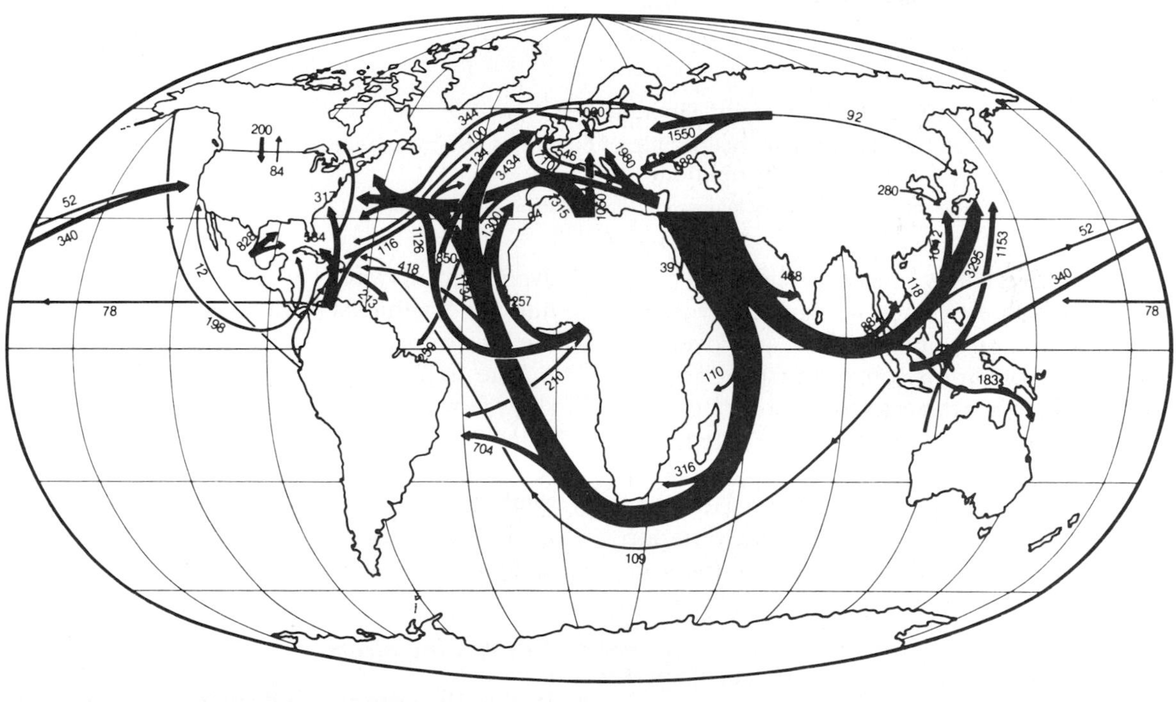

Figure 11.2. International Crude Oil Flow, 1980. (Thousand barrels per day.) No legend is used on this map but the lines are labeled to indicate magnitude of flow.

Source of map: U.S. Department of Energy, Energy Information Administration, *1981 International Energy Annual.*

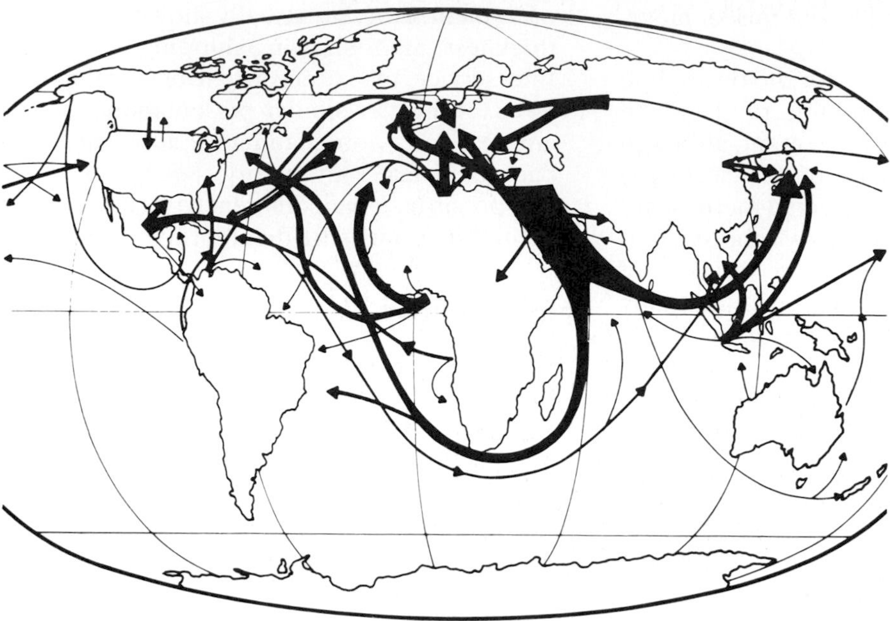

Arrows Indicate Origin and Destination
But Not Necessarily Specific Routes.

Figure 11.3. International Crude Oil Flow, 1985.
No legend appears on the map and the magnitude of flow lines are not labeled. The reader is to gain an overall impression of the pattern of distribution.

Source of map: U.S. Department of Energy, Energy Information Administration, *1986 International Energy Annual.*

be shown on the flow lines. International commodity flows are frequently mapped by quantitatively scaled lines, such as overseas movements of grains, ores, and produce. In many instances, the cartographer does not place amounts on the maps (as in a legend), but relies on the map reader to judge relative amounts visually. (See Figure 11.3.) However, direction of movement is frequently important, so arrowheads are likely part of the symbolization.

Traffic Flow Maps

Mapping traffic flows is uniquely suited to flow line symbolization and is the oldest form of this kind of map (see below). Varying line widths are used to symbolize the number of vehicles passing over portions of rail, water, road, air, or even bird flyways. (See Figure 11.4.) The number of vehicles passing a certain point over the last twenty-four hours is a typical example. A state traffic map, for another example, may show flow bands proportional to the annual average 24-hour traffic volumes. It is common to see **traffic flow maps** without directional symbols. They are often used to show the organizational and hierarchical nature of urban systems.[2]

Desire Line Maps

Desire line maps are a special case of quantitative flow maps and are unique in that they do not attempt to portray actual routes followed or the type of transportation used.[3] Typically, they illustrate social or economic interaction by use of straight lines connecting

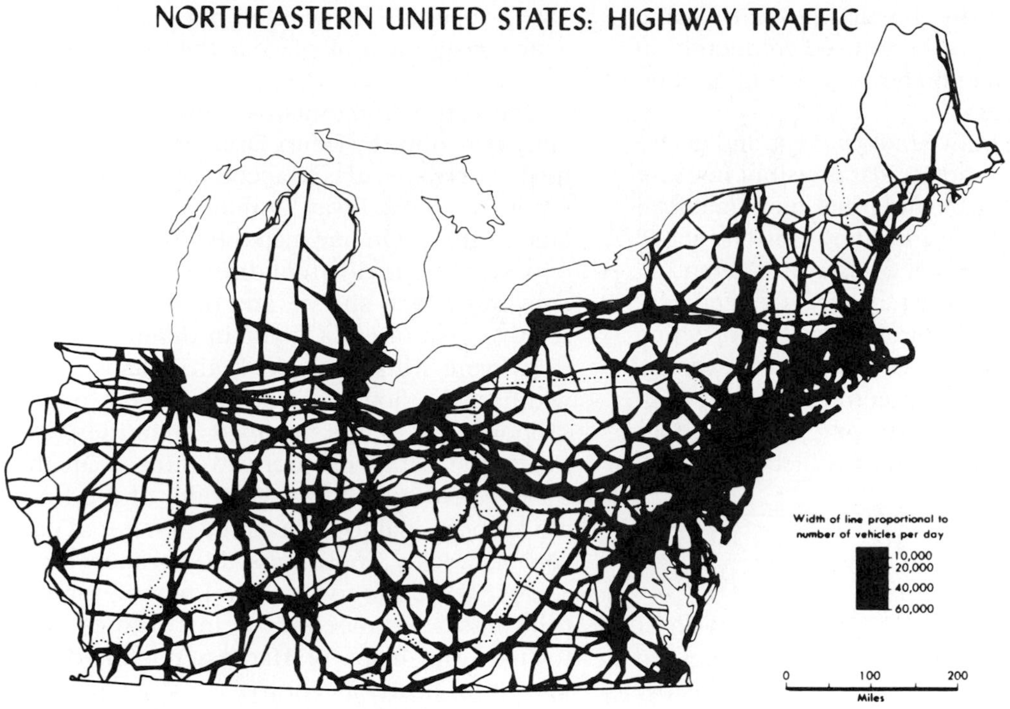

Figure 11.4. A typical traffic flow map.
In many instances such as this one direction of flow is not mapped. Width of lines shows number of vehicles passing in both directions for a specified time period. Reprinted by permission of Prentice-Hall, Inc.

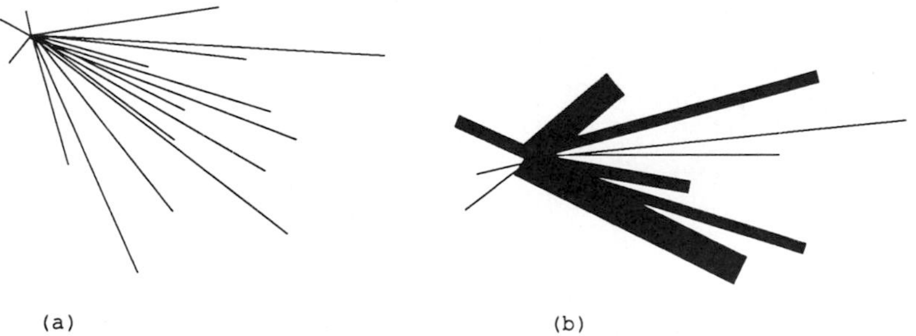

Figure 11.5. Desire line maps.
In (a) the lines are of uniform thickness (often joining origin and destination of the traveler). In (b) the desire lines vary in thickness based on aggregate interaction.

points of origin or destination. (See Figure 11.5.) Often lines represent the movement of one person between points (and a number of pairs are drawn on the same map), but they can be drawn between enumeration units and symbolize aggregated movement data. In the former, each line has the same width, and in the latter, line widths are scaled to magnitudes as in typical flow mapping.

The application of desire line maps is best found in those instances where nodal geographical patterns are to be focused on, or possibly where urban hierarchies are being stressed.[4] They have been used frequently in this latter instance to show shopping or commuting structures.

Demarcation between flow maps and desire line maps is often not a clear division. Because of limitations of scale and inherent generalization that takes place in symbolization, not to say anything about lack of data, it is often impossible to show exact routes. Generalized, often diagrammatic routes are selected, which may look like desire lines although the cartographer sets out to do something else. As with other forms of mapping, the purpose of the map will set the stage and ultimately dictate choices.

Historical Highlights of the Method

Our discussion here is limited by space and coverage, but a few remarks about the history of flow mapping are necessary to place this form of mapping in the correct historical perspective. Quantitative flow mapping began in the "golden age" of statistical cartography in western Europe, in the two decades preceding the middle of the 19th century. Never prevalent in agency mapping in the United States, this form of map was utilized heavily in economic geography text books throughout the first half of the twentieth century.

Early Flow Maps

Early statistical cartography had its start in the late 1700s, although a few such maps appeared before that time. One notable entry before the late 1700s was the famous isogonic maps made by Halley in England in 1701.[5] Another was the first contour map showing depths in the English Channel made by Phillippe Buache in France 1752.[6] By the late 1700s and early 1800s, however, a variety of statistical maps had begun to emerge. By the year 1835, western Europe had entered the "golden age" of geographic (statistical) cartography, a period lasting to roughly 1855.[7] It should be pointed out that little statistical mapping took place in the United States prior to 1850.[8]

The earliest quantitative flow maps apparently were done by Henry Drury Harnesss when he prepared the atlas to accompany the second report to the Railway Commissioners of Ireland in 1837.[9] On one map the relative number of passengers in different directions by regular conveyance was shown, and on the other, the relative quantities of traffic in different directions were illustrated. In both instances the width of the flow lines or bands were drawn proportional to the mapped quantities. Actually, the data were derived values. On the traffic conveyance map, the width of the lines are proportional to the average number of passengers weekly. These maps did not show exact routes, but showed straight lines of varying thicknesses connecting points (cities and towns). The maps of Harness remained unknown for nearly a century.[10]

Within ten years of the production of the flow maps of Harness, Belpaire of Belgium, and more notably, Minard of France began publishing flow maps of essentially similar designs to those of Harness.[11] Charles Joseph Minard was more productive than Belpaire, and Minard's interests were primarily in the areas of economic geography. Minard apparently had no contact with geographers or cartographers, although he was instrumental in popularizing the flow line technique among statisticians. By his own account, he was interested in showing quickly by visual impression numerical accuracy.[12] So much so, that he often overgeneralized other portions of his maps so that the flow lines themselves would command attention. This design aim is still relevant today. Although Minard did not invent the flow map, one contemporary authority has said that he did bring ". . . that class of cartography [flow maps] to a level of sophistication that has probably not been surpassed".[13]

Minard produced some fifty-one maps altogether, most of which were flow maps.[14] The mapped subjects of the flow maps were varied, and included such topics as people, coal, cereal, mines, livestock, and others. He mapped the distribution or flow of these commodities not only in France, but worldwide as well. His style, especially on the maps showing movement of travelers on principal railmaps in Europe (1862), is the same used today. Perhaps the most unique and provocative of his illustrations is the flow chart showing the demise of Napoleon's army in Russia; as suggested by Tufte, "It may well be the best statistical graph ever drawn."[15]

Flow Maps in Economic Geography

From the time of Harness, Belpaire, and Minard quantitative flow maps have been used to map patterns of distribution of economic commodities, people (passengers), and any number of measures of traffic densities. As suggested earlier, the flow maps of Minard reached a sophistication of technique never really matched by others since. There was a period of time, however, in the decades of the first half of this century, that a great many quantitative flow maps appeared in college economic geography text books, and in many cases with laudable design techniques.

In a recent study of flow mapping in college geography text books, several hundred flow maps were found among seventy-one books published between 1891 and 1984.[16] Although the author did not consult all books in this category in this time period (an enormous task), the study is remarkable for its breadth and attention to detail, and summarizes carefully the findings of the study and includes numerous examples of this kind of mapping. Although used extensively in this publishing medium, in general flow mapping lagged behind other forms of thematic mapping techniques.[17]

It is interesting to note from the study done by Parks that qualitative flow maps appeared in these books as early as 1891, and that this form outnumbered the quantitative form by three-to-one.[18] Qualitative flow maps are used to illustrate migration routes, explorers routes, and transportation networks. Desire line maps as a category of qualitative flow maps are used also, especially after 1960. The earliest quantitative maps, from those textbooks studied by Parks, appeared in 1912. Transportation themes are most often illustrated by the quantitative flow maps, especially international import and export of agricultural commodities.[19]

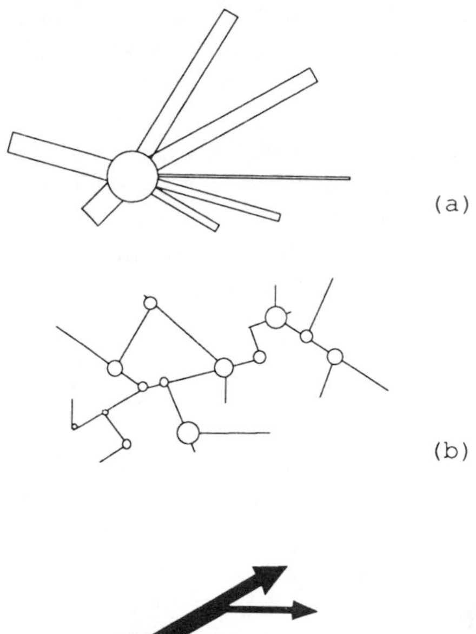

Figure 11.6. Classification of flow line maps in geography textbooks.
A radial type is illustrated in (a). The network class is represented in (b), and the distributive type in (c).

The study by Parks yields a classification of flow map design, including these three distinct patterns: radial, network, and distributive.[20] **Radial designs** are easily distinguished by a radial or spoke-like pattern, especially when the features and places mapped are nodal in form. Present day traffic volume maps fit into this category. (See Figure 11.6a.) **Network flow maps** are those that are used to reveal the inter-connectivity of places, especially evidenced by transportation or communication linkages. (See Figure 11.6b.) Airline route maps are a good example of this kind.

Flow maps in the third class include those that present the **distribution** of commodities or migration flows. Trade flows, such as shipments of wheat among countries, is a good example of this form. (See Figure 11.6c.) Maps that show diffusion of ideas or things are included in this class. Maps illustrating diffusion are often found in textbooks on cultural geography.

Topics mapped by the flow line technique are quite varied, and suggest the innovation often employed by cartographers and geographers in using this method. Railway and airline route maps are common (showing interconnectivity between places). Shipments of national gas, wheat, animals, migration of people, ore, coal, and cotton are just a few examples of the topics commonly found among flow maps. Trade flows of marine harvests is yet another. Migration routes, such as by American Indians, the French, Spanish, and English peoples in settling the New World, and others have been mapped by the flow map technique. There are few subjects that contain a from-to relation that cannot be mapped by flow symbolization.

This brief examination of flow maps in textbook cartography is intended to provide a backdrop only to the fascinating study of this form of mapping. Space will not permit a detailed presentation of hundreds of examples and the rich variety of design found among these maps. The thematic map design student is urged to explore examples from the actual textbooks themselves. A worthwhile design activity can be found in such an experience.

Designing Flow Maps

Creating effective flow maps through a careful and thoughtful design plan represents clearly one of the more difficult challenges for the map designer. There are three aspects to their design that must be considered: map organization and figure-ground (including the selection of the projection), line symbolization and data scaling, and legend design.

Map Organization and Figure-Ground

The map's hierarchical plan must be carefully considered. It seems apparent that the flow lines, or desire lines, are to be the most dominant marks on the map. As with other forms of thematic mapping, they should be placed high in the hierarchy so they clearly stand out as strong figures. (The topic of the visual hierarchy and map design is dealt with in greater detail in the next section.) Several figures in this chapter, notably Figures 11.2 and 11.3, illustrate this idea. However, achieving this hierarchy is sometimes difficult because the flow lines may stretch over several different levels on the maps. For example, lines may first be over land, then water, then land again and so on. They may intersect other flow lines, creating confusion for the reader.

As with other thematic symbol types, and as explained in the next section, flow lines should have strong edge gradients and be rendered so that visual conflict with other symbols do not result. (See Figure 11.7.) Interposition, which can be achieved easily by not having transparent symbols, is a technique often helpful in designing strong symbols. In the case here in Figure 11.7, rendering the flow lines black (or non-transparent white) is a way to improve the thematic symbols on this map.

Flow maps are ordinarily quite complex visual graphics. On most thematic maps the organization of the graphic components, land and water, symbols, titles, legends, and other marks on the maps tend to fall neatly into an easily followed plan. Land and water contrasts are developed by following figure-ground principles, and symbols usually occupy space over land areas. Title and legends are dealt with

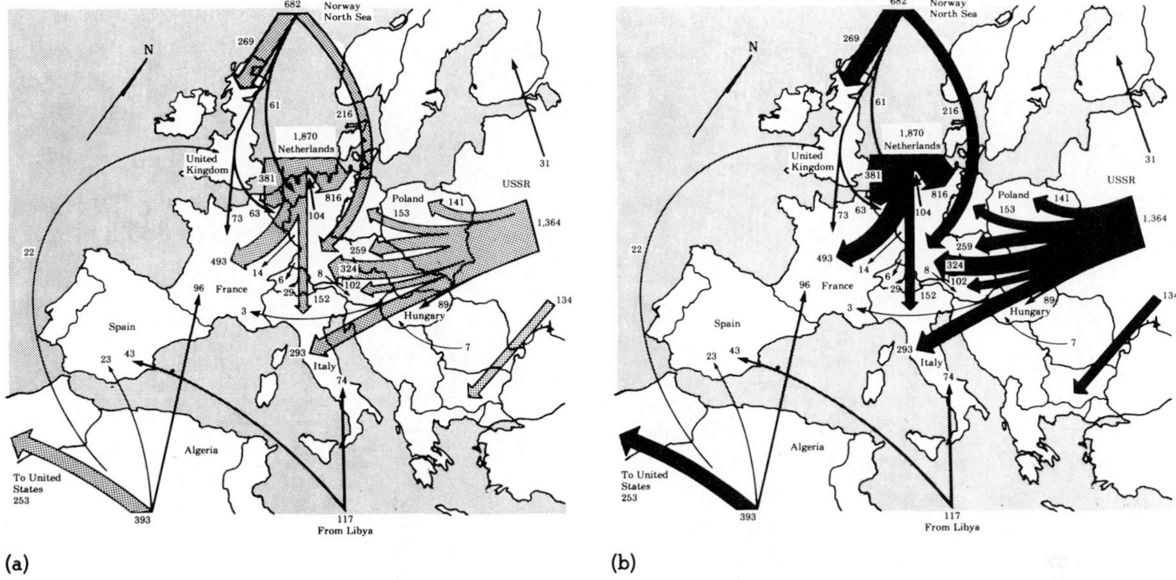

Figure 11.7. Flow lines that lack strong figure characteristics.
The flow lines in (a) could be stronger figures by either eliminating the coastlines and boundaries beneath them, or by rendering them as black symbols, as in (b).
Source of map: U.S. Department of Energy, Energy Information Administration, *1981 Annual Report to Congress, Vol. 2, Energy Statistics,* p. 112.

similarly. But on many flow maps, especially those illustrating international movements, the thematic symbols are likely to occupy spaces both over land or water, or both, and many times the lines themselves are intertwined and appear to rest in different visual levels. The nature of the visual complexity on many flow maps, then, creates unusual and challenging design problems for the cartographer.

Minimally, the cartographer should provide clear land and water distinctions on flow maps. The flow symbols must be dominate figures in perception, with strong edge gradients and clear continuity. Labelling flow lines with their values, often useful to assist the reader, should not interfere with the symbol's visual integrity. Using patterns or screens on flow symbols should not lead to confusion with other areas on the maps. The scaling of the flow lines should not cause them to be too large for the maps, which can result in too little base map information showing through. Attention to design details such as these will assist the cartographer in reaching successful results.

Projection Selection

Perhaps of equal importance to that of achieving a good visual hierarchy on the flow map is the necessity of selecting an appropriate projection. Placement of the center of the flow, if there is a center, must be strategically planned and this may require careful consideration of the projection, its center, and aspect. Placement and design of the flow lines should be done so that the map does not become an incomprehensible mesh of confusing lines. (See Figure 11.8.) For flow mapping, the equal-area and conformality attributes of projections may not be as important as other factors, such as continental shapes.[21]

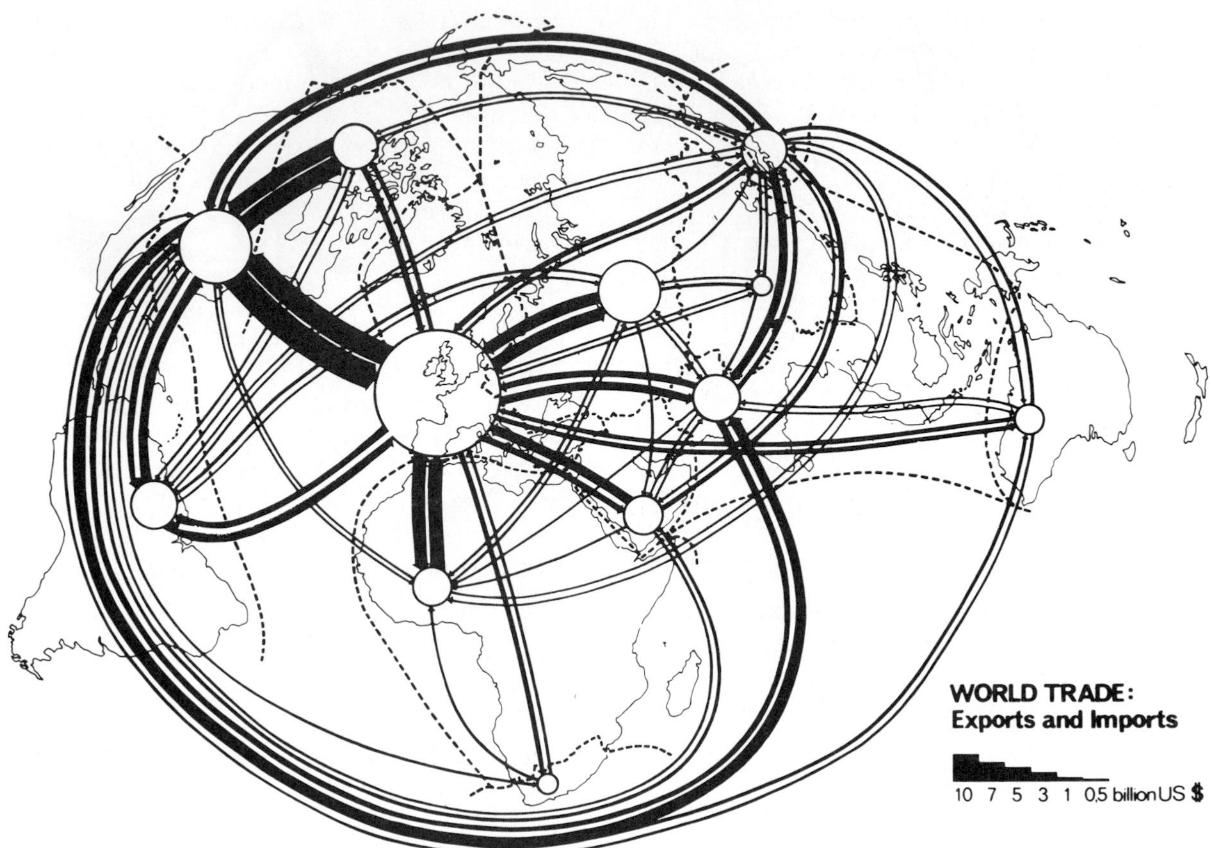

Figure 11.8. World Trade: Net-Exports, 1967.
The selection of the projection (uncommon in this case), its orientation in the map frame, and the nature of the flow distribution all contribute to a confusing array of lines.
Reprinted by permission of Van Nostrand Reinhold Company, Ltd.

Selecting a projection, adoption of its center, choosing an aspect for it, and placement in the map frame must all be considered when developing a flow map. The final map that results from these choices should be one that connotes organization and control of the mage, and deftly satisfies the purpose of the map. It is the responsibility of the cartographer to know the flow pattern he or she wishes to portray, then make decisions regarding the projection that best illustrates this pattern. For example, if the pattern is single origin to multiple destinations, or if the pattern reflects multi-nodal origins and a single destination, the employment of the projection should complement the pattern.

Balance and layout of the map's elements concludes the design activities related to achieving the visual and intellectual plans for the map. Placement of map objects and utilization of space is dealt with so that a pleasing result is reached, defined as one in which no other solution seems merited. The map's elements, as with other thematic map forms, are placed in intellectual order and treated graphically to satisfy the plan. Titles, legends, scales, source materials, and other elements are therefore treated accordingly. However, the thematic symbols, the lines themselves, remain the most important features of the map and all other elements are second in importance.

Essential Design Strategies

A summary of the essential design strategies for flow maps should include these:

1. Flow lines are highest in intellectual and therefore highest in visual/graphic importance.
2. Smaller flow lines should appear on top of larger flow lines.
3. Arrows are necessary if direction of flow is critical to map meaning.
4. Land and water contrasts are essential (if the mapped area contains both).
5. Projection, its center and aspect, are used to direct readers attention to the flow pattern important to the map's purpose.
6. All information should be kept simple, including flow line scaling.
7. Legends should be clear and unambiguous, and include units where necessary.

Line Scaling and Symbolization

On most quantitative flow maps, the width of the flow lines are proportionally scaled to the quantities they represent. Thus, a line representing 50 units will be five times the width of one symbolizing 10 units. This practice has been employed since the time of Harness, Belpaine, and Minard. Perceptually, the reader is being asked to make this visual judgment, and ordinarily this can be done by most readers in a linear fashion.[22] Scaling, therefore, appears to be straight forward and should not pose severe problems for the designer.

Chief among the concerns for the designer is the data range that must be acommodated by the widths selected. In many instances this poses considerable restraint on the project. Ordinarily the best plan is to select the widest line that can be placed on the map (and still preserve the integrity of the base map), and then determine, by looking at the data range, the narrowest line on the map. If the widest line is 1.27 cm (.5 in), and the data range is 5000 units, then the smallest line would be .000254 cm (.0001 in), obviously too small to draft on the map sheet. Either the widest line would have to be enlarged, already determined to be unacceptable, or some other solution would need to be reached. It is important to note, too, that as the linear symbols become wider, they cease to appear as such, and may take on the qualities of area symbols. This limit should be avoided.

There appears to be three ways out of this dilemma: abandon the map in favor of some other form, provide a standard line to represent values below a certain critical value (and symbolize the remaining values by conventional methods), or range-grade the values and symbolize the resulting classes by scaling the flow lines proportionally to the midpoint of the classes. In some cases, both the second and third options can be combined on the same map. (See Figure 11.9.)

Other methods in addition to proportionally scaling line widths have been used to symbolize flows. One method is a dot method in which small dots, each representing a unit of flow (for example, 100 vehicles) are placed along the route.[23] Visual impression of volume differences are quickly noted, routes are easily followed, and if necessary, the reader can count the dots to retrieve actual magnitudes. (See Figure 11.10a.)

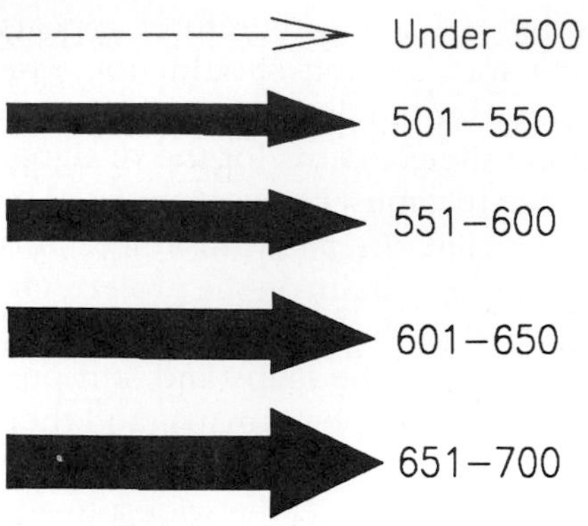

Figure 11.9. Use of a standard line in scaling flow lines.
When the data range is too great to scale all flow lines proportionally, a standard line may be selected to symbolize several values.

One other method is to render several lines of uniform thickness parallel to the route, and provide the proper number of these lines to represent the volumes. (See Figure 11.10b.) This method yields results similar to the conventional method and, in addition, the individual lines can be counted for greater precision, if necessary. One disadvantage to this method is that compilation is more difficult and time consuming.

Yet one other scaling solution can be reached by a method analagous to applying patterns to choropleth maps. In this method, broad lines of uniform thickness are drawn connecting points in the mapped space, and depending on the numerical class into which the line segment falls, are given an areal pat-

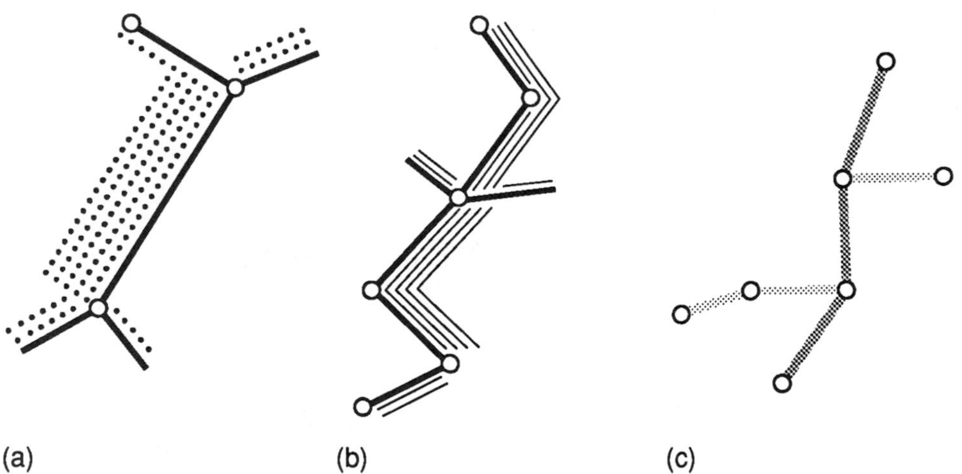

Figure 11.10. Alternative methods of symbolizing quantitative flow lines.
In (a) dots symbolizing a unit of flow are placed along side two actual routes. In (b) parallel lines are used, with each line representing a unit of flow. Finally, in (c) lines of uniform thickness are given areal patterns based on numerical classes.

tern.[24] (See Figure 11.10c.) Areal symbols are selected as they are in choropleth mapping, that is, higher values are represented by darker (higher percent area inked) patterns or screens. Color may be used instead of patterns or screens to represent the different classes. Here different color intensities of the same hue would be selected to represent the different classes. If the segments were classed into different nominal classes, however, different hues may be chosen.

This list of alternative scaling methods for flow lines is not meant to be exhaustive, but rather to be representative. Enterprising designers will no doubt add to these possibilities and newer solutions may be forthcoming. Regardless of the method, however, the scaling method must visually suggest proportional flow and do so easily.

Symbol Treatment

No rules have emerged that govern how flow lines should be treated graphically in all map cases. The only convention appears to be the way distributive flow lines are treated. (See Figure 11.11.) When mapping flows that separate into smaller flows, the widths of the individual branches should add up to the width of the trunk. This makes intuitive sense. The same applies when smaller branches come together to make a larger one.

If arrowheads are used they should be clear and be scaled proportionally to the lines to which they belong. (See Figure 11.12.) Arrowheads with small shoulders should be avoided. If flow lines overlap, smaller ones should be made to appear on top of larger ones. (See Figure 11.13.) This is accomplished by

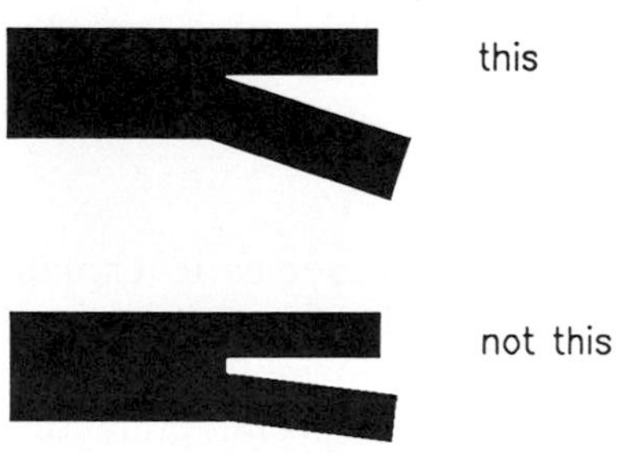

Figure 11.11. Symbolization of flow lines. When quantitative flow lines branch or unite, their widths are drawn proportionally.

Figure 11.12. Proper arrowheads improve the appearance of flow lines.

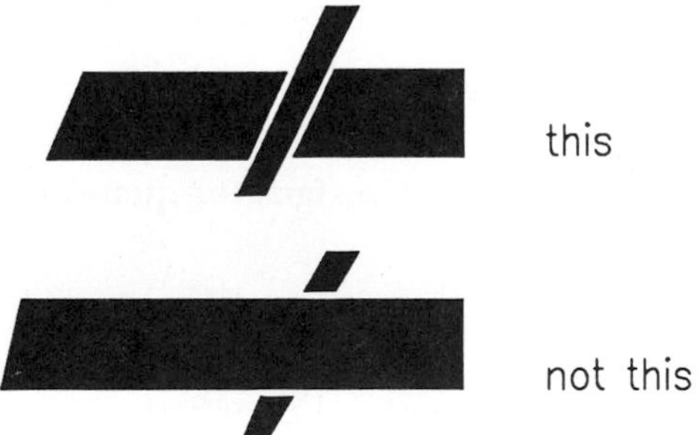

Figure 11.13. Interposition is useful in flow line symbolization. Interposition is a technique useful in making smaller lines appear to rest on top of larger ones.

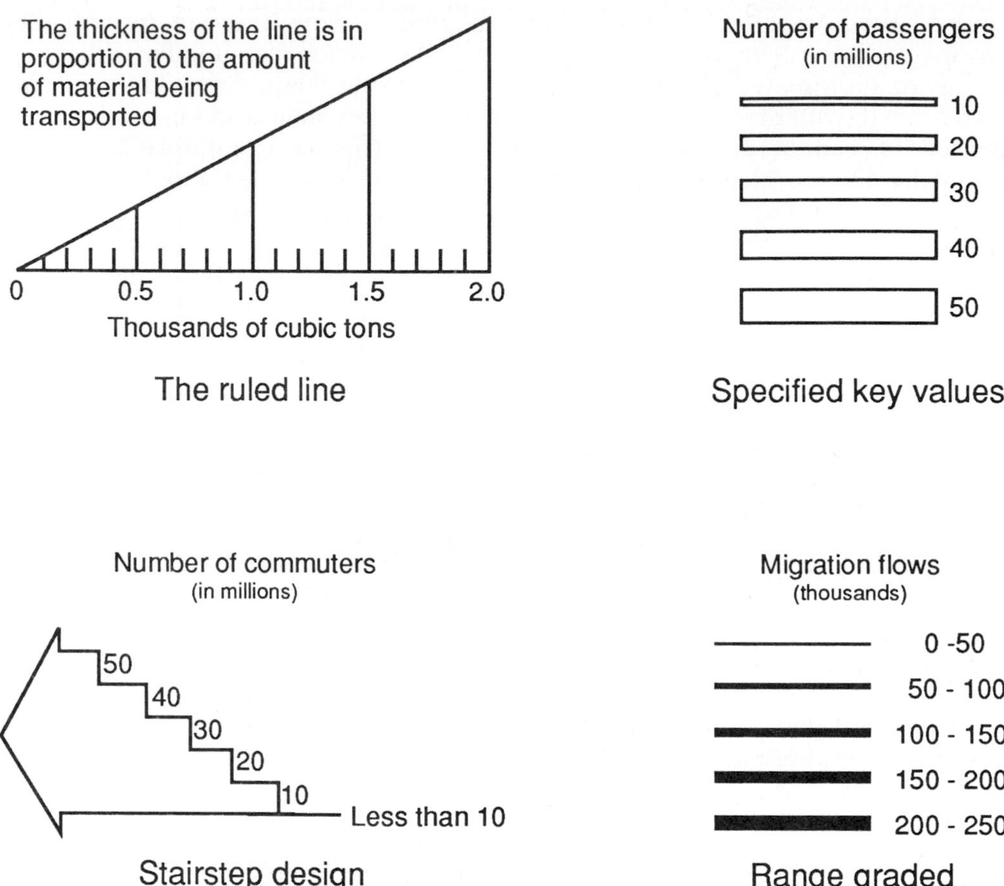

Figure 11.14. Legend designs for quantitative flow maps.

breaking the lines of the larger flow line, and having the smaller one continue uninterrupted. This is a simple and effective technique that adds a plastic and fluid dimension to the map. More will be said of interposition in the next section.

Legend Design

One of the more difficult tasks in the design of the flow map is with the preparation of the legend. The legend is the crucial link in cartographic communication between cartographer and map reader as it serves to explain carefully the symbols on the map. The legend must above all be clear and unambiguous. Units of measurements must be prominently displayed and it must be obvious how the lines are scaled, and the flow lines in the legend must appear exactly as they are on the map. If the data have been classed, the class boundaries should be clearly represented.

In one study of flow maps four general types of legend designs were identified: the ruled line, scaled bar or triangle, graduated lines, and key values (including "stairstep" designs).[25]

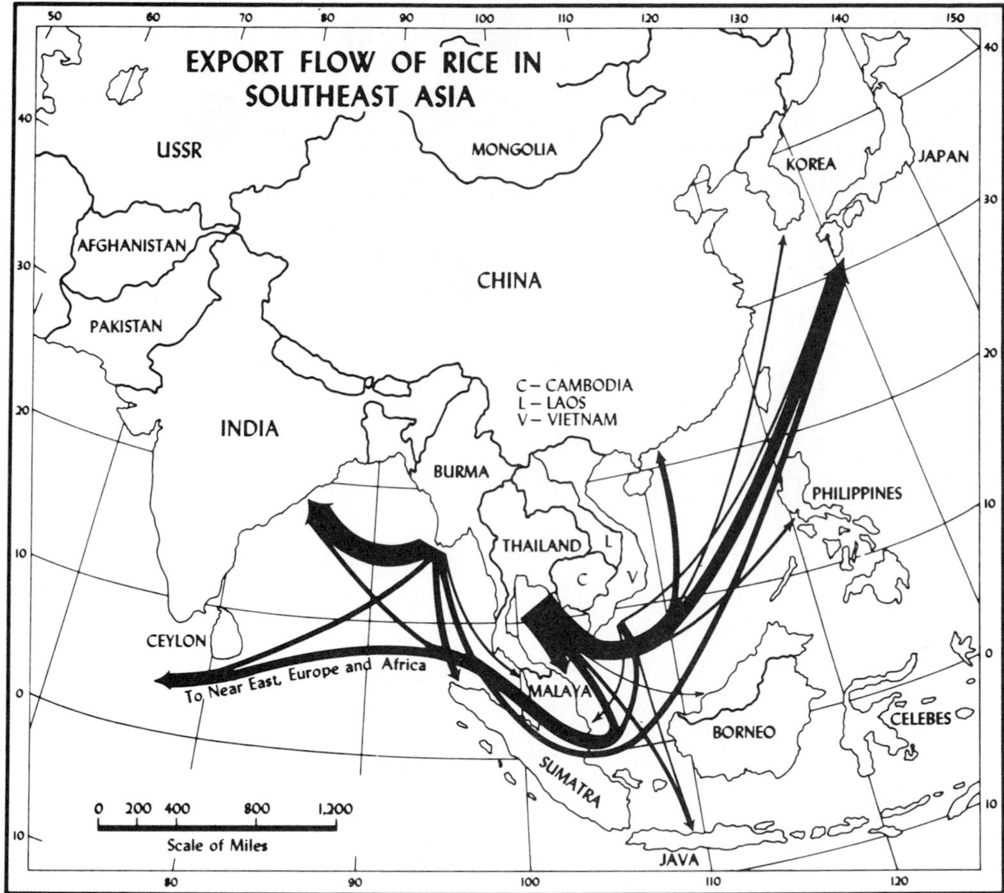

Figure 11.15. **Export Flow of Rice in Southeast Asia.**
See text for explanation.
Reprinted by permission of Prentice-Hall, Inc.

(See Figure 11.14.) There have been no exhaustive studies on flow map legend designs. In the meantime cartographers must rely on their best judgments. A good idea would be to try out the design solution on representative readers before a final decision is made.

In some cases, the values of the lines on the map may be labeled for greater precision. (See Figure 11.2.) The overall pattern of the lines shows the organization of the movement, and the labelled values provide detailed information for the reader seeking tabular data. Whether or not a legend also is used in these cases depends on the map purpose. In some respects this appears redundant. And in some cases, although the lines are scaled and drawn proportionally, no labels at the lines are provided, and no legend is included. (See Figure 11.15.) Presumably, the cartographer is wishing only to show geographical organization and pattern. It would seem imperative that these drawings be accompanied by written narrative.

Unique Solutions

A number of innovative and unique solutions for flow mapping have been used by cartographers and geographers. Up to this point in our discussion, the majority of examples have been drawn from a rather conventional pool. Other solutions do exist and can be mentioned to stimulate experimentation. One such example illustrates connectivity by using flow lines of uniform width, and quantity of shipments by proportional circle.[26] (See Figure 11.16a.) Another interesting solution, quite different, shows movements of coal by constant width flow lines connected at their ends by bar graphs scaled to the magnitudes of shipments.[27] (See Figure 11.16b.)

A kind of flow line has been used in centrographic studies. In these examinations geographical centers of gravity for different time periods are connected by lines of uniform thickness.[28] (See Figure 11.16c.) The visual result shows the geographical trend over time and the flow lines show the connectivity from time to time. These are largely descriptive studies, although centrographic analyses can be employed inferentially also.

Desire lines have been used in very unique ways. Conventional use of desire lines would include lines of uniform thickness (although they may be scaled to proportional widths, and generally without directional arrows), usually placed on a geographical base map. One unusual use has them placed on a square root transformed base map.[29] (See Figure 11.17.) In this instance the map projection origin (the "origin" of the desire lines) is located in Toronto, and the lengths of the desire lines are no longer linear. While these kinds of unique solutions may reveal patterns otherwise hidden, map readers may need, by way of an explanatory legend, some tips on map reading strategy.

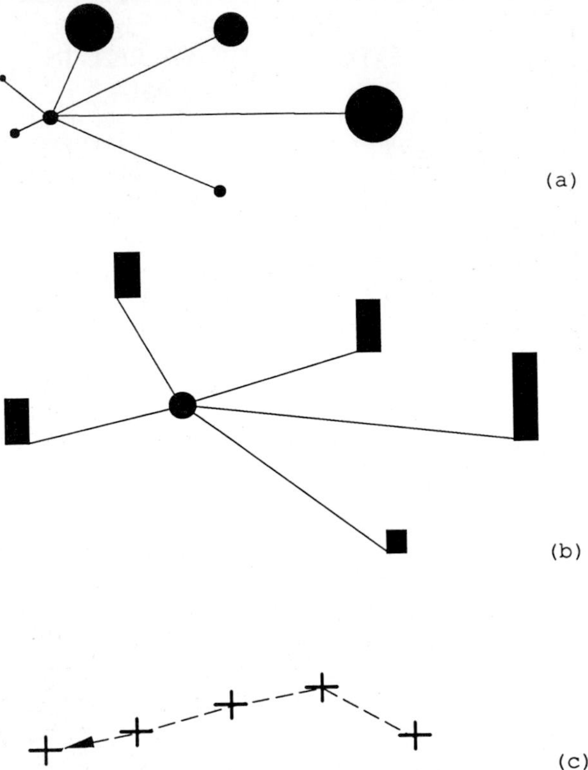

Figure 11.16. Innovative solutions to flow line representation and scaling.
See text for explanation.

Very heavily generalized and more stylized flow map solutions also have been used. (See Figure 11.18.) In the example used here showing the international trade of mineral fuels (mostly oil), the geographical base map has been abandoned in favor of a blocktype organizational chart. The author does not provide any clues about the sizes of the countries and regional rectangles, nor is there a legend de-

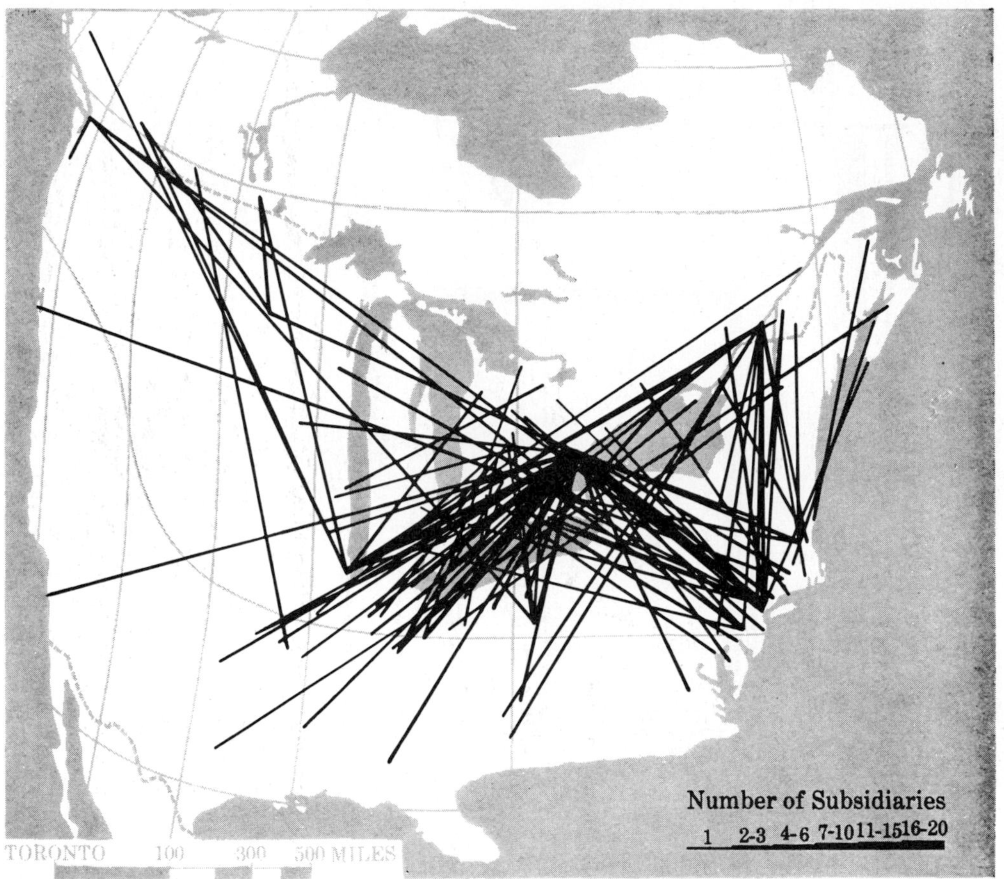

Figure 11.17. Desire lines on a square root—transformed projection, centered on Toronto, Canada.

Reprinted by permission from D. Michael Ray, "The Location of United States Manufacturing Subsidiaries in Canada," *Economic Geography* 47 (1971): 392.

scribing the widths of the flow lines. However, it is clear that Western Europe, the United States, and Japan are the three main destinations in international fuel shipping. This type of presentation illustrates organization in the activity, although at the expense of not showing actual geographical routes.

Computer Solutions

Computer solutions to flow mapping have not gone unnoticed by cartographers. In a recent study, internal migration data for the United States, in the form of a 50 × 50 (states) "from-to" table, was used in the development of a

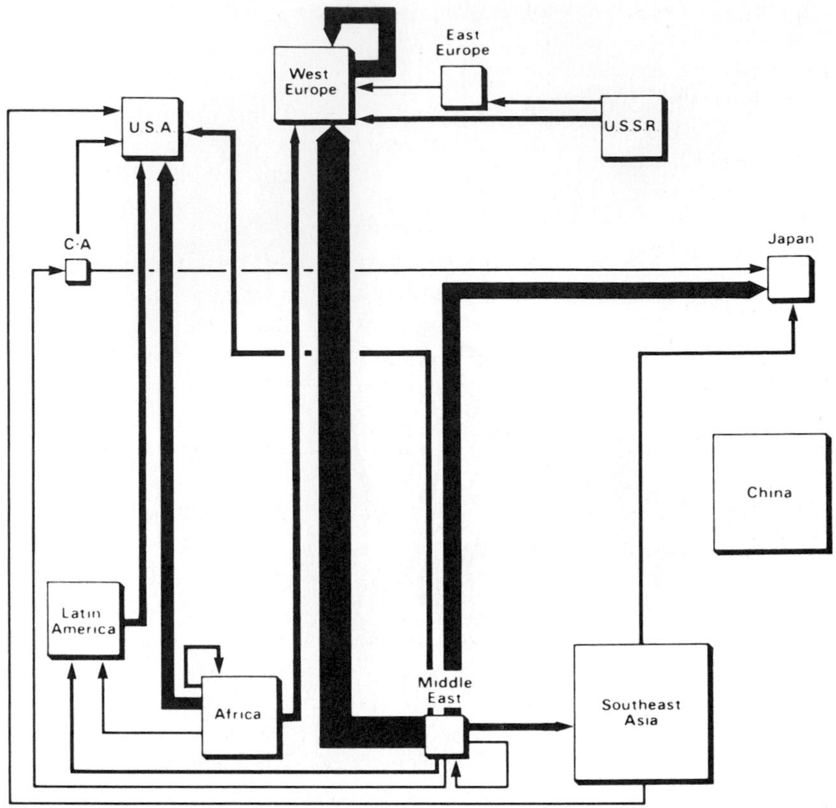

Figure 11.18. Heavily generalized quantitative flow map.
See text for explanation.
Reprinted by permission from Butterworths.

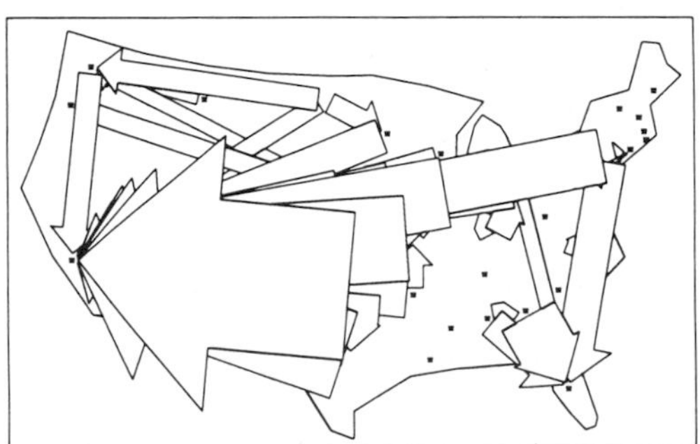

Figure 11.19. Computer solutions to a flow mapping problem.
Reprinted by permission from Waldo R. Tobler, "Experiments in Migration Mapping by Computer," *American Cartographer* 14 (1987): 159.

computer program to automatically draw flow lines or bands.[30] Line width scaling, arrowhead design, flow line overlap, and other typical graphic solutions have to be dealt with as in manual design solutions. The advantage in using the computer, however, is that alternative designs can be examined quickly. A solution is provided here in Figure 11.19.

Notes

1. Erwin Raisz, *Principles of Cartography* (New York: McGraw-Hill, 1962), pp. 218–20.
2. Stephen S. Birdsall and John W. Florin, *Regional Landscapes of the United States*

and Canada (New York: John Wiley, 1981), pp. 50–52.

3. Peter Davis, *Data Description and Presentation* (London: Oxford University Press, 1974), p. 92.
4. Brian J. L. Berry, *Geography of Market Centers and Retail Distribution* (Englewood Cliffs, NJ: Prentice-Hall, 1967), pp. 11–21; see also Brian J. L. Berry, Edgar C. Conkling, and D. Michael Ray, *The Geography of Economic Systems* (Englewood Cliffs, NJ: Prentice-Hall, 1976), p. 230.
5. H. Gary Funkhauser, "Historical Development of Graphical Representation of Statistical Data," *Osiris* 3 (1937):269–404.
6. *Ibid.*
7. Arthur H. Robinson, "The 1837 Maps of Henry Drury Harness," *Geographical Journal* 121(1955):440–50.
8. Herman Friis, "Statistical Cartography in the United States Prior to 1870 and the Role of Joseph C. G. Kennedy and the U.S. Census Office," *American Cartographer* 1(1974):131–57.
9. Robinson, "The 1837 Maps," pp. 440–50; see also Arthur H. Robinson, *Early Thematic Mapping in the History of Cartography* (Chicago: The University of Chicago Press, 1982), pp. 64, 147.
10. Robinson, *Early Thematic Mapping,* p. 147.
11. Robinson, *Early Thematic Mapping,* pp. 147–54; see also Arthur H. Robinson, "The Thematic Maps of Charles Joseph Minard," *Imago Mundi* 21(1967):95–108.
12. Robinson, "Minard," pp. 95–108.
13. *Ibid.*
14. Robinson, *Early Thematic Mapping,* p. 150.
15. Edward R. Tufte, *The Visual Display of Quantitative Information* (Cheshire, CT: Graphics Press, 1983) p. 40.
16. Mary J. Parks, "American Flow Mapping: A Survey of the Flow Maps Found in Twentieth Century Geography Textbooks, Including A Classification of the Various Flow Map Designs," unpublished Master's thesis, Department of Geography, Georgia State University, 1987, p. 14.
17. *Ibid.,* p. 34.
18. *Ibid.,* p. 39.
19. *Ibid.,* p. 43.
20. *Ibid.,* pp. 50–66.
21. Borden D. Dent, "Continental Shapes on World Projections: The Design of a Poly-Centred Oblique Orthographic World Projection," *Cartographic Journal* 24(1987):117–24.
22. George F. McCleary, Jr., "Beyond Simple Psychophysics: Approaches to the Understanding of Map Perception," *Proceedings of the American Congress on Surveying and Mapping,* 1970, pp. 189–209.
23. Gwen M. Schultz, "Using Dots for Traffic Flow Maps," *Professional Geographer* 8(1961):18–19.
24. David E. Christensen, "A Simplified Traffic Flow Map," *Professional Geographer* 8(1961):21–22.
25. Parks, "American Flow Mapping," pp. 67–68.
26. George H. Primmer, "United States Flax Industry," *Economic Geography* 17(1941):24–30.
27. Walter H. Voskuil, "Bituminous Coal Movements in the United States," *Geographical Review* 32(1942):117–27.
28. E. E. Sviatlovsky, "The Centrographical Method and Regional Analysis," *Geographical Review* 27(1937):240–54.
29. D. Michael Ray, "The Location of United States Manufacturing Subsidiaries in Canada," *Economic Geography* 47(1971):389–400.
30. Waldo R. Tobler, "Experiments in Migration Mapping by Computer," *American Cartographer* 14(1987):155–63.

Glossary

desire line map unique flow map in which actual routes between places are not stressed, but interaction is; direction of flow often not shown

distributive flow map flow map on which the distribution of commodities or migration is the principal focus

dynamic map term often used to describe the ordinary flow map

flow map map on which the amount of movement along a linear path is stressed, usually by lines of varying thicknesses

network flow map flow map that reveals the interconnectivity of places

radial flow map class of flow map that is characterized by a radial or nodal pattern

traffic flow map particular kind of flow map in which movement of vehicles past a route point is shown by scaled lines of proportionally different thicknesses

Readings for Further Understanding

Friis, Herman. "Statistical Cartography in the United States Prior to 1870 and the Role of Joseph C. G. Kennedy and the U.S. Census Office." *American Cartographer* 1(1974):131–57.

Funkhauser, Gary H. "Historical Development of Graphical Representation of Statistical Data," *Osiris* 3(1937):269–404.

Parks, Mary J. "American Flow Mapping: A Survey of the Flow Maps Found in Twentieth Century Geography Textbooks, Including a Classification of the Various Flow Map Designs." Unpublished Master's Thesis, Department of Geography, Georgia State University, 1987.

Robinson, Arthur H. *Early Thematic Mapping in the History of Cartography.* Chicago: The University of Chicago Press, 1982.

Tufte, Edward R. *The Visual Display of Quantitative Information.* Cheshire, CT: Graphics Press, 1983.

Part

III

Designing Thematic Maps

The text preceding this part dealt with forming the base map, techniques of symbolization, data manipulation, and mapping techniques. All of these subjects are important in the development toward the final map product. There are other design considerations required, however, before one completes the mapping assignment. This part of the book combines chapters that present many of the techniques of map production and reproduction, and material regarding the visual design aspects of thematic mapping. This part begins with the material presented in Chapter 12 dealing with techniques of map production and reproduction. Manual and computer-assisted developments are provided.

Cartographic designers face two-dimensional design problems much as artists do. Furthermore, the elements of two-dimensional design can be studied carefully and techniques can be learned. The basic approaches are presented in the next chapter, and advanced concepts and techniques are presented in Chapter 14, where the ideas of total map organization and figure-ground ideas are presented. Of all the visual design ideas discussed in this book, an approach to a design problem by way of having a visual hierarchy plan is the most fundamental and necessary. This plan steers the page design activity and assures success. Typographics, the study and application of language labels to maps, is a basic part of map design. Chapter 15 provides current ideas about this subject and includes modern techniques of producing type by laser printing. Chief among the design variables for the thematic cartographer is color. Color essentials, specification, application, and standards are subjects about color that cartographers need to know. These topics are presented in the last chapter of this part, and also form the closing remarks about thematic map design in this book.

Chapter 12

Graphic Tools of Map Production and Reproduction

Chapter Preview

Cartographic production and reproduction techniques are integral to the design planning of a thematic map. From the outset, the designer develops the art work specifications based on a knowledge of how the map will be reproduced. A working knowledge of the technology of printing is thus essential for the well-trained cartographic designer. Color reproduction is especially important to understand, since this form of printing requires specific detail and is considerably more complex than black-and-white duplication. Map production techniques—the execution of the design concept into final art form ready for printing—underwent extensive change during the 1950s and 1960s and more changes are likely in the decades ahead. The methods and materials include not only drafting of ink on paper or film, but also scribing or etching onto special films. Design work requires careful planning to accommodate sophisticated printing technology, and cartographers will need to specify each step in the process carefully.

This chapter is intended to introduce modern practices of thematic map production and reproduction, which play a significant role in the overall design process. One cannot adequately approach a design task without at least a rudimentary knowledge of the relevant technology. As in other fields, change is taking place rapidly in this area. Computers are instrumental in this change, and their impact and potential will be readily evident in the material that follows.

In very large cartography offices there may be a division of labor between the designer and draftsperson. In small offices a single individual may be the designer and implementer all in one. Designers should be familiar with all techniques because they may come to instruct others in this aspect of cartography. Through knowledge and experience of these processes, better design will be possible. Like many other visual and graphic activities, good design solutions in cartography may happen "at the board" or at the time of final execution. Although many professional cartographers wish to isolate themselves from drafting, really good cartographic designers do not completely divorce themselves from the execution phase. They keep informed of new techniques and recognize potential improvements.

Printing Technology for Cartographers

A distinction should be made between production of manuscript maps and of maps for reproduction, since the approaches are different for each. In **manuscript map** preparation, only one copy of the map is usually being made. The art in final form is the completed map and will not ordinarily be duplicated. Most maps today are constructed for production of multiple copies, in one form or another. This chapter deals primarily with the methods of making maps for reproduction.

Brief History of Map Printing

Map printing has paralleled the history of all printing. The development of **letterpress,** the process used in making the Gutenberg Bible in 1452, led to the first **woodcut map** in 1472.[1] From that time to the present, there has been a close working relationship between cartographers and printers. In fact, during the first 300 years of printing, the printer probably had as much to say about cartography as the cartographer did. This was no doubt due to the strong influence of the printing craft guilds. It was not until the present century that cartographers began to control their own products. Designers now have access to a broad range of printing technology and can write knowledgeable specifications to implement their designs.

During the history of printing, three major printing methods have been used by cartographers: relief, intaglio, and planar. Each has its own merits and weaknesses for map reproduction. Most maps are reproduced today by planar methods, which will be examined in detail later in the chapter.

Relief—Letterpress

Letterpress also called *relief* printing, is the oldest printing method. Although popularly thought to have been invented by Johann Gutenberg about 1450, the method was actually used by the Chinese hundreds of years earlier.[2] What Gutenberg actually invented was relief printing with movable type. In relief printing, ink is applied to a raised surface and pressed onto the paper. (See Figure 12.1a.) The relief blocks (letters) are reversed so that they will be right-reading after printing. For reproducing art and cartography products, this relief method is referred to as *woodcut.* End-grain wood was chiefly used. Portions of the image that would not be printed were chiseled out, leaving the print portions raised in relief.

Early maps reproduced from woodcuts suffered from a variety of problems inherent in the

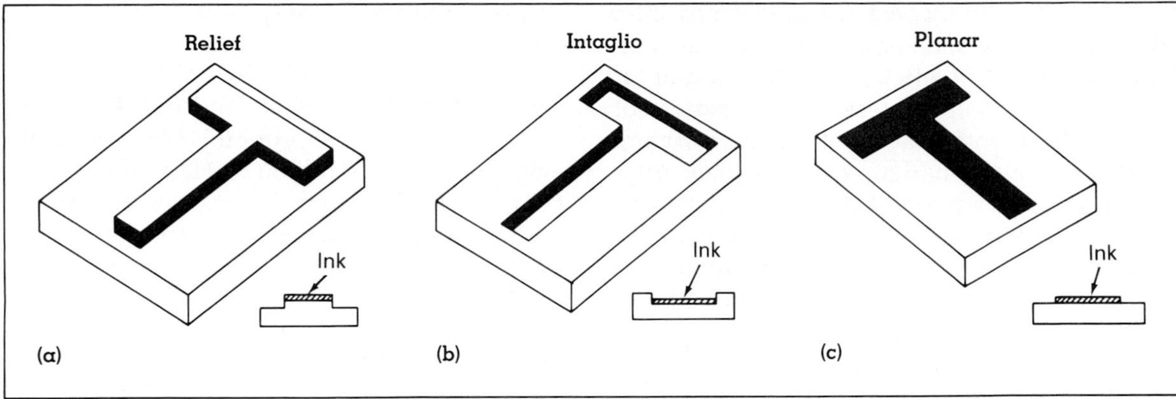

Figure 12.1. The three principal ways of printing: (a) relief, (b) intaglio, and (c) planar.

method. Lines were necessarily thick. Images were often smeared, since each impression required a new inking. The paper had to be nearly smooth and free of imperfections. In addition, it was difficult to add lettering to maps—in some cases, holes were cut in the woodcut and letter blocks dropped in. Also, large pieces of wood could not be used because of warping; large maps were often printed by piecing together the impressions made from smaller blocks. Gradation of tone was not really possible with the woodcut method, and the image had to be chiseled backwards.

Letterpress printing today is accomplished by first coating a metal plate with a light-sensitive chemical. After drying, it is exposed to a very bright light through a negative of the image to be printed. Those areas of the coated plate that are exposed will harden, and the other areas can be washed away by repeated acid baths. This process leaves the image areas raised in relief; they receive and transfer the ink during printing.

Intaglio—Engraving

Intaglio (pronounced in-tal'yo) printing is also called *etching, engraving,* or *gravure.* The elements of the image are first cut out by hand or incised by acid (etching). (See Figure 12.1b.) Ink is then applied and the plate cleaned. Some ink remains in the depressed portions; deeper depressions contain and transfer more ink than shallower depressions. When the plate is pressed onto the paper, the ink is transferred. Best results are achieved when the paper is dampened and considerable pressure applied during transfer. Actually, the paper is pressed into the depressions during printing, and afterwards there is a slight raised relief on the paper where the ink adheres. This is one way of identifying materials prepared by the intaglio process.

This form of printing was first used in the early fifteenth century and had achieved prominence by 1700. The engraved copperplate map became standard in map printing for about 150 years. In copperplate engraving, much finer lines are possible than in woodcut, then its chief competitor. As intaglio became popular, such techniques as *mezzotint* (working the surface to create tonal image), **stipple engraving** (creating tonal effects by specialized tools), and *aquatint* (special etching to create tones) were introduced. These all had wide appeal to cartographers and were especially useful for vignetting (gradation of tone or texture) at coastlines.

Modern intaglio methods include various forms of gravure, and printing on high-speed web (continuous paper) presses is possible. Techniques of platemaking for gravure include electromechanical or laser scanning of the image in order to break it into a pattern of dots used to control the mechanical engraving of small depressions (cups) in the plate. Similar techniques are used in conventional platemaking, which involves the mechanical transfer, through a carbon tissue medium, of the image onto the plate. Original art is photographed through a gravure screen. The negative is placed in contact with a photographically sensitive gravure plate and exposed. After exposure, the image is chemically etched onto the plate. The image is created on the plate by a pattern of small depressions of varying size and depth. Plates are usually of polished copper, often chromium-plated for protection. Gravure is expensive and is generally economical only for large press runs. Most maps for reproduction are not directly prepared for intaglio or gravure printing.

Planar—Lithography

Planar printing was introduced in 1796 by Alois Senefelder in Germany. Planar printing is usually referred to as lithography. The Greek word *lithos* means stone, and limestone was first used to make the plates. Today it is often called **photolithography, offset lithography,** or photo-offset lithography. This printing method relies on the fact that water and oil (grease) do not mix well. On the printing stone or plate, the image area receives the greasy ink and the non-image areas do not. When paper is pressed to the plate, only the inked areas will transfer to the paper. Unlike the relief or intaglio methods, planar printing creates no relief differences on the plate. (See Figure 12.1c.) Lithography no longer uses stones for plates except in rare circumstances—perhaps only in the graphic arts. This form of printing has come into wide use in practically all commercial applications, and most maps today are printed by this method.

As lithography was introduced to the printing industry, cartographers gradually began to see its advantages. Old copper engravings could easily be updated, transferred to stone, and then printed. Lithography also meant faster preparation of plates than copper engraving; most engraving craftspeople learned the new lithographic techniques easily. Modern lithographic techniques are examined in more detail below.

Cartographic Design and the Printer

The relationship between the cartographer and the printer has gone through different stages since 1450. Arthur Robinson of the University of Wisconsin at Madison, identifies the following periods:[3]

1. During the period when woodcut maps predominated, most cartographers did not do their own woodworking. This usually led to better map designs, because woodcutters were better craftspeople than cartographers.
2. For the most part, this relationship remained unchanged during the time when intaglio was preeminent. A few cartographers—such as Mercator—were also engravers. When national map surveys first flourished, cartographers and in-house engravers worked more closely than ever. Map design was still mainly shaped by engravers' styles.
3. When the transfer process (making printing plates from right-reading material) and lithography were introduced, anything could be reproduced. Specialists such as engravers and cartographers could be bypassed entirely, replaced by others who did not possess any cartographic knowledge. Cartographic design suffered.
4. Today, the situation is reversing. Since about 1950, printers and cartographers have once again been working closely

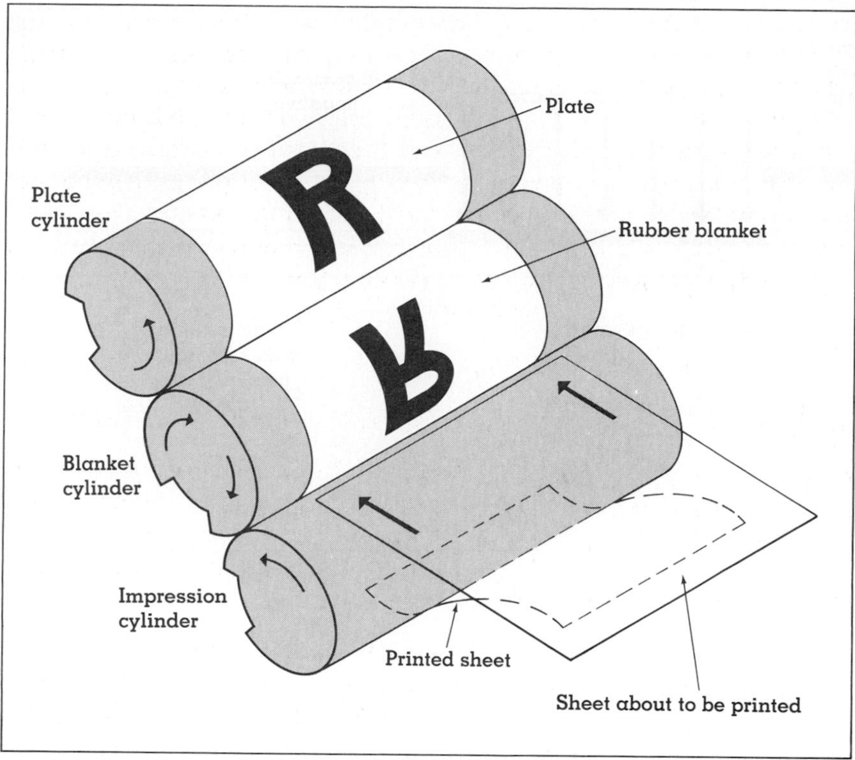

Figure 12.2. The principle of offset printing.
Notice that the image on the plate is right-reading, a feature that has made this form of printing very popular.

together. Although it is increasingly difficult for cartographers to master the full range of modern printing technology, design is surely improving in the attempt.

As the printing industry became more and more dominated by lithography (especially photolithography), the printer became primarily a duplicator. There were no craftspeople intervening to influence the look of maps. Cartographic education did not stress the aesthetic aspects of maps, except in rare cases. Today, the cartographer must assume the craftsperson's role, handling the aesthetic aspect to map design. Knowledge of the printing industry and its varied techniques gives the cartographer greater freedom of choice in design.

Modern Photolithography

Lithography, as noted above, had its beginnings around 1796. By the middle of the nineteenth century, the development of photography made it possible to put the image onto a thin metal plate that could be attached to a rotating cylinder. The impressions were made directly onto paper, requiring the image to be backwards until printed. It was not until 1905, when Ira Rubel accidentally printed an impression on a blanket cylinder, that the *offset principle* was discovered. (See Figure 12.2.) Print, art, and map materials could be right-reading through the entire preparation process, making this phase much less risky.

In producing the photolithographic plate, a light-sensitive chemical coating is applied to a

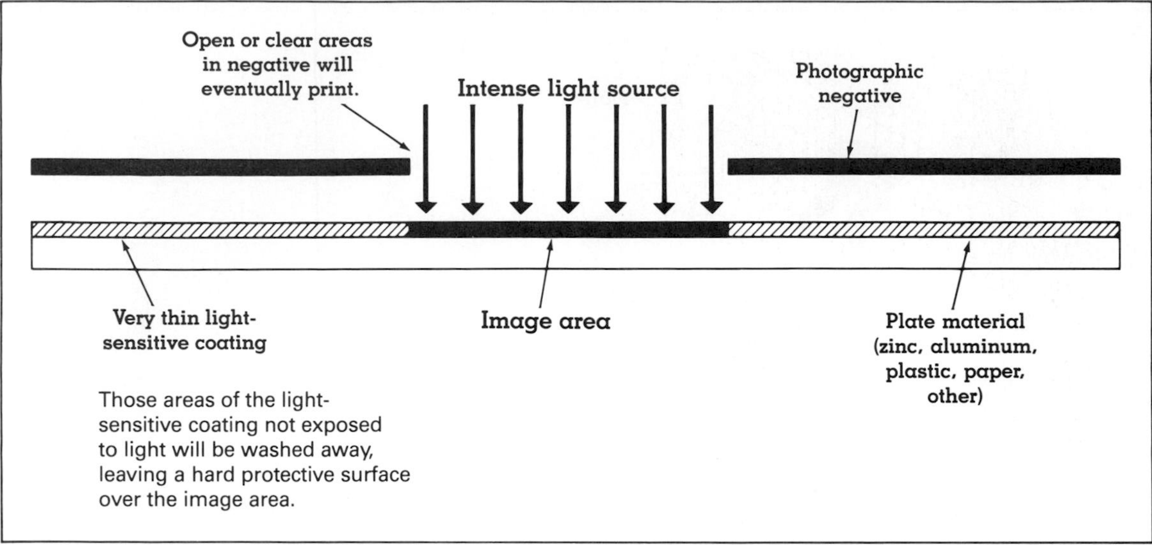

Figure 12.3. **Photolithographic negative-working plate process.**

thin metal plate. (See Figure 12.3.) Most plates today are purchased with the coating already applied. A *copy negative* of the original art is placed in contact with the plate, and both are exposed to an intense light source. After exposure, the plate is chemically washed with a developing fluid that washes away the coating where the plate has not been exposed to light. The image area, that part exposed to light, hardens and becomes ink-receptive (water-repellent) during printing; the plate is first dampened in a water bath, making the non-printing areas ink-repellent. The plate receives ink only on the printing or image areas. The ink is transferred to a rubber blanket cylinder and then to incoming paper.

Photolithography requires several steps in sequence:

1. Preparation of original art (which may be in negative form)
2. Preparation of negatives (if not done in step one)
3. Proofing
4. Platemaking
5. Printing

Although art preparation is the first step in preparation of the final printed map, we will look at the other steps first. An understanding of lithographic principles directs how the art is rendered.

Process Cameras and Special Film Screens

Offset-quality film negatives are made in precision **process cameras.** In its elemental form, such a camera comprises *copyboard, lensboard,* and *backboard* or *ground glass.* (See Figure 12.4.) The original art is placed in the copyboard, where it is usually held in place by vacuum pressure and/or film screws. The copyboard, lensboard, and backboard are in planes parallel to each other to assure accurate copy and focus controls. Most cameras contain lights (usually carbon arc or tungsten) that shine onto the art and reflect light to the backboard. The negative is exposed in this manner.

Process cameras can be quite large. In fact, the backboard and other camera components on that side of the lensboard maybe in a darkroom, with the remainder of the camera in an

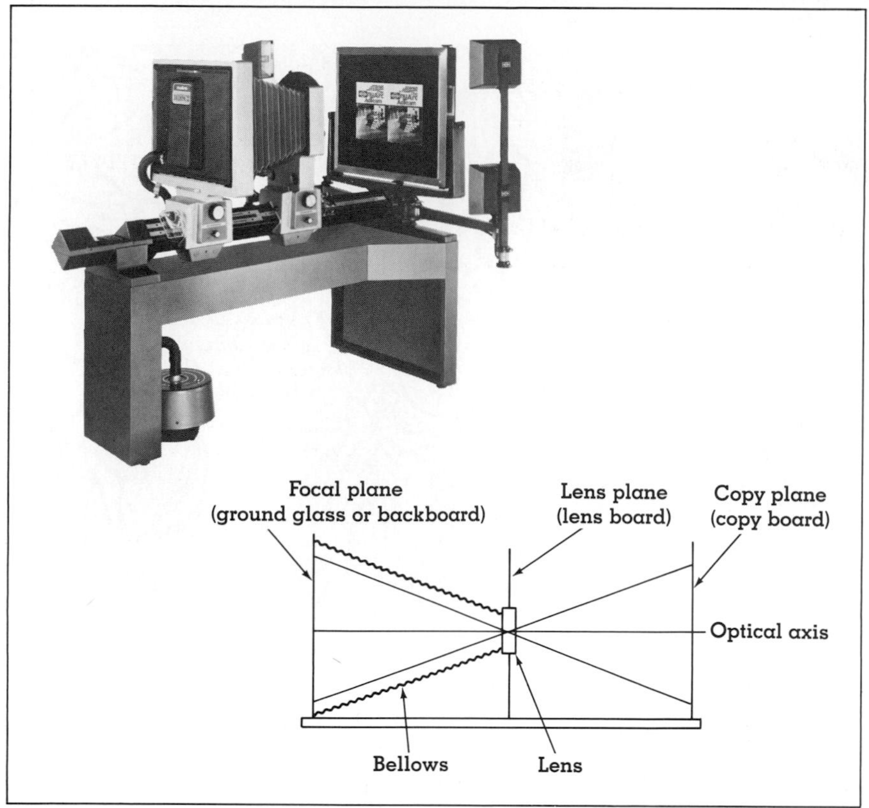

Figure 12.4. A modern horizontal process or copy camera.
Photograph courtesy of Nu Arc Company, Inc., Chicago, Ill.

adjoining room. This type of installation is for very large art and negative work; most mapping does not require process cameras of this size. A common camera size has a copyboard that is 36 × 46 cm (14 × 18 in). Smaller cartography offices do not normally have the larger cameras; they send out their artwork to professional photographers.

Film negatives. **Negative** material suitable for photolithography includes only two types: the *line negative* and the *halftone negative.* The line negative is photographic film that, after developing, is composed of opaque (black) background and open or clear areas. It does not produce the variable gray areas possible in *continuous-tone negatives.* Continuous-tone negatives are not suitable for making lithographic plates. The **half-tone negative** is really a line negative that has been broken into a pattern of very small dots by exposing it to the art through a *halftone screen.* (See Figure 12.5.) The halftone screen is a glass (or contact film negative) marked with opaque lines intersecting at right angles, producing a pattern of very small squares. When line or continuous-tone art is photographed through the screen, the gray areas produce a pattern of dots on the line negative. The resulting halftone negative can be exposed to the lithographic plate in the usual fashion.

The halftone method works because the dots are too small to be resolved by the human eye at normal reading distances. They tend to

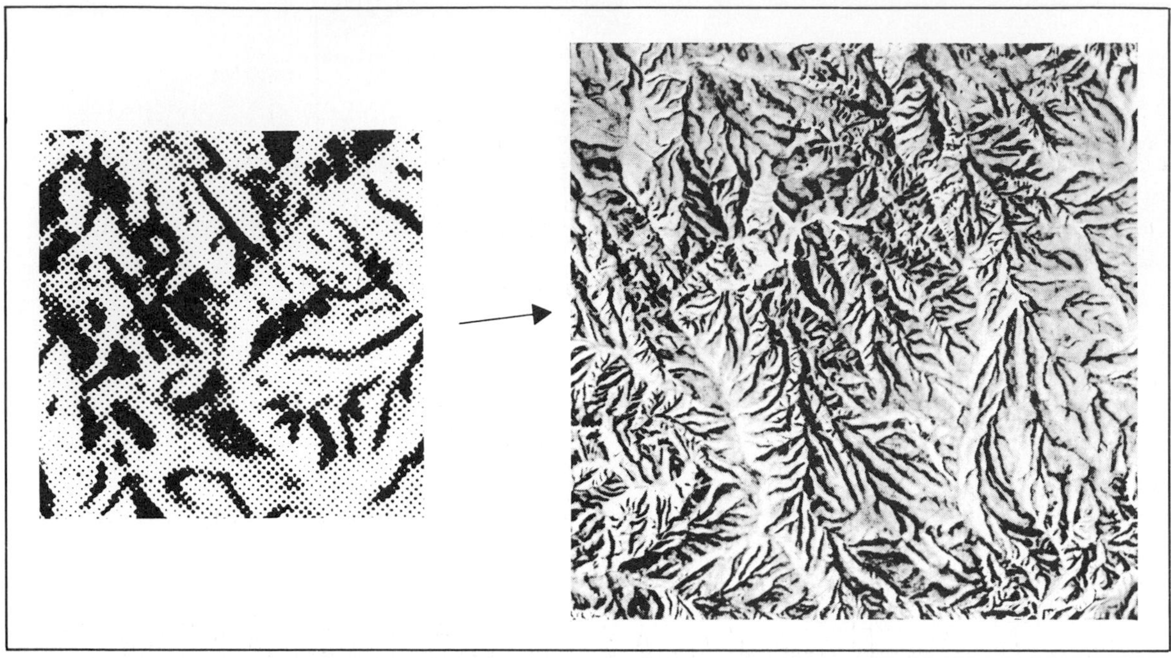

Figure 12.5. The effect of halftoning.
The halftone screen has produced this hill-shaded map, which is composed of a pattern of dots of varying size and density. A continuous-tone effect is produced because we cannot normally resolve the individual dots.

merge together, creating the impression of gray tones. The dots actually vary in size, depending on the amount of light that was reflected from the original art during exposure. Most thematic map applications do not require the use of the halftone technique. Exceptions are vertical hill-shaded terrain maps and symbols rendered for plastic effect. However, cartographic design specifications do usually call for the use of screen tints.

Screen tints. **Screen tint** sheets are film negatives composed of small dot patterns on opaque backgrounds. They are sometimes referred to as *mechanical* or *flat screens.* These screens are placed between the line negative and the light source when the plate is made. Their effect is to break up the solid-line or black areas so that when printed, they will appear gray. (See Figure 12.6.) Screen tint sheets are manufactured with two variables, the line designation (number of rows of dots per inch) and **gray tone** (approximate percentage of area covered with ink after printing). Common line designations are 65, 85, 100, 110, 120, 133, and 150 lines per inch. Coarse lines, such as 65 and 85, are not recommended in most map work because the individual dots can be seen by the human eye. This detracts from the desired flat tonal quality. The 133 line is recommended for most map work; 150 line is preferable when fine lines or lettering are to be screened. (See Figure 12.7.)

Gray tones usually range from 5 to 95 percent, in 5 percent gradations. The cartographer chooses to specify gray tone when contrast of tone is sought over the map or when a quantitative map is being symbolized to achieve a gray spectrum. Under normal conditions, no less than a 15 percent difference between two tints on the map should be specified (20 percent is preferred). The human eye finds it difficult to differentiate two tints closer in percentage than 15 percent.

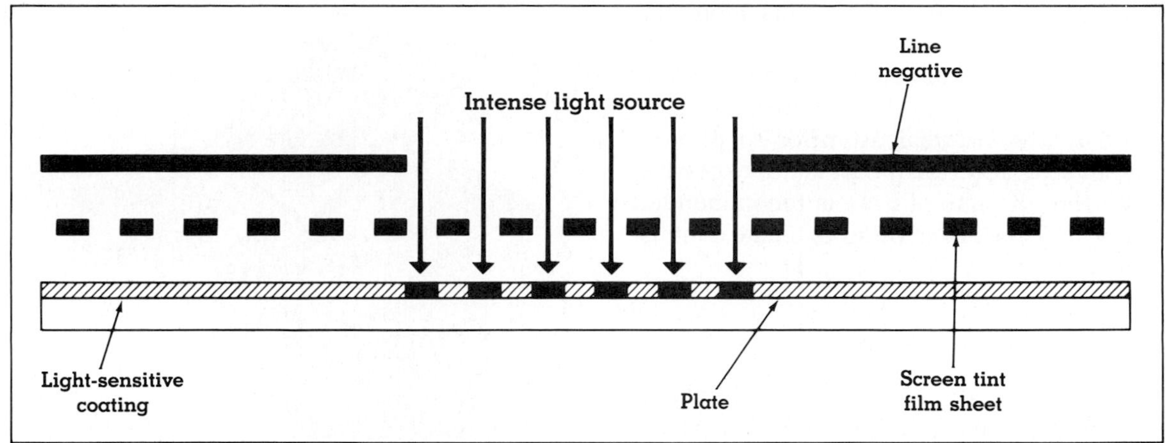

Figure 12.6. The screen tint process.
Screen tint film sheets are placed between the line negative and the printing plate during exposure. This results in the image being broken into a pattern of very small elements, usually dots.

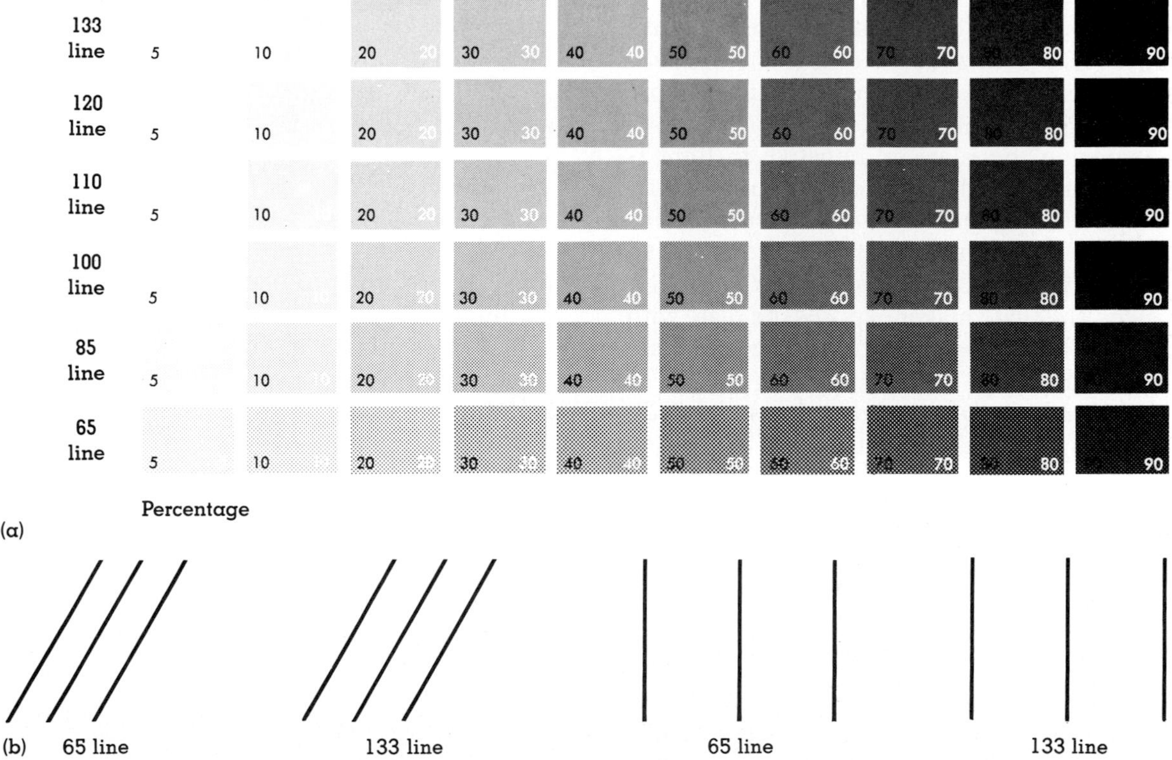

Figure 12.7. Screen tints and the effects of screening lines.
A sample of various tints is illustrated in (a). In (b), lines have been screened to show the effect of lines on at least two orientations. Note the ragged edges when coarse screens are used.

Screen tints courtesy of the ByChrome Company, Box 1077, Columbus, Ohio 43216.

Map designers use screen tints more frequently today than several years ago. The quality achieved is far superior to what can be obtained by using preprinted adhesive sheets applied to the original art prior to photography, and the additional cost is not excessive. In addition, the use of tints is recommended in many map designs in order to create contrast on the map.

Plate Preparation

There are a number of ways to prepare plates. Most modern photolithography shops have a *plate-making machine* (see Figure 12.8), which is basically a large box with a flat surface covered by a glass plate, where the plate and negatives are placed. These are subjected to a vacuum so that firm contact between the two is assured. On larger machines, this bed is rotated so that it can be exposed to a very intense light source from within the box. The duration of exposure is varied by means of timing controls on the machine.

Exposing the plate or negative to light is called **burning.** The plate can be burned more than once—*double* or *triple burning.* This occurs when more than one negative is used in the preparation of the image. In these instances, the plate maker exposes the plate with various negatives, without developing the plate before each burn.

The cartographic designer does not ordinarily get involved with this aspect of printing but should be aware that multiple burning of the plate is possible. This might facilitate art preparation and will certainly affect any photoengraving or plate specifications.

Figure 12.8. A modern plate-making machine.

Photograph courtesy of Nu Arc Company, Inc., Chicago, Ill.

Color Printing and Four-color Process

Color printing can be divided into two kinds: **flat color printing** and **process color printing.** Flat color printing is accomplished in a manner similar to black-and-white except that ink colors other than black are used. Process color printing is more complex, requiring considerably more expertise and detailed planning. It is difficult to speculate on which kind of printing is mostly used in cartographic work, but it is suspected that flat color is more common.

Discussion of color as it relates to cartographic design is relegated to another chapter, but some aspects of it must be presented here. Most of the present material deals with the objective aspects of color, leaving the more subjective and perceptual aspects for later. Of course, cartographers need to know both for effective map design.

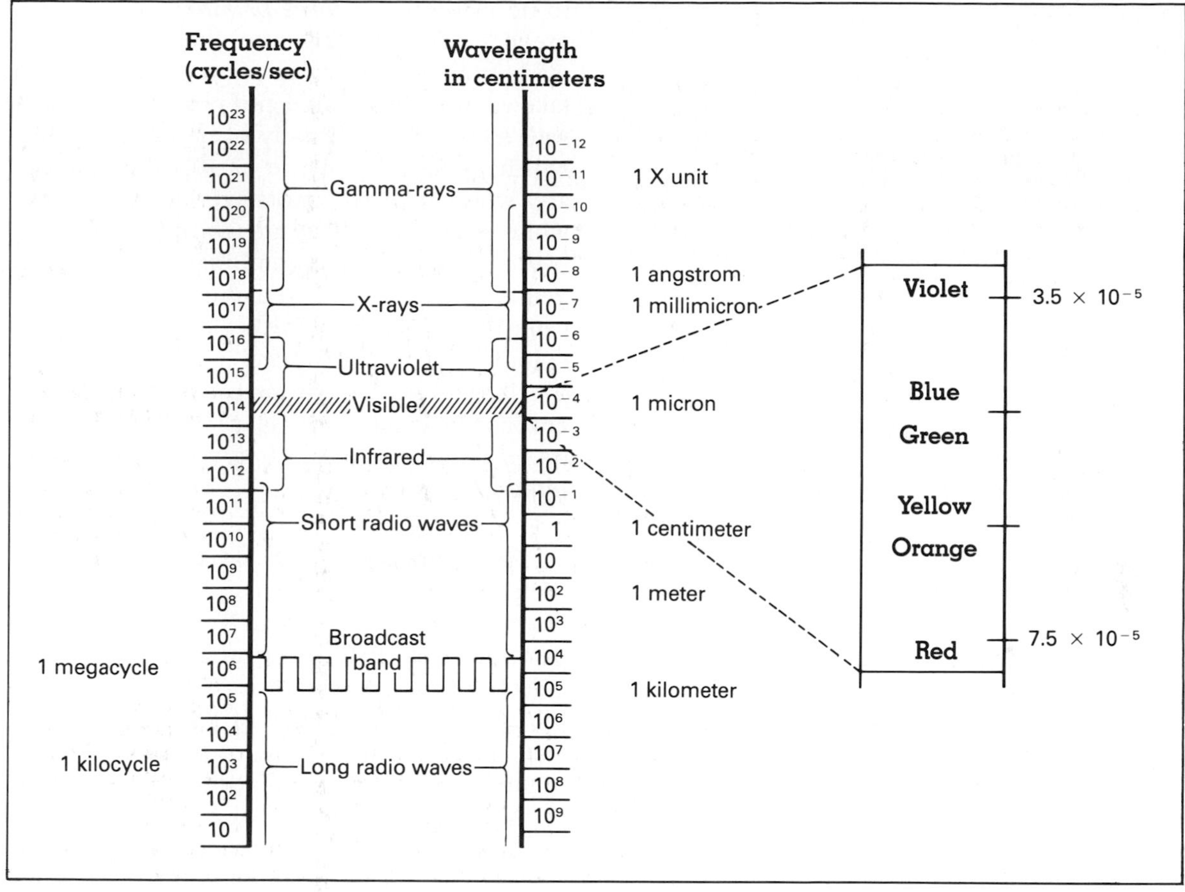

Figure 12.9. The electromagnetic spectrum.
Visible light occupies only a small portion of the entire spectrum.

Light and the color spectrum. **Light** is that part of the electromagnetic energy spectrum that is visible to the human eye. (See Figure 12.9.) The radiation spectrum is characterized by energy falling on us at different wavelengths. These wavelengths vary from very short (10^{-12} cm) to very long (10^{5} cm, or one kilometer). All visible light varies from 7.5×10^{-5} cm to about 3.5×10^{-5} cm. **Color** is simply light energy at different places along this spectrum. When our eyes detect light energy at approximately 7.5×10^{-5} cm in wavelength, we see red; when we detect wavelengths at 3.5×10^{-5} cm, we see violet. Other colors and their distinct wavelengths are in between these two.

Additive and subtractive colors. Color can be *refracted* from white light by passing light rays through a different medium. For example, we can see the different colors after passing a ray of light through a prism. Although visible light is composed of a myriad of different colors at various wavelengths, we consider white light

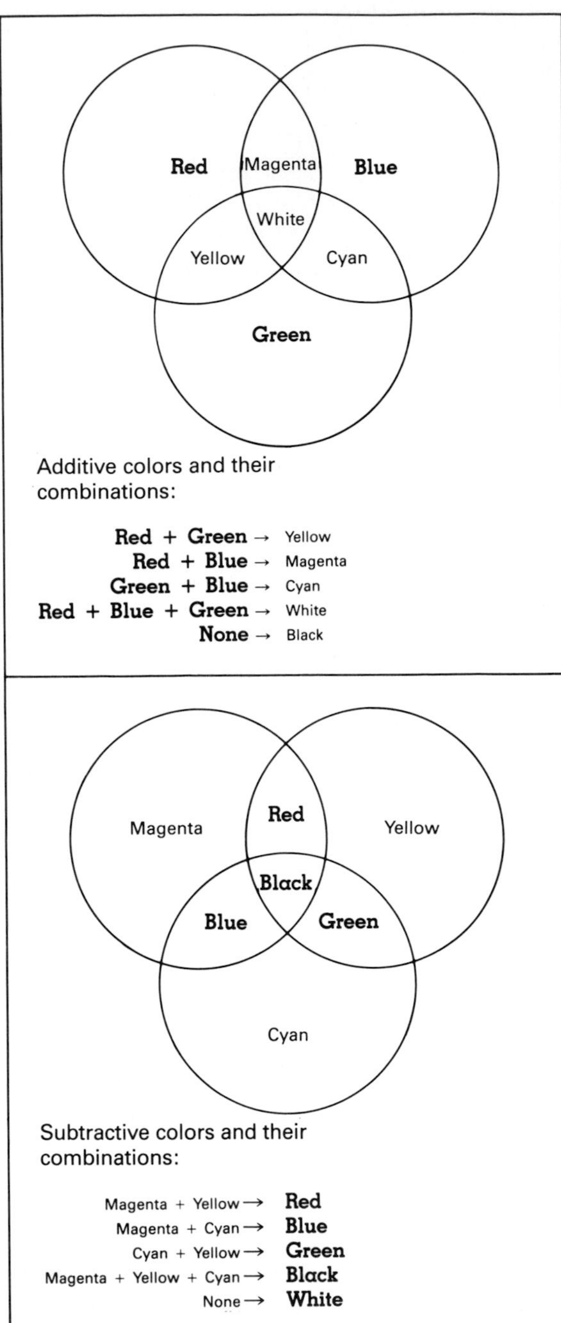

Figure 12.10. Additive and subtractive colors and their combinations.
The four-color process printing technique uses the principle of subtractive colors to produce virtually any hue. Different percentages (tints) of the process inks are used to create various hues.

to be made up of *three primary colors*—red, green, and blue—because these cannot be made from combinations of other colors. If we take each of these three and project them on a wall so that they partially overlap, two things will happen. (See Figure 12.10.) First, in that area where all three colors overlap, we will see white. Second, in those areas where two colors overlap, we will see a combination called magenta, cyan, or yellow. Red, green, and blue are called the **primary additive colors.** The new colors that are produced (in the areas of overlap) are called the **primary subtractive colors,** because each is produced by subtracting one additive primary from the white light.

Color printing. Color printing is based on the principles of subtractive coloring and the characteristics of inks used. Normally, flat color printing uses opaque printing inks; process color printing requires the use of transparent inks. Inks are composed of a liquid medium (e.g., polymer, linseed oil, or soybean oil) and pigments of organic or inorganic substances chosen to reflect certain colors.

Opaque printing ink reflects its color, and absorbs the remaining colors at its surface. (See Figure 12.11.) Red ink absorbs the blues and greens and reflects the red to the reader's eye. **Transparent inks** behave somewhat differently. The pigments of red transparent ink absorb the blues and greens, but in this case the red light is first transmitted to the paper surface before it is reflected back to the reader's eye. Thus the color of the paper or other ink is perceived by the reader. Process color printing uses the subtractive primaries of magenta, cyan, and yellow in transparent inks to create any hue or to recreate a continuous-tone color image. For example, mixing the subtractives magenta and yellow will yield red. Applying the subtractives together in different percentage screens can result in almost any color.

Cartographic designers have two choices in flat color printing. They can select the desired color of opaque ink from literally hundreds of

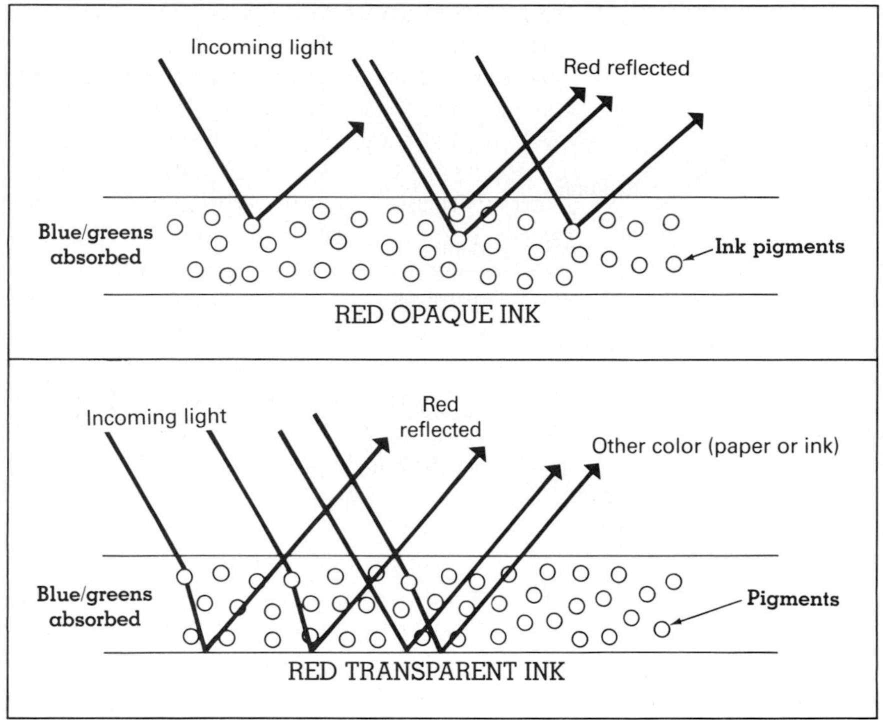

Figure 12.11. The reflecting qualities of opaque and transparent inks.
Light striking opaque inks is immediately reflected. When light strikes transparent inks, it is first transmitted to the paper surface and then reflected.

choices.[4] The alternative is to develop virtually any color by using transparent magenta, cyan, and yellow inks. The number of colors planned for the final map determines the choice. Printers normally charge by the number of different inks (regardless of whether they are process or opaque) and press time. The visual quality of a flat opaque ink color may be somewhat better, but competent printers can achieve excellent results using transparent ink overprinting. The designer must look closely at other costs in order to determine the best alternative. Regardless of what choice is made, the decision must come early on, since art preparation is dependent on it.

Process color printing in map work can begin in two ways. In one method, the cartographer produces a color manuscript map. (See Figure 12.12.) That is, the map is completely finished and looks exactly as it will appear after printing. The completed art is photographed four times, once for each of the primary colors and black.

Black is usually used (but not absolutely necessary) to overcome the limitations in faithfully reproducing the original color by the transparent inks (hence the term four-color process printing). Different filters are used to take out all colors except the one being worked. A negative and plate are made for each of the process colors, plus black. Four runs through the press are required, once for each ink, unless a multicolor press that requires only one press run is used. In this way, the original map is reproduced. This is the same general procedure followed in the color reproduction of continuous-tone color images such as photographs, illustrations, or similar art.

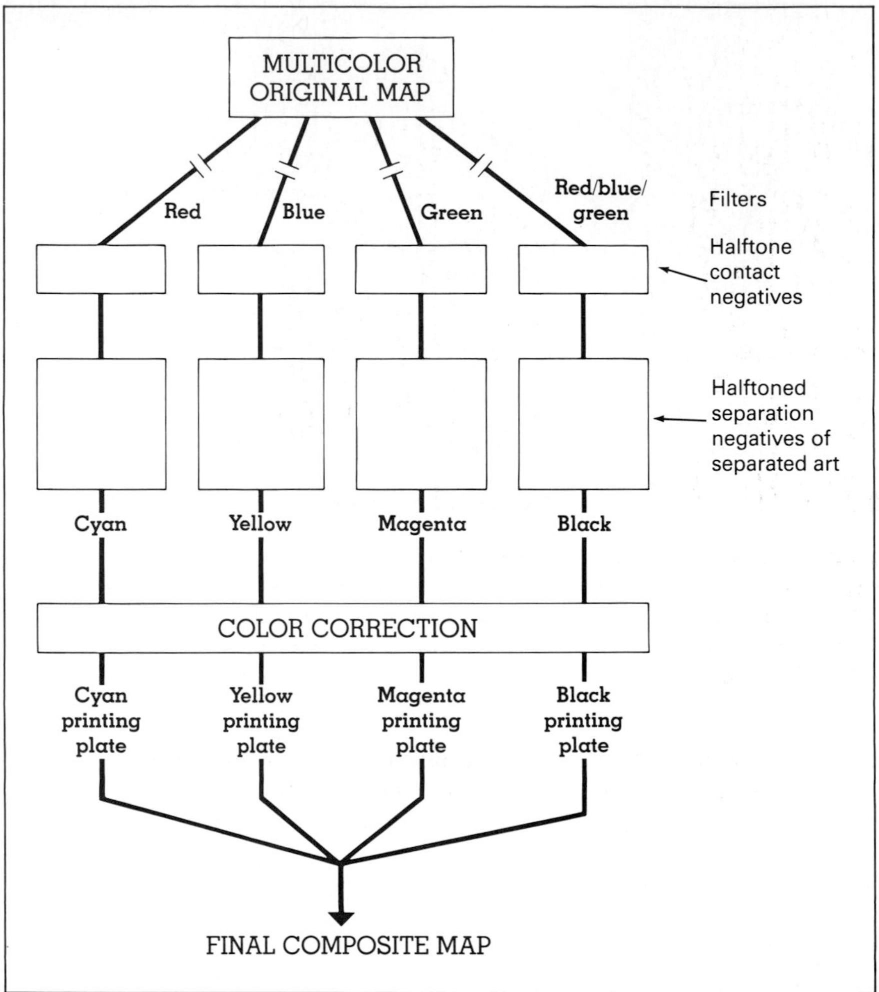

Figure 12.12. Steps in reproducing a color original map.
Color correction is the process of removing from the separation negatives some unwanted dots resulting from their original preparation.

A second way to reproduce a process color map is by developing the art originally into **color separations.** (See Figure 12.13.) In this method, the original art is produced in pieces (separations), usually four—one for each process ink and black—and negatives are made of each separation. Separate plates are again made from each negative, and the map is then printed in usual fashion.

The production of the map's camera-ready artwork begins after the designer has completed initial sketches and chosen the reproduction technique. Any consideration of the map's design must necessarily involve the reproduction aspects, since these will dictate how the map will be mechanically assembled. Over the years, *drafting* has been the word used to describe this final activity. Drafting has now

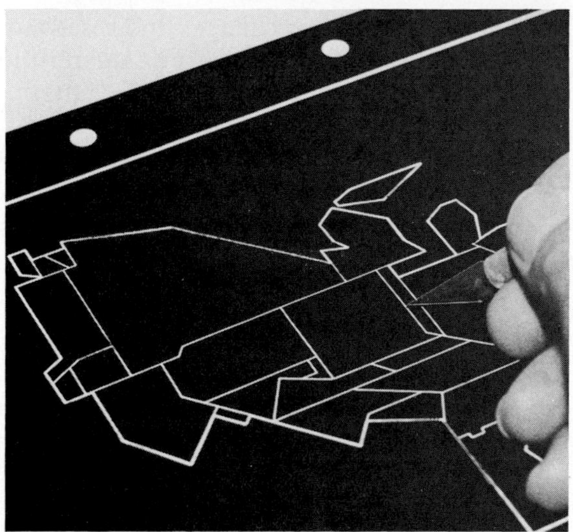

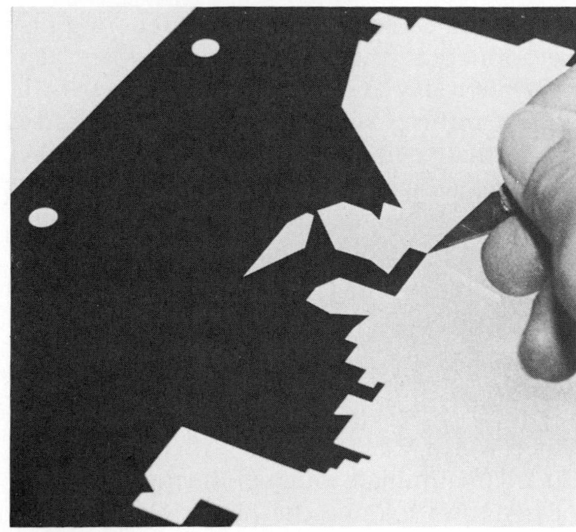

Figure 12.13. Preparing art separations for color printing.
Most maps that are to be printed in color involve the preparation of art "flaps" (or mechanicals) that will be photographed into separation negatives. In flat color printing, there is usually one separation for each color; in color printing using the process colors, several negatives may be prepared, depending on the color combinations required. Careful planning is required to ascertain that the correct separations are chosen.

taken on new dimensions, especially brought on by the introduction of new tools and techniques.

Map Production Techniques

The cartographic drafting techniques presented in this chapter are intended to serve only as an introduction to the methods currently used by professionals. They will change as technology changes, as old systems are replaced by new and better ones. The prime criteria in choosing drafting techniques are precision and economy. Beyond these considerations, the cartographer is free to adopt any technique, equipment, brand, shortcut, or personally favored approach.

Two Manual Techniques

Drafting techniques, those steps necessary to bring a sketch to final camera-ready status, have undergone many changes in the last thirty years. Modern drafting may utilize various gravers (e.g., straight-line, swivel, or turret) in addition to instruments such as ink pens and compasses. In the not-too-distant future, we may see all these instruments fall by the wayside as computer-aided design and machine drafting take over. For the present, our purpose is to introduce the drafting methods and equipment now used by the thematic map designer, though the cartographic designer must be flexible and be willing to adopt better techniques when introduced.

Non-scribing Methods

Non-scribing methods of cartographic drafting resemble engineering drawing in technique and equipment. The draftsperson works with positive images. Areas to print are black or red on a white background and are rendered right-reading. Red can be used because it behaves like black under the strong lights of the copy camera. Black drawing ink is normally used on a stable-base drafting film. Specially prepared preprinted screens, symbols, or lettering may be applied to these films, as necessary. It is possible today to reduce the inking activity to almost nothing, since line tapes can be used as an alternative to inked lines. The cartographer prepares as many separations—**mechanicals** or **flaps**—as are required for the planned project. For a multicolor or four-color process map, a separation is customarily planned for each color.

Scribing

Since World War II, a method has been introduced to eliminate much of the traditional use of pens and ink and the preparation of negatives. In **negative scribe,** the draftsperson cuts away portions of a special coating that has been applied to a stable-base film. This method was first introduced to printing earlier this century in *glass scribing.* A coating called *Flopaque* was used to coat the glass, and the image was created by stripping away this coating. Today, scribing is done on a plastic base (usually Mylar), but the process is identical.

There are two ways to begin the engraving process in negative scribing. In one method, the scribe film is placed over the manuscript map and the scribe coat is prepared by tracing the original, using specially designed **gravers** to remove the coating from the plastic base. In another method, a negative of the manuscript map is made. The image is photographically transferred to sensitized scribe film and then scribed. The portions removed make up the image that will be printed. Scribe films are usually red, rust, or yellow. All behave similarly in photographic reproduction, but yellow is often preferred because it is easier to see through during tracing, thus reducing eye strain. The USGS uses yellow scribe almost exclusively. Scribed sheets can be duplicated for large jobs. With Etchscribe film, a registered product, duplications can be done without the aid of a darkroom.

There are several advantages and disadvantages to each method of drafting. Among the advantages often listed for scribing are these:

1. Training time is often less than in non-scribing drafting methods.
2. Consistency of line weights (widths) is greater in scribing because precision-ground graver points or blades are used.
3. Corrections, editing, and updating are easier.
4. Direct preparation of copy in the desired size eliminates the need for film negatives.
5. Film dimensions are stable, and the films are durable and easy to store.

Several disadvantages for scribing have also appeared:

1. Considerable eye strain results from long hours of tracing on the scribe coat, because of the contrast between the orange and red film and the light of the light table.
2. Symbol preparation at direct printing size is often difficult.
3. Lettering cannot be placed directly onto the scribe separation. Intermediate copies or separate lettering mechanicals must be made.

Government agency drafting is now done exclusively by scribing because of the advantages outlined above. Many smaller offices are still using non-scribing methods, but modern drafting films are replacing paper. With the proliferation of other techniques for color separations, an expert non-scribing drafting can rival the precision of the engraved lines on scribe materials. Each office must weigh the costs and benefits of each system in light of its own needs. A chart comparing costs of materials is included as Table 12.1.

Table 12.1 Comparative Costs of Cartographic Drafting Materials

Item	Cost per sq ft (index = 100)
Drafting film (.005 in., matte one side)	100
Clear film (.005 in.)	45
Masking film (.005 in.)	171
Pressure-sensitive light blocking sheets	365
Scribe film (.005 in.)	256
Copy negative (commercially prepared)	1,956

Technique

Many of the fundamental techniques of cartographic drafting used for both non-scribing and scribing methods are presented as guidelines for the student. Mastery will come only with experience and practice.

The Work Station

The cartographic work station usually includes a drafting table, preferably a light table. Of course, the size of the office dictates the number of tables. Drafting tables are manufactured in a number of sizes and styles. The tops are usually hardwood, laminated to insure flatness; the better ones have perfectly straight metal edges on the right and left sides. Light tables are manufactured so that a fluorescent light source is beneath a frosted-glass drawing surface. Light tables are used in many drafting operations, especially tracing, scribing, and using special films such as masking films. The better drafting and light tables have movable tops so that they can be slanted for a more comfortable working position. Cushioned drafting stools are available in a variety of styles and colors. Drafting furniture should be purchased with quality and flexibility in mind, since initial purchase is costly and frequent replacement not advisable.

A typical work station has a variety of useful drafting instruments, as shown in Figure 12.14. As with furniture, quality is the best approach when buying these devices. Students and professionals develop their own brand preferences.

Lines

Line work is done either with inking pens or with straight or flexible gravers (in the case of scribing). There are inking pens for general use and for specific tasks; considerable advances have been made in recent years. Most ink drafting is now done with **technical pens** (Figure 12.15), which have nibs manufactured to specified line widths. They are filled with drawing ink by pouring the ink into a cartridge inside the barrel. Ink flow is downward from this cartridge to the nib. Nibs of varying quality are available, with the harder metals being more durable and expensive. Disposable technical pens are now marketed; some are good quality, but the range of nib size is small.

Using technical pens requires some practice. Consistency of line width requires three considerations. 1) The pen must be functioning properly and the nib not worn out of specification. 2) The draftsperson must hold the pen properly. 3) The ink and the paper or film used must be compatible—there are special inks for plastic drafting films. Practice will yield precision results.

Although ruling pens have for the most part been replaced by technical pens, they have the great advantage of flexibility. One pen can do the work of a whole set of technical pens, since a ruling pen can be adjusted to draw different line widths. The chief disadvantage is that considerably more skill is required to master their use; controlling line width requires greater precision. Special pens include the railroad pen

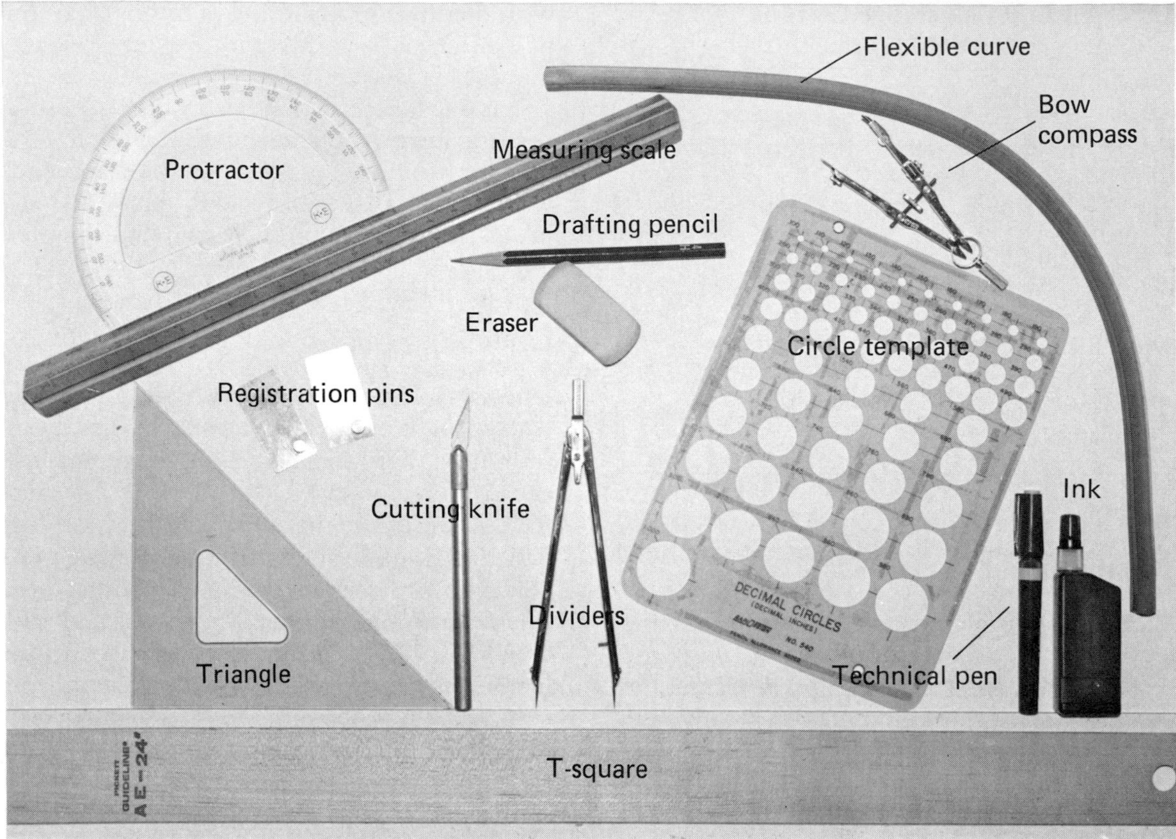

Figure 12.14. The cartographer's work station.
The major production tools are labeled for convenience.

(with two heads), the contour pen (with a swivel head), and the crow quill pen (with a point that spreads with greater pressure). Crow quill pens are used for drafting rivers and streams. Since cartographers will find use for such pens on occasion, they should learn to be comfortable with their function and use.

Line work on negative scribe film is done by special gravers (Figure 12.16) that accept special blades or points to remove the coating on the negative scribe when dragged along the surface. Blades and points can be obtained at various thicknesses to ensure accurate line widths. Careless use of these instruments can cause uneven alignment, leading to inconsistent results. Sharp blades or points and correct pressure during graving are the key elements of success.

Two kinds of gravers for line work are common: fixed, for straight lines, and flexible, for curved and irregular lines. Straight or fixed-lead gravers are moved along straight edges for tracing straight lines. The flexible head can trace irregular lines directly.

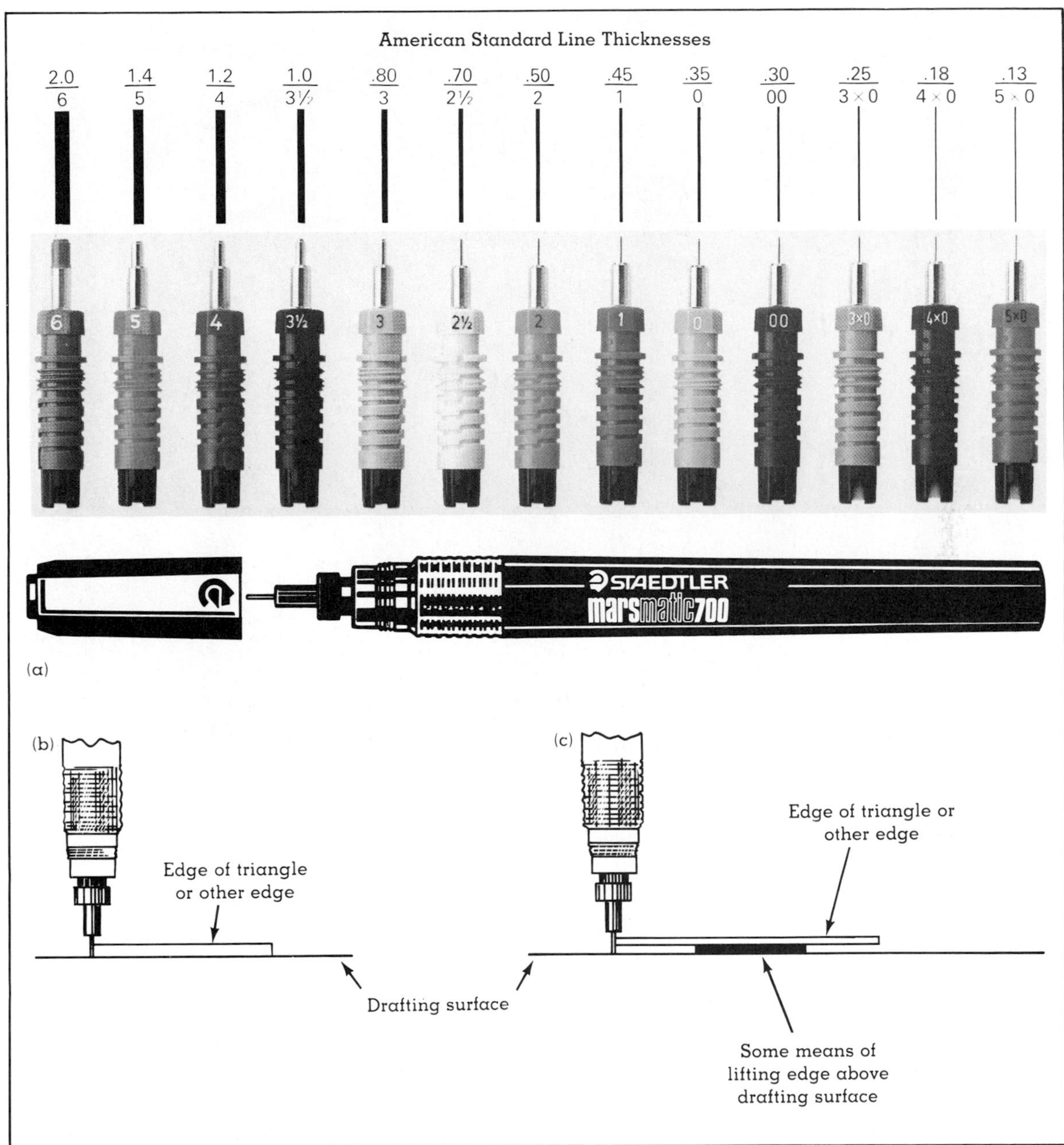

Figure 12.15. Modern drafting pens.
A typical pen and representative line weights are shown in (a). In (b), we see the incorrect way to use a pen along an edge. The correct way is represented in (c).
Pen drawings courtesy of J. S. Staedtler, Inc.

(a)

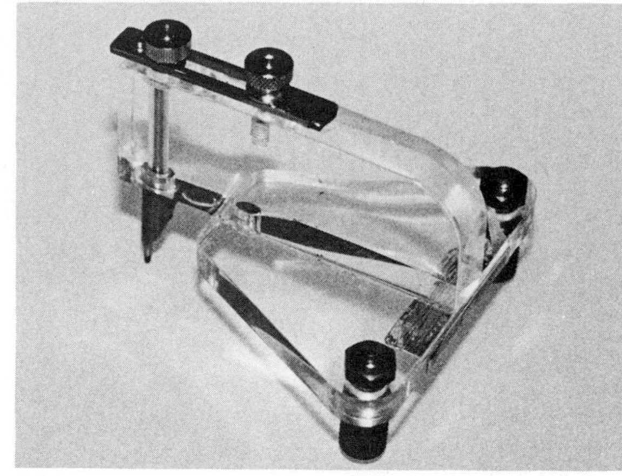

(b)

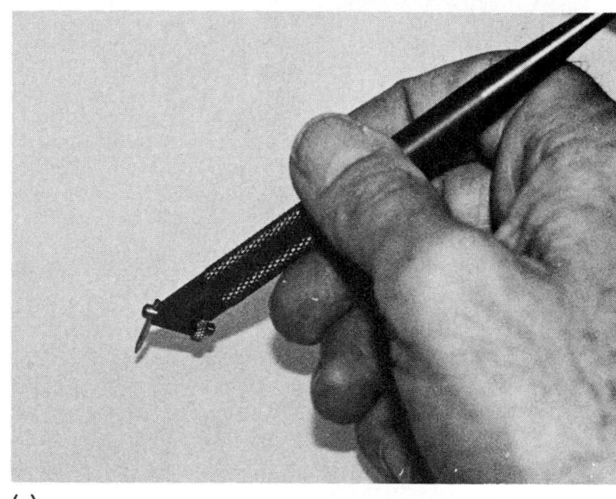

(c)

Figure 12.16. Three scribing gravers.
They are (a) rigid graver, (b) flexible graver, and (c) needle-holder graver.

Source: The instruments shown are products of the Keuffel and Esser Co., Morristown, New Jersey.

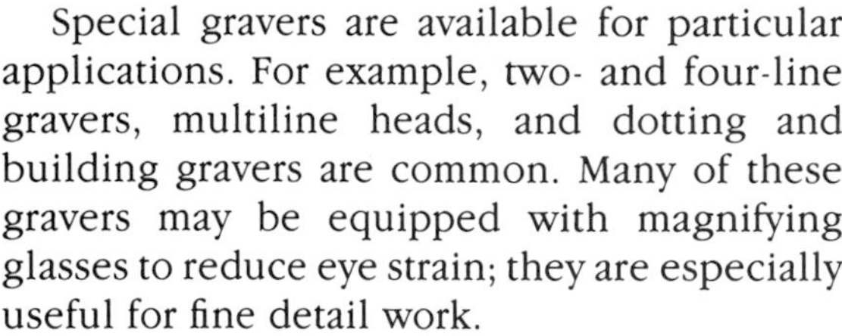

Special gravers are available for particular applications. For example, two- and four-line gravers, multiline heads, and dotting and building gravers are common. Many of these gravers may be equipped with magnifying glasses to reduce eye strain; they are especially useful for fine detail work.

Fine-line gravers are basically pin-vise holders that secure a steel needle tip. They are often angled so that the desired vertical point orientation is possible when the graver is held in normal writing fashion. They are often used for lines narrower than .002 in. and for template work. Templates are metal or polyester patterns that make the scribing of symbols far easier.

In positive art preparation, as is done in nonscribing drafting, special **border tapes** can be used instead of line inking. (See Figure 12.17.) These tapes come in a variety of widths and are obtainable in two forms: flexible (crepe) and nonflexible. Flexible tapes are especially useful for doing irregular curves and regular curves of small radius—for example, fine-line isarithms and contour lines. The use of tapes is only recommended when exact line-width specifications are not required, but their use can save time. Special care must be given to intersections and corners. If tapes are used with inklines, there can be problems in focusing the copy camera because of the thickness of the tape.

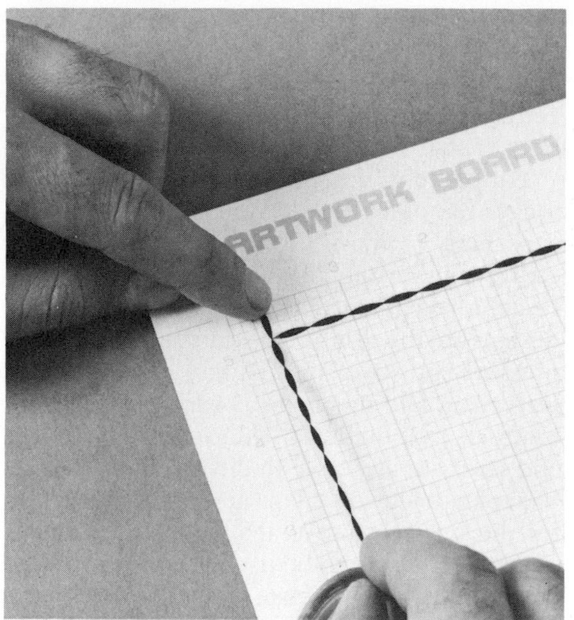

Figure 12.17. The application of adhesive-backed border tapes.
These tapes are very useful in many mapping applications.

Photographs courtesy of Graphic Products Corporation, Rolling Meadows, Ill.

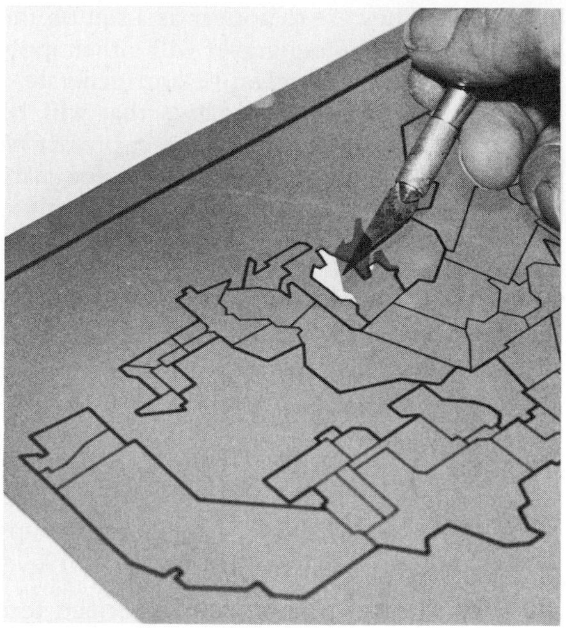

Figure 12.18. Masking film separation.
Masking films are useful in the preparation of open-window images for map artwork.

Areas and Patterns

There are several ways to prepare large areas on positive art that will eventually be screened or separated to receive color ink. Perhaps the least expensive is to develop a hand-cut, **masking film** separation, mechanical. (See Figure 12.18.) Rubylith is one such registered product name. Rubylith contains a clear plastic sheeting on which a thin red peelable emulsion layer is bound. The emulsion layer, which behaves like black in the process camera is removed (peeled) after it has been cut, leaving the clear base. A masking film mechanical is normally prepared after the line work sheets are done, so that proper registration can be maintained. Rubylith film is dimensionally stable and is used primarily with negatives.

Another method is to use red adhesive-backed film that can be cut and burnished in place on the artwork. This is more difficult to use because it is not as transparent as masking film, harder to cut, and more expensive. It is usually applied to a separate separation film, such as clear Mylar or matte film, which also adds to its cost. Adhesive-based film can be recommended when the areas to be covered are small and cutting not intricate.

A most useful product called **sensitized masking film** is widely used by cartographers.[5] A negative is prepared from the line mechanical or separation sheet and *contact-exposed* to the sensitized masking film in daylight conditions. The line image appears on the sensitized film, and areas can be removed by peeling, as with masking films. The advantage is that correct registration will result from contact-exposing the line image.

Inked areas, or emulsion layer areas of masking films or sensitized masking films, are the areas that print. Line negatives are made from these, with open (clear) areas on the negative where the image areas are to appear on the map. If they are to appear as a tint on the final map, the photoengraver will either apply a film screen to the negative and generate a single-copy composite negative that will be useful in making the plate, or apply a film screen to the original negative when the plate is made. Most printers prefer the first method because the composite negative facilitates the production of color proofs.

Cartographic drafting was considerably eased after World War II when **preprinted adhesive-backed films** with patterns were first widely manufactured. They are used in positive art preparation. There are now literally hundreds of patterns to choose from; some have become standard areal symbols, mostly by convention. There are three fundamental groups of areal patterns, as shown in Figure 12.19.

1. Dot screen patterns with specified texture (number of rows of dots per inch) and percent area inked.
2. Line patterns of various line widths and spacing between lines.
3. Irregular pattern sheets containing standard areal symbols or special screen effects.

It is important to learn the advantages and disadvantages of these patterns. For example, the percentages of area inked are not exactly as claimed by the manufacturer. To some degree, patterned films have been replaced by the precision film screens of the photoengraver. Special pattern and line screens are still widely used, however.

Generating Typographics

The preparation of type on map artwork has undergone considerable change in recent decades. The quality of map type took a turn for the worse earlier in the century when cartographers took over that responsibility from photoengravers. In part to offset the inability of draftspersons to hand-letter maps, **mechanical lettering** was introduced. Lettering could be done with small hand-held machines that followed precut grooves in templates. These forms may have been quite acceptable in engineering graphics, but their adoption by cartographers led to poorly designed maps. Lettering became rigid and lacked flow and variety of style and character.

Fortunately, the picture is much brighter today. Cartographic designers and draftspersons have many more acceptable and interesting options open to them. Individuality can be achieved at reasonable cost; more pleasing designs are the result. Type can be chosen from pressure-sensitive type films, **dry-transfer lettering,** and mechanical (dry or chemical) or photo-type methods. Hundreds of typefaces and sizes are available to designers.

Separate lettering flaps are usually prepared for the final art—several of them if different colors or tints are to be used. The type may be placed on clear acetate films or on matte drafting films. The camera operator takes special care with such art because the slightest incorrect exposure can cause imperfect results. Letters may "close up," especially if the original art has been reduced in size; the designer must bear this in mind in choosing the type style.

Registration

A critical factor in successful map printing is **registration**—the preplanned alignment of multiple images on the paper. In printing involving several line tints, pattern overlays, or separations, it becomes even more important that proper registration be achieved. Nothing is more troublesome and distracting on a map than improper registration.

The accurate positioning of multiple images on a single piece of paper is the responsibility of the designer, the photoengraver, and the printing-press operator. The designer ensures that all artwork is properly aligned; the photoengraver (or the cartographer, if doing his or her own photography) maintains the alignment during photography (especially while making composite negatives with complex screening); and the press operator controls registration during the actual printing, especially in multicolor press runs. It is difficult

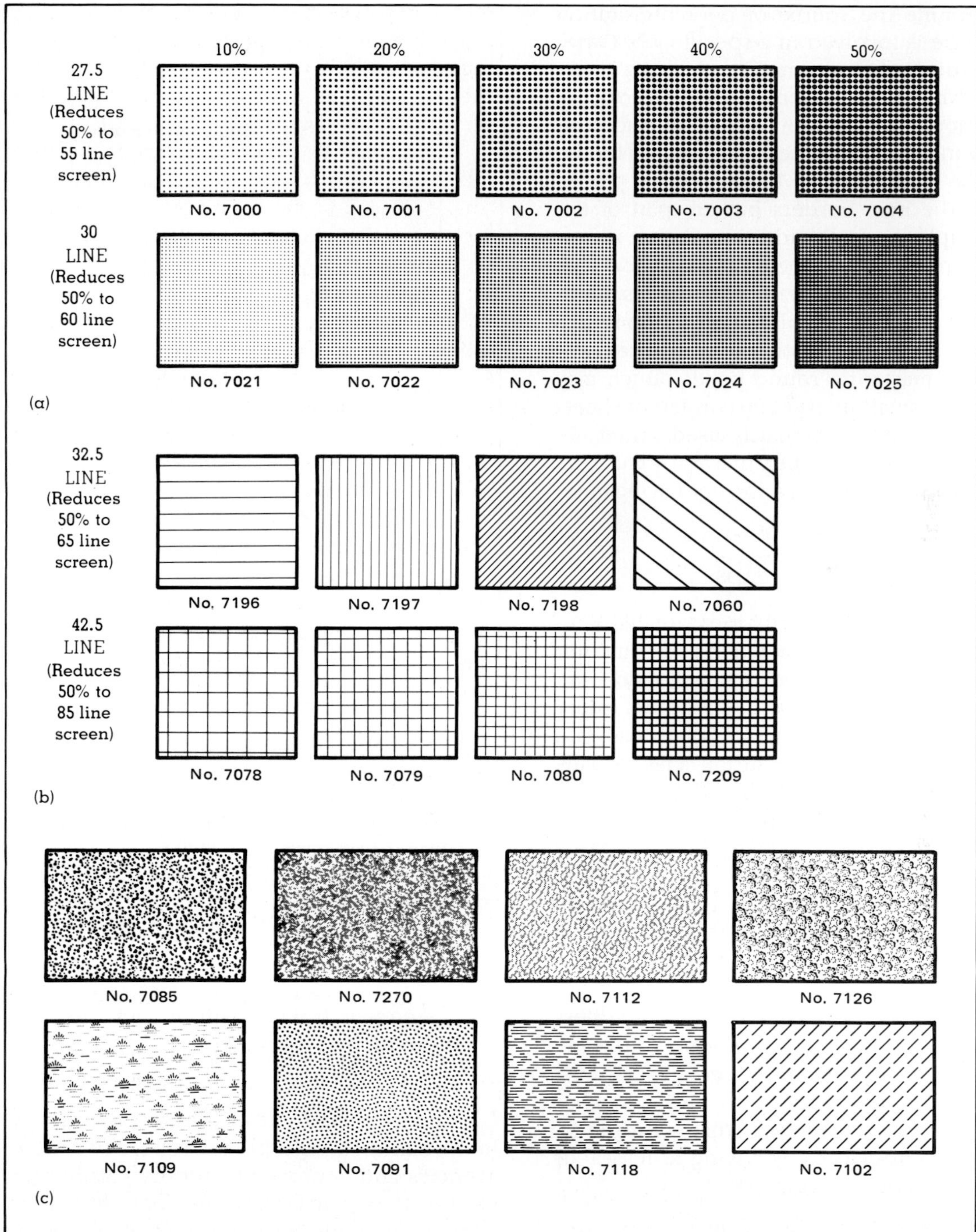

Figure 12.19. Preprinted adhesive films for positive art preparation.
(a) Several examples of dot screens are shown here, at actual size. Representative line patterns are in (b), and special pattern sheets are shown in (c).
Patterns courtesy of Graphic Products Corporation, Rolling Meadows, Ill.

to determine the source of poor registration unless one is involved in a specific job. Cartographic designers must make sure that all art is properly registered when it leaves the office.

Position or registration marks are placed on all artwork overlays as they are prepared for the job. (See Figure 12.20.) The marks are placed outside the map borders but are part of the photographed area. In addition, the art is usually pin-registered for greater accuracy. The photo-engraver will usually use this system in any event. Registration pins are uniform in size but come in a variety of formats. A standard two-hole one-quarter-in. round paper punch may be used for small map sheets; on larger sheets several holes are customarily used called pin bar registration. It is imperative that some form of registration be provided for jobs having more than one piece of art.

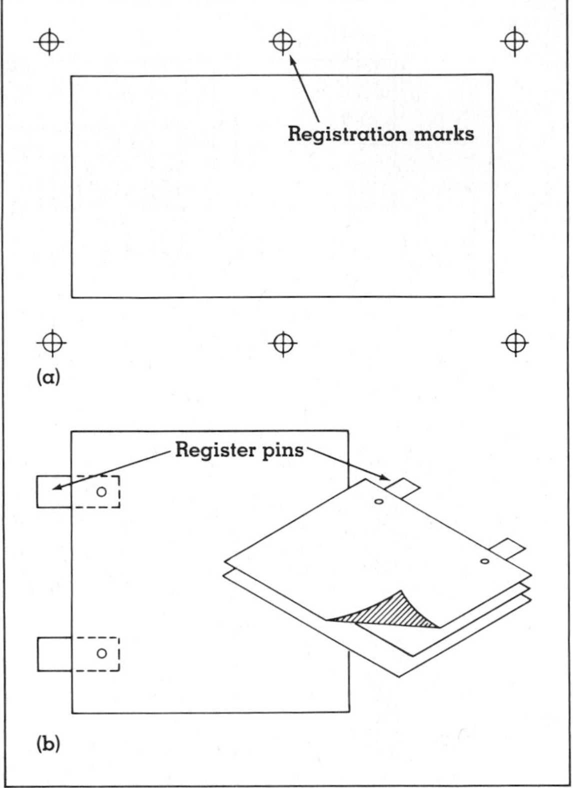

Figure 12.20. Registering art.
Registration may be accomplished by affixing special marks along the outside edges of each piece of art (a). Special pins (b) may also be used. In the latter case, each piece of film or paper is punched with holes of the exact size that will accommodate the register pins.

Computer-generated Maps

The shelves of today's libraries are likely to contain as many books on video and computer graphics as on engineering drawing or drafting. Large industrial firms now use a system called **computer-aided design (CAD)** instead of a large room of draftspeople sitting at tables. This is having an impact on cartographic drafting as well and is changing the course of cartographic production. Maps are drawn by large *laser plotters* directly onto scribe films. (See Figure 12.21.) Of course, such maps are only possible if the data are in digital form. Separation negatives or positives containing open windows can also be done this way, eliminating much tedious manual work. Regardless of how the separations are produced, the importance of map planning remains, as discussed in the next section.

Joel Morrison believes that **computer-assisted cartography** is having at least four effects on all of cartography.[6]

1. The emergence of cartography as a distinct discipline
2. The development of precision in defining cartographic terms
3. Freedom from repetitious and tedious production tasks
4. New products

Computer-assisted cartography remains mostly in the hands of large federal, state, and private agencies and is not economically feasible for most small private firms or smaller universities, although new and relatively inexpensive products are quickly challenging the more traditional means of production. (See box "The Revolution in Cartographic Production.")

Figure 12.21. Computers can be used to drive large plotting machines.
In the example shown here, the plotting head is scribing on negative scribe film.
Source: Photograph courtesy of United States Department of the Interior, Geological Survey.

Computers are already used in a variety of ways in the conventional production of maps. Computer-plotted projections are a good example. In the past, the complications of manual methods of generating projections often caused designers to choose the easiest projection, not the most desirable from the point of view of map content or design. The time formerly spent on manual work can be spent on selecting the most appropriate projection, which can lead to better design. Also, the display of maps on CRT screens can aid in the selection of class intervals, colors, and symbols.

In map production and reproduction, the products and techniques of computer-assisted cartography should be reviewed as viable alternatives to the more conventional methods of production. However, we must be alert to the

The Revolution in Cartographic Production

In the span of just a few short years, technology that replaces much of manual drafting of small-scale thematic maps has enveloped the discipline. The microcomputer and its peripheral devices, and newly developed mapping software, are what has made this possible. First came dot matrix printers capable of producing elementary choropleth maps (because filling polygons was easy), then came inexpensive single- and multi-pen plotters to do line work and line patterns (that looked like what a draftsperson or preprinted adhesive sheet would look like). Early ink-jet printing was bulky and unreliable, and was used primarily coupled to larger computers. Although used earlier on these large applications, the new technology of laser xerography (printing) hit the minicomputer world at reasonable prices at about the mid-1980s. Application of "desktop" printers has now revolutionized low-cost printing, and will for the foreseeable future have dramatic impact on cartographic production. Replacement of dot matrix and pen plotter technology by low-cost laser printers easily can be envisioned.

The emergence, in the 1980s, of device-independent page description language (pdl) for microcomputers, to organize and produce ***both*** text and graphics on plain paper (and now film), coupled with low-cost laser printers is revolutionizing this segment of the printing industry. An early entry into the field (but not the only one) is ***PostScript*** (a product of Adobe Systems, Inc.). What makes this language useful is the large number of programs that support it. Now that very powerful graphics packages have emerged that utilize *PostScript,* a whole new world is opening for the small-scale thematic map designer.

Although the graphics packages that support pdl's are not developed specifically for mapping, they have tremendous potential for such application. Two such graphics packages today are the Adobe *Illustrator 88* (a product of Adobe Systems, Inc.) and Micrografx *Designer* (a product of Micrografx, Inc.). *Illustrator 88* currently works in an Apple (Apple, Inc.) environment, and *Designer* in an IBM (or compatible) environment. In the years ahead we will no doubt see mapping applications software written to support page description languages. Or at least, existing graphics packages will be modified to accommodate more cartographic requirements.

Low-cost Laser Printers. Relatively low-cost laser printers utilize xerography image transfer technology. A rotating drum coated with photosensitive material is exposed to a light (laser) that is driven by computer control. Areas (dots) on the drum are either charged or not charged by the light source. An electronic pattern is thus composed on the drum. The turning drum picks up toner (plastic particles) of an opposite charge, and the particles adhere to the drum at the dots exposed to the light source. These particles are then transferred to plain paper, heated by rollers, and permanently sealed to the paper. Clear acetate may also be used, and on very expensive printers film negatives can be specified as output.

Resolution of course determines the appearance of the final printed product. The resolution of the pattern is specified in dots per inch (dpi) ***in both directions.*** A 300-dpi machine, therefore, can produce 300 × 300, or 90,000 dots per square inch. Cost of laser printers is determined by resolution. Low-cost machines are 300 dpi, very expensive ones exceed 2,000 dpi. However, the graphics software packages described here support printers at both low and high resolutions, without requiring any changes to the CRT image.

Equally important to the cartographer is that many of the low-cost printers also contain graphics languages that pen plotters use, making software designed for plotters only all that much more useful.

Image Creation. Maps can be created in the graphics packages in several ways. In *Designer,* for example, maps are created on the screen by mouse and draw commands, by enhancing (on the screen) images imported by desktop laser scanners, or by importing digitized files from computer-aided drafting packages (such as *AutoCad*). The image in *Designer* can be divided into "layers" of information elements to facilitate design. A layer may be composed of polygons, which can be filled with line, dot, or other patterns. Lines can be continuous, dashed, or have other characteristics, and width can be controlled in minute amounts.

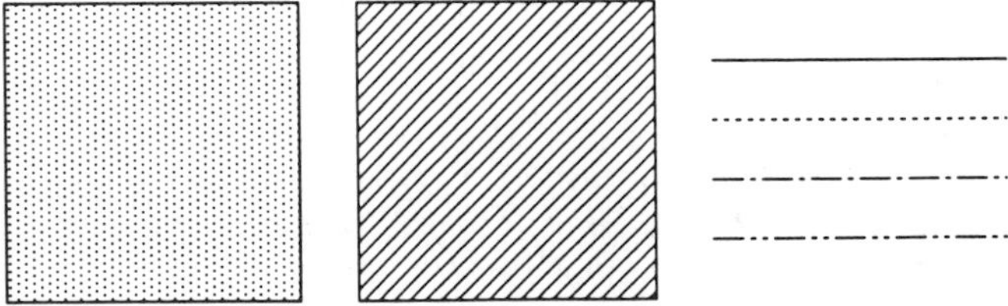

Tremendously helpful to the cartographer is that the layers can be printed out as registered separations, and be specified to the *Pantone Matching System* (a registered trademark of PANTONE, Inc.).

For the thematic map designer, scaling of quantitative symbols (areas, points, or lines) is not easy. Through various draw commands, quantitative symbols must be scaled individually, although great precision is possible.

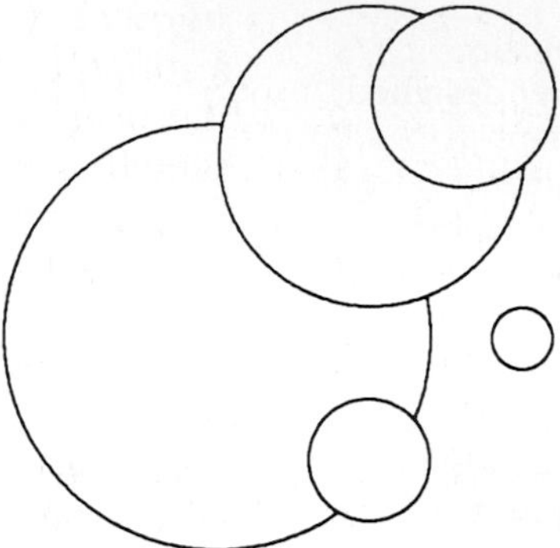

At the present time these graphics packages do not support the quantitative data handling operations of sorting, classifying, or statistical operations, that are integral to much of thematic mapping. No doubt this will change in the future.

Text creation, a severe limitation of microcomputer mapping software utilizing dot matrix or plotter output devices, is ***easily*** and handsomely handled by these graphics packages, working together with a page description language and the laser printer. Hundreds of styles and forms are available and can be placed on the image in almost any specified point size and in any orientation. Letter quality, of course, depends on printer resolution as mentioned earlier.

This is a sample of the lettering from a 300dpi laser printer.

This is a sample of the lettering from a 300dpi laser printer.

Evaluation. Critics, of course, and rightfully, can question the images created by these low-cost laser printers. In the case of text, they do not yield the clean edges of the high resolution "typeset quality" images we have grown to expect. The 300–dpi machines show raggedness, especially on slanted portions of letters and lines. Extremely high resolution (e.g., greater than 2400 dpi) printers cost thousands of dollars more, and do yield typeset quality images. Some believe that 600–dpi printers are acceptable for most text quality printing, and machines capable of this resolution are within reach of the microcomputer cartography work station. In the meantime, the low-cost 300–dpi printer, coupled with a page description language and a graphics or mapping software program, can yield very good results.

Seattle
Spokane
WA
Portland
Salem
OR

ever-present danger of allowing machines to do all the design thinking. We are cautioned, for example:

> Yet there is no assurance that electronically-produced maps will be inherently better designed than traditional, manually-produced maps. If anything, the computer seems to give many aesthetically insensitive, geographically ignorant people the opportunity to create cartographic monstrosities with unprecedented ease. Muller notes, for instance, that a number of recent maps produced with expensive hardware and complicated programs violate several, almost intuitively obvious, elementary principles of graphic symbolization.[7]

The Artwork Plan

Designing the production of a thematic map requires detailed planning so that the printed map will look in every way as the designer has intended. The cartographer necessarily deals with content, generalization and symbolization, scale, and other mapping components, but the final map will not emerge as envisioned without careful planning of the reproduction method. The purpose of this section is to present ways of organizing and managing this plan.

Getting Started

Map production requires putting the sketch map into camera-ready form, which calls for considerable detailed planning by the designer. The sketch map is the documentation of the designer's authoritative plan for the final stages of production. Detailed specifications for production must be actually on the sketch or in instructions accompanying it. The appearance of lines and any screening of them, all colors, lettering location and styles, the completed base map, and all other parts of the map must be carefully specified. In most instances, the sketch map must be accurately rendered because final art is to be traced directly from it. In other cases, the sketch map may simply be a set of detailed instructions on how to make the final art. Whichever procedure is used, a sketch map is completed first.

The sketch map can often be rendered on any suitable drawing or tracing paper, not necessarily on stable-based materials. Pencil-only (not suitable for ink) drafting or rendering paper is frequently used. It should be of suitable thickness (about .003 in.) and allow for erasing. Most such papers are made from cotton cuttings.

Flow Charting the Plan

It will be generally helpful to make a **flow chart**—a plan of the necessary steps in the preparation of the art separations. (See Figure 12.22.) Flow charts can be general or specific, depending on the nature of the job. Such a plan directs the production team in every step of the entire process and forces the designer to attend to detail. A flow chart can facilitate later editing and proofing. The flow chart is a kind of design blue-print, necessary in all but the simplest map jobs.

Map Editing

Map editing, which consists of proofreading the finished art for mistakes or omissions, is a crucial step in map production and reproduction. Editing is especially important when the map contains hundreds of place names or when many complex political boundaries are included. Even the simplest map must be checked for errors, however.

No one method is used in editing, and the activity varies depending on the size of the office and the complexity of its products. Regardless of the procedure used, some editing

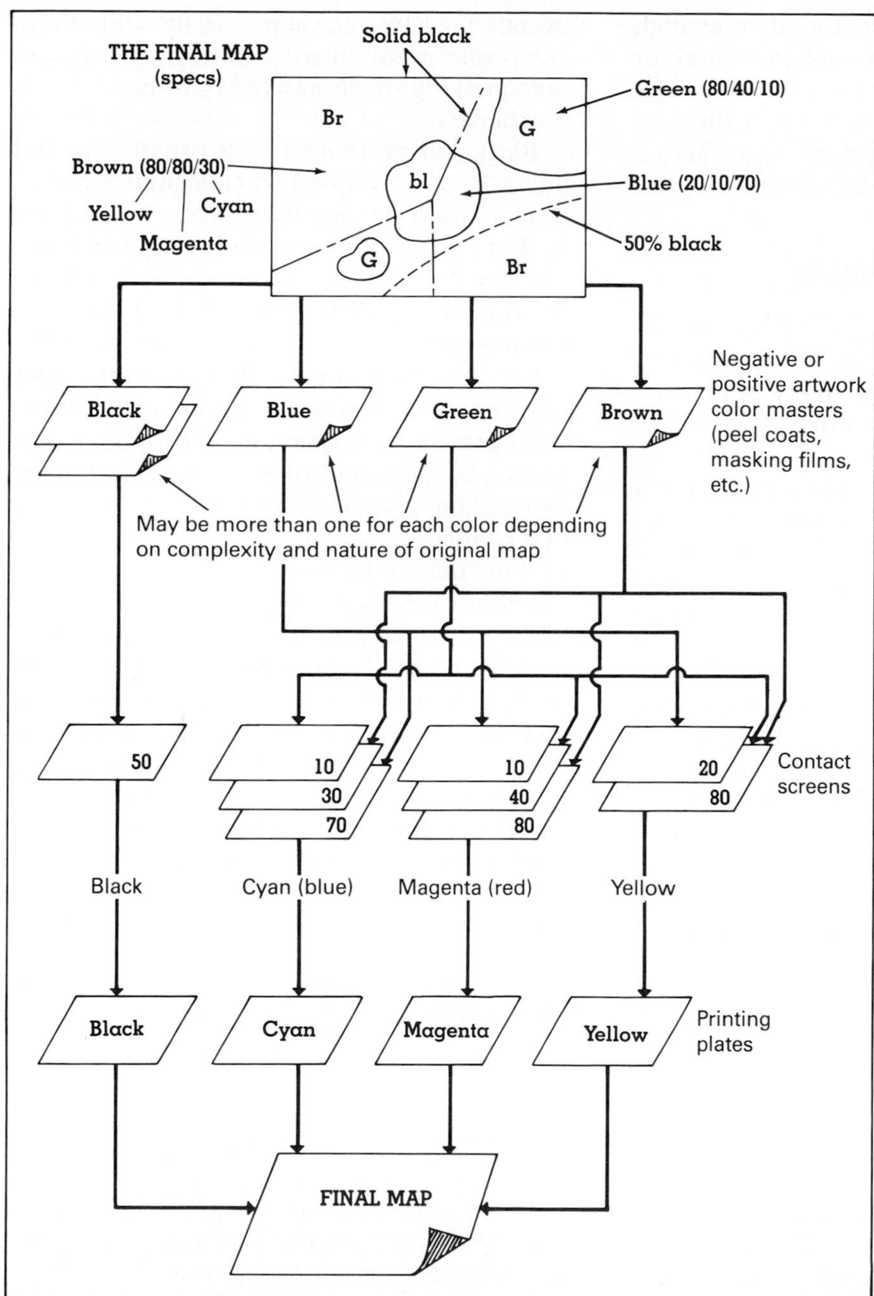

Figure 12.22 A sample map preparation flow chart.
Flow charts may be more or less detailed, and there may be several ways to chart a particular job. They should serve to direct the work flow and can be used to pinpoint economical alternatives in the plan.

must be done. The benefits usually outweigh its costs in terms of saving embarrassment or avoiding redoing the artwork. Editing should be done prior to the extensive and costly negative work and should be done again before negatives are sent to the plate maker.

Preparing Copy for Printing

Original art is most often "pasted up" on artboard for the photoengraver. The separations are affixed to a stiff artboard, in rough registration, and a tracing paper is often taped to the top to protect the art. These are called *mechanicals* or *keyline art.* The photographer will receive the whole job in composite form, facilitating preparation of accurate negatives.

Written specifications for the art (e.g., screens or masking) can be written directly on the art with a *non-photo blue pencil,* whose lead will not photograph under the harsh lights of the copy camera. Alternatively, each separation can be labeled and written instructions attached to indicate what is to be done with each flap. Photographers seem to prefer the former method. It is good practice to go over the art mechanicals with the photographers to make certain they understand the specifications. This will help avoid errors later.

Proofing Completed Art

The map designer should always strive to have proofs made of the cartographic art before it is printed, especially for color maps. These are called **pre-press proofs**—a means of representing the final map, in color, prior to printing. The two principal reasons for proofing at this stage are to evaluate the design before the expensive process of printing begins and to check on the quality of the photoengraver's work. Printers appreciate pre-press proofing as a way of assuring approval by their clients before printing. The designer can often visualize the overall final design more easily with these composite proofs than with negatives pasted onto masking sheets and liberally marked with notations.

Black-and-white art is not usually proofed unless it contains complex screen tints. In these instances, *silverprints* (blueline or brownline) or *diazo* proofs can be made by exposing special papers (or films in the case of diazo) through the copy negative and chemically processing them.

For process color or complex multicolor printing, there are two methods of pre-press color proofing: transparent and opaque. In the former, the copy negatives are used to contact-expose films chemically processed to yield one of the process colors. Proofing is done by sandwiching these together, in correct register, and examining them over a sample of the paper that will be used for printing, in bright light. Reasonably good results can be achieved, although colors will not be exact and glare is a problem. Separations can be checked and errors in the planning of the job easily detected.

Opaque proofing is similar to transparent except that the proof is on only one sheet of paper. One such system has been devised by the 3M Company and is called "3M Transfer-Key." Good results are achieved. Regardless of the brand products used, opaque color proofing is generally preferred to transparent.

Press Inspections

Press inspection, or *press proofing,* is generally done only on very large press runs involving large printing costs. A press proof is a printed copy of the map on the final paper. Press inspections are specified in printing contracts and are usually very expensive. The presses must be stopped and the cartographer summoned for the inspection, which is time consuming. Pre-press proofing, if done carefully, is normally adequate.

Notes

1. Arthur H. Robinson, "Mapmaking and Map Printing: The Evolution of a Working Relationship," in *Five Centuries of Map Printing,* David Woodward, ed. (Chicago: University of Chicago Press, 1975), pp. 1–23.
2. Michael J. Adams and David D. Faux, *Printing Technology, A Medium of Visual Communication* (North Scituate, MA: Duxbury Press, 1977), pp. 213–14.
3. Robinson, "Mapmaking and Map Printing," pp. 16–23.
4. Cartographers usually choose colors base on the *Pantone Matching System,* a color-mixing system that has nearly become standard in the printing industry. It is a product of Pantone, Inc. of Moonachie, New Jersey.
5. Sensitized Masking Film is a product of the Keuffel and Esser Company.
6. Joel Morrison, "Computer Technology and Cartographic Change," in *The Computer in Contemporary Cartography,* D. R. Fraser, ed. (New York: John Wiley, 1980), p. 10–11.
7. Mark S. Monmonier, "Geographic Information and Cartography," *Progress in Human Geography* 8 (1984):381–91.

Glossary

border tape special adhesive-backed tape that can replace the ink line; comes in a variety of widths; some forms (e.g., crepe) can be shaped to form irregularly curved lines.

burning exposure of a printing plate to a harsh light.

color light energy at different wavelengths along the visible electromagnetic spectrum; red is about 7.5×10^{-5} cm in wavelength, and violet is about 3.5×10^{-5} cm

color separation a process of creating a printed color image that looks like a continuous-tone image but is actually formed by patterns of small dots printed in black plus the three transparent inks of magenta, cyan, and yellow.

computer-aided design (CAD) interactive system that allows an operator to design drawings and develop specifications at a cathode-ray tube (CRT) screen; eliminates the need for conventional drafting methods and tools.

computer-assisted cartography cartography process that relies on the assistance of the computer; usually, the computer drives output devices (e.g., CRT image and pen plotters)

copy or process camera precision instrument used to make line negatives in modern photoengraving; consists of a copyboard, lensboard, ground glass, and backboard.

dry-transfer lettering method of applying lettering to positive art; lettters are applied to special transport films that are positioned over the art work and burnished, transferring the letters to the art

flaps art separations; the pieces of art that make up the composite image to be printed

flat color printing accomplished by using different-color opaque inks, in contrast to four-color process printing

flow chart design plan including all steps, in the proper sequence, in a cartography printing project

graver instrument used to cut away the coating of negative scribe film; a variety of forms are available, including rigid, flexible, and double

gray tone in black and white cartography, the amount of ink printed per unit area on paper surface; e.g., 10 percent or 20 percent

halftone negative photographic film containing a precision-ruled pattern of lines; used to photograph a continuous-tone negative; breaks the original image into a pattern of very small dots of uneven size and density

intaglio forms of printing in which depressions are engraved in the block or plate; the depressions receive ink, which is transferred when pressed on the paper; copperplate was a popular method of intaglio map printing during the seventeenth and eighteenth centuries

letterpress form of printing in which the printing surfaces are raised in relief from the printing block or plate; raised portions receive the ink

light that part of the electromagnetic spectrum that creates sensations in human eyes; light occupies wavelengths from 7.5×10^{-5} to 3.5×10^{-5} cm

manuscript map a single map not made for reproduction, or the working compilation map that will be used for making scribe sheets

masking film a special plastic sheet on which is applied a thin light-blocking film that can be cut and removed to make open-window art separations

mechanical lettering usually applied to art by means of special templates and scribing instruments

mechanicals pasted-up art to be sent to the photographer for line negative preparation

negative a piece of photographic film containing the reverse of the image to be printed; black areas of the original are open on the negative, and white areas of the original are black on the negative; line negatives have only black or open areas, unlike continuous-tone negatives

negative scribe drafting process that involves the cutting away of a coating applied to a tough (usually plastic) film; areas removed will be printed

offset lithography an intermediate drum or roller on the press allows the plate image to be right-reading; the image on the blanket or offset roller is backwards

opaque printing ink reflects back its color at the ink's top surface but absorbs all other colors

photolithography modern lithographic printing; so called because of the photographic preparation of the printing plate

planar printing relies on the fact that water and grease do not mix well; areas on the printing block or plate are at the same elevation, with ink either adhering or not, depending on the surface preparation between water and grease; a popular form is lithography

pre-press proof system to develop black-and-white composites of a printing project before it is printed; used for editing and checking specifications and for visual approval of a printing job

preprinted adhesive-backed film positive-working art film having any number of preprinted patterns that can be burnished to art

primary additive colors red, green, and blue; additive colors are those that, when combined with other colors, yield new ones

primary subtractive colors combinations of colors resulting from the absorption of a primary color or colors; also called secondary or complementary colors

process color printing achieved by using subtractive color combination and transparent inks; often called four-color process printing

registration any system that ensures the proper alignment of all pieces of art

screen tint photographic film used to break a line negative into a regular pattern of very small dots; after printing, the image appears as a percentage of solid color

sensitized masking film can be exposed to a line negative and areas of the film peeled away; much like peel-coat films; especially useful in cartography with difficult registration requirements

stipple and aquatint engraving special methods of engraving a copper plate in order to create the impression of tints when printed

technical pen a precision drafting pen with a specified line width; used widely, generally replacing the ruling pen

transparent inks first reflect their color to the paper or other ink on which they are printed, then reflect it back to the reader's eye; process printing requires the use of the transparent inks magenta, cyan, and yellow

woodcut map method used to print maps during the fifteenth and sixteenth centuries; based on the relief method; similar to early letterpress

Readings for Further Understanding

Adams, Michael, J. and David D. Faux. *Printing Technology, A Medium of Visual Communication.* North Scituate, MA: Duxbury Press, 1977.

Brannon, Gary R. *An Introduction to Photochemical Techniques in Cartography.* Waterloo, Ontario: Cartographic Centre, University of Waterloo, 1986.

Cardamone, Tom. *Color Separation Skills.* New York: Van Nostrand Reinhold, 1980.

Carter, James R. *Computer Mapping, Progress in the '80s.* Resource Publications in Geography. Washington, D.C.: Association of American Geographers, 1984.

"Computer Aided Designers." *Designer* (Spring 1988):8–9.

Dalley, Terence, ed. *The Complete Guide to Illustration and Design, Techniques and Materials.* Secaucus, NJ: Chartwell Books, 1980.

Earle, James H. *Engineering Design Graphics.* Reading MA: Addison-Wesley, 1977.

Felici, James, and Ted Nace. *Desktop Publishing Skills.* Reading, MA: Addison-Wesley, 1987.

A very useful primer incorporating design, typography, graphics, and hardware associated with all aspects of desktop publishing. Highly recommended for the beginning cartographer.

International Paper Company. *Pocket Pal: A Graphic Arts Production Handbook.* New York: International Paper Company, 1974.

Lang, Kathy. *The Writer's Guide to Desktop Publishing.* London: Academic Press, 1987.

Monmonier, Mark S. *Computer-Assisted Cartography, Principles and Prospects.* Englewood Cliffs, NJ: Prentice-Hall, 1982.

———. "Geographic Information and Cartography." *Progress in Human Geography* 8 (1984):381–91.

———. *Technological Transition in Cartography.* Madison: The University of Wisconsin Press, 1985.

Moore, Lionel C. *Cartographic Scribing Materials, Instruments and Techniques,* Washington, D.C.: American Congress on Surveying and Mapping, 1968.

Morrison, Joel. "Computer Technology and Cartographic Change." In *The Computer in Contemporary Cartography,* D. R. Fraser, ed. New York: John Wiley, 1980.

Olson, Judy M. "Component-Color and Final-Color Separation in Mapping." *Cartographica* 22(1985):61–69.

Peucker, Thomas K. *Compuer Cartography.* Commission on College Geography. Resource Paper No. 17. Washington, D.C.: Association of American Geographers, 1972.

Robinson, Arthur H. "Mapmaking and Map Printing: The Evolution of a Working Relationship." In *Five Centuries of Map Printing,* David Woodward, ed. Chicago: University of Chicago Press, 1975.

Schweitzer, Ricahrd H., Jr. *Mapping Urban America with Automated Cartography.* Washington, D.C.: United States Department of Commerce, Social and Economics Statistics Administration, Bureau of the Census, 1973.

Sena, Michael L. "Back to Camp—Advances in Computer-aided Map Publishing Drive New Industry." *Computer Graphics World* (March 1982):119–24.

United States Department of the Army. *General Drafting.* Technical Manual TM 5–581A. Washington, D.C.: Headquarters, Department of the Army, 1972.

———. *Offset Photolithography and Map Reproduction.* Technical Manual TM 5–245. Washington, D.C.: Headquarters, Department of the Army, 1970.

Chapter

13

Elements of Map Composition

Chapter Preview

This chapter is our first venture into the design of maps and deals with the appearance of the final map product. The basic concepts of graphic composition as they relate to maps will be presented, and they should encourage experimentation. No one best way to a design solution can be predetermined for all maps—only principles and general approaches can guide the cartographer. The ordinary thematic map can be considered to have two or more planes or levels. Thematic map design deals with the arrangement of the map's elements at each level or between levels. The map's design elements are those marks on the map that must be arranged into a graphic composition suitable to the map's communication purpose. The elements include the map's title, scale, border, symbols, and other marks. By using visualization and experimentation techniques, the composition elements are arranged to satisfy the design goals. These elements include the planar organizational elements of balance, focus, internal organization, figure and ground, and a number of contrast elements. Visual acuity plays an important part in design; cartographers must be aware of the limitations of the human eye as they plan and develop map specifications.

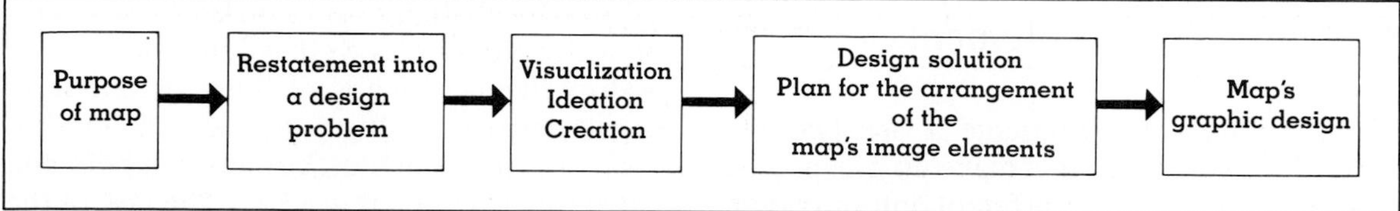

Figure 13.1. The sequential steps leading to the final graphic design of a map.

The map's graphic design can be considered the end product of a chain of events. (See Figure 13.1.) The purpose of the map appropriately comes first and can be restated into the design problem. After application of techniques of visualization and creative problem-solving, a design solution is reached. The solution involves the arrangement of the map elements to facilitate the communication goals of the map. The graphic design is the end product, the tangible map that we see.

It is always the responsibility of the cartographic designer to make certain that the map is cartographically and geographically sound. This means that the map's spatial information is correct and that distortion, other than that introduced by the chosen projection, is nonexistent or at least minimized. Once assured of this, the cartographer can render the map in any graphic way considered appropriate.

The graphic solution should seek to cause a change in the reader, as mentioned in Chapter 1. In viewing art, the reader may feel that nothing is required other than mere contemplation. When looking at effective graphic advertising, the reader feels that something must be done about it. In effective cartographic design, the reader should gain spatial knowledge and understanding.

Design Levels on the Map

It is useful at the outset to imagine the thematic map as composed of different planes or levels. (See Figure 13.2.) Usually the levels are differentiated by visual prominence. Each component of the map belongs to a specific level. More than one map element can be placed on a particular level, but *a single element should never be assigned to more than one level.* Thinking of the map in this way will facilitate the map's overall design.

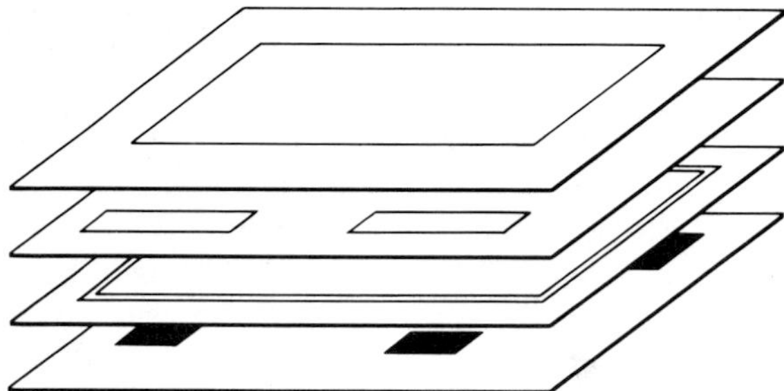

Figure 13.2. The organizational levels and the thematic map.
A typical thematic map can be considered to be made of several distinct levels or planes. In planning a map, the designer assigns the various map elements to these levels. This causes the designer to think of each element in its proper role, thus leading to a more organized design.

Map composition, the arrangement of the map's elements, takes place at each level and between levels. The arrangement at a given level may be called **planar organization,** and that between levels **hierarchical organization.** Hierarchical organization is also referred to as the visual hierarchy. The cartographer ordinarily approaches design solutions by simultaneously manipulating all elements at and between all levels.

The Map's Design Elements

Thematic maps are instruments of visual communication. The marks that make up a map are visual elements, and transfer of information takes place through them. The map designer arranges the visual elements into a functional composition to facilitate communication. This functional approach to design was first expressed by Arthur Robinson; "Function provides the basis for design."[1] That something is functional means that it has been "designed or developed chiefly from the point of view of use." Design decisions regarding the map's elements should be made on the basis of how each element is to function in the communication. The challenge is to make the map aesthetically pleasing as well as functional.

Most thematic maps contain similar **map elements:** titles, legends, scales, credits, mapped areas, graticules, borders, symbols, and place names. (See Table 13.1.) The task of the designer is to arrange these into a meaningful, aesthetically pleasing design—not an easy task. Within the **map frame** (an area on the map sheet bounded by a real or imagined border), the designer can exercise considerable freedom in arranging the map elements. The only real limitations are imposed by convention and lack of imagination. (See Figure 13.3.) The designer must blend each of the elements with the whole map's design so that a cohesive visual whole is achieved within the map frame.

Table 13.1 Elements of the Thematic Map

Name of Element	Description and Primary Function
Title (and subtitle)	Usually draws attention by virtue of its dominant size; serves to focus attention on the primary content of the map; may be omitted where captions are provided but are not part of the map itself
Map legend	The principal symbol-referent description on the map; subordinate to the title, but a key element in map reading; serves to describe all unknown or unique symbols used
Map scale	Usually included on a thematic map; it provides the reader with important information regarding linear relations on the map; can be graphic, verbal, or expressed as an RF
Credits	Can include the map's data source, an indication of their reliability, dates, and other explanatory material
Mapped and unmapped areas	Objects, land, water, and other geographical features important to the purpose of the map; make the composition a map rather than simply a chart or diagram
Graticule	Often omitted from thematic maps today; should be included if their locational information is crucial to the map's purpose; usually treated as background or secondary forms
Borders and neatlines	Both optional; borders can serve to restrain eye movement; neatlines are finer lines than borders, drawn inside them and often rendered as part of the graticule; used mostly for decoration
Map symbols	Wide variety of forms and functions; the most important elements of the map, along with the geographical areas rendered; designer has little control over their location because geography must be accurate
Place names and labeling	The chief means of communicating with maps; serve to orient the reader on the map and provide important information regarding its purpose

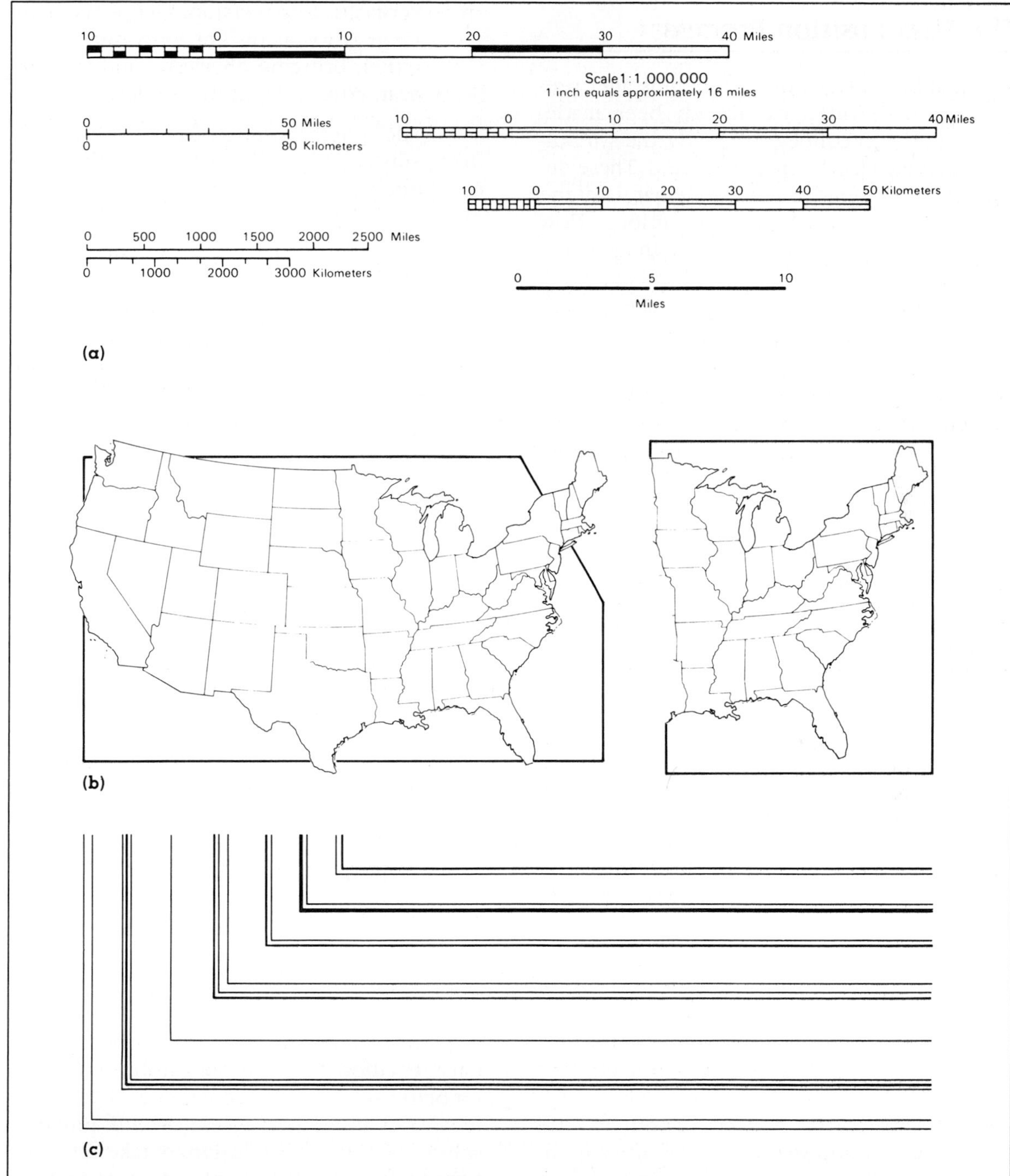

Figure 13.3. Different graphic treatments of three map elements.
In (a), several different ways of presenting graphic scales are shown. In (b), map borders need not be conventional boxes around the map but can be closely integrated with the graphic design. The illustration in (c) shows different map borders and neatlines. In graphic composition of the map, the designer must be open to many alternative ideas.

The Composition Process

The graphic design solution begins after preliminary decisions have already been made. Data have been collected; usually, the symbolization of data has been determined. These decisions have been dictated by the nature of the data and any applicable conventions. Now comes the exciting phase of design in which the final map image is conceived and prepared. Two activities are critical in this stage: visualization and experimentation.

Visualization

Our purpose here is to introduce the importance of **visualization** to the task of map design. Visualization, ideation, and creativity are necessary to solving map design problems. Map design is essentially a creative act; the designer is assembling for the first time a map's elements into a new whole, in the hope of transferring knowledge to others. This is done by creative means.

There are many descriptions of how new ideas are formed in the minds of creative people. A composite pattern might include these four stages in the creative process.[2]

1. Preparation. At this stage, a person consciously "files away" into memory visual images that can be useful for a problem at hand.
2. Incubation. The person releases all conscious hold on the problem and turns to other tasks. It is theorized that images in a person's mind are rearranged into new alignments and patterns during this truly creative stage.
3. Illumination. The solution to the problem appears suddenly, often spontaneously.
4. Verification or revision. The person consciously works out the details of the solution, bringing all efforts and skills to bear. Formal structures result.

Map designers must develop an **image pool** in the subconscious, from which creative ideas can form during the incubation stage. (See Figure 13.4.) Some psychologists believe that ideation results from realignment of images stored from earlier perceptual events.[3] ***Visualization*** is the process of putting oneself into a mental state of experiencing or seeing these new creations.

Map designers should therefore experience as much graphic art, art, and cartography as possible. It is worthwhile to spend time exploring an art museum, seeing an animated film exhibit, or going through old atlases. From such experiences, the designer builds an image inventory which can later provide creative design solutions.

Experimentation

Experimentation is necessary in testing the new idea. Sometimes ingenious solutions do not work when put onto the map sheet. Creative solutions must be worked out in detail on paper, to preview their visual effects. Do the elements of the design work effectively together? Do the details, however creative they may be, detract from the map as a whole? Are alternative solutions now apparent that were not noticeable before? A willingness to explore alternative solutions is essential to developing the best graphic design for a map.

Unfortunately, many map designs are first solutions. Time and cost are enemies of experimentation but must not inhibit the search for better solutions or refinements of good first ideas. It is altogether too easy to drift into easy solutions. Good map designers take the extra time to explore all possible ideas, even stop-

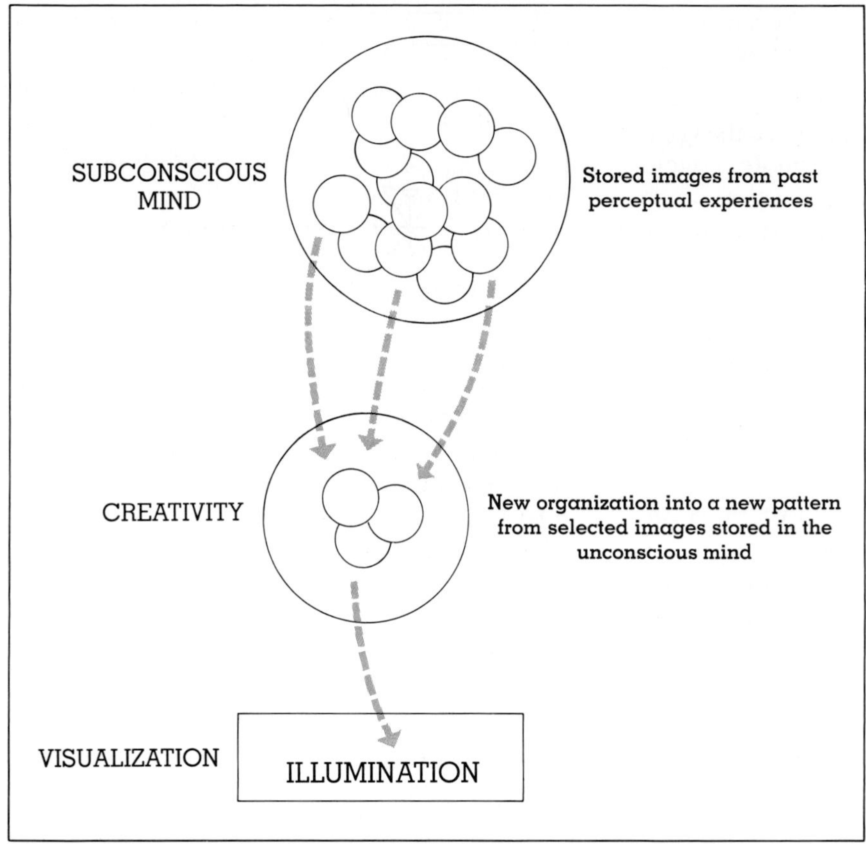

Figure 13.4. The visualization process.
It is theorized that creativity is the result of new arrangements of stored visual images in the subconscious mind. Visualization is the ability to "see" these arrangements.

ping to let incubation begin again. The process from visualization through experimentation to final solution often involves much repetition and backtracking.

Elements of Map Composition

The map's graphic composition is the arrangement or organization of its elements. The composition principles introduced below include the purpose of map composition, planar organization, figure and ground organization, contrast, and visual acuity. Knowledge of these principles and their application assists the cartographer in seeking better design solutions.

Purpose of Map Composition

Map composition is much more than layout, which is simply a sketch of a proposed piece of art, showing the relationships of its parts. The term *composition* is used here because it indicates the intellectual dimension as well as the visual. Map composition serves these ends.

1. Forces the designer to organize the visual material into a coherent whole to facilitate communication, to develop an intellectual *and* a visual structure
2. Stresses the purpose of the map
3. Directs the map reader's attention
4. Develops an aesthetic approach for the map
5. Coordinates the base and thematic elements of the map—a critical factor in establishing communication
6. Maintains cartographic conventions consistent with good standards
7. Provides a necessary challenge for the designer in seeking creative design solutions

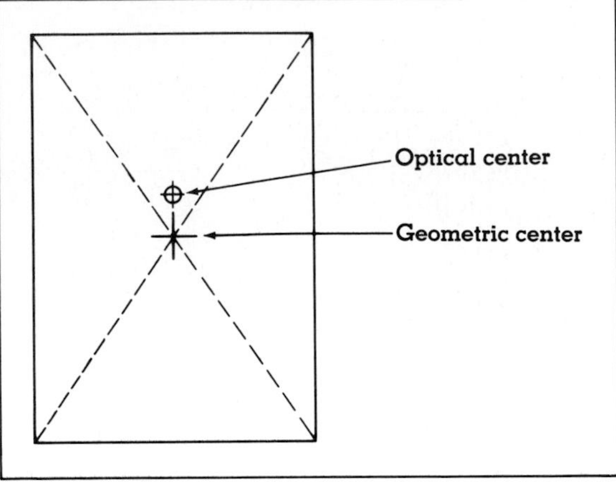

Figure 13.5. The two centers of an image space.
The designer should arrange the map's elements around the natural (optical) center, rather than the geometric center.

Planar Organization of the Visual Elements

There are three aspects of planar visual organization: balance, focus of attention, and internal (intraparallel) organization. Each is important to the designer's language, and their visual possibilities and effects must be explored.

Balance

Balance involves the visual impact of the arrangement of image units in the map frame. Do the units appear all on one side, causing the map to "look heavy" on the right or left, top or bottom? An image space has two centers: a geometric center and an **optical center** (as shown in Figure 13.5). The designer should arrange the elements of the map so that they balance visually around the optical center.

Rudolf Arnheim, a noted expert on the psychological principles of art, suggests that **visual balance** results from two major factors: weight and direction.[4] Objects in the visual field (e.g., within the borders of a map) take on weight by virtue of their location, size, and shape. Direction is also imposed on objects by their relative location, shape, and subject matter. Arnheim stresses that balance is achieved when everything appears to have come to a standstill, "in such a way that no change seems possible, and the whole assumes the character of 'necessity' in all its parts."[5] In Arnheim's view, unbalanced compositions appear accidental and transitory.

Arnheim's observations on balance resulting from visual weight and direction can be summarized as follows:[6]

1. Visual weight depends on location.
   - Elements at the center of a composition pull less weight than those lying off the tracks of the structural net. (See Figure 13.6a.)
   - An object in the upper part of a composition is heavier than one in the lower part.
   - Objects on the right of the composition appear heavier than those on the left.

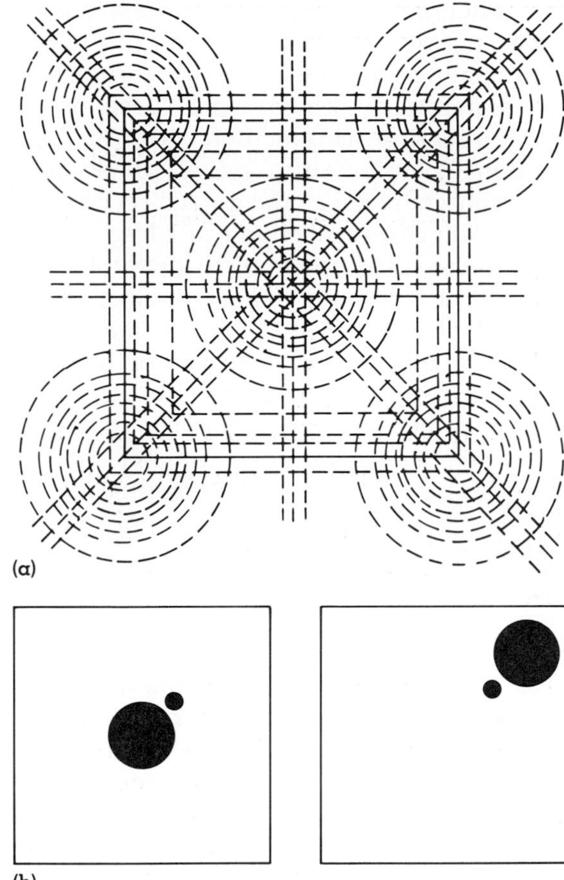

Figure 13.6. Balance in the visual field. Arnheim stresses that a structural net, as in (a), determines balance. Objects on the main axes or at the centers will be in visual balance. An object is given direction by other objects adjacent to it. In (b), the small disc's directional element is shifted as the large disc's position is changed. Each thematic map will have a unique structural net created by the locational pattern of its elements.

- The weight of an object increases in proportion to its distance from the center of the composition.

2. Visual weight depends on size.
   - Large objects appear visually heavier than small objects.
3. Visual weight depends on color, interest, and isolation.
   - Color affects visual weight. Red is heavier than blue. Bright colors appear heavier than dark ones. White seems heavier than black.
   - Objects of intrinsic interest due to intricacy or perculiarity seem visually heavier than objects not possessing these features.
   - Isolated objects appear heavier than those surrounded by other elements.
4. Visual weight depends on shape.
   - Objects of regular shape appear heavier than irregularly shaped ones.
   - Objects of compact shape are visually heavier than those not so shaped.
5. Visual direction depends on location.
   - Weight of an element attracts neighborhood objects, imparting direction to them. (See Figure 13.6b.)
6. Visual direction depends on shape.
   - Shapes of objects create axes that impart directional forces in two opposing directions.
7. Visual direction depends on subject matter.
   - Objects possessing intrinsic directional forces can impart visual direction to other elements in the composition.

Of course, Arnheim recognizes that the elements of compositional balance operate together in complex fashion. He also advises not to forsake the content of a composition simply in order to create balance: "the function of balance can be shown only by pointing out the meaning it helps to make visible."[7]

It is often difficult to achieve balance on the map. Cartography is not an expressive art form in which the graphic elements may be rearranged at will. Many of the shapes and their locations are imposed by the geographic facts.

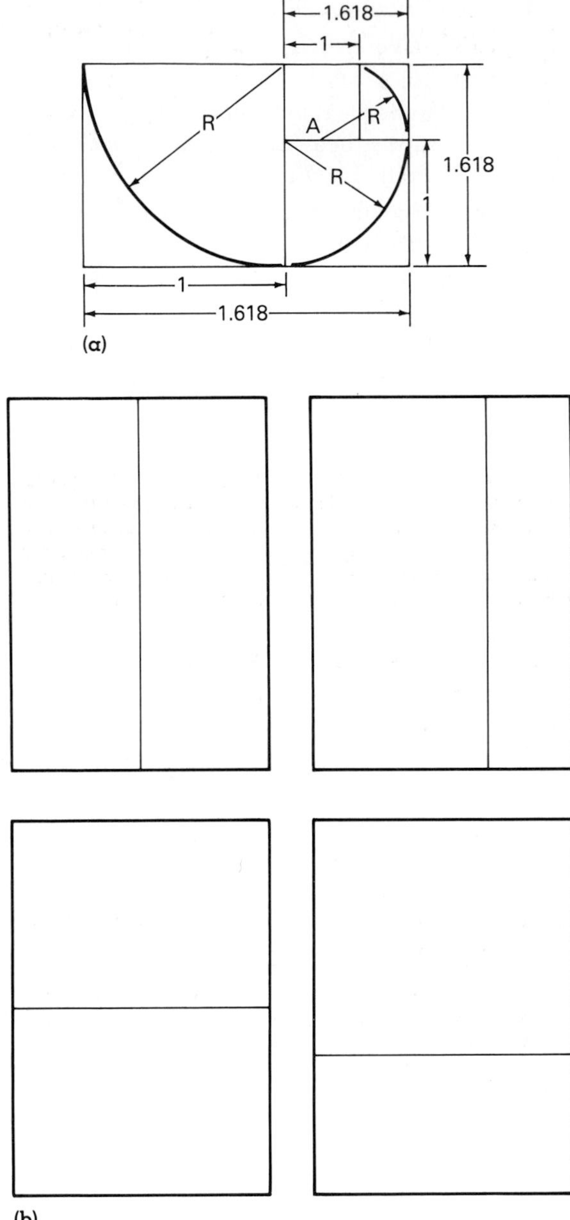

Figure 13.7. Methods of achieving balance. Balance is dynamic and will result from appropriate proportioning of the image space. The method of arriving at the golden section is illustrated in (a), and several alternatives are pictured in (b). In (b), notice that unequal divisions of space are more interesting.

The guidelines presented by Arnheim should nevertheless be applied whenever possible.

Graphic art professionals who work with two-dimensional design often speak of the **golden section.** This method of devising proportions is attributed to classic Greek architects and sculptors. In the golden section, the proportion of a smaller unit to a larger is the same as that of the larger unit to the whole. This method of sectioning can be duplicated any number of times. (See Figure 13.7a.) Proportion is the relationship that a part of the visual field has to the remainder or whole. Balance is achieved when pleasing proportions among the parts are maintained. Cartographers should bear proportion in mind in arranging the different elements of the map. Applying the golden section in cartographic design is an intuitive matter, not subject to rigid quantification.

Visual balance can be looked at from a different point of view. Writing more than fifty years ago, Richard Surrey, an advertising artist, developed important ideas about composition. Surrey observed that layout involves not only the arrangement of units (balance) but also the division of space. "In other words, instead of layout being a process of addition (putting together units), it is much more easily grasped when considered as a process of division."[8] (See Figure 13.7b.) His further thoughts on this idea may be summarized:

1. Equal divisions of space are the least interesting. Inequality and the pursuit of equilibrium make layout visually alive.
2. Small spaces struggling against large spaces are visually alive.
3. Variety, for example the division of the image space into four unequal parts, creates interest. Complex designs may be more exciting than simple ones.

To illustrate how balance can affect the impression one has when viewing a map, several different locations of the shape of Africa are included in Figure 13.8. Which appears better balanced within the map frame?

Achieving visual balance, of course, is not always so simple as in this case of Africa. Normally, thematic maps contain most of the elements mentioned earlier, and all of these must be handled in terms of balance. Visual weight caused by texture, solid black and white areas, and other elements must figure in the planning. Open spaces take up "balance space" and must be used effectively in the overall design. Complex designs require careful planning to use all spaces efficiently while retaining a visually harmonious balance. Acceptable balance is reached when the relocation of any one element would cause visual disturbance. Balance is a state of equilibrium.

In recent research using thematic maps, the balance of the map's elements is shown to have an initial affect on the way the map reader goes about looking at those elements.[9] However, the longer the reader views the map, the less importance balance seems to have on map reading behavior. Better balance also leads to less reading difficulty and to somewhat better memory of the map's message. It is not altogether clear exactly what constitutes good and poor balance in such studies because these extremes are subjective at best. Nonetheless, the balance of the map's elements is a vital concern for the cartographic designer.

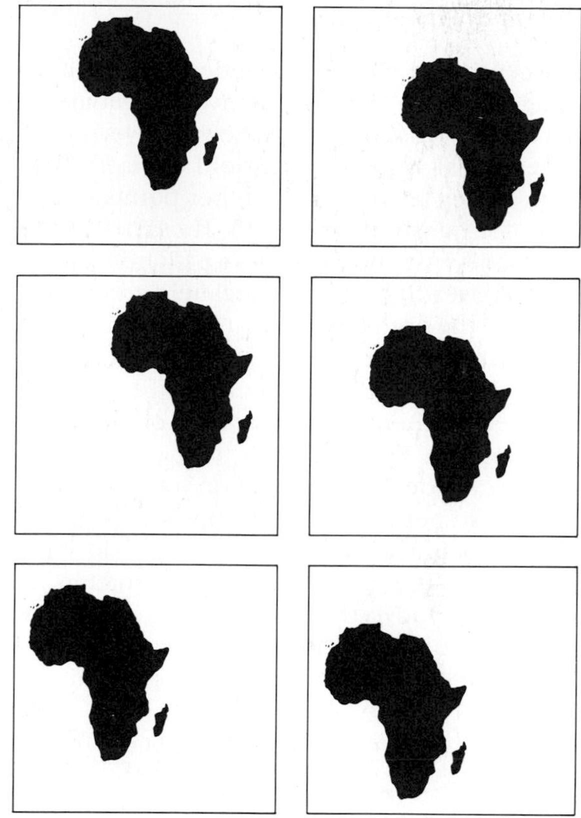

Figure 13.8. Map balance.
Position of map elements in the image space affects the balance of the map. The difference can be visually subtle, as this illustration shows. In which image does a natural visual equilibrium exist?

[E]very visual pattern is dynamic. Just as a living organism cannot be described by its anatomy, so the essence of a visual experience cannot be expressed by inches of size and distance, degrees of angle, or wave lengths of hue. These static measurements define only the "stimulus," that is, the message sent to the eye by the physical world. But the life of a precept—its expression and meaning—derives entirely from the activity of the kind of forces that have been described. Any line drawn on a sheet of paper, or the simplest form modeled from a piece of clay, is like a rock thrown into a pond. It upsets repose, it mobilizes space. Seeing is the perception of action.

Source: Rudolf Arnheim, Art and Visual Perception (Berkeley, Cal.: University of California Press, 1965), p. 6.

Focus of Attention

As mentioned above, the optical center of an image area is a point just above the geometrical center. This attracts the viewer's eye, unless other visual stimuli in the field distract attention. Surrey makes several other points significant for questions of design. He says that the reader's eye normally follows a path from upper left to lower right in the visual field and passes through the optical center.[10] (See Figure 13.9.) Furthermore, the point of greatest natural emphasis is where a line of space division intersects either the focus or field circles of attention. (See Figure 13.10.)

Surrey's ideas were based on intuitive judgments and personal observations and have not been scientifically proven. Yet they do have appeal for the designer. An examination of recent print advertisements attests to the general applicability of his ideas. Map designers can learn from these and other graphic de-

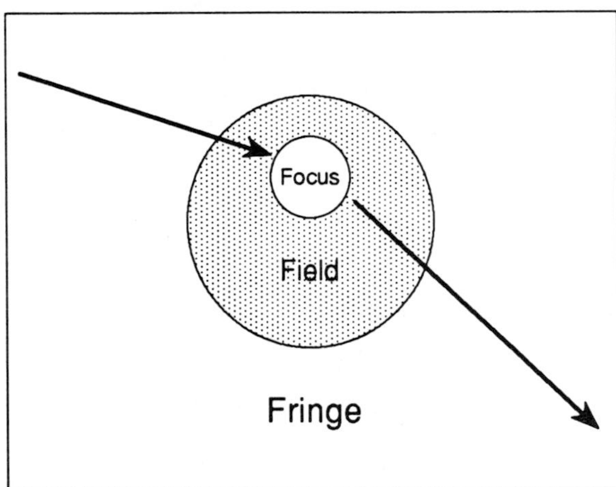

Figure 13.9. Eye movement through the image space.
In normal viewing, the reader's eyes enters the image space at the upper left, proceeds through the visual center (focus), and exits the space at the lower right. Cartographic designers may use this pattern when arranging the map's elements, so that the positions of important objects on the map correspond to the natural eye movements.

(a)

(b)

Figure 13.10. Recentering the important part of the map to align with the focus and major division lines.
In (a), San Francisco is too far removed from that important intersection. It has been moved in (b) and rests in a visually important part of the graphic arrangement. Designers will not always have the opportunity to do this but should explore the possibility in developing final designs.

signs. The map is a visual instrument, so the designer must learn what works in the visual world.

Internal Organization—Intraparallelism

The internal organization of the map's visual field relates to visual or perceptual order. Rudolf Arnheim defines order as "a wealth of meaning and form in an overall structure that clearly defines the place and function of every detail in the whole."[11] Order implies an underlying structure, graphic or intellectual, that binds the parts of the whole together. In an ordered map, the graphic elements are arranged into a composition that develops a clear visual expression of the meaning of the communication and that shows an underlying structure of the graphic elements.

One technique to give structure to the graphic elements, at least at the planar level, is intraparallelism.[12] Intraparallelism is achieved when the elements of internal structure are aligned with each other. (See Figure 13.11.) Intraparallelism reduces tension in perception; it can be introduced into map designs in subtle ways. (See Figure 13.12.) The designer must consider and experiment with a variety of visual techniques to simplify the graphic design of maps.

Figure and Ground Oganization

There is probably no perceptual tendency more important to cartographic design than **figure and ground organization.** A person's underlying behavioral tendency is to organize the visual field into categories: figures (important

Figure 13.11. Internal structure of graphic elements provided by intraparallelism.
The objects in (a) display a lack of internal structure because each has its own orientation. After realignment in (b), the objects appear to be more of an integrated unit. Lack of intraparallelism leads to tension and disunity.

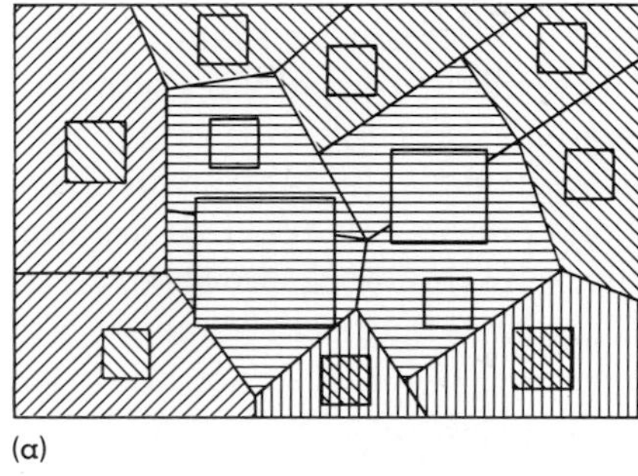

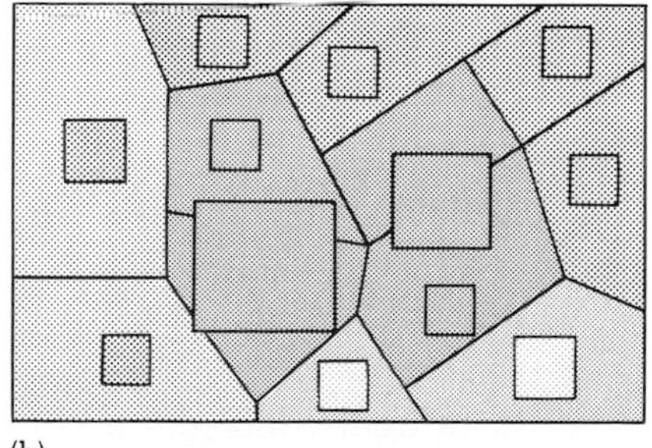

Figure 13.12. Rearranging map elements to achieve greater internal order.
The line patterns of (a), with different orientations, have been replaced in (b) by dot screens having greater internal order. The rows of repeating dots have identical orientations.

objects) and grounds (things less important). This concept was first introduced by Gestalt psychologists early in this century. Figures become objects of attention in perception, standing out from the background. Figures have "thing" qualities; grounds are formless. Figures are remembered better, while grounds are often lost in perception.

In the three-dimensional world, we see buildings in front of sky and cars in front of pavements. Likewise, we see some objects in front of others in the two-dimensional world, given wise graphic treatment. In cases such as the words printed on this page, no extraordinary measures are needed to make figures stand out from ground. In Figure 13.13(a), the graphic treatment resists the formation of figures in perception. With a few changes the square symbols leap out as figures in perception, as in Figure 13.13(b).

In the planar level of design planning, the designer should structure the field in a way that directs the reader's perception along paths commensurate with the communication goals of the map. For example, objects that are important intellectually should be rendered so as to make them appear as figures in perception.[13]

Deborah Sharpe, a color designer not from the cartographic profession, has stated very succinctly the importance of incorporating figure and ground perception into design.[14]

> In my own design work, I have found that designating the features that are to represent figure and those that are to represent ground as a first step on the job eliminates the trial and error inherent in the beginning stages of most creative tasks.

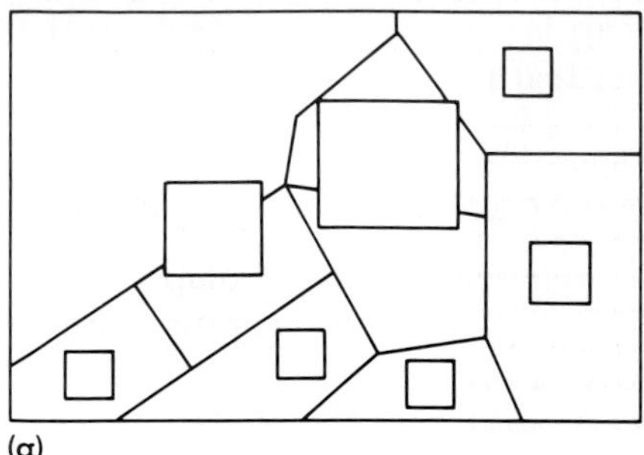

(a)

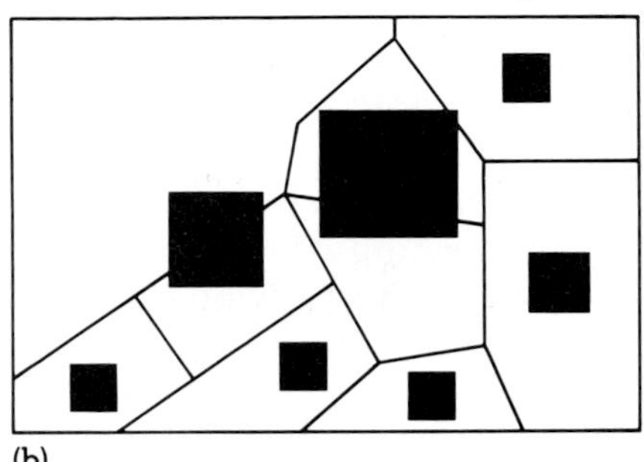

(b)

Figure 13.13. The development of figures on maps.
The graphic treatment of the squares in (a) resists figure formation. The application of black to the squares in (b) causes them to emerge as strong visual figures. In many cases, only minor graphic adjustments are necessary to develop figure formation.

Contrast

Closely associated with figure and ground organization, and of nearly equal importance, is the feature called **contrast.** Contrast is elemental in developing figure and ground but can be considered a design principle in its own right. Visual contrast leads to perceptual differentiation, the ability of the eye to discern differences. A lack of visual contrast detracts from the interest of the image and makes it difficult to distinguish important from unimportant parts of a communication. Map elements that have little contrast with their surroundings are easily lost in the total visual package. Contrast must clearly be a major goal of the designer.

Contrast can be achieved through several mechanisms: line, texture, value, detail, and color. All of these could be used in one design, but the result might be visual disharmony and tension—potentially as unrewarding as having no contrast at all.

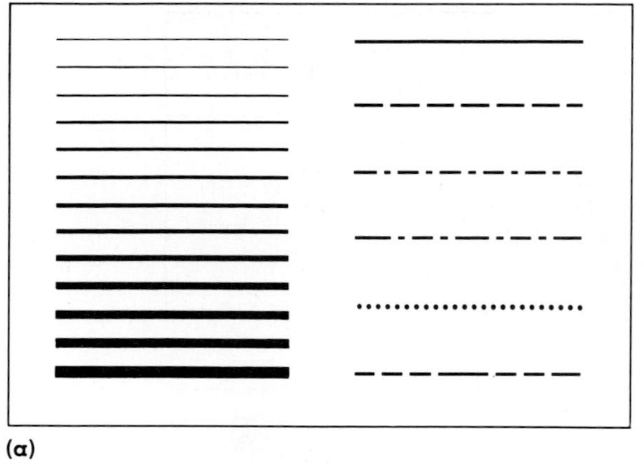

(a)

(b)

(c)

Figure 13.14. Line contrast.
In (a), several different line weights and line characters are shown. The visual effect of varying line weights on a map is illustrated by comparing (b) and (c). More visual interest is achieved with greater contrast of weight. In this case, the figure and ground organization is strengthened.

Line Contrast

Lines may be put to a variety of uses on maps. They can function as labels, borders, neatlines, political boundaries, quantitative or qualitative symbols, special symbols to divide areas, or graphic devices to achieve other goals. Line contrast can be of two kinds: character and weight. **Line character** derives from the nature of the line and its segments—its value or color. (See Figure 13.14a.) The order of visual importance of various line characters has not been well established. The subject and purpose of the map very often restrict choice of line character. On some maps, there may be no lines. For example, some recent designs use edges rather than lines to evoke a response.

The thickness of a line is its **line weight,** although no clear-cut relationship exists between thickness and visual or intellectual importance. Although a broader line generally carries more intellectual importance, very fine lines can be visually dominant also. The designer must strike a balance, keeping the map's purpose firmly in mind.

Contrast of line character and weight introduces visual stimulation to the map. A map having lines of all one weight is boring and lacks potential for figure formation. (See Figure 13.14b.) On the other hand, a map with

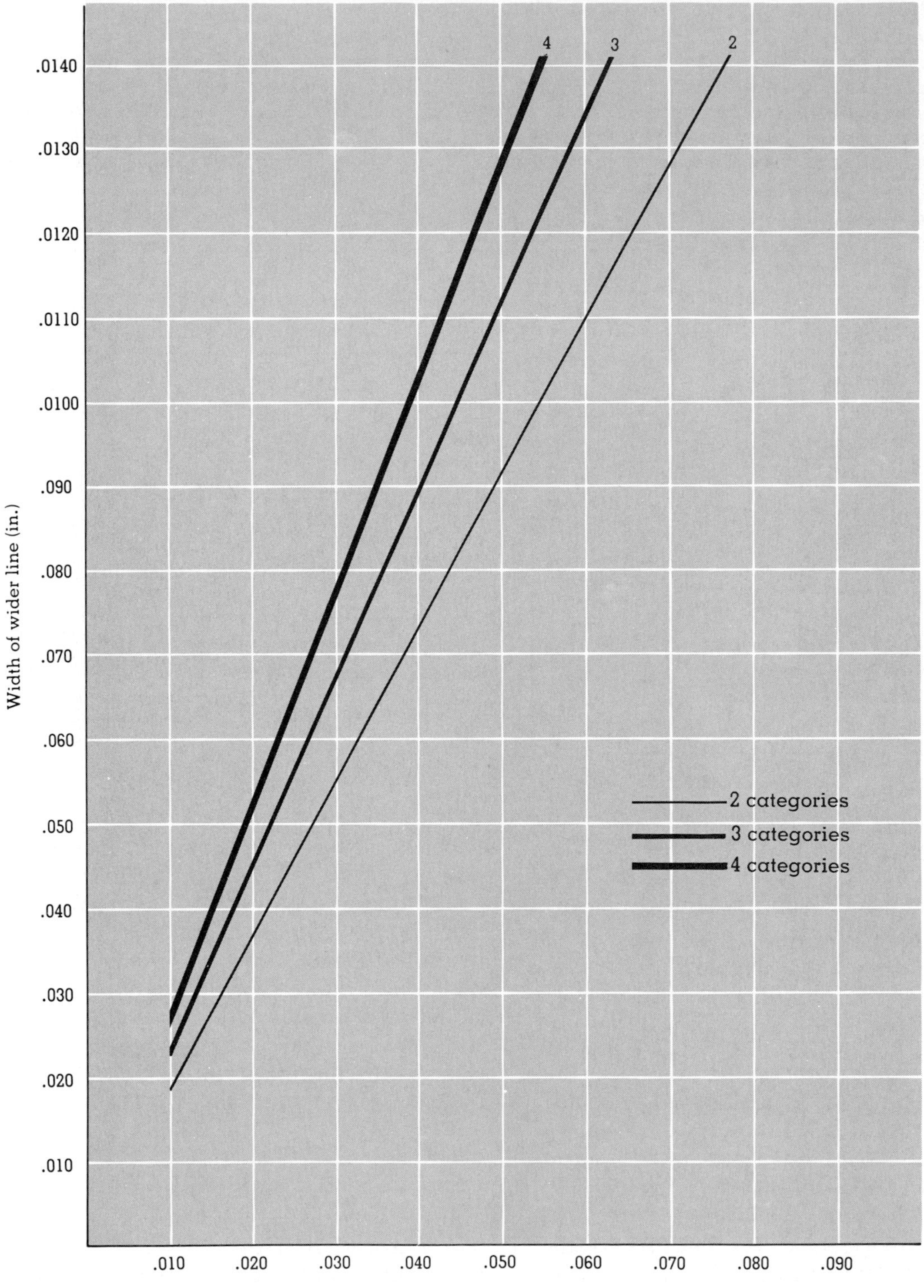
4
3
2
.0140
.0130
.0120
.0110
.0100
.090
.080
.070
.060
.050
.040
.030
.020
.010
Width of wider line (in.)
2 categories
3 categories
4 categories
.010
.020
.030
.040
.050
.060
.070
.080
.090
Width of narrower line (.000's in.)

Figure 13.15. Line width selector chart on facing page.
This chart may be used to determine line widths when two, three, or four different widths are used on maps. For the two-line category, simply determine the width of the narrower line and enter the chart along the bottom axis. Extend a perpendicular to the two-category ordinate, and from this point extend a horizontal line to the vertical axis. Read the width of the wider line directly from the vertical axis. For the three-category case, first determine the wider line of a two-line case, and then use the line width of that determination as the narrower line for the three-line case (this time using the three-category ordinate). The four-line case can be handled in a similar manner.
Source of chart: Richard D. Wright, "Selection of Line Weights for Solid, Qualitative Line Symbols in Series on Maps" (unpublished Ph.D. dissertation, Department of Geography, University of Kansas, 1967), pp. 91–92.

lines of several weights and characters focuses attention, is lively, and aids the map reader's perceptual organization of the material. Guidelines can assist the designer in choosing lines so that discrimination between weights is possible. Generally, a line-weight difference exceeding .05 in. is discernible by more than half of all map readers. A difference of .15 in. is easily noticed by practically all readers. Figure 13.15 is a chart that can aid the designer in choosing line weights to assure discernible differences.

Texture Contrast

Contrast of texture involves areal patterns and how they are chosen for the map. In this context, texture is a pattern of small symbols (e.g., dots) repeated in such a way that the eye can perceive the individual elements. Texture is often determined by the selection of quantitative or qualitative symbols for the map. Contrast considerations should be part of symbol selection. In some instances, patterns are selected and applied to the map solely to provide graphic contrast (e.g., in the differentiation of land and water). Texture is sometimes applied in order to direct the reader's attention to a particular part of the map.

Another possibility, not often used, is to use textured lettering. This differentiates labels from other lettering, enabling the designer to use more lettering in the design. Textured lettering is only possible when the letters are geometrical, not composed of many thick and thin strokes.

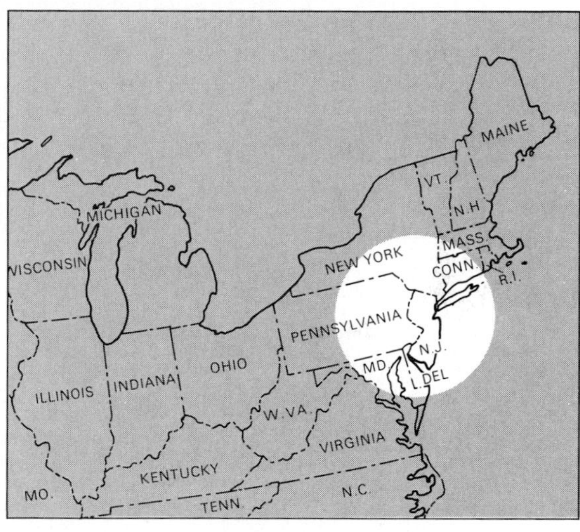

Figure 13.16. The use of differences in value to focus the reader's attention.
In this example, the absence of the pattern over New Jersey forces the eye to that part of the map. Contrast of texture can also be used in this manner.

Value Contrast

Texture is observable because the individual dots or other elements of the pattern are easily seen. Reducing such a pattern to the point where the elements are below the threshold of visual resolution acuity results in the perception of a visual tone or value. Contrast of value is another design technique used by cartographers, although some of the contrast is often dictated by the nature of the data (qualitative or quantitative). In cases not determined by the data, contrast of value can be used in ways similar to contrast of texture. (See Figure 13.16.)

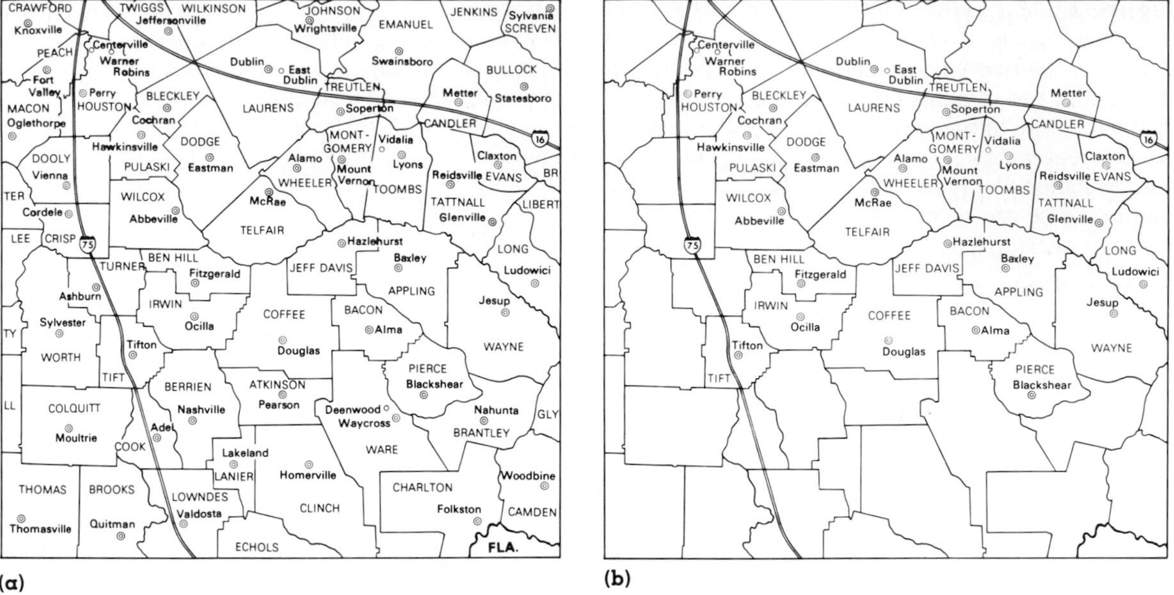

Figure 13.17. The provision of detail can direct the reader's eye.
In (a), the even distribution of place names does not focus the reader's attention. The eye is drawn to that part of the map in (b) that has the most detail—in this case, lettering.

Contrast of value leads to light and dark areas on the map. A good place to use this contrast type is in the development of figures and grounds. To stand out strongly, figures should have values considerably different than grounds. Land areas, for example, should be made lighter or darker than water areas.

Variation of Detail

Although designers seldom think of it as a positive design consideration, contrast of detail can be employed most effectively, especially in combination with other techniques. Along a continuum ranging from little detail at one end to great detail at the other, the reader's eye will be attracted to those areas of the map with the most detail.

This feature can work against the designer, however. Exquisite detail rendered to an unimportant feature can distract the reader's attention from the communication effort. By judicious use of extra detail in important areas of the map, the designer can subtly lead the reader to them. (See Figure 13.17.) Detail can also be used to strengthen figure formation.

Color Contrast

The employment of color is one of the chief techniques in the development of contrast in design. Color can differentiate areas on the map for a variety of purposes. Color as a major design ingredient is treated in detail in Chapter 16.

Vision Acuities

The map designer works in a visual medium, so all elements must be visible to the map reader. If the elements cannot be seen, the map's communicative attempt will be lost, no matter how well-designed the map may be in other respects. There are two important measures of the human ability to see visual elements: **visibility acuity** and **resolution acuity.**

Visibility Acuity

So far, it has been assumed that the map reader could see the map and all its design elements, but visibility must not be taken for granted. Fortunately, it is seldom a problem unless the designer overlooks it when preparing art for reduction.

Visibility acuity is a measure of a size threshold.[15] It should not be confused with intensity threshold, which is a measure of sensitivity. The parameter used in visibility is called a *subtense.* It is an angular measure, because the retinal image size for a 1-inch object 2 feet away is identical to that for a 10-inch object 20 feet away. Strictly speaking, acuity measures the threshold size of the retinal image, not the size of the object. For practical purposes, minimum object sizes can be prescribed so that objects will not fall below the visibility threshold, especially when art is to be reduced.

The subtense varies from about .44 seconds of arc to about 10 minutes for a black line on a white background, depending on illumination.[16] This means that at a reading distance of 46 cm (18 in), a black line should not be rendered smaller than .15 mm (.006 in) to about 1.27 mm (.05 in). The smallest size of nib manufactured by one technical pen manufacturer is .13 mm (.005 in). A minimum line thickness of .25 mm (.01 in) is a safer all-around specification for most design work.

The visibility acuity threshold for a black dot on a white background is 32 seconds of subtense. This becomes about 1.0 mm (.04 in) at a 46 cm reading distance.

Resolution Acuity

Resolution acuity is somewhat different from visibility acuity. Resolution is a measure of the detectable separation between objects in a visual field. When two objects are seen apart, the reader is said to resolve them. Again, the threshold subtense is the point at which this occurs accurately. The average threshold separation of two black dots on a white background is about 1 minute of arc.[17] This is approximately .076 mm (.003 in). This measure can become critical in map design when patterns are specified in design, especially if art is to be reduced. If the elements in a pattern are closer than this, the observer does not see the pattern but begins to perceive only a continuous tone. If the difference in patterns are important for differentiation in such a case, the design has failed in its task of facilitating communication. Map elements are not functional at all if they cannot be seen. The very best map designs will falter if minimum thresholds are not maintained.

Notes

1. Arthur H. Robinson, *The Look of Maps* (Madison, WI: University of Wisconsin Press, 1966), p. 13.
2. Mike Samuels and Nancy Samuels, *Seeing with the Mind's Eye* (New York: Random House, 1975), pp. 239–40.
3. *Ibid.,* p. 245.
4. Rudolf Arnheim, *Art and Visual Perception* (Berkeley, CA: University of California Press, 1965), p. 14.
5. *Ibid.,* p. 12.
6. *Ibid.,* pp. 14–17.
7. *Ibid.,* p 27.
8. Richard Surrey, *Layout Techniques in Advertising* (New York: McGraw-Hill, 1929), pp. 11–20.
9. James R. Antes, Kang-tsung Chang, and Chad Mullis, "The Visual Effect of Map Design: An Eye-Movement Analysis," *American Cartographer* 12 (1985):143–55.
10. *Ibid.,* pp. 21–27.
11. Arnheim, *Art and Visual Perception,* p. 45.
12. Borden D. Dent, "Simplifying Thematic Maps Through Effective Design: Some Postulates for the Measurement of Success," *Proceedings of the American Congress on Surveying and Mapping* (Fall Technical Convention, October, 1973), pp. 243–51.
13. Borden D. Dent, "Visual Organization and Thematic Map Communication," *Annals* (Association of American Geographers) 62 (1972): 79–93.
14. Deborah T. Sharpe, *The Psychology of Color and Design* (Chicago: Nelson-Hall, 1974), p. 102.

15. Albert M. Potts, ed., *The Assessment of Visual Function* (St. Louis: C. V. Mosby, 1972), p. 5.
16. *Ibid.*, p. 18.
17. *Ibid.*, p. 21.

Glossary

contrast important element of design; contrasts of line, texture, value, detail, and color are means through which maps become interesting and dynamic

figure and ground organization fundamental behavioral tendency to organize perception into figures and grounds; figures are dominant elements, and grounds serve as backgrounds for figures

focus of attention that part of the visual field which attracts the reader's eye

golden section method of dividing two-dimensional space so that the proportion of a smaller area to a larger one is identical to that of the larger area to the whole

hierarchical organization composition or arrangement of the map's visual elements as they appear between two or more visual or intellectual levels

image pool the collection of mentally stored visual images obtained from previous visual experiences

line character the internal elements that make up the distinctive qualities of a line; e.g., dot, dot-dash, dash-dash-dot, etc.

line weight the thickness of a line

map composition arrangement of the map's visual and intellectual components

map elements marks that make up the total visual image called the map, including the title, legend, scale, credits, mapped or unmapped areas, graticule, borders and neatlines, and symbols

map frame area on the map sheet bounded by a real or imagined border

optical center the place just above the geometric center in an image space; can be used to create visual balance

planar organization composition or arrangement of the map's visual elements as they appear at one visual or intellectual level

resolution acuity ability to discern a separation between objects in the visual field; usually measured by angular subtense

visibility acuity size threshold; the ability to discern an object in the visual field; usually measured by angular subtense on the retinal image, as opposed to the physical size of the object

visual balance state in which all objects in a visual image appear in equilibrium

visualization mental process in which the designer experiences whole new creations by rearranging previously stored visual images

Readings for Further Understanding

Adams, Robert. *Creativity and Communications.* London: Studio Vista Limited, 1971.

Antes, James; Kang-tsung Chang, and Chad Mullis, "The Visual Effect of Map Design: An Eye-Movement Analysis." *American Cartographer* 12(1985):143–55.

Arnheim, Rudolf. *Visual Thinking.* Berkeley, CA: University of California Press, 1971.

———. *Art and Visual Perception.* Berkeley, CA: University of California Press, 1965.

Beck, Jacob, ed. *Organization and Representation in Perception.* Hillsdale, NJ: Laurence Erlbaum, 1982.

Dember, William N. *The Psychology of Perception.* New York: Holt, Rinehart and Winston, 1961.

Dent, Borden D. "Simplifying Thematic Maps Through Effective Design: Some Postulates for the Measurement of Success." *Proceedings of the American Congress on Surveying and Mapping* (Fall Technical Convention, October, 1973), pp. 243–51.

———. "Visual Organization and Thematic Map Communication." *Annals* (Association of American Geographers) 62 (1972):79–93.

Ford, Kathryn. "Perceptual Organization of Complex Atlas Plates." Unpublished Master's thesis, Department of Geography, Queen's University, 1983.

Hanks, Kurt; Larry Belliston, and Dave Edwards. *Design Yourself.* Los Altos, CA: William Kaufmann, 1978.

Jones, Christopher. *Design Methods.* New York: John Wiley, 1981.

Lindbeck, John R. *Designing Today's Manufactured Products.* Bloomington, IL: McKnight and McKnight, 1972.

Potts, Albert M., ed. *The Assessment of Visual Function.* St. Louis: C. V. Mosby, 1972.

Robinson, Arthur H. *The Look of Maps.* Madison, WI: University of Wisconsin Press, 1966.

Samuels, Mike, and Nancy Samuels. *Seeing with the Mind's Eye.* New York: Random House, 1975.

Sharpe, Deborah T. *The Psychology of Color and Design.* Chicago: Nelson-Hall, 1974.

Surrey, Richard. *Layout Techniques in Advertising.* New York: McGraw-Hill, 1929.

Tufte, Edward R. *The Visual Display of Quantitative Information.* Cheshire, CT: Graphics Press, 1983.

White, Jan V. *Graphic Idea Notebook.* New York: Watson-Guptill Publications, 1980.

Whitfield, P. R. *Creativity in Industry.* Baltimore: Penguin Books, 1975.

Wright, Richard D. "The Selection of Line Weights for Solid, Qualitative Line Symbols in Series on Maps." Unpublished Ph.D. dissertation, Department of Geography, University of Kansas, 1967.

Chapter

14

Total Map Organization, the Visual Hierarchy, and the Figure-Ground Relationship

Chapter Preview

Effective thematic map design requires first the careful placement of the map's intellectual elements in an organizational or visual hierarchy. This planning activity involves the assignment of relative importance to each element. To make the design visually effective, the graphic solutions must follow the organizational plan, so that the appearance of each map element corresponds to its intellectual importance. The perceptual dichotomy of figure and ground, in which certain objects stand out from the remainder of a visual field, can be utilized to develop the visual hierarchy. Map objects that are clearly more important in the communication should be rendered so that they appear as figures in map reading. Attention to land-water contrast is essential to the process. Careful planning of figures and grounds leads to the emergence of a meaningful, harmonious, and unambiguous design.

The previous chapter dealt with planar organization and suggested that the total design of the thematic map could be thought to contain features of visual hierarchy organization as well. An understanding of how the visual hierarchy can work in design is critical to the planning and execution of effective thematic maps.

Visual Hierarchy Defined

Public speakers arrange their material to emphasize certain remarks and subordinate others. Professional photographers often focus the camera to provide precise detail in certain parts of the picture and leave the remainder somewhat blurred. Advertising artists organize ads to accentuate some spaces and play down others. Choreographers arrange the dancers so that some will stand out from the rest on the stage. Professional cartographers must go through similar activities in designing the map.

The **visual hierarchy** (or *organizational hierarchy*) is the intellectual plan for the map and the eventual graphic solution that satisfies the plan.[1] Each design activity should contain such a hierarchy. In this phase of design, the cartographer sorts through the components of the map to determine the relative intellectual importance of each, then seeks a visual solution that will cast each component in a manner compatible with its position along the intellectual spectrum. Objects that are important intellectually are rendered so that they are visually dominant within the map frame. (See Figure 14.1.)

Figure 14.1. The visual hierarchy. Objects on the map that are most important intellectually are rendered with the greatest contrast to their surroundings. Less important elements are placed lower in the hierarchy by reducing their contrasts. The side view in this drawing further illustrates the hierarchical concept.

Customary Positions of Map Elements in the Hierarchy

Each map has a stated purpose that controls the planning of the visual hierarchy. Mapped objects and their relative importance assume a place in the hierarchy. Although identical objects may vary in relevance, depending on the

Table 14.1 Typical Organization of Mapped Elements in the Visual Hierarchy

Probable Intellectual Level*	Object	Visual Level
1	Thematic symbols	I
1	Title, legend material, symbols and labeling	I
2	Base map—land areas, including political boundaries, significant physical features	II
3–4	Important explanatory materials—map sources and credits	II–III
4	Base map—water features, such as oceans, lakes, bays, rivers	III
5	Other base map elements—labels, grids, scales	IV

*A map object with a rank of 1 has greater intellectual importance to the map's message than one with a rank of 5. Visual levels I through IV roughly correspond with the intellectual levels 1 through 5.

map on which they are placed, there are general guidelines to follow in developing the hierarchy. (See Table 14.1.)

It is highly unlikely that the symbols on a thematic map will assume any rank other than the topmost. Those map objects customarily toward the bottom of the hierarchy may fluctuate more. For example, water is ordinarily placed beneath the land in the order but might assume a more dominant role if the purpose of the map has to do with marine or submarine features. The design activity calls for a careful examination of each element and its proper placement in the hierarchy.

An interesting activity for students is to analyze thematic maps with a view to their organizational schemes. One noticeable result will be that those maps without a visual plan are the least successful in conveying meaning and are visually confusing. (See Figure 14.2.) Unfortunately, too many such maps exist. On the other hand, a map with a carefully conceived and executed hierarchy pleases the reader, is visually stable, and does not cry out to be redesigned.

The oft-stated axiom that the best designs are not even noticed operates in cartography as well as in other disciplines.

Achieving the Visual Hierarchy

Ordering the intellectual importance of map elements is a relatively simple task, especially when guided by a clearly stated map purpose. Making the hierarchy work visually is another matter, involving knowledge of the perceptual tendencies of map readers. The designer must learn these tendencies if effective results are to be achieved.

Fundamental Perceptual Organization of the Two-dimensional Visual Field: Figure and Ground

The **figure-ground phenomenon** is often considered to be one of the most primitive forms of perceptual organization; it has even been observed in infants.[2] We tend to see objects having form as segregated from their surroundings, which are formless. Objects that stand out against their backgrounds are referred to as *figures* in perception, and their formless backgrounds as *grounds.* The segregation of the visual field into figures and grounds is a kind of automatic perceptual mechanism. Figures will not emerge from homogeneous visual fields, however.

In the everyday three-dimensional environment, we see a table on top of the floor (with the floor continuing beneath the legs of the

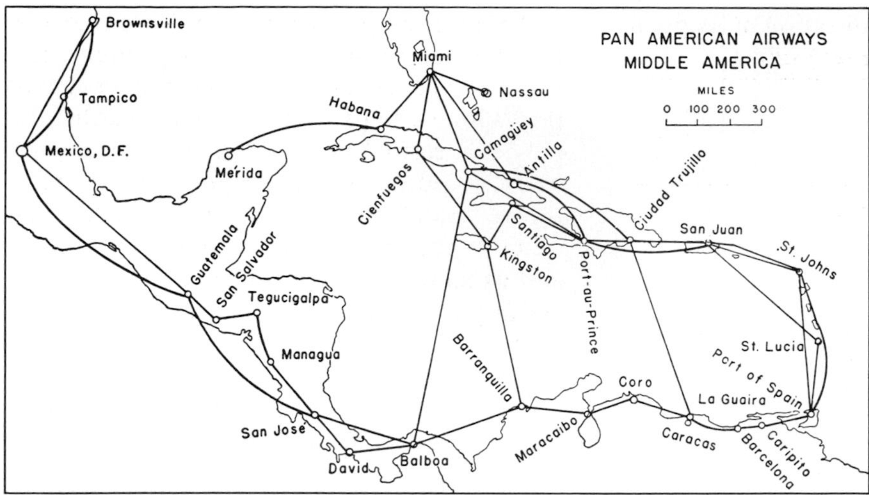

Figure 14.2. Poor organizational map plan.
In this case, it is difficult to see which areas on the map are water and which are land. Further difficulty is introduced by failure to place the thematic symbolization on the most dominant visual level. The map as a whole suffers from lack of contrast.

Illustration from Fred Carlson, *Geography of Latin America,* p. 480. Copyright Prentice-Hall, Inc. 1952. Reprinted by permission.

table), buildings in front of the sky, pictures in front of the walls on which they are hung, and so on. When we lay an eraser on a piece of paper, we see the paper continuing unbroken behind the eraser and the eraser as a complete object on the top of the paper. (See Figure 14.3a.)

Figure formation is possible in two-dimensional spatial organization as well. Figures perceived in this way are seen separately from the remainder of the visual field, have form and shape, appear to be closer to the viewer than the amorphous ground, have more impressive color, and are associated with meaning.[3] The ground usually appears to continue unbroken behind the figure, just as in the three-dimensional case. A simple example of figure from the two-dimensional world is a black disc placed within a frame. (See Figure 14.3b.)

The cartographic designer should use this perceptual tendency of figure-ground segregation as a positive design element in structuring the visual hierarchy, so that the most important map elements appear as figures in perception. With careful attention to graphic detail, all the elements can be organized in the map space so that the emerging figure and ground segregation produces a totally harmonious design. Visual confusion is eliminated, and the intent of the message becomes clear. Fortunately, psychological researchers have examined many of the mechanisms that lead to figure formation, so designers have guidelines for organizing the map's graphic elements.

Perceptual Grouping Principles

Several perceptual grouping principles have been found to be primary mechanisms for figure formation. In **perceptual grouping,** the viewer spontaneously combines elements in the visual field that share similar properties, resulting in new forms or "wholes" in the visual experience. From the designer's point of view, these groupings can act in two opposing ways: as positive mechanisms to use if the map's elements will permit it, or else as mechanisms to avoid if the map's elements are arranged in such a fashion that the spontaneous grouping

(a)

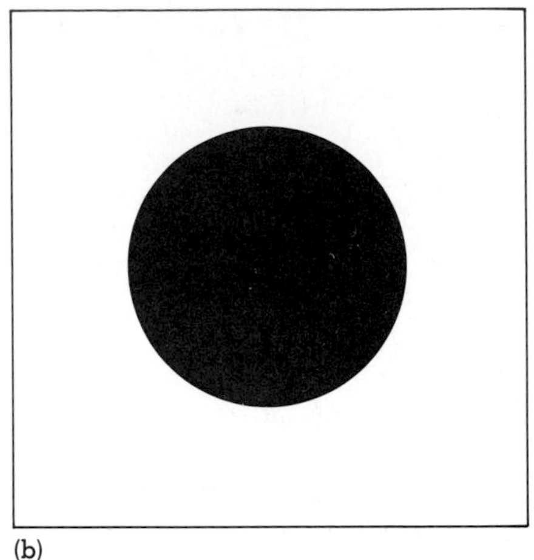
(b)

Figure 14.3. Figures and grounds.
In (a), we see familiar objects appearing as whole objects in perception. Grounds appear to continue unbroken beneath the figural objects. In the graphic two-dimensional world, as in (b), figures dominate perception and appear to rest on top of seemingly unchanging grounds. The black disc in (b) is usually seen as figure and the surrounding white area as ground.

will detract from the planned hierarchy. Their potential as design elements is explained and exemplified below.

Grouping by similar shape. Objects in the visual field possessing similar shapes are usually combined into a new group that appears distinct from the remainder. (See Figure 14.4a.) This perceptual feature undoubtedly comes into play when map readers view a map containing several different qualitative map symbols. (See Figure 14.4b.) In fact, if we could not visually combine identical symbols, it would be difficult to "see" the geographical pattern of one symbol type as distinct from others.

Grouping by similar size. Viewers tend to group similarly sized objects in the visual field into new perceptual structures. (See Figure 14.5a.) This tendency is especially important to cartographic design in at least two areas: the reading of different type sizes on maps and the visual assimilation of geographical pattern from maps containing range-graded proportional symbols (explained in Chapter 8). Designers, although they may not be aware of it, depend on this perceptual grouping feature when they work with these cartographic elements. (See Figure 14.5b.)

Recent research on the perception of graduated point symbol maps (discussed in Chapter 8) has uncovered a perceptual characteristic that is relevant to the matter of grouping by similar size. It appears that the perceived size of symbols is affected by their immediate environments. Consequently, the perceptual grouping of similarly sized symbols could possibly be affected by this condition. Patricia Gilmartin concluded her study with these findings:[4]

1. When a circle (the point symbol tested) is among circles smaller than itself, it appears larger (an average of 13 percent) than when it is surrounded by circles larger than itself. Isolated circles not surrounded by others (larger or smaller) were judged to be of an intermediate size.

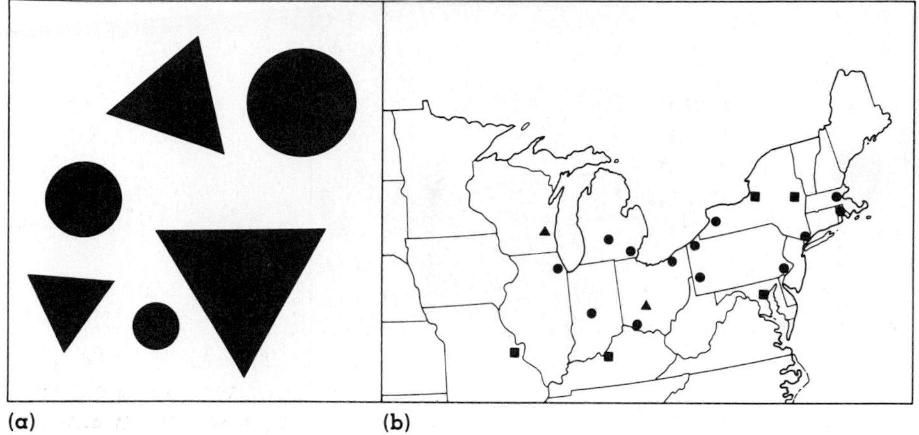

Figure 14.4. Grouping by similarity.
In perception, we tend to group similar objects. In (a), the triangles belong visually to a group distinct from the group formed by the circles. This perceptual tendency is fundamental in some cartographic situations, such as (b). We often use this perceptual grouping phenomenon to communicate thematic messages on maps.

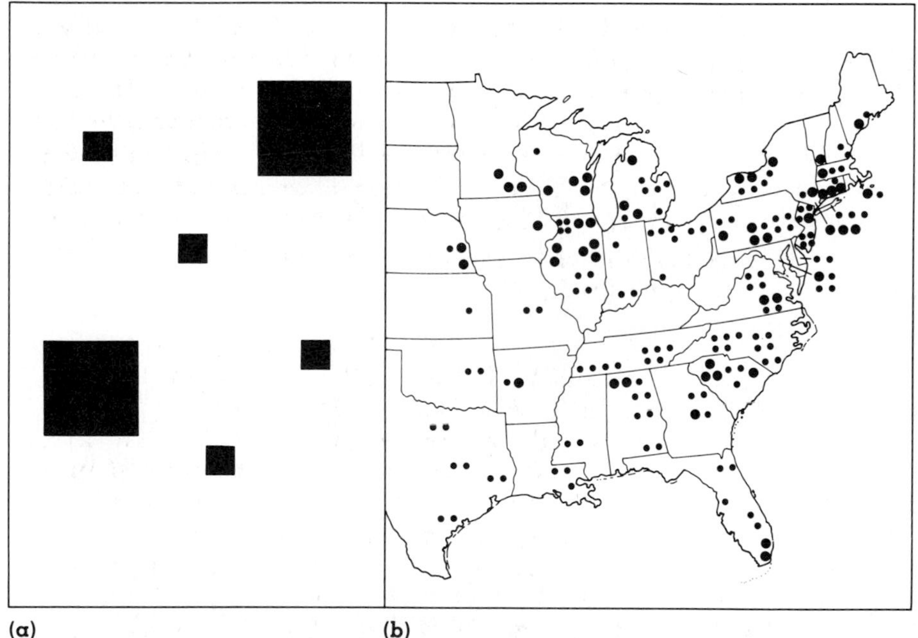

Figure 14.5. Grouping by similar size.
In (a), objects in the visual field that are similar in size tend to be grouped together in perception. It is difficult to make the experience take on a different visual structure. Try to place a small square and a large square together into a coherent new group. This perceptual tendency to group similarly sized objects is used frequently by cartographers in design of maps, as in (b).

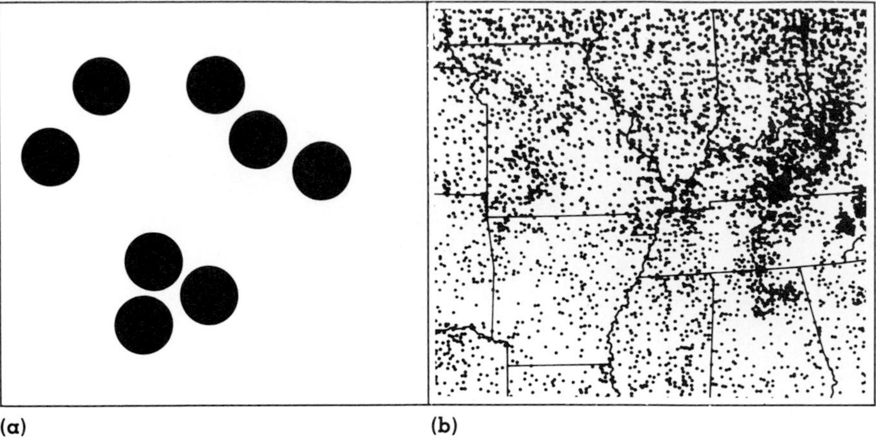

Figure 14.6. Perceptual grouping by proximity.
In (a), visual recombination of the groups into a new organization would be difficult. Cartographers use this perceptual grouping tendency in map design, as in (b).

Dot map in (b) is part of a larger map taken from U.S. Bureau of the Census, *Census of Agriculture,* 1969, Volume V, Special Reports, Part 15, Graphic Summary (Washington, D.C.: USGPO, 1973), p. 16.

2. The effect of larger or smaller circles on a surrounded circle can be reduced if internal borders (boundaries around the area represented by each graduated symbol) are used on the map.
3. There is some evidence that smaller circles are more susceptible to being judged differently because of their environments than large circles.

It is clear from such perceptual studies that the designer's task is not an easy one, or even an altogether clear one. Perceptual tendencies can be confusing and difficult to manage in cartographic design. However, if the designer strives to create patterns that are unambiguous in perception, the design is likely to be successful.

Grouping by proximity. Another strong perceptual grouping tendency is that of proximity. Elements in the visual field that are closer to other elements tend to be seen as a unit that stands out from the remainder. (See Figure 14.6a.) In fact, the words on this page are formed by the proximity principle—letters are visually grouped in words. The cartographer relies on this visual tendency when depicting geographical distributions, especially those containing clusters. (See Figure 14.6b.) If the eye did not combine elements in this way, it would be exceedingly difficult for designers to accomplish their task.

Figure Formation and Closure

Closure is an important perceptual principle. It refers to the tendency for the perceiver to complete unfinished objects and to see as figures objects that are already completed. A contour or edge is usually associated with closure. (See Figure 14.7a.) Broken contours are spontaneously completed so that we see the "whole" object. Therefore, map designers need to provide strong, completed contours around figural objects. Sometimes these edges are "broken" by the diagram to provide for lettering or other map elements. (See Figure 14.7b.) Unless skillfully executed, this design solution can have negative effects on figural areas.

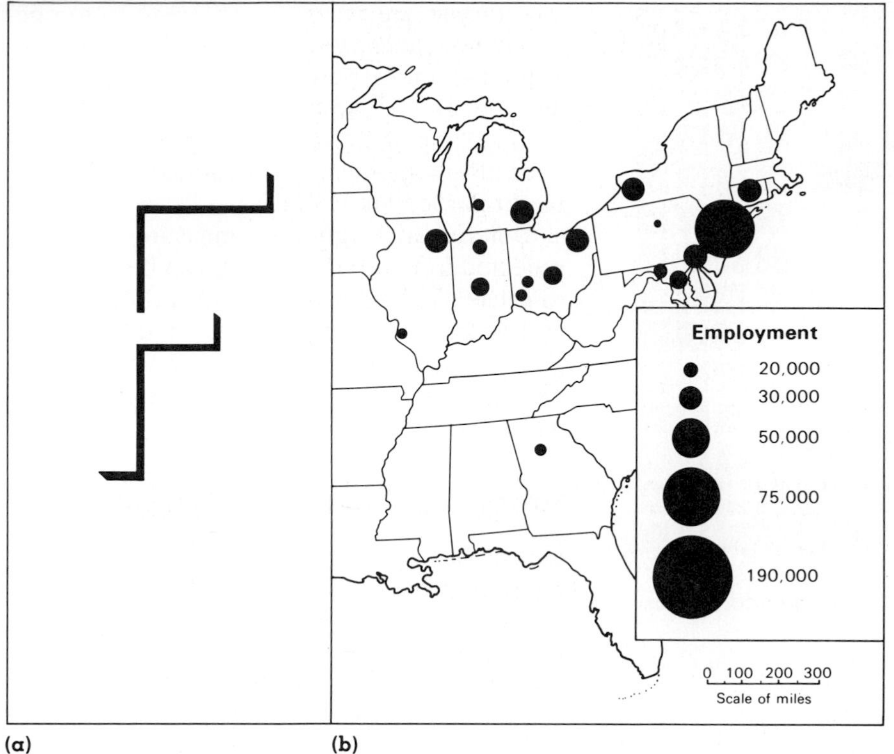

Figure 14.7. Closure.
We tend to close figural objects to form more simple structures, as in (a). The mind supplies missing elements without pausing to consider. This occurs in many cartographic situations, and the communication does not suffer. In (b) the person reading the map provides the missing coastlines and is not really bothered by their physical absence.

Using Texture to Produce Figures

Texture and differences in texture can be used to produce figures in perception.[5] It appears that orientation of the textural elements is more important in figure development than is the positioning of the elements. (See Figure 14.8.) This feature is related to the quality called **intraparallelism,** which is the similar alignment of elements in the visual field to achieve order and harmony in the experience.[6]

The cartographic literature also provides evidence that texture and texture discrimination lead to the emergence of figures. In a study to determine the effects of different graphic representations of land and water, it was found that a representation containing a textured pattern over one surface produced the least amount of ambiguity during perception.[7] Unfortunately, the subjects in the experiment also considered that particular configuration to have a low aesthetic value. This negative response may have been due to the specific textured patterns selected for the experiment, not to the inherent visual differentiation caused by texture as such.

Using texture to evoke figures in perception is closely related to the feature of **articulation** in the visual field: inequality of detail from place to place in the field. Both texture and

Figure 14.8. Textural elements and figures in perception.
Although subtle, as in this case, texture can provide just enough visual contrast to segregate the visual field into figures and grounds.

articulation are ways of achieving heterogeneity in visual experience, a necessary requirement for figure perception.[8] Adding texture and detail to shapes on the map that are planned as figures is a reliable graphic method of ensuring their visual emergence during map reading. (See Figure 14.9.)

Differential brightness, measured by light reflected from the map sheet, can be also used to cause figural development.[9] A difference of at least 20 to 25 percent in reflected light is required to produce the desired result. Brightness as it relates to color and figure formation will be dealt with in Chapter 16.

Strong Edges and Figure Development

One of the principal ways of producing a strong figure in two-dimensional visual experience is to provide crisp edges to figural objects. Conversely, figural dominance can be weakened by reducing edge definition. Edges result from contrasts of brightness, reflection, or texture.

(a)

(b)

Figure 14.9. The map and the use of textural elements.
Textural elements should be applied subtly. Texture can be placed on either figure or ground areas. The illustration here shows screened lines.

These characteristics have special significance and utility in cartographic design, particularly in coordinating the graphic elements in the planned visual hierarchy. For example, thematic symbols and other elements high in the hierarchy can be rendered in solid hue (black or color), and subordinate features can be screened, thus reducing the sharpness of their edges. Screening also reduces the intensity of the less important elements. (See Figure 14.10.)

Screening to reduce edge sharpness and intensity are photomechanical techniques used to produce **aerial perspective** on a flat map sheet. This phenomenon is part of our three-dimensional world; it accounts for the haziness and lack of clarity of distant objects. It is often called a *depth cue.* Landscape painters employ this real-world phenomenon by rendering distant objects with less detail and less intense color.

The Interposition Phenomenon

A most useful way of causing one object in the two-dimensional visual field to appear on top of or above another is to interrupt the edge or contour of one of the objects. This phenomenon is usually referred to as **interposition** and is frequently cited as a depth cue in perception.[10] In cartographic design, this technique can be used to strengthen the dominance

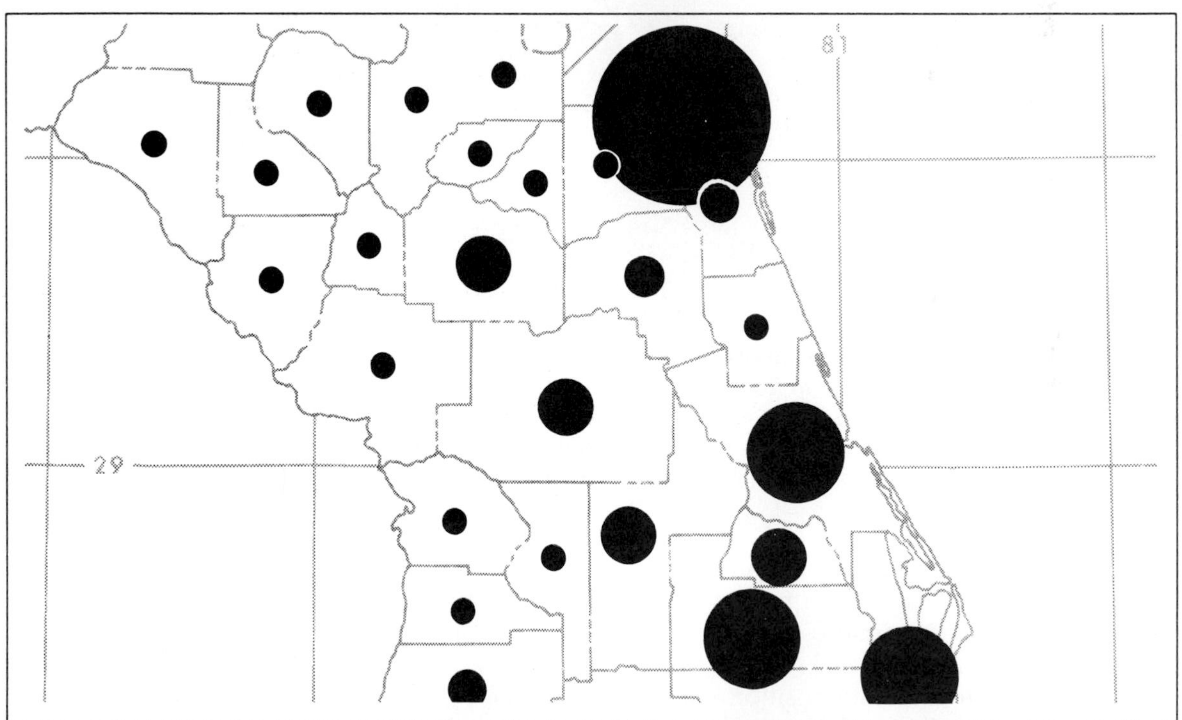

Figure 14.10. The use of screening to develop the visual hierarchy.
Less important elements (coastlines, political boundaries) have been screened to reduce the sharpness of their edges. The graticule has been screened even more, further reducing its contrast and thereby placing it on an even lower visual level. Important elements (symbols, legends) are rendered in total black, which tends to emphasize their importance in the overall organization.

Redrawn from Borden D. Dent, "Visual Organization and Thematic Map Communication," *Annals* (Association of American Geographers) 62 (1972):85. Redrawn by permission.

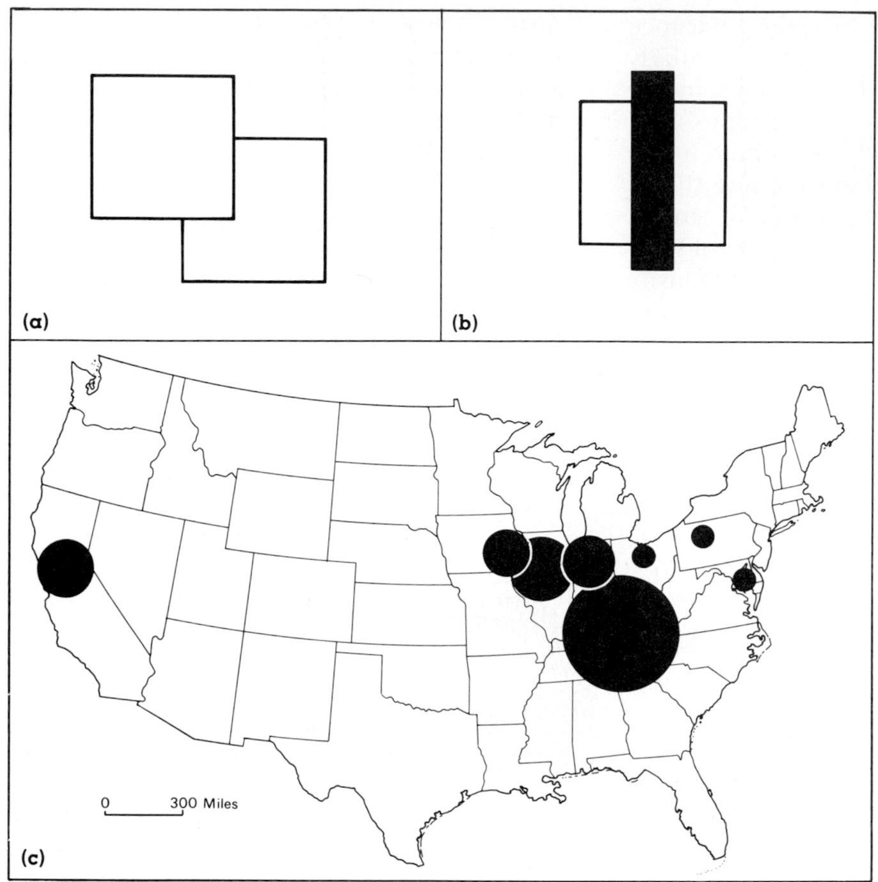

Figure 14.11. Interposition to assist in figural development.
In (a), we tend to see one square in front of the other one. In (b), the black rectangle is seen "over" the square. The objects whose edge contours continue unbroken are the ones seen as being on top. Interposition can be used on maps, as in (c). The black circles break coastlines or state shapes, and appear "on top" of the land. This enhances their figural properties.

of certain objects in the visual hierarchy. (See Figure 14.11.) The result is an impression of depth on the map—an interesting, dynamic, and fluid solution.

It is possible to use interposition cues to produce stacking effects of clustered graduated point symbols (these symbols are discussed in Chapter 8). Although this method results in a map that appears three-dimensional (thus adding interest to the design), it makes it difficult for the map reader to see the quantitative differences of the scaled symbols.[11] On the average, errors in symbol reading increase with the amount of the symbol obscured by an overlapping one. Because of this, the use of interposition is not recommended in cases where one quantitative point symbol overlaps another. Otherwise, interposition can be used whenever it enhances the overall hierarchical plan for the map.

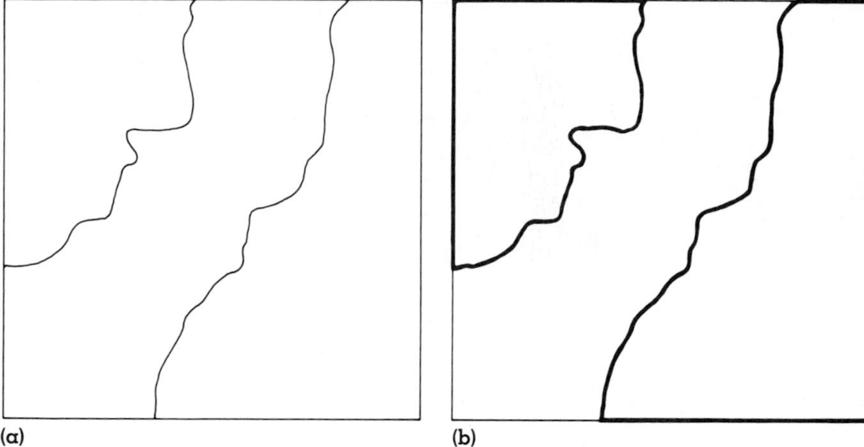

Figure 14.12. Figural development of land areas.
In (a), the configuration of land and water areas and the graphic treatment of the map border hinders the land from becoming figure. In (b), the borders joining the land areas are treated as part of the land. This forms enclosing contours and strengthens the perception of the land areas as wholes. Cartographic designers will need to apply different solutions to varying designs.

Figures and Grounds in the Map Frame

Within a bounded space in the two-dimensional visual field, areas that are smaller and completely enclosed will tend to be viewed as well-defined figures. Cartographic designers have considerable choice in such elements as texture, articulation, edging, and interposition to accentuate objects as figures, but less freedom in the manipulation of figural size and enclosedness. Constraints imposed by location, scale, and map size can preclude any adjustment to enhance figural areas of the map. In cases where these restrictions do not limit design choice, there are guidelines to assist in selecting the proper size ratios of figures and grounds.[12] In one study, acceptable size ratios ranged from 1:2.18 to 1:3.56: Nonfigural areas may be from 2.18 to 3.56 times larger than figural areas without interfering with figure formation.

One approach to assure that the figural portion of the map is completely enclosed, in the case of land-water differentiation, is to provide an enclosing contour by continuing the shoreline as the map border.[13] (See Figure 14.12.) This solution, although not preferred by the respondents in the study, did reduce ambiguity somewhat.

The Special Case of the Land-Water Contrast

The principal concern of the thematic map designer is to communicate a spatial message effectively. To a great extent, the success of this effort depends on how the message is presented or arranged for the map reader. If the graphic material is presented in a clear and unambiguous manner, success in communication will probably result. On the other hand, an unclear and confusing graphic picture will make

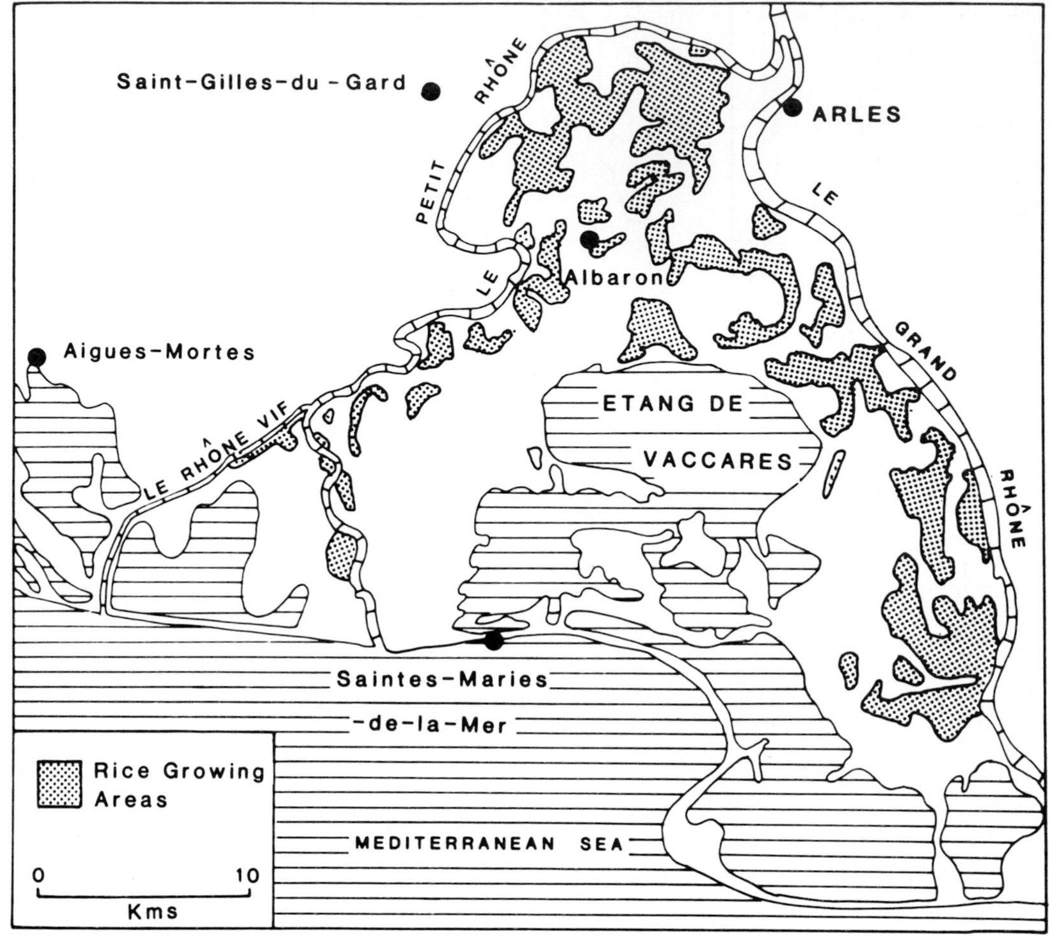

Figure 14.13. Ambiguous land-water contrast.
Some coastlines are very complex, causing difficult design solutions. If looked at long enough, land and water areas alternate in perception.
Map from A. Scarth, "Rice and the Ecological Environment in the Camargue," *Geography* 71 (1986), p. 157. Reprinted by permission of the Geographical Association.

the reader frustrated and unreceptive to the message. (See Figure 14.13.) The graphic elements of the map must therefore be arranged in such a way as to reduce any possible reader conflict. One way of eliminating possible confusion is to provide clues on the map to help the reader in determining geographical location.

A significant geographical clue is the differentiation between land and water, if the mapped area contains both. This distinction has been suggested as the first important process in thematic map reading.[14] Maps that present confusing land-water forms deter efficient and unambiguous communication of ideas. (See Figure 14.14.) Design solutions should never

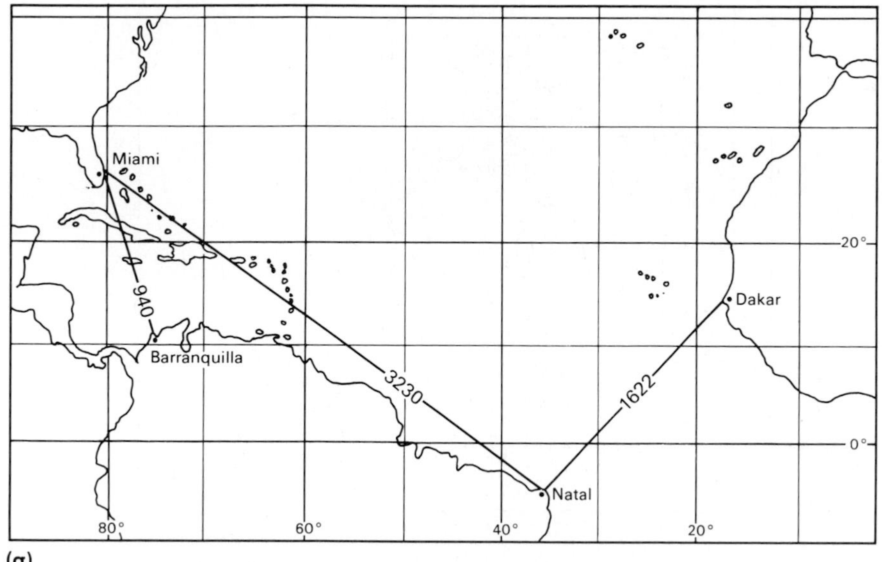

(a)

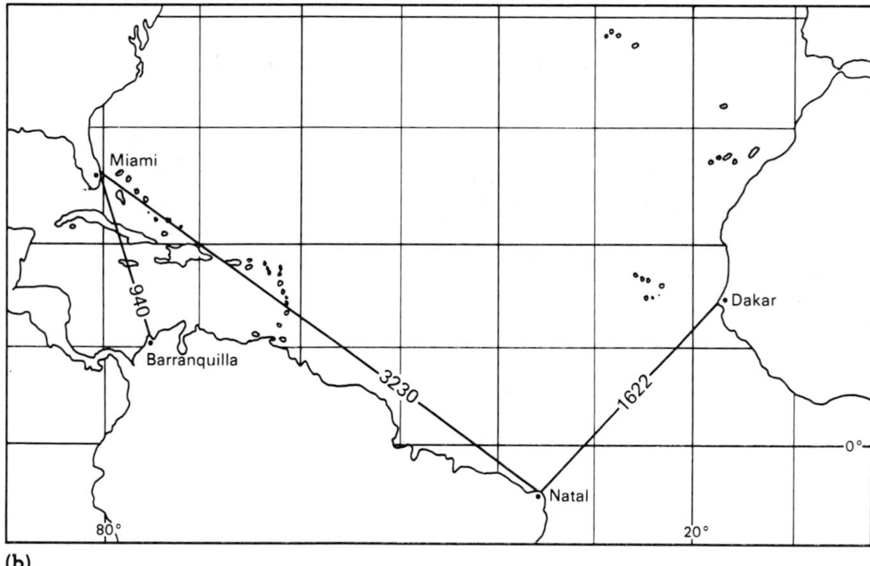

(b)

Figure 14.14. Land and water contrasts.
In (a), the contrast between land and water is not clearly evident. The graticule creates a similar texture over the entire map, resisting figure formation. A very simple correction can overcome this. In (b), the graticule has been eliminated over land areas, causing them to be enhanced as figures. In many cases, only simple corrections are necessary to solve problems of insufficient contrast between land and water.

Illustration (a) from Fred Carlson, *Geography of Latin America,* p. 491. Copyright Prentice-Hall, Inc., 1952. Reprinted by permission.

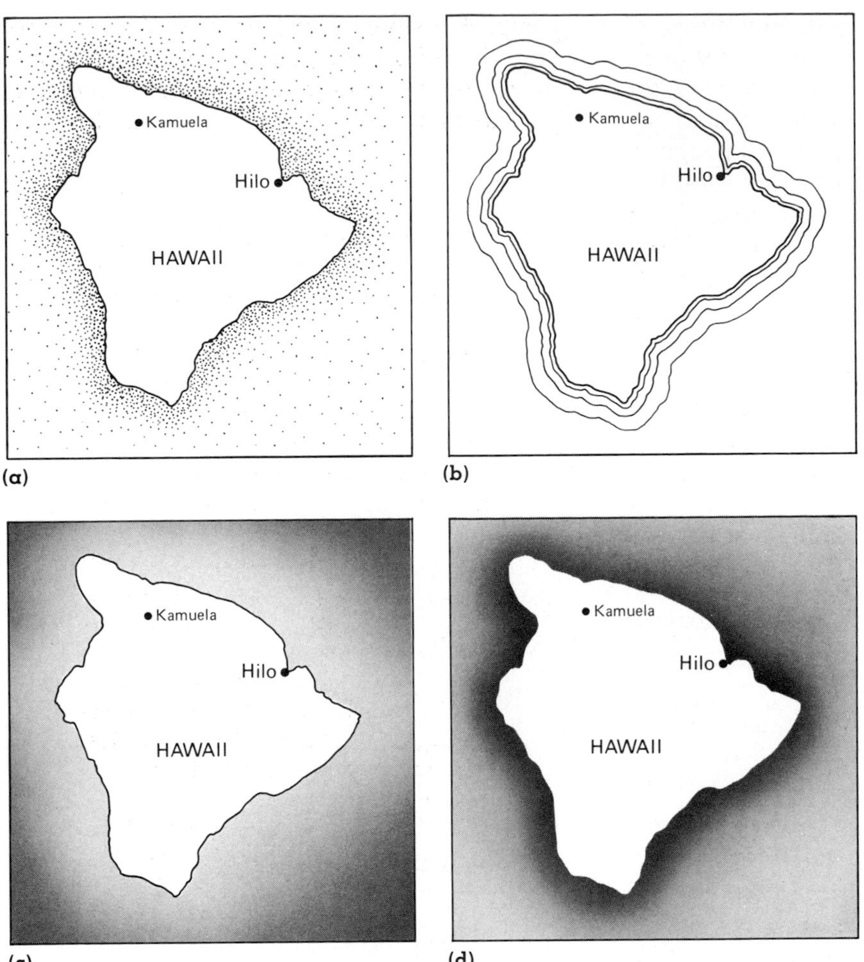

Figure 14.15. Different methods of coastal vignetting.
In (a), stippling on the water side does not clearly show the land as figure but can be used because it is conventional. In (b), form lines are used; these are not aesthetically preferable. In (c), a continuous tone of decreasing brightness is employed. This is not particularly good because it lacks contrast at the coastline. The continuous tone of increasing brightness shown in (d) is the best choice of the ones illustrated, because it enhances land as figure and is aesthetically preferable.
These figures are redrawn from Richard Lindenberg, "Coastal Vignetting and Figure—Ground Perception in Small-Scale Black and White Cartography" (unpublished Master's thesis, Department of Geography, Georgia State University, 1975) p. 76.

be visually distracting or lacking in clarity and stability in the intended order. Land-water differentiation usually aims to cause land areas to be perceived as figures and water areas as ground. In unusual cases, water areas are the focal point of the map and would therefore be given graphic treatment to cause them to appear as figures.

Vignetting for Land-Water Differentiation

Historically, engravers, draftsmen, and cartographers solved the land-water differentiation problem by use of **vignetting** at the coastline. Vignetting is any graphic treatment emerging from an edge or border and resulting in a continuous gradient of brightness. Vignetting of the coast is most common, although it has been used on maps at political or other borders for visual differentiation between land areas.

Some of the popular ways of coastal vignetting are stippling, form lines, and continuous tones (both decreasing or increasing in brightness away from the coastline). (See Figure 14.15.) In recent years, these methods have come under close scrutiny to determine if they in fact enhance the figure and ground organization of the map. Most research has compared them to other forms of figure enhancement (e.g., heterogeneity, texture, articulation, and graticule) in terms of effectiveness.

One surprising result of these studies, reported by at least two researchers working independently, is that stippling on the water at coastlines is ineffective.[15] In one study, the stippling caused the water to be seen as land! In the other study, stippling on the water side was ranked sixth among eight in figural goodness and also in aesthetic preference. It appears that stippling should be used very cautiously, if at all.

In a study by Lindenberg, a continuous tone of increasing brightness was ranked best overall by map-reading judges in developing the figural goodness of land *and* in aesthetic preference. The form line method was not judged very effective or aesthetically preferred. Both studies support the idea that development of heterogeneity in the visual field through differences in surface texture is as effective in developing land-water contrasts as is coastal vignetting.

Summarizing the various methods of achieving land-water differentiation, it would seem that, if vignetting is used, a continuous tone of increasing brightness should be applied to the water at the coastal interface. If no vignetting is used, at least surface texture differences between land and water should be developed. One last suggestion is to arrange the shapes in the map frame to reduce the possibility of ambiguous or reversible figures.

In sum, "The cartographer is responsible for structuring a complete synthesis which will be comprehensible to the user."[16] This can be done by effectively planned visual hierarchy, carried out by incorporating known perceptual principles of the figure and ground dichotomy.

Notes

1. Borden D. Dent, "Visual Organization and Thematic Map Communication," *Annals* (Association of American Geographers) 62 (1972):79–93.
2. William D. Dember and Joel S. Warm, *Psychology of Perception,* 2d ed. (New York: Holt, Rinehart and Winston, 1979), p. 251.
3. Ralph Norman Haber and Maurice Hershenson, *The Psychology of Visual Perception* (New York: Holt, Rinehart and Winston, 1973), p. 184.
4. Patricia P. Gilmartin, "Influences of Map Context in Circle Perception," *Annals* (Association of American Geographers) 71 (1981):253–58.
5. John P. Frisby, *Seeing: Illusion, Brain, and Mind* (Oxford: Oxford University Press, 1980), pp. 115–16; see also Dent, "Visual Organization and Thematic Map Communication."
6. William R. Sickles, "The Theory of Order," *Psychological Review* 49 (1942):403–21.
7. C. Grant Head, "Land-Water Differentiation in Black and White Cartography," *Canadian Cartographer* 9 (1972):25–38.
8. E. G. Weaver, "Figure and Ground in the Visual Perception of Form," *American Journal of Psychology* 38 (1927):194–226.
9. Clifford H. Wood, "Brightness Gradients Operant in the Cartographic Context of Figure-Ground Relationship," *Proceedings* (American Congress on Surveying and Mapping, March, 1976):5–34.
10. Julian E. Hochberg, *Perception* (Englewood Cliffs, NJ: Prentice-Hall, 1964), pp. 87–88; see also Gaetano Kanizsa, *Organization in Vision: Essays on Gestalt Perception* (New York: Praeger, 1979), pp. 94–97.
11. Richard E. Groop and Daniel Cole, "Overlapping Graduated Circles: Magnitude Estimation and Method of Portrayal," *Canadian Cartographer* 15 (1978):114–22.
12. P. V. Crawford, "Optimum Spatial Design for Thematic Maps," *Cartographic Journal* 13 (1976):134–44.

13. Head, "Land-Water Differentiation."
14. *Ibid.*
15. *Ibid.;* and Richard E. Lindenberg, "Coastal Vignetting and Figure-Ground Perception in Small-Scale Black and White Cartography" (unpublished Master's thesis, Department of Geography, Georgia State University, 1975).
16. Willis Heath, "Cartographic Perimeters," *International Yearbook of Cartography* 7 (1967):112–20.

Glossary

aerial perspective the diminution of detail with increasing distance

articulation providing detail in one part of the visual field; objects which are more articulated are frequently perceived as figure

closure spontaneous perceptual tendency to close contours or edges around objects to make them whole

figure-ground phenomenon a fundamental perceptual tendency in which visual fields are spontaneously divided into outstanding objects (figures) and their surroundings (ground)

interposition phenomenon perceptual tendency for one object to appear behind another because of interrupted contour

intraparallelism similar alignment of elements in a textural surface; can lead to figure formation in perception

perceptual groupings mechanisms identified by psychologists as leading to figure formation; include grouping by shape, size, and proximity

vignetting graphic treatment at an edge or border, resulting in a continuous gradient of brightness

visual hierarchy the intellectual plan for the map and the subsequent graphic solution to satisfy the plan; may also be called the *organization hierarchy*

Readings for Further Understanding

Arnheim, Rudolf. *Art and Visual Perception.* Berkeley, CA: University of California Press, 1965.

Crawford, P. V. "Optimum Spatial Design for Thematic Maps." *Cartographic Journal* 13 (1976):134–44.

Dember, William N., and Joel S. Warm. *Psychology of Perception,* 2d ed. New York: Holt, Rinehart and Winston, 1979.

Dent, Borden D. "Visual Organization and Thematic Map Communication." *Annals* (Association of American Geographers) 62 (1972):79–93.

Ferens, Robert J. "Design of Page-Size Maps and Illustrations." *Surveying and Mapping* 28 (1968):447–55.

Frisby, John P. *Seeing: Illusion, Brain, and Mind.* Oxford: Oxford University Press, 1980.

Gilmartin, Patricia P. "Influences of Map Context on Circle Perception." *Annals* (Association of American Geographers) 71 (1981):253–58.

Groop, Richard E., and Daniel Cole. "Overlapping Graduated Circles: Magnitude Estimation and Method of Portrayal." *Canadian Cartographer* 15 (1978):114–22.

Haber, Ralph Norman, and Maurice Hershenson. *The Psychology of Visual Perception.* New York: Holt, Rinehart and Winston, 1973.

Head, C. Grant. "Land-Water Differentiation in Black and White Cartography." *Canadian Cartographer* 9 (1972):25–38.

Heath, Willis. "Cartographic Perimeters." *International Yearbook of Cartography* 7 (1967):112–20.

Hochberg, Julian E. *Perception.* Englewood Cliffs, NJ: Prentice-Hall, 1964.

Kanizsa, Gaetano. *Organization in Vision: Essays on Gestalt Perception.* New York: Praeger, 1979.

Lindenberg, Richard E. "Coastal Vignetting and Figure-Ground Perception in Small-Scale Black and White Cartography." Unpublished Masters Thesis, Department of Geography, Georgia State University, 1975.

Sickles, William R. "The Theory of Order." *Psychological Review* 49 (1942):403–21.

Weaver, E. G. "Figure and Ground in the Visual Perception of Form." *American Journal of Psychology,* 38 (1927):194–226.

Wood, Clifford H. "Brightness Gradients Operant in the Cartographic Context of Figure-Ground Relationship." *Proceedings* (American Congress on Surveying and Mapping, March, 1976), pp. 5–34.

Wood, Michael. "Human Factors in Cartographic Communication." *Cartographic Journal* 9 (1972):123–32.

———. "Visual Perception and Map Design." *Cartographic Journal* 5 (1968):54–64.

Chapter

15

Making the Map Readable through Intelligent Use of Typographics

Chapter Preview

Map lettering is an integral part of the total design effort and should not be relegated to a minor role. Lettering on the map functions to bring the cartographer and map reader closer together and makes communication possible. To employ lettering properly, the cartographic designer should be familiar with letterform characteristics, sizes, letterspacing, type personalities, and legibility. Size is a most critical element in design, and lettering style and personality can affect the appearance of the map. Type classification is important for the cartographer in selecting a suitable typeface. Map typography includes lettering placement; overall lettering harmony can be achieved through adherence to established conventions of placement. Experimental studies expand knowledge of map typography and reveal the need for greater research in this area of cartographic design. Today, most map lettering is generated by cold type, especially pressure-sensitive materials and photographic image production. Low-cost laser printers driven by typesetting software have tremendous potential for the cartographic designer.

For many cartographic designers, the planning and application of map lettering remains the last task. This is unfortunate—in most instances, the appearance and mood of the entire map can be set by its lettering, so lettering should not be treated lightly in design. Although only a handful of cartographic researchers have examined map lettering in any detail, there is extensive general literature on type, its design, history, application, and production. This chapter deals with the fundamentals of lettering and attempts to illustrate the importance of lettering in thematic map design. Four major topics—the function of map lettering, the elements of type, map typography and design, and the production of map lettering—will be examined.

Functions of Map Lettering

All map lettering, whether on general-reference maps or thematic maps, serves to bring the cartographer and map user together, making communication possible. Only in the highly unusual situation where cartographer and map user discussed the map in person could written language be ignored.

On general-reference maps, lettering serves mainly to *name places* and to identify or *label things* (e.g., scales, mountains, oceans, straits, and graticule elements). On thematic maps, lettering is also provided for titles, legends, and other explanatory marginal materials necessary to make the map content more comprehensible. Because all written language elements (letters and words) are symbols for meaning, they serve the same function on maps.

Map lettering should be viewed first as a *functional symbol* on the map, only secondarily, as an aesthetic object. Nevertheless, map lettering, if not done well, hinders communication. Therefore, the cartographer should approach the employment of lettering with an appropriate regard for both function and form.

Map lettering in this context refers to the *selection* of lettering type and its *placement* on the map. Questions of which words to use in titles, labels, or other marginalia are not addressed here.

Lettering can express the nature of a geographic feature by its *style,* the feature's importance by its *size,* the feature's location by its *placement,* and the feature's extent by its *spacing.*[1] The variables of style and size are most important in the design of titles and legends; size, spacing, and placement are particularly important on the body of the map. These distinctions set apart map lettering from general text lettering and should be borne in mind by the designer.

The Elements of Type

The selection of typeface and the placement of lettering are the two chief concerns of the designer. Proper selection can be made only if the cartographer fully understands the fundamentals of *type design.* As there are hundreds of individual typefaces from which to choose, the designer must recognize the elements and characteristics of their design, how they may be classified to make selection and use easier, and the fact that different typefaces can make different impressions on the reader. At first, these aspects of type can be bewildering, but familiarity comes easily with practice and experience.

Typeface Characteristics

Typeface design, type size, and letterforms are the principal characteristics of type with which the designer works. Letter and word spacing may be added, although strictly speaking these are not typeface elements. They are treated in this section because they can have such a tremendous impact on the perception of individual letters or words.

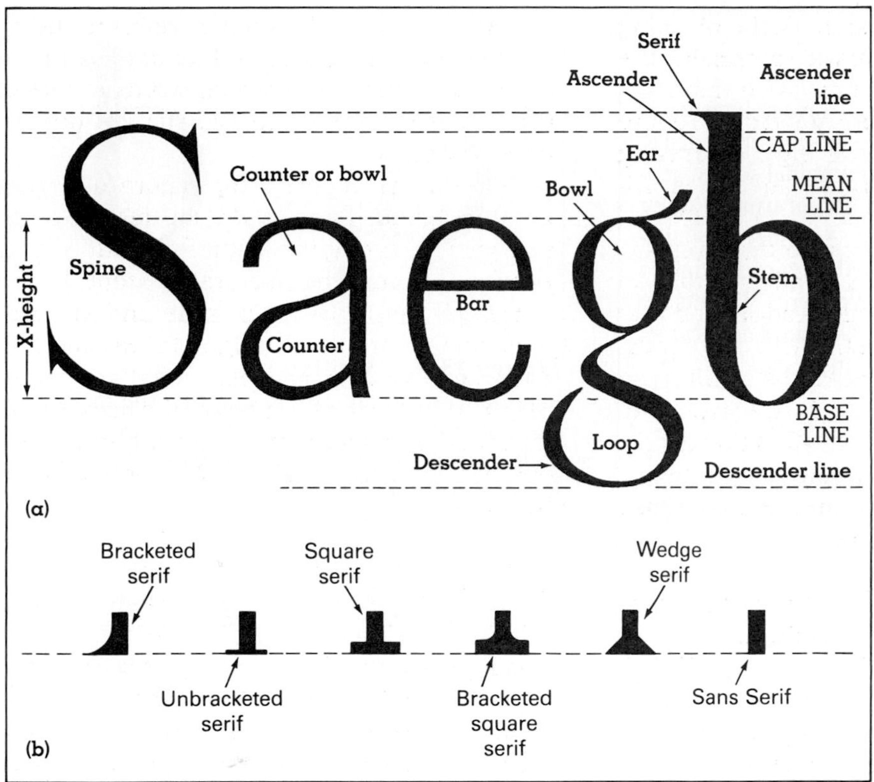

Figure 15.1. Type elements.
The principal elements of letterforms are shown in (a), and (b) illustrates the different serif forms.

Letterform Components

All **typefaces** have elements in common regardless of the letter. (See Figure 15.1.) Letters, both capital and lowercase, are begun on the **base line.** The height of the body of lowercase letters is referred to as the **x-height,** an important dimension in letter design because it often determines the readability of the type. Letter strokes that are higher than the x-height are **ascenders;** all ascenders in a lowercase alphabet will terminate at a common **ascender line. Descenders** are letter strokes that fall beneath the base line and terminate at the **descender line.** In most typefaces, capital letters are shorter than ascenders; there is therefore a **cap line** that defines the vertical dimension of capital letters.

A major element of some letterforms is the **serif.** Serifs are finishing strokes added to the end of the main strokes of the letter. Not all designs have serifs, but it is important to note that in running text such as you are now reading, letterforms with serifs are easier to read. (See Figure 15.2.) Lettering styles that do not contain such finishing strokes are called **sans serif** styles. Serifs have different appearances, depending on the way they are joined to the main strokes. The serif may or may not be supported by a **bracket** (or *fillet*).

TIMES ROMAN

Most geographical analyses involve point patterns (or centers of areas) that have weights attributed to them. Except at the simplest nominal scales, geographical point phenomena do not usually occur everywhere with equal value. An especially interesting study is to plot weighted means over time to discover a spatially dynamic pattern. Weights are most often socioeconomic data such as income, production, sales, or employment data.

HELVETICA REGULAR

Most geographical analyses involve point patterns (or centers of areas) that have weights attributed to them. Except at the simplest nominal scales, geographical point phenomena do not usually occur everywhere with equal value. An especially interesting study is to plot weighted means over time to discover a spatially dynamic pattern. Weights are most often socioeconomic data such as income, production, sales, or employment data.

Figure 15.2. Two lettering text blocks.
Times Roman, a roman, serifed type, is easier to read in running text than lettering set in sans serif style, such as Helvetica Regular.

Counters, bowls, and loops of letters should be examined carefully when choosing type. **Counters** are the partially or completely enclosed areas of a letter, and **bowls** are the rounded portions of such letters as o, b, d, and the upper part of the lowercase g. The lower part of the lowercase g is referred to as a **loop.** These spaces can close up during the photographic reproduction of the map, possibly because of reduction and the original art, poor photographic negatives, or excessive ink in printing. This reduces their discernibility and can therefore impede communication. Map designers should choose typefaces with letterforms that are open and have few light areas that can lead to such problems.

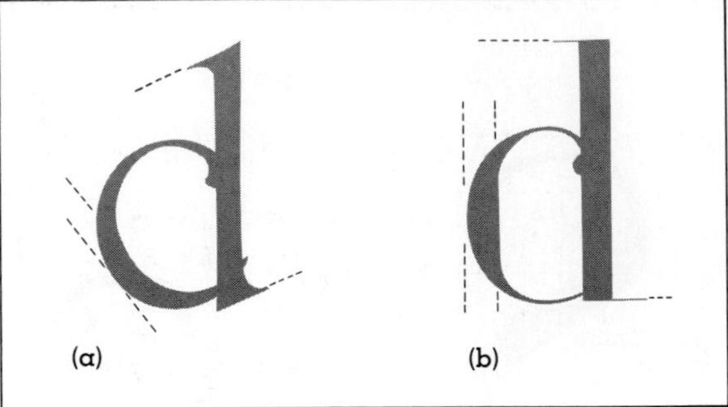

Figure 15.3. Shading.
Shading is the term used to describe the main slant of letterforms. Oblique slants, as in (a), characterize most Oldstyle typefaces. In Modern faces, such as (b), the emphasis is on strong verticals and horizontals.

Another aspect of letterforms is **shading,** a term used to describe the position of the maximum stress in curved letters.[2] (See Figure 15.3.) Shading has a definite impact on the appearance of letters and is especially important in the readability of type for text. Broadly speaking, letters with extreme vertical shading are more difficult to read than those with softer or more rounded shading. Shading also characterizes the historical development of type, a matter addressed below.

Cartographic requirements. Map lettering differs considerably from book and text lettering. Type on lines with identical background is characteristic of book typesetting. Letters are **set solid** (no letterspacing), and the **leading** (spacing) between lines of type is an important concern for the book designer. In cartography, however, letters are often spread out, placed on changing backgrounds (tints, patterns, colors), oriented in a variety of ways, and interrupted by lines or other symbols. The letters themselves may be rendered in different tones, patterns, colors, thereby causing considerable variation. As a result of these awkward situations, the chief criterion of the cartographic designer in selecting a typeface is that the individual letters be easily identifiable.

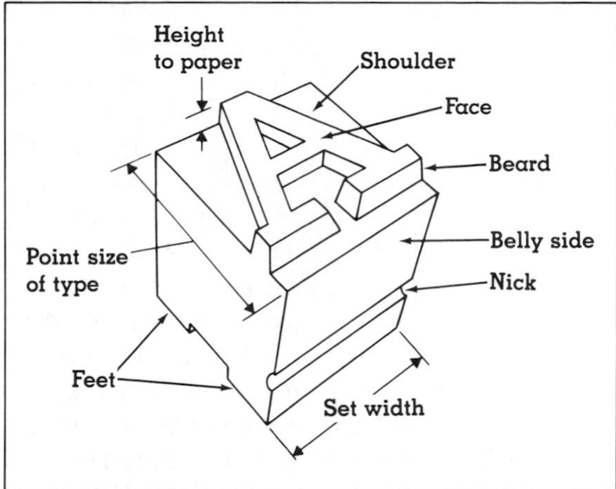

Figure 15.4. Foundry type character and common terminology to describe its parts. Notice that the point size refers to the size of the body height, not the height of the face on the body.

One cartographer has identified the following considerations in selecting type for maps.[3]

1. The legibility of individual letters is of paramount importance, especially in smaller type sizes. Choose a typeface in which there is little chance of confusion between c and e or i and j.
2. Select a typeface with a relatively large x-height relative to lettering width.
3. Avoid extremely bold forms.
4. Choose a typeface that has softer shading; extreme vertical shading is more difficult to read than rounder forms.
5. Do not use decorative typefaces on the map; they are difficult to read.

It might be added that serif letter forms are normally preferable in meeting the unusual requirements of cartographic lettering. Individual letter recognition is vital on maps. The serif can play an important role, since it tends to complete letters optically and ties one letter visually to another.[4] Aesthetic considerations often affect the choice between serif and sans serif forms also.

Type Size

Type size designation is related to the way type was originally produced on metal, or foundry, blocks. (See Figure 15.4.) The body or height of the block specifies the type size, although the actual letter on the block may not extend the full height. The system of specifying type size in the United States and Britain is based on division of the inch into 72 parts called **points.** A point equals .0138 in (.351 mm). Thus, 72 points equal .9962 in. All type is measured in point sizes. Because the face of a particular type style may not occupy the whole height of the foundry body, the actual point size of the printed letter may be different from the nominal size given for a particular typeface.

Type size specification based on foundry production of type is still used today, although the actual production of type may be by photo or computer methods. The important point is that a given style and size of type is standard. The cartographer normally chooses a type based on actual size of the typeface and specifies the nominal type size as listed in the manufacturer's catalogue.

The designer must specify type that is large enough to be read easily. A number of studies have been conducted to determine the minimal size. Professional cartographers rarely use type smaller than 4 or 5 points; safe practice is to set the lower limit at 6 points (.0828 in). Type environment can also play a role in the choice of size. Care must be exercised when original art is to be reduced, so that type sizes smaller than 4–6 points will not result.

Letterforms and Type Families

Typographic nomenclature includes the word **font,** which is a complete set of all characters of one size and design of a typeface. A font of type normally includes numerals and special characters such as punctuation marks, in addition to the letters of the alphabet. Type designers often design variations of a basic font, making up the **family** of that font. (See Figure 15.5.) A normal type family will include these abcdefghijklmnopqrstu Book

abcdefghijklmnopqrsu Medium

abcdefghijklmnopqrs Bold

abcdefghijklmnopqr Heavy

abcdefghijklmnopqrstuv Book italic

abcdefghijklmnopqrstuv Medium italic

abcdefghijklmnopqrstu Bold italic

abcdefghijklmnopqrst Heavy italic

Figure 15.5. Variants within a type family.
Both roman and italic forms, and different weights in each, may be included in a family. A heavier weight in each is also available.

This type is called ITC Cushing, designed by Vincent Pacella for the International Typeface Corporation. Reproduced with permission.

letterform variants: weight, width, roman, and italic. **Type weight** refers to the relative blackness of a type, although this variant is usually not standard among different typeface designs.[5] It is customary to find three weights: normal, lightface, and boldface. Type may be available in *condensed* or *extended* versions, called **type widths.**

The principal variants in a type family are the **roman** and **italic** forms. Roman is the basic, upright version of the design, and italic is a slanted version of its roman counterpart. Historically, an italic letterform was often slightly different from its roman partner, especially in the serifs. Today's type designers usually make the italic form identical except for its slant. In text, italic forms are frequently used for emphasis, but on maps they are most often used to label hydrographic features: oceans, seas, bays, straits, coves, lagoons, lakes, rivers, and so on.

Letter and Word Spacing

Strictly speaking, letterspacing and word spacing are not elements of typeface design. However, these features are extremely important, especially in the visual perception of letter pairs and words. Entire map projects can be weakened by poor spacing of letters and words; no other feature is so obviously incorrect at first glance. Practice and a critical eye are essential in providing correct spacing.

Letterspacing involves the appropriate distribution of spaces between letters.[6] Letterspacing is usually required for lines of type in all capitals, but not in lines of lowercase letters. Letters are usually set solid (without letter spacing) in lines of lowercase. **Word spacing** is the proper distribution of spaces between words to achieve harmonious rhythm in the line. Words set in all capitals or all lowercase require word spacing.

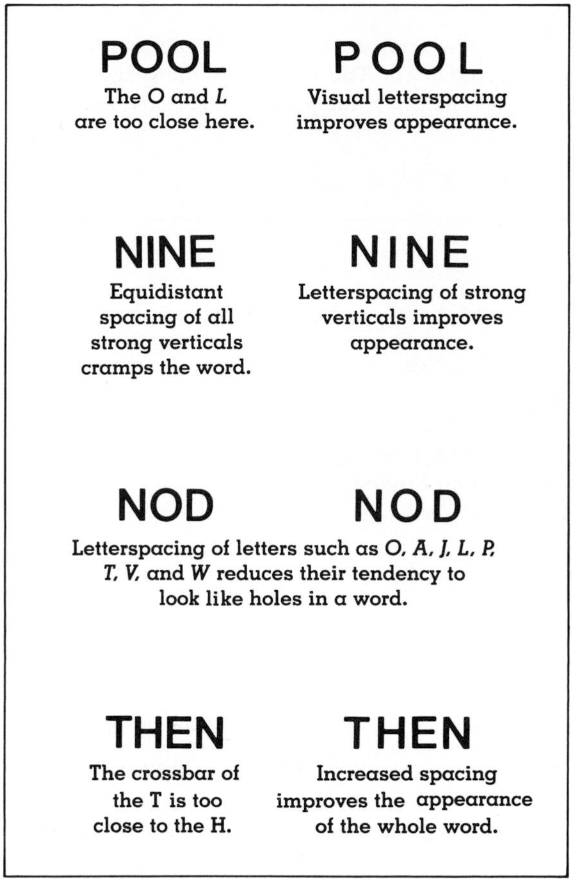

POOL The O and *L* are too close here.	P O O L Visual letterspacing improves appearance.
NINE Equidistant spacing of all strong verticals cramps the word.	N I N E Letterspacing of strong verticals improves appearance.
NOD	N O D
Letterspacing of letters such as O, *A, J, L, P, T, V,* and *W* reduces their tendency to look like holes in a word.	
THEN The crossbar of the T is too close to the H.	T H E N Increased spacing improves the appearance of the whole word.

Figure 15.6. Letterspacing.
Letterspacing is important in cartographic design. These examples point out where common mistakes are made.

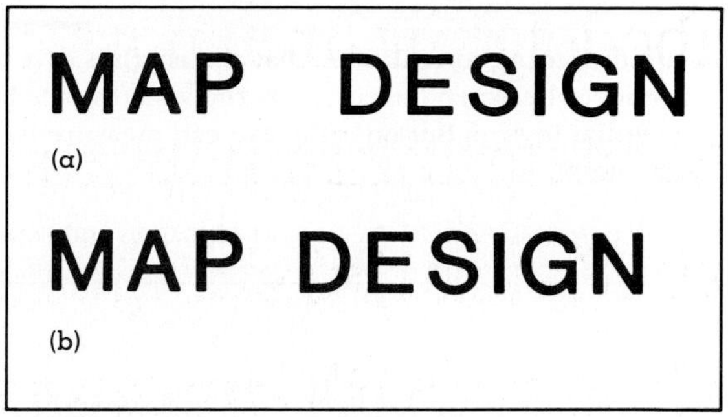

MAP   DESIGN
(a)

MAP DESIGN
(b)

Figure 15.7. Wordspacing.
The space between the words in (a) is too great. Improved word spacing, as in (b), is achieved by using approximately the same space as the letter I (plus its normal letterspacing) between words.

Spacing of both letters and words is especially important in cartography because so much of the lettering is stretched across map spaces to occupy geographical areas. Some names cause little difficulty for the designer because they are set solid: town and city names and other labels applied to point phenomena.

The real culprits are capital letters that cause open spaces, such as A, J, L, P, T, U, W, and O. Notice that these contain few vertical strokes and have mostly oblique or rounded strokes. Jan Tschichold, a noted German typographic designer, suggests that spacing of capital letters can be facilitated by the **neutralizing rule.**[7] Figure 15.6 demonstrates this, utilizing the O and the L. If the O is too close to the L (or other letter), a hole appears in the word. Letterspacing the O and the L causes the space to be visually equal to the inner part of the O, and the hole in the word disappears. Tschichold's neutralizing rule states that minimum spacing between capitals should always be equal to this optical value (or visually equal space) for all capital letters, not just for words containing an O. Letterspacing can be greater than this minimum value with no disruption of visual harmony, although word recognition will be more difficult.

As a general rule, continues Tschichold, the designer should strive to letterspace the capitals so that each letter's outline is more visually dominant than its inner space. Practice with spacing of capital letters will demonstrate this simple but revealing principle.

For some letter combinations, notably AV, AT, AW, AY, LT, LV, LW, and LY, **mortising** may be required. Mortising is fitting letters closer together to achieve proper visual balance in letterspacing.[8] Mortising usually requires that other letterspacing rules be applied as well.

Word spacing of names set in all capitals also deserves attention. One rule is to provide a distance between words equal to the letter I, including the letterspacing that is appropriate for it. (See Figure 15.7.) This principle applies to spacing of words set in capitals along a line of

> **The first thing to realize is that the rhythm of a well formed word can never be based on equal linear distances between letters. Only the visual space between letters matters. This unmeasurable space must be equal in size. But only the eye can measure it, not the ruler. The eye is the judge of all visual matter, not the brain.**
>
> Source: *Jan Tschichold,* Treasury of Alphabets and Lettering (*New York: Reinhold, 1966*), p. 29.

text, where it is undesirable for the words to form "islands" in the line. Cartographers need to exercise considerable caution here. Too much word spacing disrupts the continuity of multiple words, yet the cartographer often attempts to stretch words to occupy a given geographic space. Experimentation is required to achieve an acceptable balance in these cases.

Typeface Classification

The cartographer who has a knowledge of the history of type design and sees how different typefaces relate to each other will be better prepared to select type for maps. Type classification is one way to begin. The system discussed here is one among many; it is selected for its simplicity and historical emphasis.

Prior to the invention of movable type and the printing of Gutenberg in Germany, writing in Europe was accomplished by hand, usually by a select group of clerical scribes. The alphabets had come to them from Latin alphabets, as modified through time. Manuscript letterforms were called **textura** by the Italians.[9] Early movable type designs simply attempted to replicate the manuscript forms. A system of type classification begins with the first movable type and places all subsequent typefaces into classes, based on when they were created. A modern typeface is placed in one of these classes according to the match of its style characteristics with those of faces designed earlier.

A relatively simple classification, developed by Alexander Lawson, contains eight major classes. (See Table 15.1 and Figure 15.8.) For the most part, **Black Letter** is the term used for at least three or four different styles of early movable type that attempted to replicate manuscript forms. Two principal forms were *lettre de forme* and *lettre de somme.*[10] **Lettre de forme** typifies the design used by Gutenberg in his Bible. It is characterized by angular, sharp strokes with considerable vertical stress. **Lettre de somme** is less formal and more rounded. Another term used to describe Black Letter is *gothic,* but some confusion exists because many in the United States used this name for the early sans serif styles developed in the 1830s. Two styles used today are Cloister Black (Old English) and Goudy Text, which closely resembles the type of Gutenberg's first Bible.

Table 15.1 Typeface Classes

1. Black Letter
2. Oldstyle
   a. Venetian
   b. Aldine-French
   c. Dutch-English
3. Transitional
4. Modern
5. Square serif
6. Sans serif
7. Script-cursive
8. Display-decorative

Source: *Alexander Lawson,* Printing Types: An Introduction (*Boston: Beacon Press, 1971*), p. 27.

There are few uses of Black Letter in modern cartography. The style is very difficult to read, especially set in text, and is compatible with little else on the map. Perhaps it might find some use on decorative maps with historical content. Diplomas and newspaper logos continue to use Black Letter.

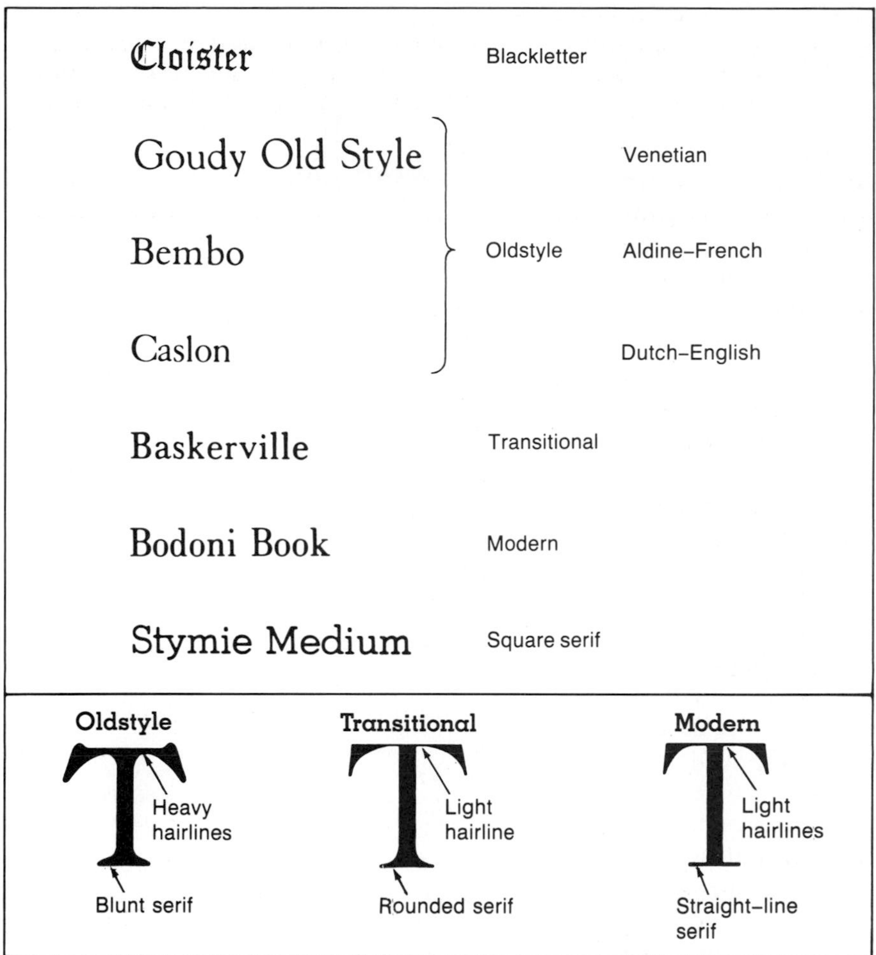

Figure 15.8. Examples of various typefaces for the classification used in this text. The chief differences in strokes and serifs among the three main classes are illustrated at the bottom.

Oldstyle, sometimes referred to as *Old face* types, may be subdivided into three groups. *Venetian* is a style first developed in northern Italy some twenty years after Gutenberg. Venetian Oldstyle is characterized by concave serifs and little contrast between thick and thin strokes. (See Figure 15.8.) A revival of the Venetian Oldstyle in late nineteenth and early twentieth centuries resulted in the introduction of such type designs as Cloister Oldstyle, Centaur, and Goudy Old Style.[11]

Another class of Oldstyle is *Aldine-French.* This group is modeled after the Venetian but was originally characterized by a greater difference between thick and thin strokes and modifications of the serifs. This style influenced French types of the sixteenth century and dominated style efforts in Europe for nearly 150 years.[12] Modern designs include Bembo and Palatino. (See Figure 15.8.)

Dutch-English forms the third group in the Oldstyle class. The French type forms were popular in Europe but soon gave way to the enterprising Dutch, who in many cases redesigned type to facilitate printing. The contrast of thick and thin strokes was increased, and serifs were straightened and bracketed in the lowercase forms. Early Dutch designs were adopted in England and enjoyed enormous popularity because of William Caslon.[13] (See Figure 15.8.) Caslon's foundry and others in England produced virtually all the type seen in America until the Revolution. Caslon forms abound today; many manufacturers offer the style. The Dutch-English tradition is at the end of the development of the Oldstyle class of letterforms.

Transitional type forms make a historical bridge between the Oldstyle and Modern type faces. John Baskerville, a printer and type founder in England around 1750, designed the type that bears his name,[14] often thought to be an excellent example of the Transitional style. The style is characterized by less serif bracketing than letterforms of the Oldstyle period. (See Figure 15.8.)

The **Modern** style of letterform began in the late eighteenth century and was considerably influenced by the Transitional forms. Baskerville's designs led Bodoni of Italy and Didot of France to design faces that have become identified with the modern movement. (See Figure 15.8.) The development of typeface designs in the modern period was in response to several other parallel achievements, notably in printing, manufacturing, and advertising. The strokes of the copper engraver led to hairline serifs and thick cross strokes. There was an emphasis on vertical shading. The development of **display types** (those too large for book-text printing) was important because of the Industrial Revolution and the advertising that it brought. Type of the Modern period was still characteristically roman in style, but brought to the extreme in geometrical regularity.

Square serif type styles, also called *slab serif* or *Egyptian,* were first introduced around the beginning of the nineteenth century. These type forms are characterized by even stroke widths and usually have unbracketed slablike serifs. The Clarendon typeface appeared in the mid-nineteenth century and was a more graceful form of the square serif. Stymie was produced in the early part of this century and

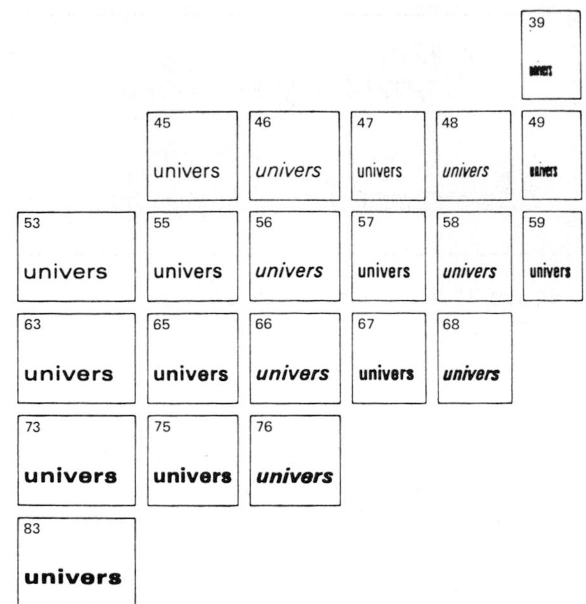

Figure 15.9. The complete Univers palette. A unique feature of this type family is the use of numbers instead of descriptive terms to designate weight.

The idea for representing the Univers family this way comes from Alexander Lawson, *Printing Types* (Boston: Beacon Press, 1971), p. 100. Redrawn with modifications from same source.

carries on the tradition of this class of type. (See Figure 15.8.)

The most notable letter form to appear in the history of type design is ***sans serif,*** a letter having no serifs. Previous innovations in design had at least carried along the Latin alphabet's tradition of serifs, but the new sans serif styles marked a radical departure. They have remained in use since their introduction and have enjoyed widespread popularity. A Swiss type designer named Adrian Frutiger styled a complete family of sans serif alphabets called ***Univers,*** one of the most popular such forms ever designed. Unique to this family was his use of number series, rather than descriptive names, to designate weights. Those in the 40s are lightface, in the 50s medium, in the 60s bold, and in the 70s extra bold. Both roman and italic forms are included in each number series. (See Figure 15.9.)

Script type face

Decorative type face

Figure 15.10. Script and decorative typefaces.
Thematic map designs do not employ these typefaces.

Two other classes of lettering styles round out this classification. **Script-cursive** and **display-decorative** are similar in that they are not useful for ordinary text typography but find their place in decorative or advertising display. They are difficult to read in most circumstances and have had little use in ordinary cartography. (See Figure 15.10.)

Nearly any type class may be used effectively on the thematic map except Black Letter, script or decorative forms. Certain designs may be more effective than others or may impart a more appropriate mood to the map. Certain principles regarding lettering class combinations and use of families are recommended and outlined below. First, however, let us look at a key quality of letter designs—their personalities.

The Personality of Typeface

Most graphic artists and designers seem to agree that typefaces are capable of creating moods, and that there is such a thing as **type**

Some types are versatile enough to be appropriate for almost any job. Others are more limited in what they can do. But all types have some special qualities that set them apart. Art directors are not in agreement about these qualities, but here are a few familiar faces, along with descriptions of the moods they seem to cover.

1. Baskerville—beauty, quality, urbanity.
2. Bodoni—formality, aristocracy, modernity.
3. Caslon—dignity, character, maturity.
4. Century—elegance, clarity.
5. Cheltenham—honesty, reliability, awkwardness.
6. Franklin Gothic—urgency, bluntness.
7. Futura—severity, utility.
8. Garamond—grace, worth, fragility.
9. Standard—order, newness.
10. Stymie—precision, construction.
11. Times Roman—tradition, efficiency.

These qualities if indeed they come across at all to readers, come across only vaguely. Furthermore, a single face can have qualities that tend to cancel out each other. (Can a type be both tradition-oriented and efficient?) While the art director should be conscious of these qualities and make whatever use he can of them, he should not feel bound to any one type because of a mood he wants to convey.

Assume he is designing a radical, militant magazine like Ramparts. Baskerville seems an unlikely choice for such a magazine, and yet it has been used effectively by Ramparts. As it has been used by Reader's Digest!

Sometimes the best answer to the question, "What type to use in title display?" is to go with a stately, readable type—like Baskerville—and rely upon the words in the title to express the mood of the piece.

Source: *Roy Paul Nelson,* Publication Design (*Dubuque, Iowa: William C. Brown, 1972*), p. 94.

personality. There is no question that the selection of type can alter the appearance of a map, but it is exceedingly difficult to categorize each type design by the kind of response it is likely to produce. Certainly, little is known about typeface and moods in cartographic design; more detailed knowledge is needed.

In general, Oldstyle designs, with their emphasis on diagonal shading, create more mellow and restful designs.[15] Bibles are more likely to be printed in Caslon or Bembo than Bodini. On the other hand, Modern faces, characterized by strong vertical and horizontal strokes, create abrupt, dynamic, and unsettling lettering. Lettering set in a sans serif face is dazzling; it is difficult for the eye to settle down. These observations are especially true in text typography but are also observable in display settings, such as in most cartographic applications.

As a starting point, the student might wish to consider the contents of Table 15.2. These are only generalizations; the appearance of type in a cartographic environment may be significantly altered in the map's content and other style features. Nonetheless, an awareness of type's influence on mood is important for the designer.

Legibility of Type

Although the designer may be interested in the aesthetic qualities of letters and alphabets, the real concern is readability. If type on the map cannot be read easily, its employment has been a wasted effort. **Type legibility,** the term most often used in research dealing with readability, has been defined this way:

> Legibility deals with the coordination of those typographical factors inherent in letters and other symbols, words, and connected textural material which affect ease and speed of reading.[16]

In cartography, **type discernibility**—the perception and comprehension of individual words not set in text lines—is more critical than legibility.

Table 15.2 Personality of Typefaces

Dignity	Oldstyle typefaces
Power	Bold sans serif
Grace	Italics or scripts
Precision	New sans serifs and slab serifs
Excitement	Mixtures of typefaces

Source: *Roy Paul Nelson,* Publication Design (*Dubuque, Iowa: William C. Brown, 1972*), pp. 34–35.

There is extensive literature dealing with text-type legibility, but little that deals specifically with display type discernibility. Nonetheless, certain important facts are known to apply to cartographic lettering.

The individual lowercase s, q, c, and x are difficult to distinguish, and the letters f, i, j, l, and t are frequently mistaken for each other.[17] The lowercase e, although a frequent letter in print, is often mistaken. In general, these legibility patterns have been recorded:

dmpqw	high legibility
jrvxy	medium legibility
ceinl	low legibility

The legibility of individual letters becomes important in cartography because the map may be embedded with different backgrounds, textures, or colors that can make letters difficult to distinguish. In addition, legibility may be affected by the way the letters are shown; they can be screened or rendered in color.

Lettering in all capitals, more than any other factor, slows reading. This is apparently caused by a larger number of eye fixation pauses; because capital letters occupy more space, the result is a reduction in the number of words perceived. Lowercase letterforms lead to quicker recognition. "Lower case words impress the mind with their total silhouette while capitals are mentally spelled out by letter."[18] (See Figure 15.11.) Also, the upper halves of lowercase letters contribute more recognition cues than the lower halves. Unfortunately, many of the tops of the letters in the sans serif forms are identical or nearly so, creating problems of legibility.

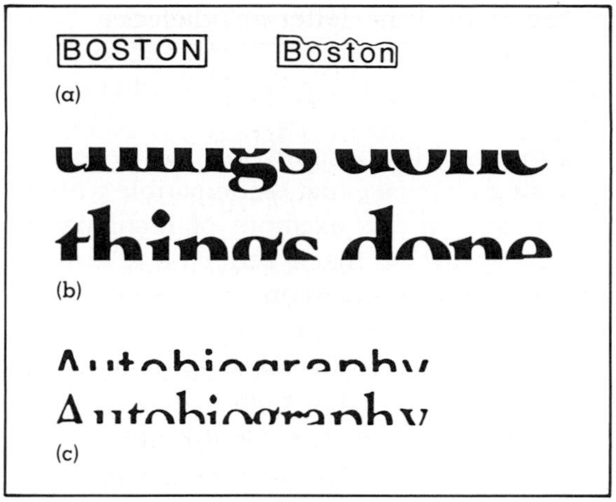

Figure 15.11. Reading type.
Lowercase letters lead to easier word recognition because of the silhouette they produce (a). It is easier to distinguish words by the tops of letters than by their bottoms (b). Serif letterforms are frequently easier to read, because the letters are more distinct, especially along the upper halves of lowercase letters (c).

In the cartographic context, placement of letters relative to their environments and the selection of letterforms can either add to or detract from overall legibility.

In the styles commonly used in text printing, legibility is not influenced by typeface style. This also applies to sans serif styles. Aesthetic preference might lead the designer away from sans serif, even though it might be equivalent in legibility.[19] It is interesting to note that the United States Geological Survey has recently selected a sans serif typeface (Univers) to appear on many of the 1:100,000 series maps, as well as others they produce.[20]

A significant research finding is that lettering in all italics slows reading, or at least is not preferred by most readers.[21] Italic letterforms have historically been used for labeling hydrographic features, and this practice is not likely to change. Of course, the legibility studies of italic forms were in a text format. Similar findings might not apply in cartographic contexts.

Thematic Map Typographics and Design

The selection of map type and its proper placement have been developed mostly by tradition. There are few experimental studies that pertain solely to map design. This section presents the highlights of conventional practices and the few experimental results available.

Lettering Practices

General guidelines, selecting typefaces, the use of capitals and lowercase, and lettering placement have all developed through empirical use. For the most part, the trained eye and aesthetic judgment of the cartographer have produced the design strategies that are now standard.

General Guidelines

There are four major goals in approaching the lettering of a map.[22]

1. Legibility
2. Harmony
3. Suitability of reproduction
4. Economy and ease of execution

Legibility research traditionally deals with the characteristics of letter design and how they affect the speed and accuracy of reading. Because lettering on maps usually occurs in a much more complex environment, legibility is achieved not only by choosing the appropriate typeface, but also by attending to good placement, adequate spacing of letters and words, and lettering environments such as background textures and confusing linework. Care and advanced planning help to avoid problems.

Lettering harmony involves several features related to the selection of a typeface. This goal of typographic design will be treated in detail below. The third aim, suitability for reproduction, pertains to the character of the typeface and how well it stands up to reduction (if any) and printing. Typefaces containing small bowls and counters tend to "close up" during printing, especially if excessive ink is used. Styles with unusually thin strokes and serifs may not photograph well; these letter features may disappear during printing. The designer should evaluate how well each letter in a proposed style will reproduce in the job at hand.

Finally, the designer needs to evaluate the cost of lettering. The various means by which cold type can be produced introduce options that should be investigated. If possible, unit costs (cost per word) should be compared for the different means. For many mapping projects, the cost of the lettering exceeds any other material costs.

Selecting Typefaces and Lettering Harmony

Many professional cartographers hold that only one typeface should be used on the map. To achieve contrast and harmony, the cartographer may, however, select several variants of a single type family.[23] The preferred way of mixing type is to vary roman and italic forms in the same family and to vary weight (light, medium, or bold). The least preferred is to mix different typefaces; if different typeface designs are used, do not select those that contrast markedly.

The following have also been suggested as fairly good combinations if more than one face is used: a strong geometric modern with a sans serif; and a square serif with a sans serif.[24] It is also good practice to choose different types (from the same period) that are not too similar; each face should be distinct in style to avoid confusion.

Using only type of different weights in one family should not be considered a limitation in design. The different weights, if used effectively, can be very expressive. Weights should be chosen in accordance with the importance of the feature the lettering identifies. Name and label weights may be lightened to deemphasize lettering that is large for other design reasons.

Typographic harmony is achieved by choosing a typeface that is compatible with the map's content. An example of incompatible choice would be Black Letter on a map depicting a twentieth-century geographical theme. Harmony means using a typeface with a personality that befits the map, in type sizes that correspond to the map's intellectual hierarchy. If the type size is carefully planned, the hierarchy is apparent to the map reader.

Very small lettering (4 or 5 points) is difficult to read at a normal reading distance of 30–46 cm (12–18 in). As a general rule, the larger or more important the feature, the larger its label should be.[25] This does not suggest that label sizes are proportional to the sizes of features, but rather that there should be an ordinal association. Rarely should there be more than four to six different sizes.

The selection of the typeface for a map requires careful inspection of each letter in the font. Typefaces with great contrast in thick and thin strokes, large x-heights, open bowls and loops, and dissimilar top halves of lowercase letters should be favored over others. Serif forms are good when letters are spread out, because they help tie the letters together. Display types should be investigated for their usefulness as these can be more appropriate for a given map than ordinary text faces.[26]

The Use of Capitals and Lowercase

Good lettering design on the map can be achieved by contrast of capitals and lowercase. A map that contains only one form or the other is exceptionally dull and usually indicates a lack of planning. In general, capitals are used to label larger features such as countries, oceans, and continents, and important items

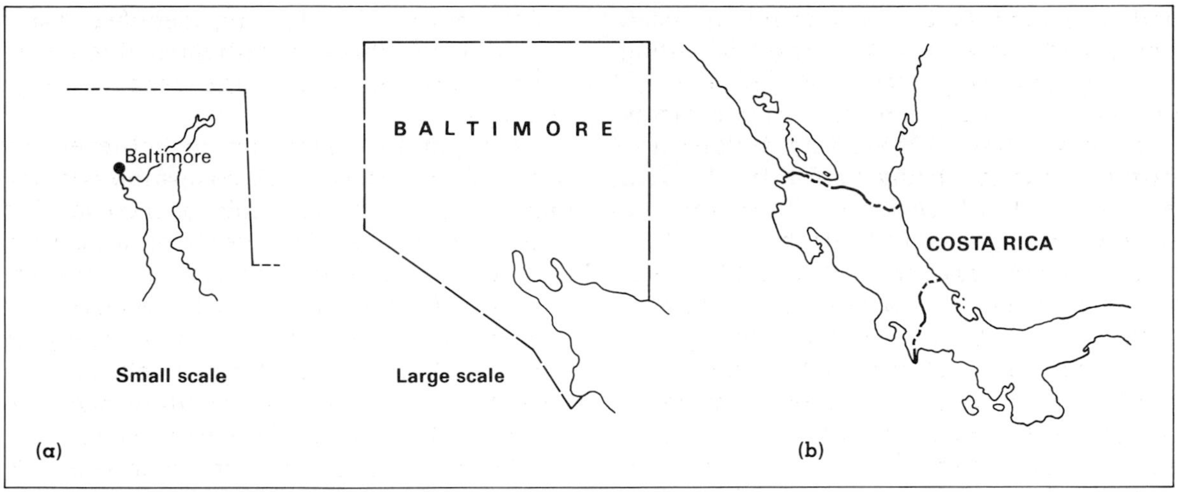

Figure 15.12. The use of capital and lowercase letterforms.
Point phenomena (a) are customarily labeled with lowercase. At larger scales where inside lettering can be accommodated, capitals are suggested. Because of space problems, areal phenomena such as countries may have their labels outside, as in (b). In these cases, the convention of using capitals is retained.

such as large cities, national capitals, and perhaps mountain ranges.[27] Smaller towns and less important features may be labeled in lowercase with initial capitals.

One practice is to label features represented by point symbols with lowercase lettering. In contrast, features that have large areas should be labeled in capitals. (See Figure 15.12a.) Scale determines the choice. At small scales, the city of Baltimore is represented by a dot and labeled in lowercase; at a much larger scale, in capitals. An exception is made for country names, normally set in capitals, which are still set in capitals when placed outside the country at smaller scales. (See Figure 15.12b.) The general rule is, however: inside the feature, capitals; outside, lowercase with initial capital.

The Placement of Lettering

Careful lettering placement enhances the appearance of the map. There are several conventions, supported by a few experimental studies. Most decisions regarding lettering placement fall into one of the following categories: labeling point symbols, designating linear features, naming open areas, and placement and design of titles and legend materials.

Point symbol labeling. Most professional cartographers agree that point symbols should be labeled with letters set solid (no letterspacing). On small-scale maps, these should follow the line of the parallels (if apparent). They should be set horizontally on large-scale maps.[28] (See Figure 15.13a.) The preferred location of the label is above and to the right of the symbol. (See Figure 15.13b.) The signature location, to the right and below, is next preferred. Never locate the label on the same horizontal line with the symbol, and if possible do not locate it to the left of the symbol.

Research has shown that distractors (other graphic marks) can impede the finding of names on a map when located anywhere near the beginning of a word.[29] Consequently, never

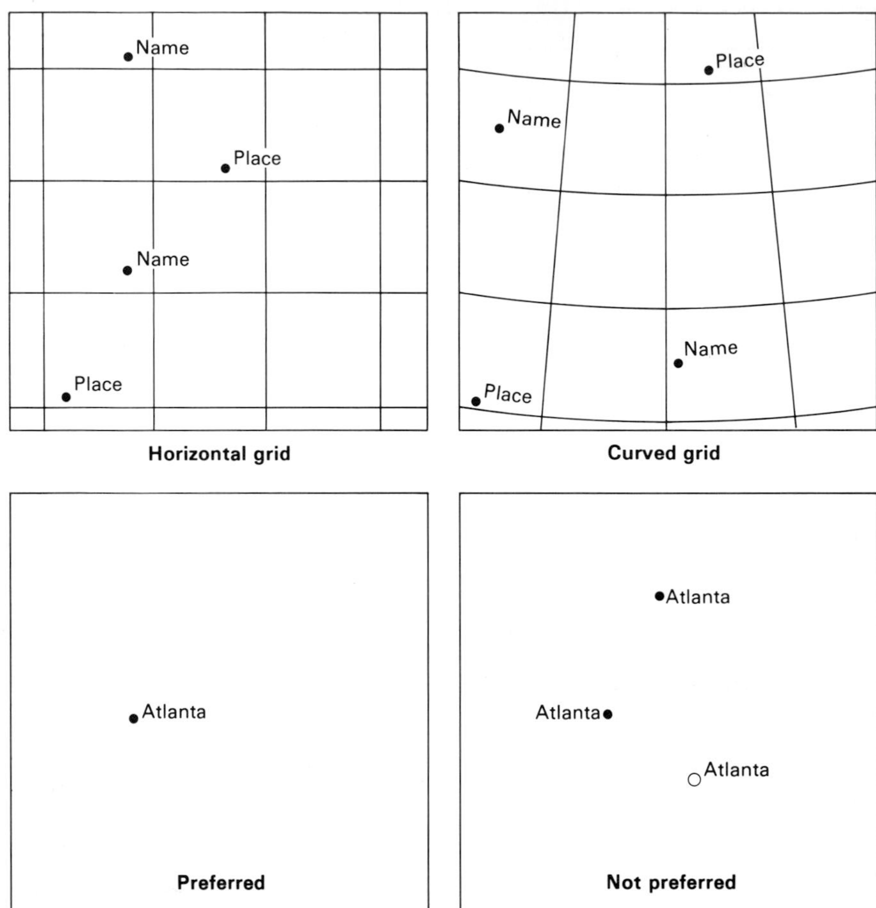

Figure 15.13. Lettering placement conventions.
Labels should be aligned with the grid, as in (a). There are preferred ways of labeling point phenomena, as in (b). Above and to the right of the symbol is most preferred. The labels should never be set on the same horizontal lines as the symbol. Symbols that can be confused with letters (such as an open circle) should never be used.

use open point symbols that can be confused with letters of the label. Other marks, such as lines and texture, can also distract the reader. With careful planning, these situations can be avoided.

Names of ports and harbor towns should be placed seaward, if possible.[30] Names at coastlines or rivers should never be placed so they overlap the coastline. Likewise, town names should be placed on the side of the river on which the town is located.

Designating linear features. Linear features on thematic maps include rivers, streams, roads, railroads, streets, paths, airlines, and many linear quantitative symbols. The general rule is that their labels should be set solid (no letterspacing) and repeated as many times along the feature as necessary to facilitate its identification.

The ideal location of the label for a linear feature is above it, along a horizontal stretch if possible. (See Figure 15.14.) Do not crowd the

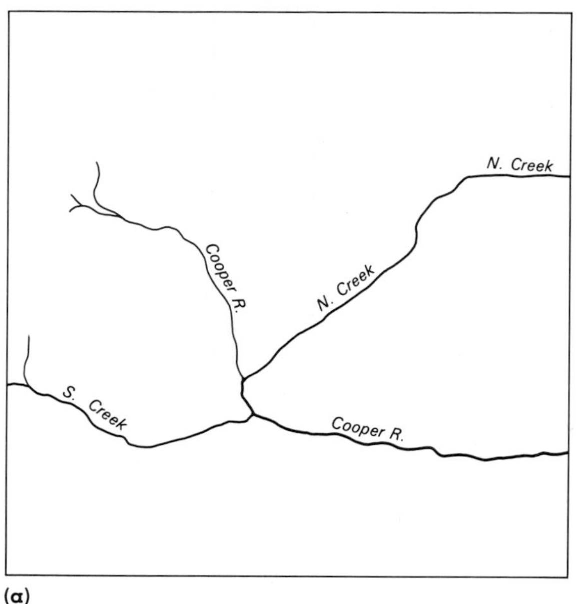

(a)

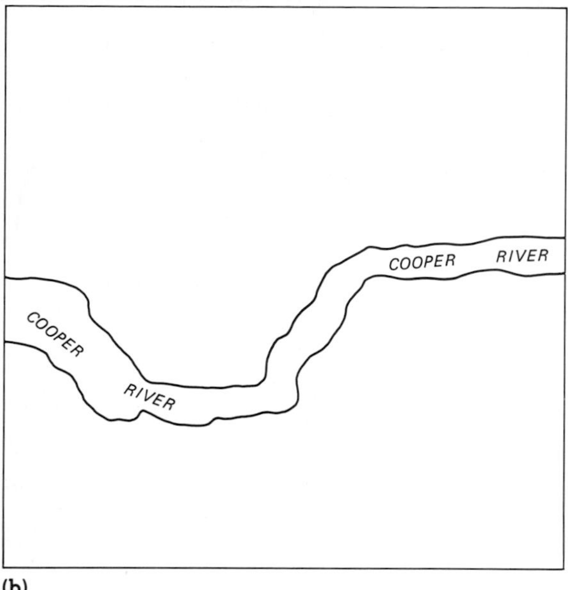

(b)

Figure 15.14. Labeling linear features.
Names outside a feature should be set solid, never letterspaced. Labels may be repeated for clarity. Inside a feature such as a river at large scale, the name should be set in all capitals as in (b), but not spread out.

label into the feature. Room must be reserved for lowercase descenders, if any. If at all possible, place a river name so that its slant is in the direction of the river's flow (assuming the label is italic, which is preferred).

Names in areal features. If a space on the map is large enough to accommodate lettering completely inside it, it may be designated an areal feature. Examples include oceans and parts of oceans, large bays, lakes, continents, countries, states, forests, and geographical subdivisions. The general rule is to letterspace the words so as to reach the boundaries of the feature.[31] The extent of the feature is then obvious from the letterspacing of its label. (See Figure 15.15.)

Each map presents unique problems in the placement of areal labels, but the following simple rules are helpful:

1. Curved lines of letters should be gentle and smooth, and the curve should be constant for the entire word. (See Figure 15.16a.)
2. Do not hyphenate names and labels.
3. If a line of lettering is not horizontal, make certain it deviates significantly from the horizontal so that its placement will not look like a mistake.
4. Do not locate names and labels in a way that the beginning and ending letters are too close to the feature's borders.
5. Choose a plan for the lettering placement of the whole map in accordance with the normal left-to-right reading pattern.[32] (See Figure 15.16b.) Never place labels and names vertically.
6. Never position names so that parts of them are upside down. (See Figure 15.16c.)

Placement and design of titles and legends. Map titles, legends, and other explanatory information, like any other graphic

Figure 15.15. Areal labels.
Names set inside areal features should be letterspaced to occupy the entire feature. Such labels may be curved to suggest the shape of the feature. Notice that the big labels are not letterspaced.

Source: Paul E. Lydolph, *Geography of the U.S.S.R. (New York: John Wiley, 1964). Reprinted by permission of John Wiley and Sons, Inc.*

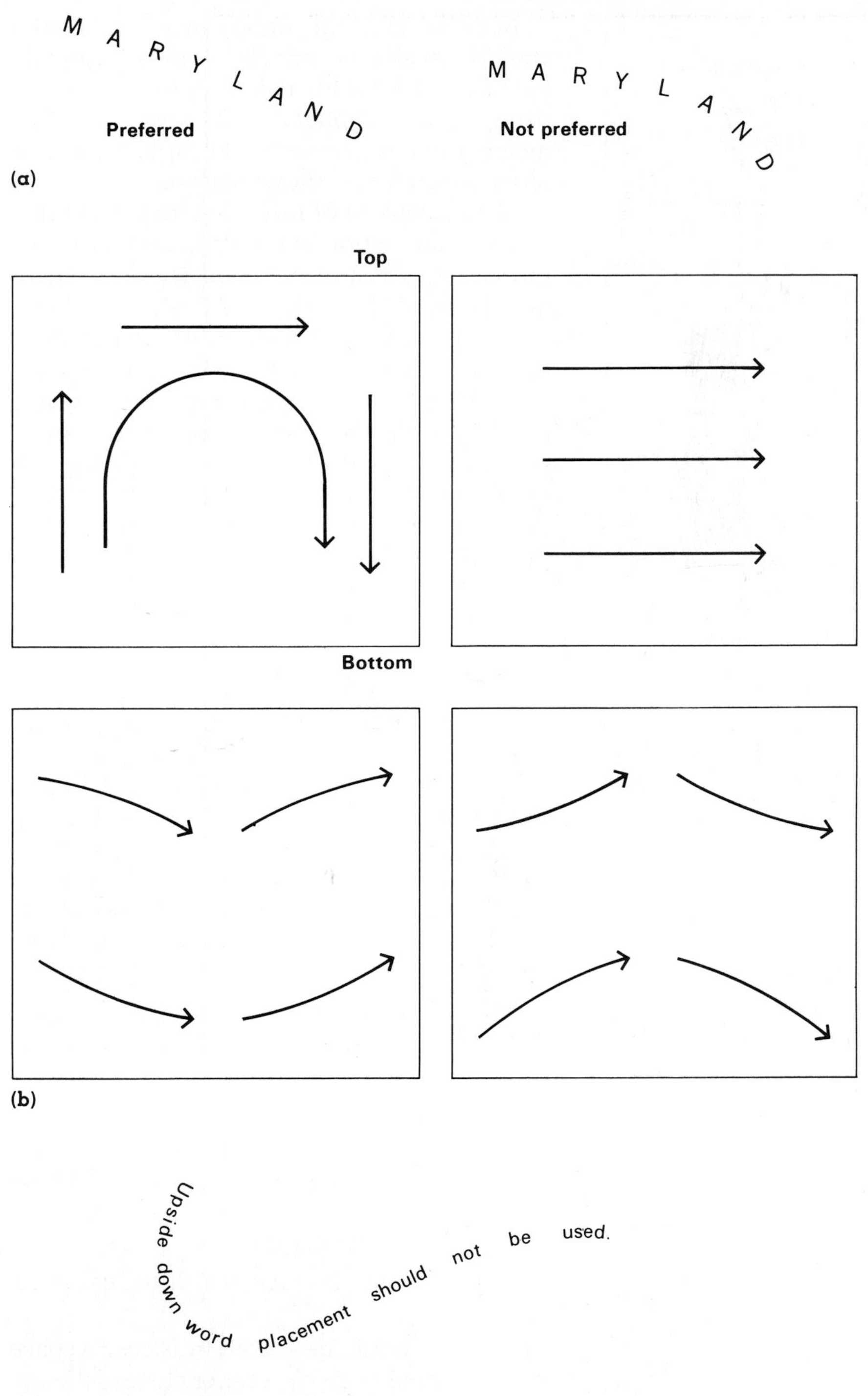

Figure 15.16. The flow of names and labels on the map.

SUBTITLE SUBTITLE
TITLE TITLE
SUBTITLE SUBTITLE

(a)

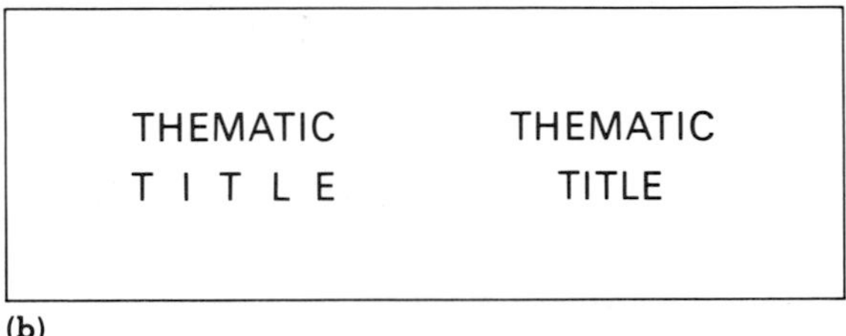

(b)

Figure 15.17. Planning title designs.
If the bottom line of a three-line title is to appear nearly equal to the top line, it should be in a smaller point size, as on the right in (a). Never spread letters to fill in a legend box or imaginary rectangle, as on the left in (b). It is better to use consistent letter spacing and center the titles.

elements, need to fit into the whole plan for the map. Titles are generally the most important intellectually and should therefore be largest in type size. Any subordinate titles should have somewhat smaller point sizes. Legend materials are important elements on the thematic map but are subordinate intellectually and visually. Their lettering size should reflect their position in the hierarchy. Map sources, explanatory notes, and the like are the smallest in point size.

Spacing in title or legend boxes requires attention from the designer. When three lines of type are used and the middle line is to be largest, the lower line must be a bit smaller than the top line if it is to appear the same size. (See Figure 15.17a.) One type designer notes that "lines consisting of letters of the same size in the same arrangement must, under all circumstances, have the same letterspacing."[33] (See Figure 15.17b.) This has particular significance when placing titles within boxes or title frames. Short words should never be letterspaced to achieve a real or imagined rectangle.

Because of their importance, map titles should usually be rendered in all capitals. Subtitles can be made to appear subordinate by using initial capitals and lowercase. Remember that words set in all capitals must be letterspaced for aesthetic reasons.

The placement of titles, legends, and other written information will be dictated in part by the overall layout of the map. These elements must be worked into the total design to achieve proper balance and proportion. Their treatment must be subtle so that they command the attention they deserve without overpowering the map. Title lettering or title boxes can be located so that they overlap other map features. (See Figure 15.18.) This helps to position them in the visual hierarchy and establish their intellectual rank.

Experimental Studies

Few experimental studies have dealt with lettering in thematic map design, but some significant findings have come to light. In one study, researchers investigated the effect of map typography on the speed of searching for place names on a map—a task similar to finding a city or town name on a general-reference map. The findings that have direct bearing on typographic design for the map are as follows:[34]

1. Names set in lowercase with an initial capital are easier to find than names set in all capitals of the same point size. Lowercase names are recommended.
2. Boldface type is no more legible than type of normal weight. It should be avoided because it has a "cluttering" effect on the map.
3. Choice of typeface has little effect on legibility.
4. Type should be placed in as clear a space as possible. Avoid clutter close to the initial letter of a word.
5. Typographic coding by color or point sizes can reduce search time; irrelevant coding can increase it.

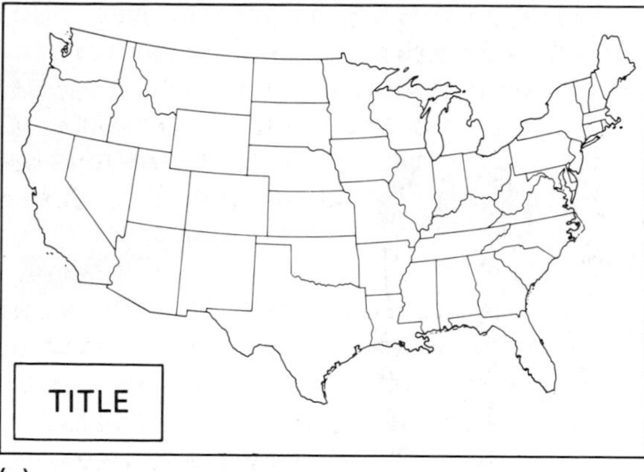

(a)

(b)

Figure 15.18. Using titles and title boxes in map design.
Positioning titles and title boxes over map areas may make them more integral to the overall visual and hierarchical plan of the map.

Another researcher studied the map search abilities of seventh- and eighth-grade students, reaching the following conclusions:[35]

1. Ordinary type styles are about equal in "searchability." A 4-point size variation had no significant effect on search times.
2. On a map containing three type styles, search is slowed if the reader does not know beforehand in which face the target will be. On the other hand, search time is reduced if the reader knows beforehand the target's typeface. This supports the design practice of replicating map elements (including symbols and lettering) in legend materials.
3. Text reading is not similar to map reading, at least in terms of type.

A common practice in thematic map design is to suggest an ordinal classification of phenomena with labels of different size. Thus, cities having small populations are labeled with lettering of relatively small point size, and larger cities are identified by labels in larger point sizes. Barbara Shortridge investigated this design practice to determine what lettering sizes and differences between them would best serve to indicate this kind of ordinal ranking. Her general conclusions were these:[36]

1. Lettering size differences of 34 percent are easily discriminated. Thus, a difference of 2 to 2½ points in the range of 5½ to 15 points is above this threshold.
2. Lettering size differences of less than 22 percent should not be used.

The Production of Cold Type Map Lettering

Cold type production should be distinguished from hot type methods. **Hot type** is any type production process that involves a raised (relief) surface.[37] The term is a carryover from earlier periods in printing when individual type was cast from molten metal. **Cold type** composition involves the preparation of letter images for photographic reproduction. Virtually all cartographic lettering today is produced by cold type methods. There are four main processes: hand lettering (including lettering machines), pressure-sensitive materials, photographic image production, and low-cost laser printers.

Hand Lettering

Hand lettering, still a versatile and economical way of producing lettering for maps, is rarely used nowadays, especially for maps that

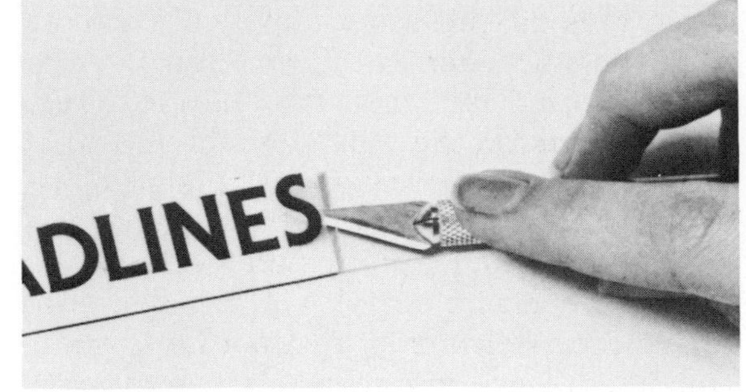

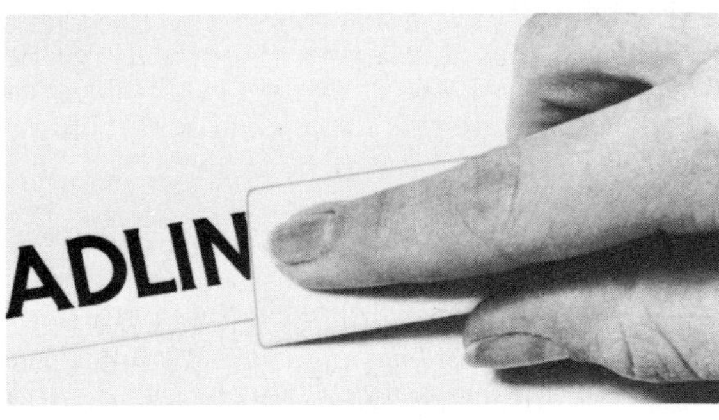

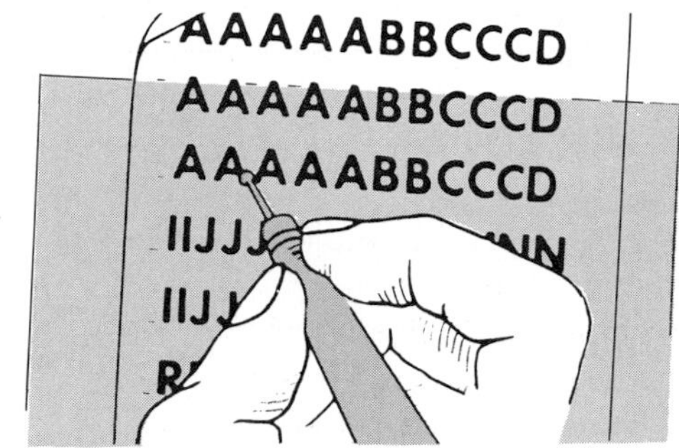

Figure 15.19. Two forms of cold-type image transfer.
In (a) through (c), three steps of applying pressure sensitive letters are illustrated. Dry transfer is shown in (d).
Photographs Courtesy of Graphics Products Corporation, Rolling Meadows, Illinois.

are to be reproduced in quantity. Any lettering that is to be reproduced must be done with pen and india ink. Graphite pencil images will not photograph adequately. Special lettering pens may be used to create serif style lettering, or technical pens may be used in certain circumstances. Consistent lettering stroke, slant, and size must be achieved—tasks not easily mastered by most cartographers. It is desirable to make artwork larger than reproduction size. Reduction can eliminate many inconsistencies.

Lettering machines, notably the Leroy lettering system, became exceedingly popular during the middle decades of this century. A special scriber containing an ink tip, or one capable of holding a technical pen, is drawn along a grooved template. Different slants are possible by resetting the scriber. Unfortunately, many of the early templates contained only a sans serif style, thus limiting the choice in design. Later template styles included serif letterforms. These machines offered the cartographer freedom from hand lettering, gave consistent results with clean, crisp lettering, and therefore became quite popular. The use of lettering machines has begun to wane.

Pressure-sensitive Materials

Two kinds of image-transfer systems have become very widely used in cartographic production. One such method involves a thin plastic sheet, on one side of which appears the lettering image. On the other side is an adhesive layer. A letter is cut from the sheet which contains one or several of each letter of the alphabet, it's carefully placed on the cartographic art and positioned, and then burnished (rubbed) in place. (See Figure 15.19a.)

Removal is difficult if not impossible without destroying the letter. Pattern and tints are manufactured on these sheets as well as lettering.

Dry transfer, the second method, is more popular than the above kind. In this system, the image is carried on a special transparent film and transferred to the art by merely rubbing the letter. (See Figure 15.19b.) It must be positioned accurately before the transfer is made. Hundreds of lettering styles and sizes are available in these pressure-sensitive materials. Their images are usually sharp and clear and photograph well. Because of their versatility, reasonable cost, availability, and wide design selection, they have become very popular.

One disadvantage in the use of pressure-sensitive lettering is time. For one or a few maps containing little lettering, the element of time is not significant. In the preparation of a large number of maps containing hundreds of place names and other labels, the use of pressure-sensitive materials involves formidable amounts of time. Fortunately, in these circumstances, the cartographer can have lettering set by a typographer onto transfer sheets as a special order. Whole names and words can then be transferred to the art and need not be done letter by letter. Special spacing of letters and words can be specified. Special orders such as these cost more and are recommended only for rather large jobs.

Another disadvantage is that the type is easily damaged once it has been transferred to the map unless it is sprayed with a fixative or other precautions are taken.

Photographic Image Production

Cartographers often use type of display size rather than text size. Display type is usually 12 points or larger. For the production of such type, several photographic machines are available; two kinds are currently popular. One kind produces the final art by passing light through a negative carrier containing the letters onto photosensitive film or paper. The paper or film is affixed by wax or glue to the cartographic art.

Another kind of cold type machine operates fundamentally like a typewriter with a carbon ribbon. Lettering is placed on a transparent adhesive tape carrier by "strike-on." A carbon ribbon is placed between a raised relief letter on the font wheel; when struck, the ribbon transfers the image to the front of the tape. Whole words or running text can be produced in this manner. Each word is positioned on the cartographic art and burnished in place. This type is also easily damaged if not treated with a fixative, preferably before it is placed on the artwork. Initial investment in such cold type machines is high ($900–$5000). Only the larger production facilities can justify their purchase.

Low-cost Laser Printer Typographics

Photographic typesetting is the traditional method for achieving high-quality map lettering. Modern digital typesetting machines expose characters onto silver-rich photosensitive paper by either laser or CRT technologies, using images composed of from 600 dots per inch (dpi) resolution up to 5,000 dpi.[38] Photographic typesetting machines, however, are very expensive and most cartographic designers send such typesetting work out to typesetting shops.

Laser printers, essentially a form of xerography, came into use in the mid-1970s and were used for printing in large computer facilities. Decreasing cost and size, improved technology, and coupled with micro-computer page description languages (pdl's), low-cost laser printers—those having 300 dpi—today are offering the cartographer another means of type production.

The laser printing process includes three steps. First, the image (type or graphics) is directed by laser beam controlled by computer software onto a rotating drum. Next, toner (dry ink) adheres to the drum where it was "etched" by the laser beam. Finally, the ink is permanently affixed to the paper by heat.

Type quality varies according to the resolution of printer "engine." A resolution of 300 dpi is discernible by the human eye, and a

Baltimore
Baltimore
Baltimore
Baltimore
Baltimore
Baltimore
Baltimore
Baltimore

Baltimore
Baltimore
Baltimore
Baltimore
Baltimore
Baltimore
Baltimore
Baltimore

Figure 15.20. Low-cost laser printer typographics.
Helvetica typeface on the left is printed actual size at 300 dpi, and on the right reduced by 25 percent.

ragged edge may be seen. (See Figure 15.20.) On higher resolution images of 1,200 dpi or more, the individual dots of ink are rarely noticed. Some cartographers use the 300–dpi printer images as draft-quality only, although some are using them as final art. They can be reduced photographically to achieve better results, although care must be used. (See Figure 15.20.)

One advantage of the low-cost laser printers for the cartographer is their enormous flexibility. Software programs that drive them contain many type styles, and almost any type size can be specified. As printers containing resolutions of 600 dpi or more emerge at lower prices, this form of type production will become commonplace.

Notes

1. Erwin Raisz, *Principles of Cartography* (New York: McGraw-Hill, 1962), pp. 54–55.
2. John R. Biggs, *The Use of Type* (London: Blandford Press, 1954), p. 18.
3. John S. Keates, "The Use of Type in Cartography," *Surveying and Mapping* 18 (1958):75–76.
4. R. A. Gardiner, "Typographic Requirements of Cartography," *Cartographic Journal* 1 (1964):42–44.
5. Alexander Lawson, *Printing Types: An Introduction* (Boston: Beacon Press, 1971), p. 29.
6. Jan Tschichold, *Treasury of Alphabets and Lettering* (New York: Reinhold, 1966), p. 29.
7. *Ibid.*
8. Biggs, *The Use of Type,* p. 40.
9. Lawson, *Printing Types,* p. 47.
10. Daniel Berkeley Updike, *Printing Types, Their History, Forms, and Use: A Study in Survivals,* 3rd ed., two vols. (Cambridge: Harvard University Press, 1966), pp. 61–64.
11. Lawson, *Printing Types,* p. 59.
12. *Ibid.,* p. 61.
13. Updike, *Printing Types,* pp. 101–6.
14. *Ibid.,* pp. 107–16.
15. Biggs, *The Use of Type,* pp. 18–20.
16. Miles A Tinker, *Legibility of Print* (Ames, IA: Iowa State University Press, 1963), p. 8.
17. Herbert Spencer, *The Visible Word* (New York: Hastings House, 1968), p. 25.
18. Tschichold, *Treasury of Alphabets,* p. 35.
19. Tinker, *Legibility of Print,* p. 64.
20. Clarence R. Gilman, "1:100,000 Map Series," in "Map Contemporary," *American Cartographer* 9 (1982):173–77; see also David Woodward, "Map Design and the National Consciousness: Typography and the Look of Topographic Maps," *Technical Papers* of the American Congress on Surveying and Mapping (Falls Church, VA: American Congress on Surveying and Mapping, 1982), pp. 339–47.
21. Spencer, *The Visible Word,* p. 31.
22. A. G. Hodgkiss, "Lettering Maps for Book Illustration," *Cartographer* 3 (1966):42–47.
23. Arthur H. Robinson, *The Look of Maps: An Examination of Cartographic Design* (Madison, WI: University of Wisconsin Press, 1966), p. 39.
24. *Ibid.*
25. Raisz, *Principles of Cartography,* p. 54; also Hodgkiss, "Lettering Maps for Book Illustration," pp. 42–47.
26. Gardiner, *"Typographic Requirements,"* pp. 42–44.
27. Robinson, *The Look of Maps,* p. 37.
28. Eduard Imhof, "Positioning Names on Maps," *American Cartographer* 2 (1975):128–44; see also Robinson, *The Look of Maps,* p. 47.

29. Liza Noyes, "The Positioning of Type on Maps: The Effect of Surrounding Material on Word Recognition Time," *Human Factors* 22 (1980):353–60.
30. Imhof, "Positioning Names on Maps," pp. 128–44.
31. *Ibid.*
32. *Ibid.*
33. Tschichold, *Treasury of Alphabets,* p. 41.
34. Richard J. Phillips, Elizabeth Noyes, and R. J. Audley, "Searching for Names on Maps," *Cartographic Journal* 15 (1978):72–76.
35. Barbara S. Bartz, "Experimental Use of the Search Task in an Analysis of Type Legibility in Cartography," *Cartographic Journal* 7 (1970):103–12.
36. Barbara Gimla Shortridge, "Map Reader Discrimination of Lettering Size," *American Cartographer* 6 (1979):13–20.
37. J. Michael Adams and David D. Faux, *Printing Technology: A Medium of Visual Communication* (North Scituate, MA: Duxbury Press, 1977), p. 66.
38. James Felici and Ted Nace, *Desktop Publishing Skills: A Primer for Typesetting with Computers and Laser Printers* (Reading, MA: Addison-Wesley, 1987), p. 9.

Glossary

ascender letter stroke extending above the x-height or body of a lowercase letter, as in h, d, b, etc.

ascender line horizontal line marking the maximum extent of all ascenders

base line bottom horizontal line from which all capital letters and lowercase bodies rise

Black Letter name of the type style first used in movable type; attempted to replicate textura or manuscript forms

bowls rounded portions of such letters as o, b, and d

bracket a fillet that helps join serifs to the main stroke

cap line horizontal line marking the height of all capital letters; usually lower than the ascender line

cold type preparation of letter images for photographic reproduction

counters partially or completely enclosed areas of a letter

descender letter stroke extending below the x-height or body of a lowercase letter, as in y, g, q, etc.

descender line horizontal line marking the maximum extent of all descenders

display-decorative special typefaces used in art and advertising; not ordinarily useful in cartography

display types type sizes not ordinarily used in running text; usually 12 points or larger

dry transfer method of producing letters by rubbing images from an intermediate film carrier onto cartographic art; very common form of lettering production

family a series of lettering styles all related to the basic style; a family contains variations of the basic style in weights, widths, roman, and italic

font a complete set of all characters of one size and design of a typeface

hand lettering forming letters by hand or by lettering machines

hot type any type production process that involves a raised (relief) surface; traditionally refers to type cast from molten metal

italic slanted version of a type style; minor letterform differences may exist between an italic and its roman counterpart

leading spacing between horizontal rows of continuously running text

letterspacing additional horizontal spacing placed between letters; required when all capital letters are to be set, but optional in lowercase

lettre de forme angular shaped Black Letter with extreme vertical stress; very difficult for modern readers

lettre de somme more rounded version of Black Letter; less formal, easier to read than lettre de forme

loop rounded portions of lowercase descenders such as j, g, and y

Modern type design developed in early eighteenth century; these styles have a pronounced vertical shading which distinguishes them from earlier forms

mortising fitting letters closer than in the ordinary (geometrical) arrangement, to achieve a good visual result

neutralizing rule a letterspacing principle based on visual assessment of letterspacing rather than geometrical

Oldstyle early type designs that did not try to replicate manuscript styles; Venetian (Italy), Aldine-French (mostly French), and Dutch-English are three variations of Oldstyle

point unit of measurement for type; 72 points to the inch (0.138 in. per point)

roman basic, upright version of a type style

sans serif a letterform containing no serifs

script-cursive type style that replicates handwriting; not ordinarily useful in cartography

serif finishing stroke added to the end of the main strokes of a letter

set solid no letterspacing

shading the position of maximum stress or slant of a letterform; Oldstyle type has greater shading than do Modern faces

square serif bold and squared off, with thickness similar to that of main letter strokes; also called *slab serif* or *Egyptian*

textura early manuscript alphabet form common at the time of Gutenberg's first printing of the Bible

Transitional type design that bridged the gap between Oldstyle and Modern; Baskerville (English) best example of period

type discernibility perception and comprehension of individual words not set in runnning text; perhaps of greater theoretical use in cartographic design than type legibility

typeface particular style or design of letterforms for a whole alphabet

type legibility type design characteristics that can affect ease and speed of reading

type personality informal way of describing the impressions that type can elicit; e.g., dignity, power, grace, precision, and excitement

type weight relative blackness of a type; normal, lightface, boldface

type width different width of a letter design, such as condensed or extended

Univers a popular sans serif typeface designed by Adrian Frutiger in the 1930s

word spacing space between words set horizontally

x-height height of the body of lowercase letters

Readings for Further Understanding

Adams, J. Michael, and David D. Faux. *Printing Technology: A Medium of Visual Communication.* North Scituate, MA: Duxbury Press, 1977.

Bartz, Barbara S. "An Analysis of the Typographic Legibility Literature: An Assessment of Its Applicability to Cartography." *Cartographic Journal* 7 (1970):10–16.

———. "Experimental Use of the Search Task in an Analysis of Type Legibility in Cartography." *Cartographic Journal* 7 (1970):103–12.

Biggs, John R. *The Use of Type.* London: Blandford Press, 1954.

Cook, Rick, "Page Printers." *Byte* 12 (September, 1987):187–97.

Dailey, Terrence, ed. *Illustration and Design: Techniques and Materials.* Secaucus, NJ: Chartwell Books, 1980.

Felici, James, and Ted Nace. *Desktop Publishing Skills: A Primer for Typesetting with Computers and Laser Printers.* Reading, MA: Addison-Wesley, 1987.

Gardiner, R. A. "Typographic Requirements of Cartography." *Cartographic Journal* 1 (1964):42–44.

Gilman, Clarence R. "1:100,000 Map Series." In "Map Commentary." *American Cartographer* 9 (1982):173–77.

Hodgkiss, A. G. "Lettering Maps for Book Illustration." *Cartographer* 3 (1966):42–47.

Imhof, Eduard. "Positioning Names on Maps." *American Cartographer* 2 (1975):128–44.

Keates, John S. *Cartographic Design and Production.* New York: John Wiley, 1973.

———. "The Use of Type in Cartography." *Surveying and Mapping* 18 (1958):75–76.

Lang, Kathy. *The Writer's Guide to Desktop Publishing.* London: Academic Press, 1987.

Lawson, Alexander. *Printing Types: An Introduction.* Boston: Beacon Press, 1971.

Lewis, John. *Typography: Design and Practice.* New York: Taplinger, 1978.

Monmonier, Mark Stephen. *Technological Transition in Cartography.* Madison: The University of Wisconsin Press, 1985.

Nelson, Roy Paul. *Publication Design.* Dubuque, IA: William C. Brown, 1972.

Noyes, Liza. "The Positioning of Type on Maps: The Effect of Surrounding Material on Word Recognition Time." *Human Factors* 22 (1980):353–60.

Phillips, Richard J., Elizabeth Noyes, and R. J. Audley. "Searching for Names on Maps." *Cartographic Journal* 15 (1978):72–76.

Poulton, E. C. "A Note on Printing to Make Comprehension Easier." *Ergonomics* 3 (1960):245–48.

Raisz, Erwin. *Principles of Cartography.* New York: McGraw-Hill, 1962.

Robinson, Arthur H. *The Look of Maps: An Examination of Cartographic Design.* Madison, WI: University of Wisconsin Press, 1966.

Robinson, Arthur H., Randall Sale, and Joel Morrison. *Elements of Cartography,* 4th ed. New York: John Wiley, 1978.

Schlemmer, Richard M. *Handbook of Advertising Art Production.* Englewood Cliffs, NJ: Prentice-Hall, 1976.

Shortridge, Barbara Gimla. "Map Lettering as a Quantitative Symbol: A Preliminary Investigation." Unpublished Ph.D. Dissertation, Department of Geography, University of Kansas, 1979.

———. "Map Reader Discrimination of Lettering Size." *American Cartographer* 6 (1979):13–20.

Spencer, Herbert. *The Visible Word.* New York: Hastings House, 1968.

Tinker, Miles A. *Legibility of Print.* Ames, IA: Iowa State University Press, 1963.

Tschichold, Jan. *Treasury of Alphabets and Lettering.* New York: Reinhold, 1966.

Updike, Daniel Berkeley. *Printing Types, Their History, Forms, and Use: A Study in Survivals,* 3rd ed. Two volumes. Cambridge: Harvard University Press, 1966.

Woodward, David. "Map Design and the National Consciousness: Typography and the Look of Topographic Maps." *Technical Papers* of the American Congress on Surveying and Mapping. Falls Church, VA: American Congress on Surveying and Mapping, 1982, pp. 339–47.

Zachrisson, Bror. *Studies in the Legibility of Printed Text.* Stockholm: Almquist and Wiksell, 1965.

Chapter

16

Principles for Color Thematic Maps

Chapter Preview

The application of color to the thematic map is one of the most exciting aspects of cartographic design, yet perhaps the least studied by cartographers. Color is produced by physical energy, but our reaction to it is psychological. Color has been found to have three dimensions: hue, value, and chroma. Our perception of color is influenced by its environment and by the subjective connotations we attach to colors. Map readers have preference for certain colors, but these change throughout life. Our understanding and use of color, especially in color research, is enhanced by learning how color can be specified. Color theorists may use the Munsell, Ostwald, or CIE specification systems. In cartographic practice, color is often specified by a matching system, such as the Pantone. Cartographic designers are aware that colors can function in certain ways in design. Design strategies to achieve figure and ground, the proper degree of contrast, and color harmony improve the use of color on the thematic map.

Introducing color into the design of a thematic map can be both exciting and troublesome. On the one hand, color provides so many more design options that designers often quickly seize the opportunity to include it. Yet the inclusion of color invites many potential problems. Costs increase, design solutions require greater precision, and production problems can begin to mount. The designer is plagued by uncertainty as to how the reader will respond to the map. However, if allowed to choose, with no budgetary restraints, most map designers choose color mapping because of its inherent advantage of greater design freedom.

Color in thematic mapping is perhaps the most fascinating and least understood of the design elements. Color is subjective rather than objective:

> The concept of color harmony by the color wheel is an objective conclusion arrived at by intellectual activity. The response to color, on the other hand, is emotional; thus there is no guarantee that what is produced in a purely intellectual manner will be pleasing to the emotions. Man responds to form with his intellect and to color with his emotions; he can be said to survive by form and to live by color.[1]

Another problem in dealing with color is that it is difficult, if not impossible, to set color rules. Certain standards for color use have been adopted for some forms of mapping, notably USGS topographic maps. No standards or rules for color use, except for a few conventions, exist for thematic maps. This situation will demand greater attention in the years ahead as the production of temporary color maps, such as those generated on color CRT screens, can be designed from a palette of thousands of colors. On the other hand, rigid rules can also be restrictive in design.

Color is a complex subject and can be studied in many different ways. Physicists, chemists, physiologists, psychologists, and artists approach color from different perspectives, with distinct purposes.[2] The physicist looks at the electromagnetic spectrum of energy and how it relates to color production. Chemists examine the physical and molecular structures of **colorants,** the elements in substances that cause color through reflection or absorption. The physiologist treats the mechanisms of color reception by the eye-brain pathway, and the psychologist deals with the meaning of color to human beings. Finally, the artist works with aesthetic qualities of color, using information gained from the physiologist and psychologist.

This chapter introduces the characteristics of color that are important in thematic mapping: color perception, color specification systems, cartographic conventions, and design strategies. The focus throughout is on the psychological and aesthetic aspects of color. The student interested in further investigation will find ample fascinating reading listed at the end of this chapter.

Color Perception

Reading thematic maps is a perceptual process involving the eyes and brain of the map reader. Light reflected off the map is sensed by the eyes, which report sensations to the brain, where cognitive processes begin. The sensing and cognitive processing of color is called **color perception.** Aspects of color perception treated in this section are the physiology of the human eye, physical properties of color production, and the dimensions of color.

The Human Eye

The human eyes are wondrous organs, often considered to be external linkages to the brain itself. The processing of light that falls on the human eye is like a vast, complex data-analyzing system. Each eye is a spherical body about one inch in diameter. (See Figure 16.1.)

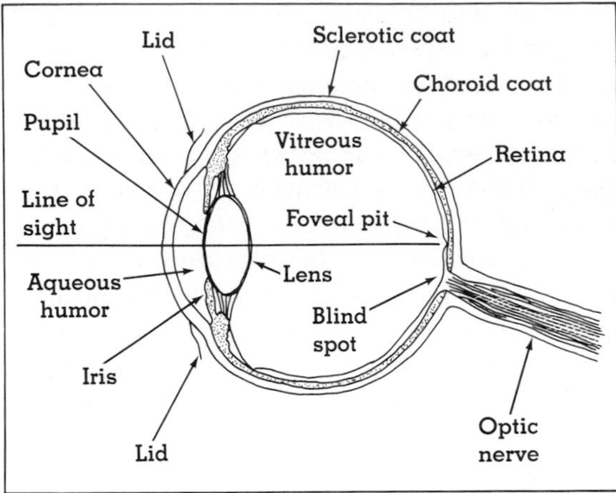

Figure 16.1. The parts of the human eye.

Light enters through the **cornea,** a transparent outer protective membrane. The amount of light entering the eye is controlled by the **iris,** a diaphragm-like muscle at the center of the eye. Light is then focused by the transparent **lens** and passed onto the back wall of the eye chamber. The inside of the back of the eye chamber is covered by a thin tissue called the **retina.** A fluid (vitreous humor) fills the eye and serves to keep the distance between lens and retina nearly constant.

Light-sensitive cells make up the retina. **Rod cells,** numbering about 120 million, provide for only achromatic sensations and have no color discrimination. **Cone cells,** numbering about 6 million, are of three types. Some have peak sensitivity to blue, some to red, and others to green portions of the electromagnetic spectrum. As these rod and cone cells are triggered, electric impulses pass through a complex system of specialized cells to the **optic nerve,** which is a bundle of nerve cells that transmits the electrical impulses to the brain. Color perception takes place in the brain and involves cognitive processes that add meaning and substance to the light patterns sensed by the eye. Color does not exist in the environment, or in objects. Color is psychological, a product of the mind.

The cartographic designer does not ordinarily need to pay particular attention to the details of the color-sensing apparatus of the map reader, except to make sure that graphic marks do not exceed the lower limits of visual acuity. Illumination is very important. At low levels, only achromatic sensing is possible; color detection becomes possible as illumination increases. Most map reading situations exceed the minimal illumination threshold for color discrimination. When objects subtend an angle on the retina of less than about 12 minutes of arc, discrimination of color becomes difficult. Absolute color identification is generally poor for objects subtending less than 10 minutes of arc.[3] It is advisable to limit color use to map objects that subtend an angle on the retina of 20 minutes of arc or more. At a reading distance of 30 cm (12 in), this would be about .15 cm (.06 in).

Physical Properties of Color Production

The production of color requires three elements: a light source, an object, and the eye-brain system of the viewer. This defines the **object mode** of viewing. There is also an **illuminant mode,** when a viewer looks directly into the light source. Color production in cartography is normally the object mode.

Visible light is that part of the electromagnetic energy spectrum to which our eyes respond. Sir Isaac Newton showed that the visible portion of the energy spectrum is composed of the various colors, each having a different wavelength. Physicists plot energy against wavelength on charts to produce a

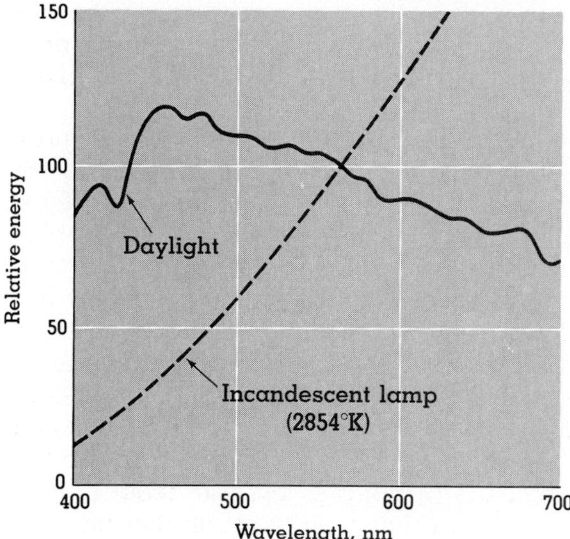

Figure 16.2. Spectral energy.
The figure shows spectral energy, or relative energy at each wavelength, for two light sources: typical daylight and an incandescent lamp at 2854°K (color temperature).

spectral energy distribution curve. Different light sources, such as the sun, a tungsten light bulb, or a fluorescent bulb, each generate a different spectral energy distribution curve. (See Figure 16.2.)

When we attempt to describe the physical properties (physical dimensions) of color, we must specify what source is used. Each produces a characteristic spectral energy distribution curve, so that the colors of the viewed objects vary. In a paint store, color chips are often displayed in such a way that the customer can switch on either incandescent or fluorescent lights, to see how the colors behave in different light. It is no wonder that color specification and color choice are so difficult in cartography; color maps may be designed in one environment and viewed in another.

The physical characteristics of color are also affected by the quality of the object's surface. Some surfaces transmit all light falling on them, such as a white sheet of paper. These are called **transparent objects.** Most objects are **reflective** of portions of the visible spectrum; that part of the spectrum that is reflected defines the color of the surface. Surfaces that absorb all light are **opaque** and appear black. The amount of light that is reflected from surfaces can be plotted on a diagram that is called a **spectral reflectance curve.** This type of diagram defines the physical dimensions of reflected light from objects. (See Figure 16.3.)

Cartographers do not usually delve into the spectral reflectance curves of the papers and inks of their printed maps. The combinations of printed color inks and different papers are so numerous that measurement of this kind is prohibitive, except in large map-making organizations such as the USGS. Research cartographers, however, need to be especially mindful of these characteristics, so that their work can be replicated and extended. The complexities of this kind of research have had a negative impact on color experimentation in cartography.

Color Dimensions

Cartographic designers are directly concerned with the psychological dimensions of color—those that describe what we see in the form of reflected light. These dimensions are much more difficult to explain because of the range of variation in human behavior, the difficulty of reporting visual perception objectively, and the fact that we often associate colors with behavior moods. No set of exactly tailored rules are available to the designer; a general understanding and sharp intuition must suffice.

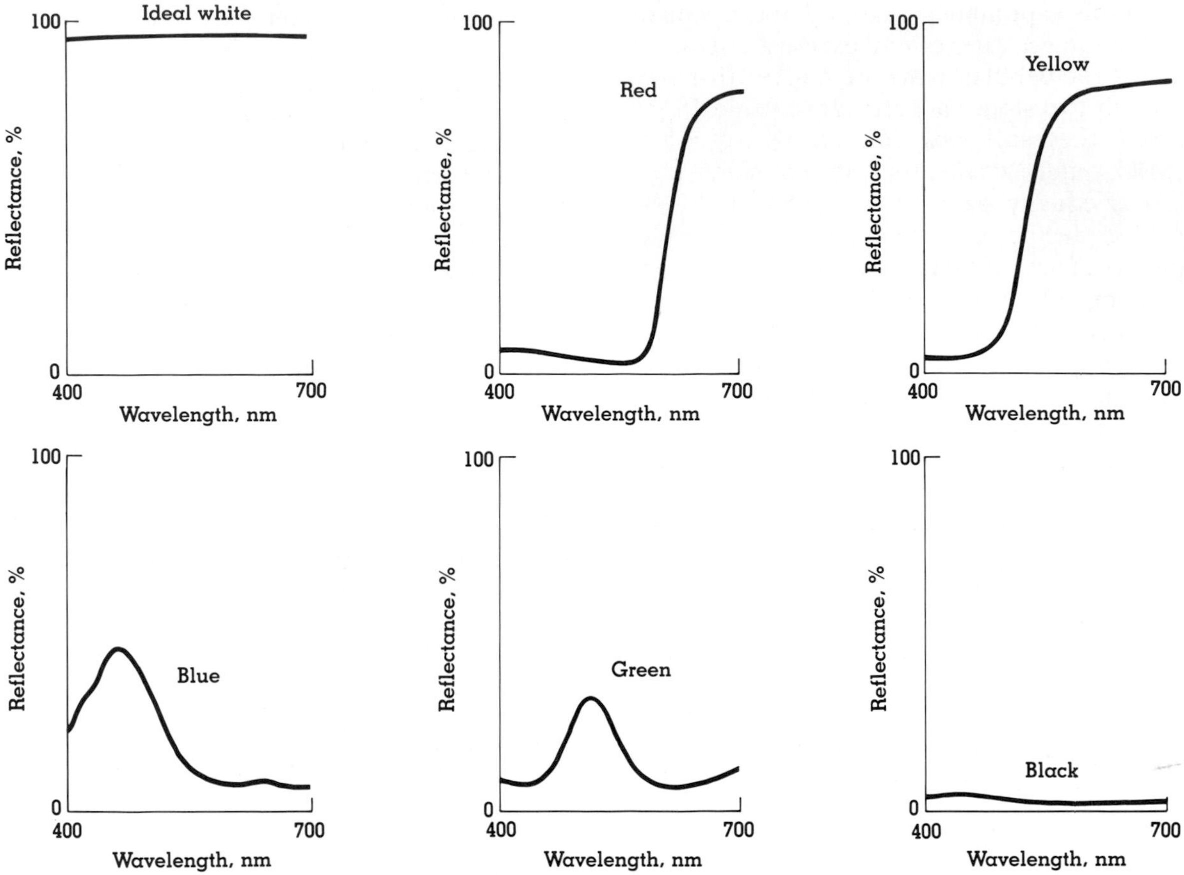

Figure 16.3. **Spectral reflectance (*R*) curves of several colored materials.**
Notice that white reflects the most light and black the least. Also, the similarity in the curves of colors close to each other (red and yellow or blue and green) is quite remarkable.

The Desert Island Experiment

A subject is asked to imagine that he or she is on a desert island that is covered by many large pebbles of different colors.[4] The task is to arrange (classify) the pebbles by color, using any system. One subject first decides to divide the pebbles into two groups—those with color **(chromatic)** and those that are white, gray, or black, or without color **(achromatic).**

The next step in classification by this subject might be to arrange the achromatic pebbles (the whites, grays, and blacks) into an array from white to black, with the grays in between. This array, based on the single quality *lightness,* is called a **value scale.**

To deal with the sorting of the chromatic pebbles, the subject decides to make piles of red stones, blue stones, green stones, and so on. If greater precision is needed, there could be groups of blue-green pebbles, red-orange, and so forth. The chromatic pebbles are now divided into **hue** categories.

Within each pile, the subject now decides to array the colored pebbles by lightness or value. Thus, for example, the red stones could be arrayed from light red to dark red, from pinks to the deepest cherry reds.

In this experiment, a last judgment is made by the subject. After careful examination of, say, all the red pebbles, it may be noticed that even though two stones are similar in value (lightness), they still *look* different. A red radish looks somehow different from a tomato whose red is equally dark. After careful study, the subject decides that the two reds can be compared in terms of their neutral (or middle) gray content; all reds of the same value can be further divided into yet another spectrum. This is sometimes referred to as *saturation,* or **chroma.** A hue having less gray in it is more saturated.

This completes the desert island experiment. The subject has classified colors into groups based on appearance or **psychological dimensions.** Because these dimensions are based on perception, they have particular significance in thematic map design.

Hue

Hue is the name we give to the various colors we perceive: the reds, greens, blues, browns, red-oranges, and the like. Each hue has its own wavelength in the visible spectrum. Hue is conventionally illustrated by a **color wheel,** as in Figure 16.4 and Plate 4. Theoretically, a color wheel can contain an almost infinite number of hues, but most include no more than 24. Eight or twelve are more common, twelve being especially useful for artists.[5] The structure of any color wheel, especially as used by the artist, is tied to the so-called **primary** (or pure) colors of red, yellow, and blue. These hues cannot be obtained by mixing other hues, but all other colors can be obtained by mixing the primaries. The color wheel is usually organized so that the primaries are positioned at the vertices of an imaginary triangle contained within the wheel.

Although the color wheel normally contains twelve hues, the human eye can distinguish millions of different colors. In ordinary situations, however, we are not called on to do so;

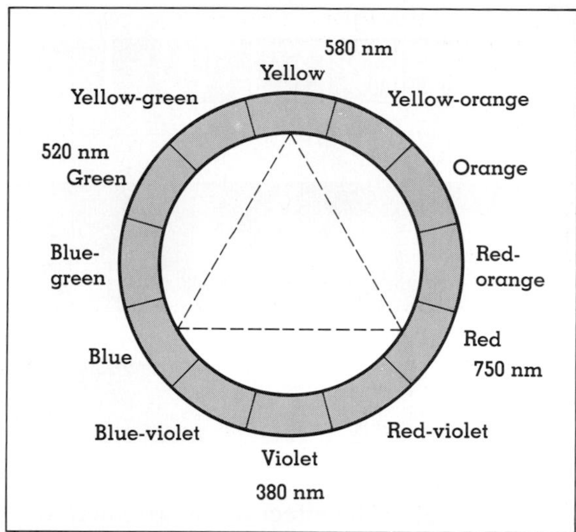

Figure 16.4. The twelve-part color wheel, showing approximate wavelengths of the various hues.
Typical color wheels will contain eight or twelve hues. The primary hues (red, yellow, and blue) normally occupy the vertices of an enclosed imaginary triangle. The abbreviation nm stands for nanometers; one nanometer is one-billionth of a meter.

in cartographic design, our assortment of hues is certainly far less. In fact, it would be difficult to imagine any case where a map reader would have to identify more than a dozen (this is common on soil or geologic maps). The United States Census has recently done 16-color two-variable choropleth maps,[6] but these will probably not find wide use because of the complexities inherent in reading them.

Value

Value is the quality of *lightness* or *darkness* of achromatic and chromatic colors. Conceptually easy to grasp, value is difficult to deal with in practice. Sensitivity to value is easily influenced by environment, and apparent lightness or darkness is not proportional to the reflected light of achromatic surfaces. A given gray, or

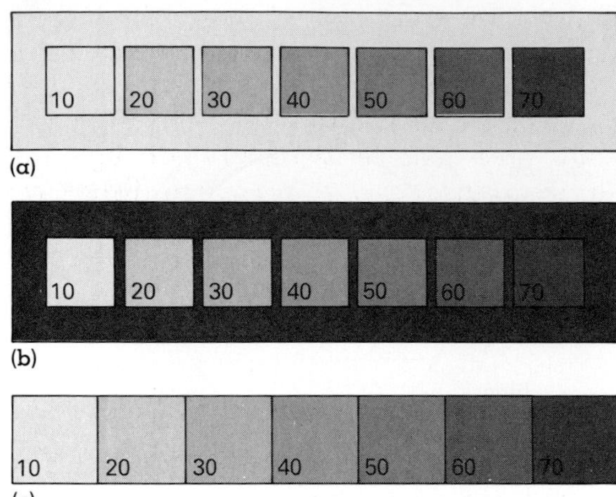

Figure 16.5. The effect of environment on the perception of value.
The surrounding environments of (a) and (b) modify the appearance of the group (otherwise identical) in the strips they surround. Induction results when color patches of different values are closely juxtaposed, as in (c).

hue of a specified value, looks one way when examined individually against a white background but differs when viewed in an array of grays against different backgrounds.

People distinguish better among grays if all are similar to their backgrounds. Against a light background, light grays are more easily distinguished from each other; against a dark background, dark grays are better distinguished. (See Figure 16.5.) Also, relative to the whole value spectrum, greater differences at each end (both white and black) are required for better discrimination.[7] These conditions are compounded when the value scales are chromatic. People cannot easily discern values from an array containing a number as large as ten, unless they have training. Five is preferable for cartographic purposes.

In art, value is controlled by the addition of white or black pigment to a hue. If white is added to a hue, a **tint** results. When black is added, a **shade** is produced. A **tone** results from adding equal amounts of a hue, white, and black.[8]

Controlling value on printed color maps is difficult. One suggested method is discussed below, but it will be better understood if the chroma dimension is first explained.

Chroma

Chroma is also called *saturation, intensity,* or *purity.* This color dimension can best be understood by comparing a color to a neutral gray. With the addition of more and more pigment of a color, it will begin to appear less and less gray, finally achieving a full saturation or brilliance. For any given hue, chroma varies from 0 percent (neutral gray) to 100 percent (maximum color). At the maximum level, the color appears pure and contains no gray. Chroma levels vary with hues; the most intense yellow appears brighter than the most intense blue-green. Achromatic colors are said to have zero chroma.

Color Interaction

When we look around us we see a world of many colors, patches of color surrounded by other colors, and colors adjacent to other colors. In many cases, maps are composed of more than one color. Reaction to a color patch is always modified by its environment. Seldom is the reader asked to look at a single color—even simple color maps usually contain variations of one hue, combined with black. There are at least three **color interactions**—simultaneous contrast, successive contrast, and color constancy—that must be considered in cartographic design.

Simultaneous Contrast

When the eye spontaneously produces the complementary color (opposite on the color wheel) of a viewed color, the effect is called **simultaneous contrast.**[9] A color, if surrounded by another color, begins to appear tinged by the complementary color of the surrounding color. A simple experiment demonstrates this effect. Place a small gray patch inside a box of pure color. After you gaze at the surrounding color for a brief period, the gray square begins to look like the surrounding color's complementary color.

Simultaneous contrast (Figure 16.5) causes adjacent colors to be lighter in the direction of the darker adjacent colors, and darker in the direction of the lighter colors. This is sometimes referred to as induction. Induction is particularly bothersome in map design when several different values of the same color are juxtaposed on the map or in the legend. The effects can be reduced by separating color areas by white or black outlines.[10]

An interesting feature of induction, or any simultaneous contrast, is that it cannot be photographed; it is produced by the eye alone.

Successive Contrast

Successive contrast results when a color is viewed in one environment and then in another in succession. The color will be modified relative to these new surroundings. It may appear darker or lighter, more or less brilliant compared to each new environment. For example, a gray will look darker against a lighter background and lighter against a darker background. An orange will appear more red and darker on a yellow background and lighter and more yellow on a red background. New contrasts are therefore established in each new situation. Cartographers cannot control all color environments, but positive design attempts to reduce the number of successive contrasts on the map.

Color Constancy

Although it is not of central concern to the map designer, **color constancy** is a feature of color perception that should at least be noted. We tend to judge colors based on presumed illumination. For example, the shadow areas of the folds of a red drape are gray rather than red, but because we assume the drapes to be lit by a common source (e.g., sunlight) we perceive these gray areas to be red also. This kind of judgment is made during map reading when color areas fall in shadows. Although the shadows cause actual color areas to change, we perceive them in constancy. If it were not for this, reading color maps would be almost impossible.

Subjective Reactions to Color

Human beings react to color for a number of reasons. Simultaneous and successive contrast are caused by our physiological system (eye-brain); most people have nearly identical response mechanisms to these. More variable, and much more difficult to control in design, are psychological reactions to color: color preferences, the meanings of color, and behavioral moods and color. Taken together, these are called the **subjective reactions** to color.

Color Preferences

There has been an enormous amount of research dealing with color preferences. Most of the formal research has come from the fields of psychology and advertising. Little research has been conducted by cartographers on color preference, so our understanding must be borrowed from the other disciplines.

There is abundant evidence that color preference is somewhat developmental. We must first define terms. **Warm colors** are those of the longer wavelengths (red, orange, and yellow), and **cool colors,** at the opposite end of the spectrum, have shorter wavelengths (violet, blue, and green). Warm and cool, of course, are psychological descriptions of these color ranges. Children about the age of four or five years prefer warm colors. Red is most popular, with blues and greens next.[11] Young children also prefer highly saturated colors, but this preference begins to drop off after about the sixth grade.

The findings regarding children's color choices have prompted one cartographic researcher to comment on color design for children's maps:[12]

1. As children are only aware of small ranges in hue, colors should be chosen from within the basic spectrum colors—blue, green, yellow, orange, and red.
2. Because school-age children begin to reject fully saturated colors, a step or two down the saturation range is more desirable.
3. Children appear to dislike dull unattractive colors, so color choice should avoid these. Stay close to the spectral hues.
4. Children generally reject achromatic color schemes—the gray scale. These reduce the attractiveness of the map.
5. Choosing colors that have greater compatability with the expected yields greater comprehension. Thus, for example, blue is better than red for water.

As we mature and leave childhood, our color preferences change. Generally, we tend to favor colors at the shorter wavelengths. Color preferences among North American adults are blue, red, green, violet, orange, and yellow, in that order. The greenish-yellow hues are the least liked by both men and women. Women show a slight preference for red over blue and yellow over orange, whereas men slightly prefer blue over red and orange over yellow. Both sexes choose saturated colors over unsaturated ones.

Although color preference yields some insight into the broader realms of color psychology, a simple list of preferences may not be enough. Color preference tests performed by psychologists are conducted for the sake of color choice only and are not product-related. Most color experts would agree that other variables, such as color environment, product name, packaging, context, and merchandising schemes, are also important in color choice.[13]

Parallels may be drawn for color use in cartographic design. Color conventions should overrule broader color preferences; cartographic design should otherwise seek to follow the preferences of the marketplace. Until detailed studies are provided, map designers need to follow the leads provided by merchandising, art, and related psychological literature.

Colors in combination. In one important study,[14] ten different hues, each with three different values and chromas, were selected as object colors. A random sample of 25 background colors was selected, and the subjects were asked color combination preferences. The general findings were these (see Table 16.1):

1. The most pleasant combinations result from large differences in lightness. Lightness (or value) contrast is necessary in pleasant object-background combinations.
2. A good background color must be either light or dark; being intermediate in lightness makes it poor.
3. Consistently pleasant object colors are hues in the green to blue range, or other hues containing little gray.
4. Consistently unpleasant object colors are in the yellow to yellowish-green range, or other hues containing considerable gray.
5. To be pleasant, an object color must stand out from its background color by being definitely lighter or darker. This is the single most important finding of the study for the cartographer.
6. Vivid colors combined with grayish colors tend to be judged as pleasant.
7. Good and poor combinations are found for all sizes of hue differences (that is, distances apart on the color wheel).

These findings were not the result of a cartographic inquiry, so they can be used only as general guidelines. Nonetheless, they can serve the cartographer as a starting point for color selection.

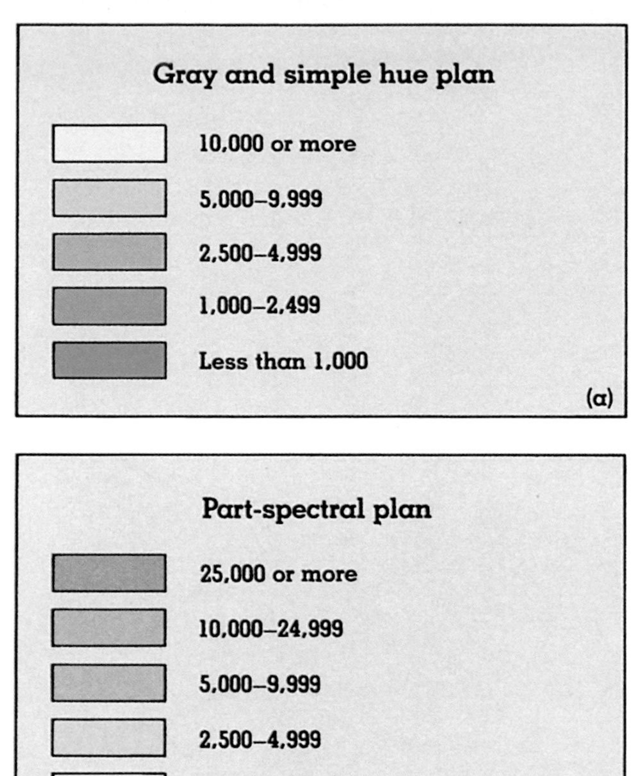

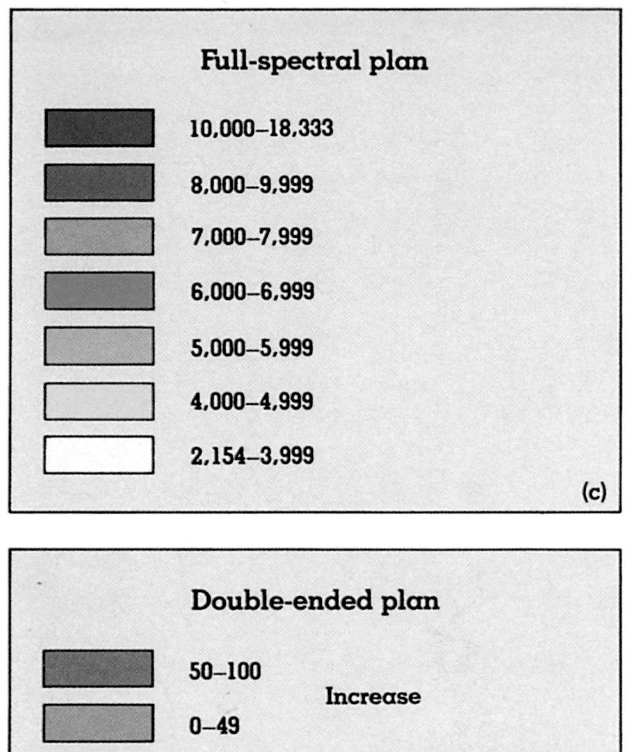

Plate 1 Different quantitative color plans.
See text for explanation. The color plan for (b) is used on Bureau of the Census GE-50 Series map number 56 (Median Family Income for 1969); the color plan for (c) is used on GE-50 Series map number 60 (Occupied Housing Units, 1970).

Plate 2 Color and figure-ground development.
Effective use of hue differences can enhance the figure-ground relationship on maps. In the example shown, the proportional symbols in (b) are clearly more visually dominant than those in (a).

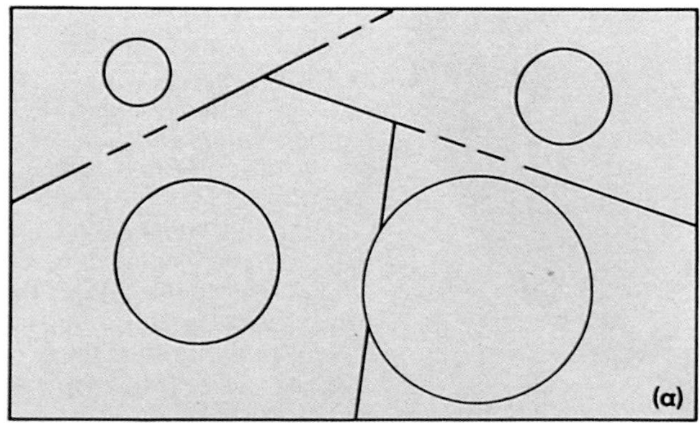

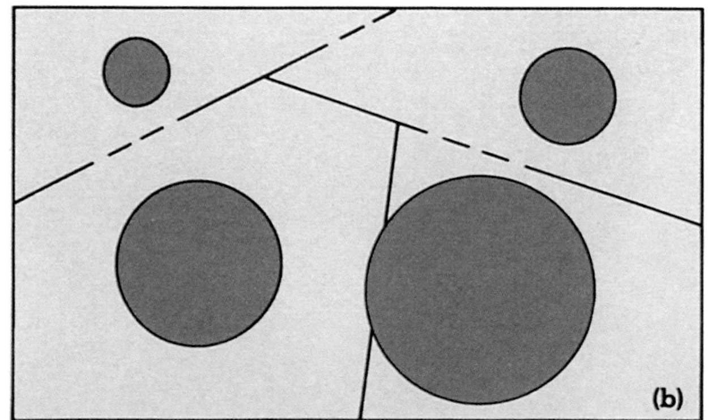

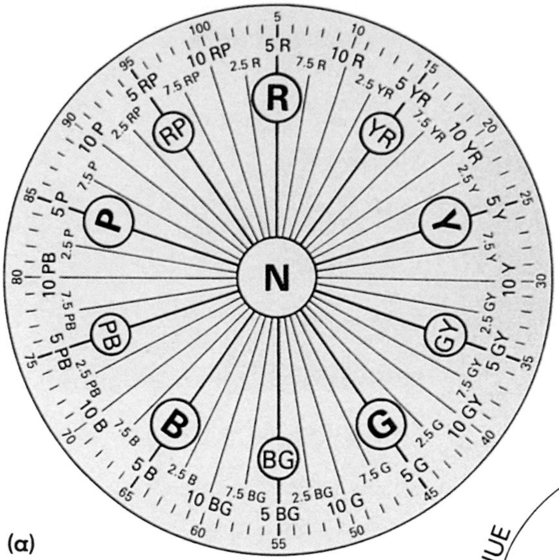

Plate 3 The Munsell color solid.

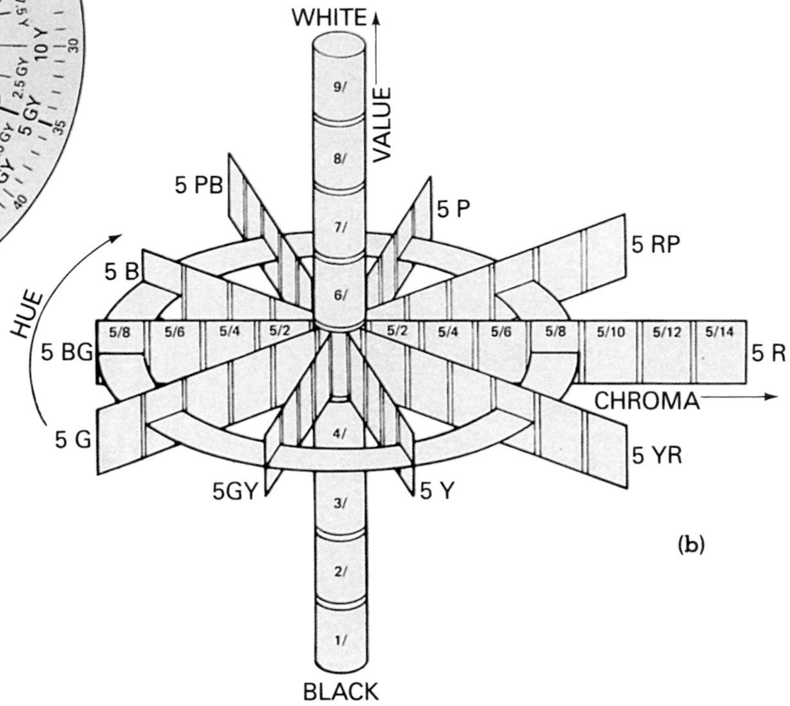

The hue symbols and their relation to one another on the color wheel are represented in (a). Hue, value, and chroma dimensions of the solid are shown in (b). A photograph of a Munsell solid is represented in (c). Courtesy Munsell Color, 2441 N. Calvert St., Baltimore, MD 21218.

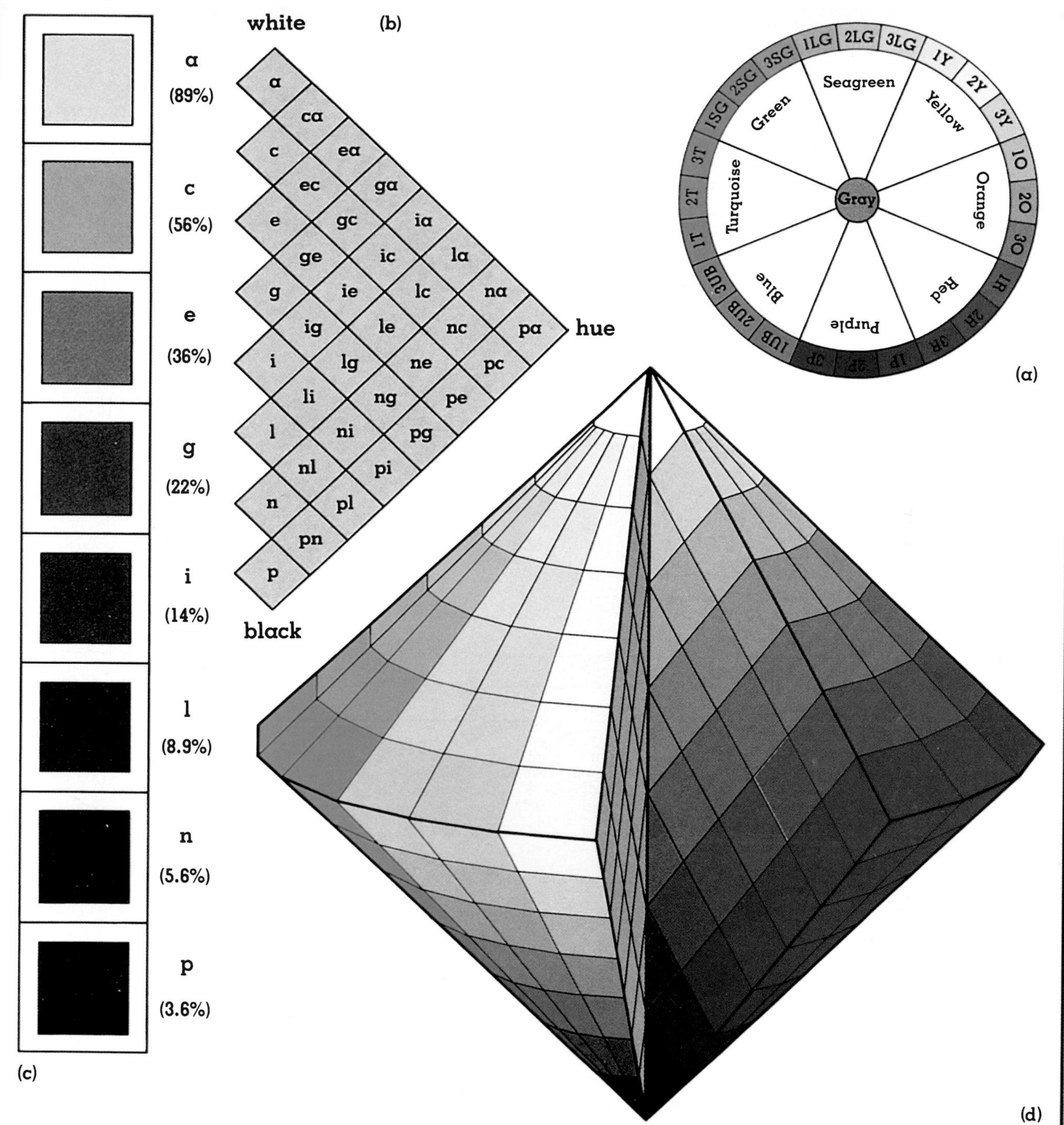

Plate 4 Features of the Ostwald color solid.
The hue circle (a) contains 24 hues. A triangle of the solid (b) contains an arrangement of hues (tints and tones) caused by mixing different amounts of white and black to each hue. The equal value gray scale used by Ostwald determined the appearance of the tints and tones (c). Figures in parentheses show white content; black content in remainder equals 100 percent. An approximation of a completed Ostwald cone is represented in (d). The screen tint percentages used here to represent the solid yellow, magenta, and cyan are modelled after those used by Allan Brown in "A New ITC Colour Chart Based on the Ostwald Colour System," ITC Journal *(1982) 2:109–118. Used with permission.*

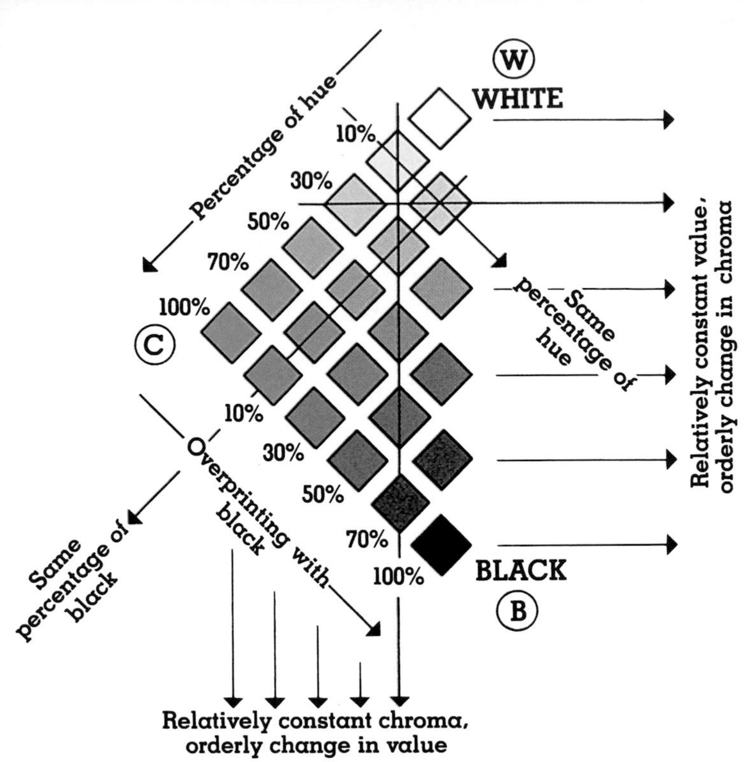

Plate 5 Color scaling for quantitative maps based on the Ostwald color triangle.
The percentages shown along the hue and black sides yield perceptual steps of approximate equality. A sequence for scaling may be selected from any vertical, horizontal or diagonal row, depending on the designer's choice of varying value (select a vertical row), chroma (select a horizontal row), or both (select a diagonal row). No absolute rules exist to assist in this selection. This diagram is based on Henry W. Castner, "Printed Color Charts: Some Thoughts on Their Construction and Use in Map Design," Proceedings (Annual Meeting of the ACSM-ASP, St. Louis, 1980), p. 378.

Plate 6 Color clarifies map elements.
The red overprint symbol on this map highlights the Appalachian Trail and generally makes the map more legible. Green forest land and blue water features further distinguish these map elements. This map is reproduced from the "Appalachian Trail," a brochure produced for the Chattahoochee National Forest by the Southern Region Forest Service of the United States Department of Agriculture, 1971.

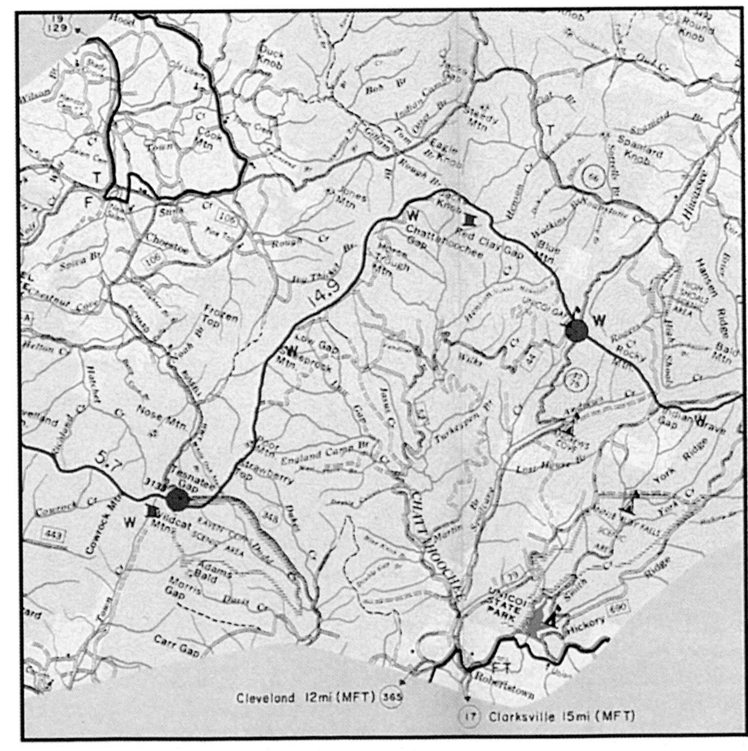

Table 16.1 Object-Background Preferred Color Combinations

Object Hues			Highest Ranked Favorable Background Hues*			Highest Ranked Unfavorable Background Hues**		
Munsell Notation		ISCC-NBS Color Names***	Munsell Notation		ISCC-NBS Color Names	Munsell Notation		ISCC-NBS Color Names
5R	5/8	Moderate red	5.0G	2/2	Very dark green	10.0RP	4/12	Vivid purplish red
5YR	5/8	Brownish orange	5.0G	2/2	Very dark green	10.0RP	4/12	Vivid purplish red
5Y	5/6	Light olive	5.0G	2/2	Very dark green	5.0R	4/14	Vivid red
5GY	5/8	Strong yellow green	5.0G	2/2	Very dark green	5.0G	5/4	Moderate green
5G	5/6	Moderate green	5.0R	8/2	Pale to grayish pink	2.5PB	5/4	Grayish blue
5BG	5/6	Moderate bluish green	N	10/	Black	5.0G	5/4	Moderate green
5B	5/6	Moderate greenish blue	5.0Y	9/12	Vivid yellow	5.0G	5/4	Moderate green
5PB	8/2	Very pale blue	5.0R	2/2	Dark grayish red-very dark red	5.0GY	7/10	Strong yellow green
5P	5/2	Grayish purple	N	10/	Black	5.0Y	5/4	Light olive
5RP	5/10	Moderate purplish red	N	10/	Black	2.5YR	5/14	Vivid orange
N	5/	Medium gray	5.0R	8/2	Pale pink	2.5PB	5/2	Grayish blue

*The background color shown for each object color represents only the top-ranked of 5 favorable ones selected by test subjects.

**The background color shown for each object color represents only the top-ranked of 5 unfavorable ones selected by test subjects.

***ISCC-NBS (Inter-Society Color Council—National Bureau of Standards.) Color names are identified for each Munsell Color Notation.

Source: H. Helson and T. Lansford, "The Role of Spectral Energy of Source and Background Color in the Pleasantness of Object Colors," *Applied Optics* 9 (1970): 1513–1562; color names from National Bureau of Standards, *The ISCC-NBS Method of Designating Colors and a Dictionary of Color Names* (Circular 553—Washington, D.C.: USGPO, 1955).

Connotative Meaning and Color

Perhaps the most interesting design aspect of color is that people react differently to spectral energies. Two people look at the same red wavelength, but their responses may be entirely different. This presents the cartographic designer with a considerable challenge, so the subjective and connotative meanings of color should be studied.

Connotative responses to colors vary considerably. The literature, both in psychology and advertising, is sometimes vague and contradictory. Nonetheless, some generalization may be made. (See Table 16.2.) Psychologists suggest that reds, yellows, and oranges are usually associated with excitement, stimulation, and aggression; blues and greens with calm, security, and peace; black, browns, and grays with melancholy, sadness, and depression; yellow with cheer, gaiety, and fun; and purple with dignity, royalty, and sadness.[15] Advertising people are quick to point out that color context and copy also suggest the connotative meaning of color.

The implications of this in cartographic design are not altogether clear. There is a paucity of cartographic research into color meaning and map design. Until adequate research can lead the way, it appears that we must borrow from the psychologists and advertising people. In general, the map designer should

Table 16.2 Connotative Meanings and Color

Color	Connotations
Yellow	Cheerfulness, dishonesty, youth, light, hate, cowardice, joyousness, optimism, spring, brightness; strong yellow—warning
Red	Action, life, blood, fire, heat, passion, danger, power, loyalty, bravery, anger, excitement; strong red—warning
Blue	Coldness, serenity, depression, melancholy, truth, purity, formality, depth, restraint; deep blue—silence, loneliness
Orange	Harvest, fall, middle life, tastiness, abundance, fire, attention, action; strong orange—warning
Reddish browns, russets, and ochres (earth colors)	Warmth, cheer, deep worth, and elemental root qualities; can be friendly, cozy or dull, reassuring or depressing
Green	Immaturity, youth, spring, nature, envy, greed, jealousy, cheapness, ignorance, peace; mid-greens—subdued
Purple (violet)	Dignity, royalty, sorrow, despair, richness; maroon—elegant, and painful
White	Cleanliness, faith, purity, sickness
Black	Mystery, strength, mourning, heaviness
Grays	Quiet and reserved, controlled emotions; can be used to create sophisticated atmosphere

Source: James E. Littlefield and C. A. Kirkpatrick, *Advertising: Mass Communication in Marketing* (Boston: Houghton Mifflin, 1970), p. 204; Al Hackl, "Hidden Meanings of Color," *Graphic Arts Buyer 12* (1981): 41–44; and Marshall Editions, *Color* (New York: Viking Press, 1980), pp. 138–41.

attempt to select colors by carefully taking into account their connotative meanings in association with the map's message.

Advancing and Retreating Colors

The eye of the reader seems to perceive colors as **advancing** or **retreating.** The colors with the longer wavelengths, notably red, appear closer to the viewer when seen along with a color of shorter wavelength. There is some evidence of a physiological basis for this claim. The lens of the eye "bulges" when it refracts red rays. This same convex shaping of the lens takes place when we view objects close up. For this reason, red may have come to be associated with proximity. In general, warm hues (long wavelengths) advance, cools recede. For brightness or value, high values advance, and low values recede. In terms of saturation, deep or highly saturated colors advance, and less saturated colors recede.

There is at least one case in which this aspect of colors has an effect on color selection. The development of figures and grounds on maps can be enhanced by choosing colors with their advancing and retreating characteristics in mind. Advancing colors should be applied to figural objects.

Color Specification Systems

Color specification—the exact naming of colors—has been examined by many artists, scientists, and others ever since Newton showed that color could be generated by passing light through a prism. Yet because of the large number of colors available to the senses, it was not easy to agree upon such a naming system. It was not until the early part of this century that standards began to appear. The Munsell system, named after its American inventor, Albert H. Munsell (1858–1918), has become the standard in the United States. Wilhelm Ostwald (1853–1932), a Latvian by birth, invented a similar system that has become commonly used throughout Europe.

Two other systems, the ISCC-NBS (the Inter-Society Color Council and the National Bureau of Standards), and the CIE (Commission International de l'Eclairage, or the International Commission on Illumination) have evolved since the work of Munsell and Ostwald. ISCC-NBS uses a unique naming system applied to the Munsell color solid, and CIE specifications are based on the physical spectrometry of reflected light.

Cartographers, especially researchers, are called on more and more to be specific in their color designations for papers and inks used in printing. Research into color CRT display in mapping is going on also. Understanding the terminology of color specification has therefore become increasingly important to map designers as they attempt to communicate about color. The value of these systems, especially Munsell's and Ostwald's, is in the way they help us to conceptualize interrelationships of the three dimensions of hue, value, and intensity of colors. Practical color specification deals mainly with ink selection.

Munsell Color Solid

Munsell recognized the three dimensions of color to be hue, value, and chroma (or saturation) and incorporated these into the **Munsell color solid.**[16] A color solid is a representation of **color space,** a way of integrating and showing the relationships of the three color dimensions. (See Plate 3.) The solid is only roughly a sphere, because of the unevenness of chroma.

The color solid of Munsell, and the quantitative specifications derived from it, is based on the whole array of colors available for human sensation. Steps or divisions between colors are based on psychological intervals, measured through extensive testing. New colors may be added as they are developed in science and industry. This is a principal advantage of this color solid.

Ten hues make up the color band around the solid; each hue has a letter designation. Ten number divisions are associated with each hue—the "5" position designating the pure color of each hue. There are therefore 100 different hues on the color wheel; for example, 7R, 2Y, and 8B each specify a particular hue on the circle.

Psychophysical testing has determined that people do not perceive color value in linear fashion. Munsell's value scale, called an *equal-value scale,* records the reflectance percentages of what appear to be equal steps from white to black. Value on the solid is represented along a vertical scale (the axis) of the color solid, perpendicular to the plane of the color wheel. The top of this value scale is white (10) and the bottom position is black (0). Between these two "poles" are nine divisions, graduating from white to black, that yield different grays. The middle gray (5) is neutral. Notation of value is by the position on this vertical scale, followed by a slash (7/, 2/, etc.).

With these two dimensions, dark red or middle red, for example, can be exactly specified. A designation of 10BG2/ would indicate a blue-green closer to blue and dark (nearly black) in value.

Munsell also incorporated chroma designation in his color solid. Chroma or saturation is represented by steps going inward toward the value axis for each hue band. The chroma scale is also psychologically stepped. As a color gets closer to the axis, it becomes weaker (more gray); conversely the farther out on the chroma scale a color is, the stronger and more saturated it becomes. Chroma designation is a number that follows the value slash, such as

/7. A complete Munsell color designation would thus be, for example, 8p 7/5.

Maximum chroma differ for the variety of hues and values. For example, for 5R, maximum chroma at a value of 8/ is 4, but for a value of 4/ it is 14. For 5PB, maximum chroma at value 8/ is 2, and for 2/ it is 6. These give the color solid an uneven and unbalanced appearance, causing it to deviate significantly from a perfect sphere. (See Plate 1.)

Ostwald Color Solid

Like the model of Munsell, the **Ostwald color solid** displays an orderly arrangement of the colors humans can sense. The Ostwald solid differs in several important respects, however. It is shaped like a double cone, with the bases touching. (See Plate 4.) The color wheel which forms the outer ring or circumference of the cones, unlike Munsell's, is composed of 24 separate hues, 3 in each of 8 primary hues. Ostwald's solid does not recognize the dimensions of hue, value, and chroma, but achieves distinct colors by hue, white, and black.[17]

The color solid is formed by the introduction of a monochromatic triangle of 28 *tones* (distinct colors achieved by adding percentages of white and black to a given hue). A neutral gray scale becomes the axis of the solid and is the innermost tier of tones in a given hue's triangle. (See Plate 4.) The neutral gray scale is a psychological gray scale—its divisions or percentages of gray are based on equally appearing amounts of gray. This scale differs from Munsell's.

A whole range of *tints* (hue plus white), *shades* (hues plus black) and *tones* (hues plus white and black) are achieved simply on the Ostwald solid. Specifications follow letters that indicate certain percentages of white and black, and hue is designated by its number from the color wheel. For example, 8 *pa* designates a relatively pure red (*p* indicates an absence of black and *a* an absence of white). A soft, grayish red would be 8 *le.*

Ostwald's color solid has one major limitation: its inflexibility. A new color having stronger brilliance (because of the invention of new dyes and colorants) cannot be added without changing the whole solid. The Munsell solid is not limited in this way, since new chromas may be added at the peripheries as they are introduced.

The Ostwald system of color specification finds its greatest utility among printers, artists, ink manufacturers, and others who develop colors by the mixing of colorants, white, and black pigments. Cartographers likewise find this system appealing, as explained below.

The ISCC-NBS Method

One interesting approach to color designation deserves comment. The ISCC (Inter-Society Color Council) and the NBS (National Bureau of Standards) color name charts were originally devised as color standards to be used by the United States Pharmacopoeia.[18] In this system, all hues are preceded by modifiers that describe the color. (See Figure 16.6.) Such adjectives as pale, light, brilliant, very pale, weak, or dusky are used to designate such colors. The adjectives are used throughout all hues, but their locations vary relative to the value and chroma scales of the Munsell system, to which they are matched.

CIE Color Specification

The CIE system of color designation describes the physical properties of color and has become an international standard for this kind of color description. This system is based on additive color mixing, generated by reflecting red, green, and blue colored lights from a white screen in an attempt to match the color of a test lamp.[19] These colored lights are considered *primary* lights, and the amounts of the lights necessary to produce a test color are called the test color's **tristimulus** values. Because this system also considers the reflective character-

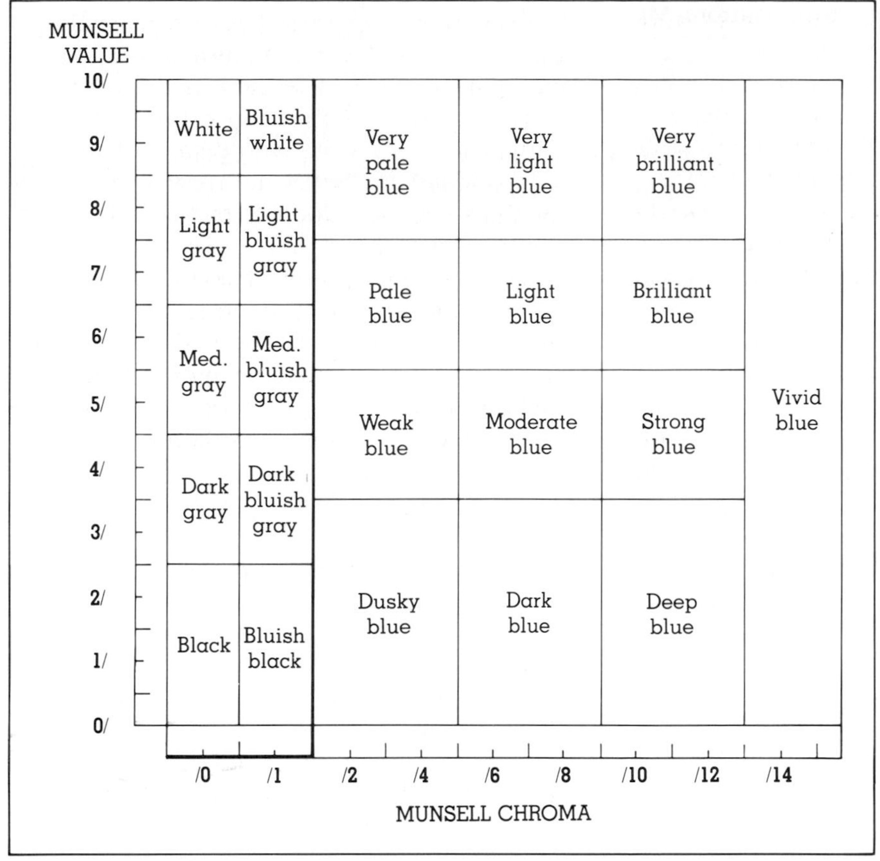

Figure 16.6. An example of the ISCC-NBS color specification system.
Each color is described by adjective identifiers and related to the chroma and value scales of the Munsell solid.

istics of the surface and the quality of the source(s) of illumination, color specification becomes quite complex. Sophisticated recording machines are required, such as a spectrophotometer that measures reflectance from a colored surface at different wavelengths.

Color charts, or physical samples such as color chips, are not produced by the CIE system. For this reason, the practicing cartographer will find little direct utility in this color system. The cartographic researcher pursuing objectives in color investigation, however, will find it useful because of its specificity in describing the physical attributes of colors.

Printing Color Specifications

The foregoing systems of color specification have considerable merit in introducing the idea of color space, and how the arrangement of the dimension of color may be shown by geometrical solid figures. In practice, the cartographer does not specify color with these systems but uses a *color matching system* adopted more or less through consensus by the printing industry. One such system, the *Pantone Matching System,* has become standard today, so much so that cartographers dealing in color should become familiar with it.

Table 16.3 Sample Colors from Pantone Mixing Guide

Color	Mixture	Percentage
Color: Pantone 165C	8 parts Pantone yellow	50.0%
	8 parts Pantone warm red	50.0%
		(Ratio 1:1)
Color: Pantone 164C	4 parts Pantone yellow	25.0%
	4 parts Pantone warm red	25.0%
	8 parts Pantone white	50.0%
		(Ratio 1:1:2)
Color: Pantone 168C	8 parts Pantone yellow	40.0%
	8 parts Pantone warm red	40.0%
	4 parts Pantone black	20.0%
		(Ratio 2:2:1)
Color: Pantone 163C	2 parts Pantone yellow	12.5%
	2 parts Pantone warm red	12.5%
	12 parts Pantone white	75.0%
		(Ratio 1:1:6)

Source: James P. DeLuca, *Pantone Matching System and Pantone Matching System Formula Guide* (Moonachie, N.J.: Pantone, Inc., 1980), p.4.

The Pantone Matching System is a registered trademark of PANTONE, Inc., Moonachie, N.J. 07074.

Eight basic colors, plus black and white, form the elementary colors of the PMS: yellow, warm red, rubine red, rhodamine red, purple, process blue, reflex blue, and green. By specifying different mixtures of these colors, over 500 different printing ink hues are possible. The cartographer selects a color from color charts produced by Pantone and provides the printer with the number of the color. The printer mixes the ink by using a *Formula Guide* that shows the percentages of the elementary colors needed to achieve the desired color. (See Table 16.3.) Mixing is by weight. Four-color process printing is done with Pantone process colors.

A variety of helpful materials are provided by the manufacturer: a *Color Tint Selector* (which shows the printed effects of a variety of screens on nearly 500 Pantone Colors), a *Color and Black Selector* (which shows the effects of mixing screened black with about 81 different colors), and a *Four-Color Process Guide* (which shows the different hues achievable by four-color process printing with Pantone process inks). There are other products not mentioned here. Some computer desktop publishing software now comes equipped with color palettes in which screen colors can be specified in Pantone numbers.

Although there are other color-matching systems available, most printers use the PMS. Other manufacturers are licensed by Pantone to produce products matched to its colors. Examples include self-adhesive overlay sheets, paper, colored markers, and the like. These materials can assist the cartographer in visualizing the final color map during various stages of production.

Color specification is possible by using such a system as Pantone for ordinary color use in cartography. Since the colorants used are relatively standard (the eight elementary colors) and mixing guides standard, mixed hues are reasonably consistent. Variability comes from printing presses, impressions, and paper. These sources of inconsistency will probably always plague the cartographer, making exactly accurate color matches difficult.

Color in Cartographic Design

Through the use of a variety of design strategies several functional uses of color can be achieved on the map, as discussed in this final section of our introduction of the use of color in cartographic design.

The Functions of Color in Design

Arthur Robinson of the University of Wisconsin–Madison has written succinctly on the various functions of color in mapping,[20] summarized as follows:

1. Color functions as a *simplifying* and *clarifying* agent. In this regard, the use of color can be useful in the development of figure and ground organization on the map. Color can unify various map elements to serve the total organization of the planned communication. (See Plate 6.)
2. Color affects the general perceptibility of the map. Legibility, visual acuity, and clarity (of distinctiveness and difference) are especially important functional results of the use of color.
3. Color elicits subjective reactions to the map. People respond to colors, especially the hue dimension, with connotative and subjective overtones. Moods can be created with the use of many colors.

Thus, *structure, readability,* and the reader's *psychological reactions* can be affected by the use of color. It might be added that color functions to clarify thematic symbolization.

Lessons learned by the advertising industry regarding the function of color in printed media are also instructive for the cartographic designer. The use of color advertisements is not hit-or-miss, as it once was. Marketing studies have shown that color is an effective attention-getter and mood-setter and can be used as a strong product identifier. Some advertising experts estimate that readership can be increased by 15 percent by using color.[21] The goals of thematic mapping are vastly different from those of most print advertising, but the two disciplines do share common ground in their ways of using color.

The functions of color in print advertising have been listed as follows:[22]

1. To attract buyer's attention
2. To stimulate interest
3. To identify products
4. To relate successive advertising
5. To provide emphasis
6. To illustrate, interpret, and prove
7. To contribute to structural motion; lead the reader's attention
8. To embody prestige and a certain mood or atmosphere
9. To relate components in the ad
10. To show structure, design, and installation of product

The functions of color in mapping are indeed similar to these.

Design Strategies for the Use of Color

Map designers use several strategies to use color to its fullest potential in map communication. Five will be treated here, with figure and ground development first.

Developing Figures and Grounds

The organization of the map into figures and grounds can be enhanced by the use of color. Color provides contrast—a necessary component in figure formation. Perceptual grouping is also strengthened by the use of color, notably by similarity. For example, similar hues are grouped in perception (although they may in fact be of different wavelengths). On a world map, for example, continents are more easily grouped as land masses if they are rendered in similar hues. Colors of similar brightness or

Table 16.4 Color Combinations Useful in Developing Figure and Ground Organization on the Thematic Map

Figure Colors		Ground Colors
Yellow	Best	Black
White		Blue
Black		Orange
Black		Yellow
Orange		Black
Black		White
White		Red
Red		Yellow
Green		White
Orange		White
Red	Worst	Green

Source: Deborah Sharpe, *The Psychology of Color and Design* (Chicago: Nelson-Hall, 1974), p. 107.

dullness are also grouped, as are warm colors, cool colors, or other like tints and shades.[23] Tints and shades are grouped with their primary color. Perceptual grouping of colors is a strong tendency and should be a positive design element.

Generally, warm colors (reds, oranges, and yellows) tend to take on figural qualities better than cool colors (greens, blues, and purples), which tend to make good grounds. (The map "Per Capita Retail Sales: 1963" on page 221 of *The National Atlas of the United States of America* is a good example of the use of colors to bring out figures and grounds.) This may be partially explained by the tendency of the warm colors to advance and the cool colors to recede.

The **power factor** concept can be applied in the context of developing figures and grounds. The power factor of a color is defined as the product of its Munsell value and chroma numbers.[24] Small areas of higher power factor are more likely to be seen as figures.

Color combinations also affect figure and ground development. (See Plate 2 and Table 16.4.) Yellow on black is noted to be the most visible color combination—yellow tends to be perceived as figure. The least visible and therefore worst combination, in terms of figure and ground development, is red on green. An example of the weakness of red on green in developing figures and grounds can be found on page 161 of *The National Atlas of the United States of America.* The colors listed in Table 16.4 are to be used only as starting points for selection, since any modification of chroma and value will affect the results. The cartographic designer needs to balance other design elements with these color combinations to achieve an overall solution.

The Use of Color Contrast

Contrast is the most important design element in thematic mapping. Contrast in the employment of color can lead to clarity, legibility, and better figure-ground development. A map rendered in color with little contrast is dull and lifeless and does not demand attention. Even in black-and-white mapping (or one color other than black), contrasts of line, pattern, value, and size are possible. With color, additional possibilities exist. One colorist lists seven possible color contrasts:[25]

1. Contrast of hue
2. Contrast of value (light-dark)
3. Contrast of cold and warm colors
4. Complementary contrasts
5. Simultaneous contrast
6. Contrast of saturation (intensity or chroma)
7. Contrast of extension

Hue contrast can be used in cartographic design as a way to affect clarity and legibility and to generate different visual hierarchical levels in map structure. Some feel that hue is the most interesting dimension in color application in mapping, more so than value or chroma.[26] Contrast of hue demands an overall color plan for the map and requires that some thought be given to color balance and harmony, topics not very well researched by cartographers.

Saturation and value are two contrasts that provide visual interest and, depending on the nature of the map, can carry quantitative information. Contrast of value is a fundamental necessity in structuring the color map's visual field into figures and grounds. Objects that are high in value (relatively light) tend to emerge as figures, provided other components in the field do not impede figure formation. It is difficult to talk of saturation and value in isolation; they are usually a part of any hue selection. Both are essential in the design of quantitative maps.

Contrast of cold and warm colors can also be used to enhance figure and ground formation on the map. Artists use this contrast to achieve the impression of distance; faraway objects are rendered in cold (blue/green) colors and nearby objects in warmer tones. Figures on maps should be rendered in colors of the warmer wavelengths, relative to the ground hues.

In most mapping cases, the designer should avoid simultaneous contrast, because this effect is visually troublesome and distracting. Simultaneous contrast can be minimized, if not eliminated altogether, by separating color areas with black or white lines.

Many artists believe that providing complementary colors in a composition establishes stability. Complementary colors are opposite on the color wheel. The primary complementaries are these: yellow-violet, blue-orange, and red-green. The eye will spontaneously produce a complementary of a color fixated on if it is not present. Because of this, it is said that harmony can be achieved in a composition by a balance of complementary colors. The eye does not then need to produce complementaries on its own, so the image is more stable. The contrast of complementaries appears to be a feature of color use that needs further investigation.

Contrast of extension relates to the much broader topic of color balance and harmony, dealt with in more detail below.

Developing Legibility

The legibility of colored objects, especially lettering, is greatly influenced by their colored surroundings. Symbols in color must be placed on color backgrounds that do not affect their legibility. Black lettering on yellow (object on background) is the most legible combination, and green lettering on red is the least. (See Table 16.5.) The difficulty is exacerbated because the lettering is usually spread over several different background colors. Black lettering may become illegible as it crosses dull or gray colors. The designer must pay careful attention to lettering and all color environments in which it is placed. An otherwise good design can fail if caution is not exercised in this respect.

Table 16.5 Legibility of Colored Lettering on Colored Backgrounds

Color Combination (object on background)	
Black on yellow	Most legible
Green on white	
Blue on white	
White on blue	
Black on white	
Yellow on black	
White on red	
White on orange	
White on black	
Red on yellow	
Green on red	
Red on green	Least legible

Source: Al Hackl, "Hidden Meanings of Color," *Graphic Arts Buyer* 12 (1981): 41–44.

Color Conventions in Mapping

Conventional uses of color in mapping may be separated into qualitative and quantitative conventions. On most color thematic maps, these conventions must be observed because breaking a convention can be extremely disconcerting to the map reader.

Qualitative conventions. Colors used on maps in a qualitative manner are those applied to lines, areas, or symbols that show kind or quality, not amount. Qualitative color conventions use color hue and chroma to show nominal (and in some cases ordinal) classifications. Many conventions are quite old—such as showing water areas as blue—and the logic of their use well established.

Arthur Robinson has itemized many of the conventional uses, as follows:[27]

1. blue for water
2. red with warm and blue with cool temperature, as in climatic and ocean representations
3. yellow and tans for dry and little vegetation
4. brown for land surfaces (representation of uplands and contours)
5. green for lush and thick vegetation

The color dimension most logically used on the qualitative map is hue. Hue shows nominal classification well and is especially appropriate because hue is difficult to associate psychologically with varying amounts or quantity of data. Caution must be exercised in using contrast of value to show nominal characteristics, because people tend to assign quantitative meanings to value differences.

Quantitative conventions. Conventions associated with color use on quantitative thematic maps operate in terms of color choice or color plan. No conventions exist for color choice on quantitative maps (e.g., population density maps are always blue, income maps are always green, and so on). **Color plan** is the way the designer chooses to use the color dimensions of hue, value, and chroma to symbolize varying amounts of data on the map. (See Plate 1.) Four color plans have been identified:[28]

1. *Gray and simple-hue plan.* More of a single ink is applied to areas that represent greater amounts. This is achieved by *screening.* An excellent example of this method can be seen at the top of page 166 of *The National Atlas of the United States of America.*
2. *Part-spectral plan.* Colors adjacent on the color wheel show variations in amount: blue, green, and yellow (blue represents high amounts, yellow low); red, orange, and yellow (red represents high amounts, yellow low). Usually, the colors are produced by two inks, with the middle color achieved by overprinting.
3. *Full-spectral plan.* A separate hue is used to represent different amounts of data on the map. Red, orange, yellow, green, and blue may be used, with red usually chosen to represent the higher amounts. Colors are achieved by overprinting or by separate inks. This plan is the most commonly used for hypsometric layer tints on maps that show elevation.
4. *Double-ended plans.* In schemes that illustrate both positive and negative aspects on the same map, dark red at the positive end may grade to light red, and light red to light blue to dark blue at the negative end, or some other hue choices.

The color dimensions of value and chroma (saturation), operating singularly or together, are preferred in symbolizing quantitative data on maps. There appears to be general agreement among cartographers that the color dimension of value be used to symbolize an ordered array of data magnitudes, with lighter values for smaller magnitudes, and darker values for larger magnitudes. For further study the student can examine several of the GE-50 Series choropleth maps produced by the United States Bureau of the Census. (See Table 16.6.) Hue differences alone should not be used, but if they are, part-spectral schemes are preferred over the full-spectral ones.[29]

Three ways are customarily used to devise single-hue plans for quantitative maps in which gradations of value or chroma (or both) are to be achieved through screening or multiple screening. (See Figure 16.7.)

1. Gradation is achieved by screening solid ink with successively higher or lower screen values. For example, 10-, 25-, 50-, and 75-percent screened areas (with solid ink as the highest) may be used. Experimental results show that the higher the value level of the initial hue,

Table 16.6 Color Plans Used on Several GE-50 Series Choropleth Maps

GE-50 Series No.	Title	Color Plan (1)	Hues (2)	Number of Classes	Printing Method (3)
29	Size of Farms, 1964	part spectral	light beige to dark blue	7	A
42	Population Trends 1940–1970	double-ended	dark blue to dark red	6	B
6	Families with Incomes under $3000 in 1959	full spectral	red, brown, orange, yellow green, blue	6	C
38	Population Density, 1970	part spectral	beige to red	4	D
47	Number of Negro Persons, 1970	part spectral	light yellow to dark blue	6	A
56	Median Family Income for 1969	full spectral	yellow-beige, light green, green, blue, orange, red, purple	7	A

Notes:

(1) A part spectral plan may be considered an ordered hue array.

(2) In each case, the first color listed represents the lowest data magnitude, and the last color listed the highest data magnitude.

(3) Method A = 3 solid inks, and screened inks overprinted
Method B = 2 solid inks, each screened successively
Method C = 6 solid inks
Method D = 2 solid inks, and screened inks overprinted
No black screen tints are used in any of these color plans.

Source: U.S. Census Bureau, GE-50 Series Maps listed

the more the screened gradations vary in chroma than in value.[30]

2. Gradation is achieved by *overprinting* a solid color ink with successively higher (or lower) screened black ink. Thus, for example, 10-, 25-, 50-, and 75-percent black screens are overprinted on the selected color ink. In terms of sensation, the color's value or chroma (or both) are altered by this method.
3. The two above methods are combined.

Cartographic designers need guidelines in the production of such graded series. Fortunately a solution has been suggested by Professor Henry Castner using the Ostwald color

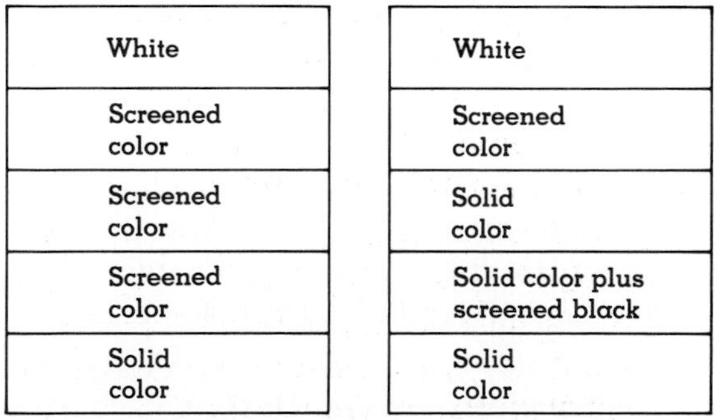

White
Screened color
Screened color
Screened color
Solid color

White
Screened color
Solid color
Solid color plus screened black
Solid color

Figure 16.7. Two commonly used systems of screening and overprinting to achieve a graded series in color on quantitative maps.

solid and the concepts of tint and shade.[31] The plan is patterned after a triangular slice from the Ostwald solid and shows the relationship of screened hues, overprinting black screens, and their effects on saturation and value. (See Plate 5.) Although not based on experimental psychological testing, this model is very helpful in devising the color plan for symbolization on quantitative maps.

Recent research by Mersey with color specification for choropleth maps yields important findings for the designer, and of significance is that the color plan should be selected based on the task placed before the map reader.[32] If the task is to extract specific information (that is, to look for an area containing a given data amount), the color plan should be based on hue variations, with few data classes. If the task is to gain overall knowledge of the spatial variation of the data mapped, the color plan should be the simple hue plan having regular value increments ordered to the data amounts (larger magnitudes are symbolized with a hue of darker value and smaller magnitudes with the same hue of lighter value). Map readers in this latter case appear to do equally well with choropleth maps having either a few data classes, or many. If the designer is developing a choropleth map that encompasses multiple purposes the best compromise is a hue-value plan in which regular value amounts are combined with a corresponding ordered hue array. Again, this appears to work equally well with choropleth maps containing a few or many data classes.

Little study has been conducted on color plan selection for other types of quantitative maps. One notable exception is one done by Lindenberg on the perception of color graduated symbols. His conclusions show that estimation of size differences of graduated circles are not affected significantly by different hues.[33]

Color Harmony in Map Design

Color harmony is traditionally felt to belong to the realm of the artist, therefore not a matter of concern to the cartographic designer. This is far from the truth, but this element of design has been overlooked in cartographic research. Some aspects of this important subject are introduced below.

Color harmony, as it applies to maps, is more than the suitable and pleasing relationship of hues. Color harmony relates to the overall color architecture for the entire map. Harmony includes these components:

1. Effectiveness of the *functional uses* of color on the map
2. Appropriateness of the *conventional uses* of color on the map
3. Overall appropriateness of *color selection* relative to map content
4. Effective use of the *quantitative color plan*
5. Effective employment of the *relationship of hues*.

Effectiveness of functional uses relates to how well the designer has employed color as a simplifying and clarifying agent. A harmonious design can be also judged on its appropriate use of color convention. Has every care been taken to provide conventional color wherever possible? Do departures from convention restrict the ease of communication?

Another important component of overall color harmony is the proper selection of color relative to map content. Maps illustrating January temperatures should not be rendered in warm hues, because of convention and connotative aspects, and deserts having sparse vegetation should not be shown in green.

The quantitative color plan (graded series of colors to show varying amounts) should be designed so that either color value or chroma differences correspond with numerical gradations. If hues are used in the graded series, the part-spectral scheme is the best choice. On maps of nominal data classes, symbols should be rendered in different hues.

Achievement of an effective relationship of hue balance of all colors on the map requires great skill and careful planning. Any multicolor map has its different colors occupying areas of varying sizes. **Color balance** is the result of an artful blending of colors, their dimensions, and their areas so that dominant colors occupying large areas do not overpower the remainder of the map. Dominant colors are those that contrast greatly with their environment; any color can be dominant, depending on its surrounding colors. *Contrast of extension* is another expression for color balance.

Brilliance (or light value) and extent (occupied area) are the two elements of **color force.**[34] This concept is similar to the power factor of a color, mentioned earlier. The light values of colors have been measured by colorists: yellow—9, orange—8, red—6, violet—3, blue—4, and green—6. For example, if a composition contains yellow and violet, the proportionality would be 9:3. To compute the areas and to maintain equal dominance, the reciprocal must be used. As yellow is three times stronger than violet, it should occupy one-third the space.

Cartographic designers have their hands full when dealing with balance. In most instances, areas are fixed by geography, so only choice of color is possible when planning balance. Consideration must be given to the other elements of color harmony as well. The designer needs experience with color and a knowledge of how colors behave in different environments to solve balance problems.

Color harmony, as artists and colorists use the term, refers to the pleasant combination of two or more colors in a composition. Harmony also refers to order—the arrangement of colors. According to Ostwald, the simpler the order, the better.[35] However, what is pleasant to one map reader may not be to another. Experimentation with the ideas of color harmony as developed by Munsell, Ostwald, Birren, Itten, and others would reveal differences.

In considering harmony, perceived color can be categorized into six compartments: pure hue, white, black, tint, shade, and tone. (See Figure 16.8.) White, black, and the three primaries produce the secondaries, tint, shade, tone:

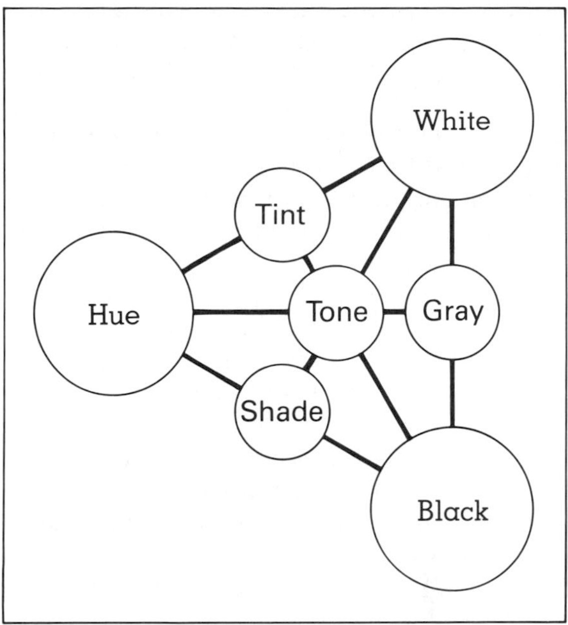

Figure 16.8. A color triangle devised by Faber Birren.
As sensation, all colors fall into one of these seven forms.
Source: Faber Birren, *Selling Color to People* (New York: University Books, 1956), p. 170.

hue + white = tint
hue + black = shade
hue + black + white = tone

The following thoughts regarding these relationships of harmony are derived from Birren:[36]

1. Most people like pure hues rather than modifications (reds, not purples).
2. Every color should be a good example in its category. If a red is chosen, it should not be possible to mistake it for a purple. Pure colors, if selected, must be brilliant and saturated.
3. Whites should be white, and blacks should be black.

4. Tints, shades, and tones should be easily seen as such, not confused with pure hues, blacks, or whites.

To these may be added the following general harmony rules:

5. Colors opposite on the color wheel tend to be harmonious.
6. Harmony can also be achieved by using colors adjacent on the color wheel.
7. Harmony can also be achieved by combining colors of the same hue but with different tints, shades, and tones.

The professional cartographer must explore in considerable depth the relationships of colors. The task is made more difficult because cartography is not a purely expressive art. Color selection and employment are constrained by the map's other elements.

Notes

1. Deborah Sharpe, *The Psychology of Color and Design* (Chicago: Nelson-Hall, 1974), p. 123.
2. Johannes Itten, *The Elements of Color* (New York: Van Nostrand Reinhold, 1970), p. 12.
3. Mari Riess Jones, "Color Coding," *Human Factors* 4 (1962): 355–65.
4. This experiment is retold in many books on color. A good version can be found in Fred W. Billmeyer, Jr., and Max Saltzman, *Principles of Color Technology* (New York: John Wiley, 1966), pp. 15–17.
5. Howard T. Fisher, "An Introduction to Color," in James M. Carpenter, *Color in Art* (Cambridge, MA: Fogg Art Museum, 1974), p. 25.
6. Judy M. Olson, "Spectrally Encoded Two-Variable Maps," *Annals* (Association of American Geographers) 71 (1981): 259–76.
7. Arthur H. Robinson, *The Look of Maps: An Examination of Cartographic Design* (Madison, WI: University of Wisconsin Press, 1966), p. 90.
8. Faber Birren, *Selling Color to People* (New York: University Books, 1956), p. 171.
9. Itten, *The Elements of Color,* p. 52.
10. Robinson, *The Look of Maps,* p. 94.
11. Sharpe, *The Psychology of Color and Design,* p. 18; see also Faber Birren, *Selling with Color* (New York: McGraw-Hill, 1945), p. 21.
12. Patrick Sorrell, "Map Design—With the Young in Mind," *Cartographic Journal* 11 (1974): 82–91.
13. Watson S. Dunn and Arnold M. Barbau, *Advertising: Its Role in Modern Marketing* (Hinsdale, IA: Dryden Press, 1978), pp. 429–32.
14. H. Helson and J. Lansford, "The Role of Spectral Energy of Source and Background Color in the Pleasantness of Object Colors," *Applied Optics* 9 (1970): 1513–1562; see also Deane B. Judd, "Choosing Pleasant Color Combinations," In *Contributions of Color Science by Deane B. Judd,* David L. MacAdams, ed. United States Department of Commerce, National Bureau of Standards—Washington, D.C.: USGPO, 1979.
15. Sharpe, *The Psychology of Color and Design,* p. 55.
16. Faber Birren, *Munsell: A Grammar of Color* (New York: Von Nostrand Reinhold, 1969), pp. 17–27.
17. Wilhelm Ostwald, *The Color Primer* (New York: Van Nostrand Reinhold, 1969), pp. 17–18.
18. United States Department of Commerce, National Bureau of Standards, *The ISCC-NBS Method of Designating Colors and a Dictionary of Color Names* (Circular 553—Washington, D.C.: USGPO, 1955), p. 1.
19. Fred W. Billmeyer, Jr., and Max Saltzman, *Principles of Color Technology* (New York: John Wiley, 1966), p. 31.
20. Arthur H. Robinson, "Psychological Aspects of Color in Cartography," *International Yearbook of Cartography* 7 (1967): 50–59; see also Arthur H. Robinson, *The Look of Maps,* pp. 75–97.
21. Jack Engel, *Advertising: The Process and Practice* (New York: McGraw-Hill, 1980), p. 503.
22. James E. Littlefield and C. A. Kirkpatrick, *Advertising: Mass Communication in Marketing* (Boston: Houghton Mifflin, 1970), p. 205.

23. Sharpe, *The Psychology of Color and Design*, p. 105.
24. A. Jon Kimerling, "Color in Map Design," paper delivered at the workshop on Map Perception and Design, ACSM Cartography Division, Annual Meetings, St. Louis, 1980.
25. Itten, *The Elements of Color*, pp. 33–64.
26. Robinson, "Psychological Aspects," pp. 50–59.
27. *Ibid.*
28. David J. Cuff, "The Magnitude Message: A Study of the Effectiveness of Color Sequences on Quantitative Maps," unpublished Ph.D. dissertation, Department of Geography, Pennsylvania State University, 1972, p. 21.
29. David J. Cuff, "Shading on Choropleth Maps: Some Suspicions Confirmed," *Proceedings* (Association of American Geographers, April, 1973): 50–54.
30. David J. Cuff, "Value Versus Chroma in Color Schemes on Quantitative Maps," *Canadian Cartographer* 9 (1972): 134–40.
31. Henry W. Castner, "Printed Color Charts: Some Thoughts on Their Construction and Use in Map Design," *Proceedings* (American Congress on Surveying and Mapping, March, 1980): 370–78.
32. Janet E. Mersey, "The Effects of Color Scheme and Number of Classes on Choropleth Map Communication," unpublished Ph.D. dissertation, Department of Geography, University of Wisconsin–Madison, 1984, pp. 212–16.
33. Richard E. Lindenberg, "The Effect of Color on Quantitative Map Symbol Estimation," unpublished Ph.D. dissertation, Department of Geography, University of Kansas, 1986, p. 121.
34. Itten, *The Elements of Color*, p. 29.
35. Ostwald, *The Color Primer*, p. 65.
36. Birren, *Selling Color to People*, pp. 169–85.

Glossary

achromatic having no hue (such as red, green, blue), but having characteristics (such as white, gray, or black)

advancing colors a hue of higher wavelength (notably red), a color of high value, or a highly saturated hue appears closer to the viewer; apparently caused by both physiological and learned mechanisms

chroma the saturation, intensity, or purity of a color; one of the three color dimensions

chromatic having the quality of hue, such as red, green, and blue

colorant the elements in substances that cause color, either through absorption or reflection

color balance the result of artful blending of colors and their areas so that dominant colors occupying large areas do not overpower the remainder of the composition; sometimes referred to as contrast of extension

color constancy the tendency to judge colors as being identical under different viewing conditions, such as different illumination

color force light values given colors by artists; brilliance

color harmony the pleasant combinations of two or more colors in a composition; can be developed by the functional, conventional, and color selection plans

color interaction the way we perceive colors together; always modified by their environment

color perception cognitive process involving the brain, where meaning and substance are added to light sensation

color plan choosing the color dimensions of hue, value, and chroma to symbolize varying amounts of data

color space a conceptual three-dimensional space used to illustrate the color dimensions of hue, value, and chroma

color wheel the organization of hues of the visible spectrum into a circle; many distinct hues can be shown, but twelve are customary, especially among artists

cone cells cells in the retina with peak sensitivity to blue, red, or green light energy

cool colors colors of the shorter wavelengths, such as violet, blue, and green

cornea transparent outer protective membrane over the lens of the eye

hue the quality in light that gives it a color name such as red, green, or blue; a way of naming wavelength; one of the three color dimensions

illuminant mode of viewing the production of color in which only the light source and the eye-brain system of the viewer are present

iris diaphragm-like muscle that controls the amount of light coming into the eye

lens crystal-like tissue that focuses light onto the back of the inside of the eye

Munsell color solid a three-dimensional geometrical figure, roughly equivalent to a sphere, designed to show the interrelationships of hue, value, and chroma

object mode of viewing the production of color in which a light source, an object, and the eye-brain system of the viewer are present

opaque objects absorb all light striking them; appear black

optic nerve bundle of nerve cells connecting the retina to the brain; conveys electrical impulses that carry light information

Ostwald color solid a three-dimensional geometrical figure roughly equivalent to two cones whose bases touch; designed to show the relationships of color dimensions; particularly useful in showing how colors can be achieved by blending hue, white, and black

power factor the product of a color's Munsell value and chroma numbers; related to color force

primary colors red, yellow, and blue hues; cannot be made or mixed from other colors

psychological dimensions of color hue, value, and chroma

reflective objects reflect some or all of the light striking them

retina membrane that lines the back of the inside of the eye; contains light-sensitive cells

retreating colors a hue of lower wavelength (blues), a color of low value, or a poorly saturated hue appears farther away to the viewer; apparently caused by both physiological and learned mechanisms

rod cells cells in the retina sensitive to achromatic light

shade the result of mixing a hue with black

simultaneous contrast a color interaction; the eye produces the complementary color of the one being viewed

spectral energy distribution curve a graphic plot of the energy and wavelength of a light source; different sources produce different curve characteristics

spectral reflectance curves graphic plots of light reflected from objects or surfaces

subjective reactions to color involve color preferences, meanings, and behavioral moods produced by colors

successive contrast a color appears different to the eye on different backgrounds, especially when viewed successively

tint the result of mixing a hue with white

tone the result of mixing a hue, white, and black

transparent objects transmit all light passing through them

tristimulus values amounts of red, green, and blue light necessary to match a test lamp color; used in CIE color specification

value scale an array of color based on lightness and darkness qualities; one of the three color dimensions

warm colors colors of the longer wavelengths such as red, orange, and yellow

Readings for Further Understanding

Arnheim, Rudolf. *Art and Visual Perception.* Berkeley: University of California Press, 1974.

Billmeyer, Fred W., Jr., and Max Saltzman. *Principles of Color Technology.* New York: John Wiley, 1966.

Birren, Faber. *Munsell: A Grammar of Color.* New York: Van Nostrand Reinhold, 1969.

———. *Selling Color to People.* New York: University Books, 1956.

———. *Selling with Color.* New York: McGraw-Hill, 1945.

Burton, Philip Ward, and William Ryan. *Advertising Fundamentals.* Columbus, OH: Grid Publishing, 1980.

Carpenter, James M. *Color in Art.* Cambridge, MA: Fogg Art Museum, Harvard University, 1974.

Castner, Henry W. "Printed Color Charts: Some Thoughts on Their Construction and Use in Map Design." *Proceedings* (American Congress on Surveying and Mapping, March, 1980): 370–78.

Color. New York: Viking Press, 1980.

Color Compass: An Illustrated Guide for Color Mixing and Selection, Color Theory and Harmony. New York: M. Grumbacher, Inc., 1972.

Cuff, David J. "Color on Temperature Maps." *Cartographic Journal* 10 (1973): 17–21.

———. "Value Versus Chroma in Color Schemes on Quantitative Maps." *Canadian Cartographer* 9 (1972): 134–40.

———. "The Magnitude Message: A Study of the Effectiveness of Color Sequences on Quantitative Maps." Unpublished Ph.D. Dissertation, Department of Geography, Pennsylvania State University, 1972.

———. "Shading on Choropleth Maps: Some Suspicions Confirmed." *Proceedings* (Association of American Geographers, April, 1973): 50–54.

———. "Impending Conflict in Color Guidelines for Maps of Statistical Surfaces." *Canadian Cartographer* 11 (1974): 54–58.

Deluca, James P. *Pantone Matching System and Pantone Matching System Formula Guide.* Moonachie, NJ: Pantone, 1980.

Doslak, William Jr., and Paul V. Crawford. "Color Influence on the Perception of Spatial Structure." *Canadian Cartographer* 14 (1977): 120–29.

Dunn, Watson, S., and Arnold M. Barban. *Advertising: Its Role in Modern Marketing.* Hinsdale, IA: Dryden Press, 1978.

Engel, Jack. *Advertising: The Process and Practice.* New York: McGraw-Hill, 1980.

Fisher, Howard T. "An Introduction to Color." In James M. Carpenter, *Color in Art.* Cambridge, MA: Fogg Art Museum, 1974.

Hackl, Al. "Hidden Meanings in Color." *Graphic Arts Buyer* 12 (1981): 41–44.

Helson, H., and T. Lansford. "The Role of Spectral Energy of Source an Background Color in the Pleasantness of Object Colors." *Applied Optics* 9 (1970): 1513–62.

Itten, Johannes. *The Elements of Color.* New York: Van Nostrand Reinhold, 1970.

Jones, Mari Riess. "Color Coding." *Human Factors* 4 (1962): 355–65.

Judd, Deane B. "Choosing Pleasant Color Combinations." In *Contributions of Color Science by Deane B. Judd,* David L. MacAdams, ed. United States Department of Commerce, National Bureau of Standards. Washington, D.C.: USGPO, 1979.

Keates, J. S. *Cartographic Design and Production.* New York: Halsted Press, 1973.

Kimerling, A. Jon. "Color in Map Design." Paper delivered at the workshop on Map Perception and Design, ACSM Cartography Division, Annual Meetings, St. Louis, 1980.

———. "Color Specifications in Cartography," *American Cartographer* 7 (1980): 139–53.

Kueppers, Harold. *Color Atlas: A Practical Guide for Color Mixing.* Woodbury, NY: Barron's, 1982.

"The Language of Color." 16-mm film produced by Pantone, Inc., Moonachie, NJ, 1983. Available through Modern Talking Picture Service, St. Petersburg, FL.

Lindenberg, Richard E. "The Effect of Color on Quantitative Map Symbol Estimation." Unpublished Ph.d. Dissertation, Department of Geography, University of Kansas, 1986.

Littlefield, James E., and C. A. Kirkpatrick. *Advertising: Mass Communication in Marketing.* Boston: Houghton Mifflin, 1970.

McCormick, Ernest J., and Mark S. Sanders. *Human Factors in Engineering and Design.* New York: McGraw-Hill, 1982.

Mersey, Janet E. "The Effects of Color Scheme and Number of Classes on Choropleth Map Communication." Unpublished Ph.D. Dissertation, Department of Geography, University of Wisconsin–Madison, 1984.

Meyer, Morton A., Frederick R. Broome, and Richard H. Schieweitzer, Jr. "Color Statistical Mapping by the U.S. Bureau of the Census." *American Cartographer* 2 (1975): 100–117.

Olson, Judy M. "Spectrally Encoded Two-Variable Maps." *Annals* (Association of American Geographers) 71 (1981): 259–76.

Ostwald, Wilhelm. *The Color Primer.* New York: Van Nostrand Reinhold, 1969.

Robinson, Arthur H. *The Look of Maps: An Examination of Cartographic Design.* Madison, WI: University of Wisconsin Press, 1966.

Sargent, Walter. *The Enjoyment and Use of Color.* New York: Cover Publications, 1964.

Sharpe, Deborah. *The Psychology of Color and Design.* Chicago: Nelson-Hall, 1974.

Sorrell, Patrick. "Map Design—With the Young in Mind." *Cartographic Journal* 11 (1974): 82–91.

United States Department of Commerce, National Bureau of Standards. *The ISCC-NBS Method of Designating Colors and a Dictionary of Color Names.* Circular 553. Washington, D.C.: USGPO, 1955.

United States Department of the Interior, Geological Survey. *The National Atlas of the United States of America.* Washington, D.C.: USGPO, 1970.

Appendix A

Geographical Tables

The table in this appendix lists the lengths of the degrees of latitude and longitude on the earth's surface, using the 1866 Clarke reference ellipsoid. In the case of latitude, the values are the lengths of the arcs extending half a degree on each side of the tabulated latitudes.

Length of a Degree of Latitude and Longitude

	Degree of Latitude				Degree of Longitude				
Lat. (°)	Nautical miles	Statute miles	Feet	Meters	Nautical miles	Statute miles	Feet	Meters	Lat. (°)
0	59.701	68.703	362 753	110 567	60.109	69.172	365 226	111 321	0
1	.702	.704	756	568	60.099	69.161	365 171	111 304	1
2	.702	.704	759	569	60.072	69.129	365 003	111 253	2
3	.703	.705	762	570	60.026	69.077	364 728	111 169	3
4	.705	.707	772	573	59.963	69.004	364 341	111 051	4
5	59.706	68.709	362 782	110 576	59.881	68.910	363 845	110 900	5
6	.708	.711	795	580	59.781	68.795	363 238	110 715	6
7	.711	.714	808	584	59.664	68.660	362 523	110 497	7
8	.713	.717	825	589	59.528	68.503	361 696	110 245	8
9	.717	.721	844	595	59.373	68.325	360 758	109 959	9
10	59.720	68.724	362 864	110 601	59.201	68.128	359 715	109 641	10
11	.724	.729	887	608	59.011	67.909	358 560	109 289	11
12	.728	.734	913	616	58.803	67.670	357 297	108 904	12
13	.732	.739	940	624	58.578	67.410	355 925	108 486	13
14	.737	.744	969	633	58.335	67.130	354 449	108 036	14
15	59.742	68.750	363 002	110 643	58.074	66.830	352 864	107 553	15
16	.748	.757	035	653	57.795	66.509	351 168	107 036	16
17	.753	.763	068	663	57.498	66.168	349 367	106 487	17
18	.760	.770	107	675	57.185	65.807	347 461	105 906	18
19	.766	.777	143	686	56.854	65.427	345 453	105 294	19
20	59.773	68.785	363 186	110 699	56.506	65.026	343 337	104 649	20
21	.780	.793	228	712	56.140	64.605	341 115	103 972	21
22	.787	.801	271	725	55.758	64.165	338 793	103 264	22
23	.794	.810	317	739	55.359	63.705	336 365	102 524	23
24	.802	.819	363	753	54.953	63.227	333 839	101 754	24
25	59.810	68.828	363 412	110 768	54.510	62.729	331 207	100 952	25
26	.818	.837	461	783	54.060	62.211	328 474	100 119	26
27	.827	.847	514	799	53.594	61.675	325 646	99 257	27
28	.835	.857	566	815	53.112	61.121	322 717	98 364	28
29	.844	.868	622	832	52.614	60.547	319 688	97 441	29

(continued)

Length of a Degree of Latitude and Longitude

	Degree of Latitude				Degree of Longitude				
Lat. (°)	Nautical miles	Statute miles	Feet	Meters	Nautical miles	Statute miles	Feet	Meters	Lat. (°)
30	59.853	68.878	363 675	110 848	52.099	59.955	316 562	96 488	30
31	.863	.889	734	866	51.569	59.345	313 340	95 506	31
32	.872	.900	789	883	51.023	58.716	310 023	94 495	32
33	.882	.911	848	901	50.462	58.070	306 611	93 455	33
34	.891	.922	907	919	49.885	57.407	303 107	92 387	34
35	59.902	68.934	363 970	110 938	49.293	56.725	299 508	91 290	35
36	.911	.945	364 029	956	48.686	56.027	295 820	90 166	36
37	.922	.957	091	975	48.064	55.311	292 041	89 014	37
38	.932	.968	154	994	47.427	54.578	288 173	87 835	38
39	.942	.980	216	111 013	46.776	53.829	284 216	86 629	39
40	59.953	68.993	364 281	111 033	46.110	53.063	280 171	85 396	40
41	.963	69.005	344	052	45.430	52.280	276 040	84 137	41
42	.974	.017	409	072	44.737	51.482	271 827	82 853	42
43	.984	.029	472	091	44.030	50.668	267 530	81 543	43
44	.995	.041	537	111	43.309	49.839	263 150	80 208	44
45	60.006	69.054	364 603	111 131	42.575	48.994	258 691	78 849	45
46	.017	.066	669	151	41.828	48.135	254 154	77 466	46
47	.027	.078	731	170	41.068	47.260	249 534	76 058	47
48	.038	.090	797	190	40.296	46.372	244 843	74 628	48
49	.049	.103	862	210	39.511	45.468	240 072	73 174	49
50	60.059	69.114	364 925	111 229	38.714	44.551	235 230	71 698	50
51	.070	.127	990	249	37.905	43.620	230 315	70 200	51
52	.080	.139	365 052	268	37.084	42.676	225 328	68 680	52
53	.090	.151	115	287	36.253	41.719	220 276	67 140	53
54	.100	.162	177	306	35.409	40.748	215 151	65 578	54
55	60.111	69.174	365 240	111 325	34.555	39.765	209 961	63 996	55
56	.120	.185	299	343	33.691	38.770	204 708	62 395	56
57	.130	.197	358	361	32.815	37.763	199 390	60 774	57
58	.140	.208	417	379	31.930	36.745	194 012	59 135	58
59	.150	.219	476	397	31.036	35.715	188 576	57 478	59

(continued)

Length of a Degree of Latitude and Longitude

Lat. (°)	Degree of Latitude: Nautical miles	Statute miles	Feet	Meters	Degree of Longitude: Nautical miles	Statute miles	Feet	Meters	Lat. (°)
60	60.159	69.229	365 531	111 414	30.131	34.674	183 077	55 802	60
61	.168	.241	591	432	29.217	33.622	177 526	54 110	61
62	.177	.251	643	448	28.294	32.560	171 916	52 400	62
63	.186	.261	696	464	27.362	31.488	166 257	50 675	63
64	.194	.270	748	480	26.422	30.406	160 545	48 934	64
65	60.203	69.280	365 801	111 496	25.474	29.314	154 780	47 177	65
66	.211	.290	850	511	24.518	28.215	148 973	45 407	66
67	.219	.298	896	525	23.554	27.105	143 117	43 622	67
68	.226	.307	942	539	22.583	25.988	135 215	41 823	68
69	.234	.316	988	553	21.605	24.862	131 273	40 012	69
70	60.241	69.324	366 030	111 566	20.620	23.739	125 289	38 188	70
71	.247	.331	070	578	19.629	22.589	119 268	36 353	71
72	.254	.339	109	590	18.632	21.441	113 209	34 506	72
73	.260	.346	148	602	17.629	20.287	107 113	32 468	73
74	.266	.353	184	613	16.620	19.126	100 988	30 781	74
75	60.272	69.359	366 217	111 623	15.606	17.959	94 826	28 903	75
76	.276	.365	247	632	14.588	16.788	88 638	27 017	76
77	.282	.371	280	642	13.565	15.611	82 425	25 123	77
78	.286	.376	306	650	12.538	14.428	76 181	23 220	78
79	.290	.381	332	658	11.507	13.242	69 918	21 311	79
80	60.294	69.385	366 355	111 665	10.472	12.051	63 629	19 394	80
81	.298	.389	375	671	9.434	10.857	57 323	17 472	81
82	.301	.393	394	677	8.394	9.659	51 001	15 545	82
83	.303	.396	411	682	7.350	8.458	44 659	13 612	83
84	.306	.399	427	687	6.304	7.255	38 304	11 675	84
85	60.308	69.402	366 440	111 691	5.256	6.049	31 939	9 735	85
86	.310	.403	450	694	4.207	4.842	25 564	7 792	86
87	.311	.405	457	696	3.157	3.633	19 180	5 846	87
88	.312	.406	463	698	2.105	2.422	12 789	3 898	88
89	.313	.407	467	699	1.052	1.211	6 394	1 949	89
90	60.313	69.407	366 467	111 699	0.000	0.000	0	0	90

Reproduced from Nathaniel Bowditch, *American Practical Navigator,* vol. II. Defense Mapping Agency, Hydrographic-Topographic Center (Washington, D.C.: USGPO, 1975), Table 6, pp. 124–125. Note: Values in this table are slightly different than those for the GRS80 reference ellipsoid newly adopted for use in North America.

Appendix

B

Census Geography Definitions

The cartographer planning to use any of the products of the United States Census Bureau is cautioned to read all introductory materials (especially definitions) of the volumes being used. The current census always contains the most recent definitions used when the census was taken. The material in this Appendix describes the Census Bureau area classifications as applied to the 1980 Census of Population.

Reproduced directly, in whole and in part, from United States Bureau of the Census, *1980 Census of Population,* vol. 1, Characteristics of the Population (Washington, D.C.: USGPO, 1982); and United States Bureau of the Census, *Census '80: Continuing the Factfinder Tradition,* by Charles P. Kaplan, Thomas Van Valey and Associates (Washington, D.C.: USGPO, 1980), pp. 144–145, 146–147.

States

The 50 States and the District of Columbia are the constituent units of the United States.

Counties

In most states, the primary divisions are termed counties. In Louisiana, these divisions are known as parishes. In Alaska, which has no counties, the county equivalents are the organized boroughs together with the "census areas" which were developed for general statistical purposes by the State of Alaska and the Census Bureau. In four states (Maryland, Missouri, Nevada, and Virginia), there are one or more cities which are independent of any county organization and thus constitute primary divisions of their States. That part of Yellowstone National Park in Montana is treated as a county equivalent. The District of Columbia has no primary divisions, and the entire area is considered equivalent to a county for census purposes.

County Subdivisions

Statistics for subdivisions of counties or equivalent areas are presented as follows:

1. Minor civil divisions (MCDs) in 29 States. MCDs are primary divisions of counties established under State law. These MCDs are variously designated as townships, towns, precincts, districts, wards, plantations, Indian reservations, grants, purchases, gores, locations, or areas. In some States, all incorporated places are also MCDs in their own right. In other States, incorporated places are subordinate to or part of the MCD(s) in which they are located, or the pattern is mixed—some incorporated places are independent MCDs and others are subordinate to one or more MCDs.
2. Census county divisions (CCDs) in 20 States. CCDs are geographic areas which have been defined by the Census Bureau in corporation with State and county officials for the purpose of presenting statistical data. CCDs have been defined in States where there are no legally established MCDs, where the boundaries of MCDs change frequently, and/or where the MCDs are not generally known to the public. Using published guidelines, the CCDs have usually been designed to represent community areas focused on trading centers, or to represent major land use areas, and to have visible, permanent, and easily described boundaries.
3. Census subareas in Alaska. For the 1980 census, census subareas have been delineated cooperatively by the Census Bureau and the State of Alaska for statistical purposes. These areas replace the subdivisions used for the 1970 census.
4. Quadrants in the District of Columbia.

Places

Two types of places are recognized in the census reports—incorporated places and census designated places—as defined below.

Incorporated Places

Incorporated places recognized in the reports of the census are those which are incorporated under the laws of their respective States as cities, boroughs, towns, and villages, with the following exceptions: boroughs in Alaska and New York, and towns in the six New England States, New York, and Wisconsin. The towns in the New England States, New York, and Wisconsin, and the boroughs in New York are recognized as MCDs for census purposes; the boroughs in Alaska are county equivalents.

Census Designated Places

As in the 1950, 1960, and 1970 censuses, the Census Bureau has delineated boundaries for closely settled population centers without corporate limits. In 1980, the name of each such place is followed by "(CDP)," meaning "census designated place." In the 1970 and earlier censuses, these places were identified by "(U)," meaning "unincorporated places." To be recognized for the 1980 census, CDPs must have a minimum 1980 population as follows:

Area	Minimum CDP population
Alaska	25
Hawaii	300
All other states:	
Inside urbanized areas:	
With one or more cities of 50,000 or more	5000
With no city of 50,000 or more	1000
Outside urbanized areas	1000

Hawaii is the only State with no incorporated places recognized by the Bureau of the Census. All places shown for Hawaii in the 1980 census reports are CDPs. Honolulu CDP essentially represents the Honolulu Judicial District. The city of Honolulu, coextensive with the county of Honolulu, is not recognized for census purposes.

Census designated place boundaries change with changes in the settlement pattern; a place which has the same name as in previous censuses does not necessarily have the same boundaries.

Urban and Rural Residence

As defined for the 1980 census, the urban population comprises all persons living in urbanized areas and in places of 2500 or more inhabitants outside urbanized areas. More specifically, the urban population consists of all persons living in (1) places of 2500 or more inhabitants, incorporated as cities, villages, boroughs (except in Alaska and New York), and towns (except in the New England States, New York, and Wisconsin), but excluding those persons living in the rural portions of extended cities; (2) census designated places of 2500 or more inhabitants; and (3) other territory, incorporated or unincorporated, included in urbanized areas. The population not classified as urban constitutes the rural population.

Extended Cities

Since 1960 there has been an increasing trend toward the extension of city boundaries to include territory essentially rural in character. The classification of all the inhabitants of such cities as urban would include in the urban population persons whose environment is primarily rural in character. For the 1970 and 1980 censuses, in order to separate these people from those residing in the closely settled portions of such cities, the Bureau of the Census classified as rural a portion or portions of each such city that was located in an urbanized area. To be treated as an extended city, a city must contain one or more areas that are each at least 5 square miles in extent and have a population density of less than 100 persons per square mile. The area or areas must constitute at least 25 percent of the land area of the legal city or include at least 25 square miles. These areas are excluded from the urbanized area.

Those cities designated as extended cities thus consist of an urban part and a rural part.

Urbanized Areas

Definition

The major objective of the Census Bureau in delineating urbanized areas is to provide a better separation of urban and rural population in the vicinity of large cities. An urbanized area consists of a central city or cities, and surrounding closely settled territory ("urban fringe").

The following criteria are used in determining the eligibility and definition of the 1980 urbanized areas:

An urbanized area comprises an incorporated place and adjacent densely settled surrounding area that together have a minimum population of 50,000. The densely settled surrounding area consists of:

1. Contiguous incorporated or census designated places having:
   a. A population of 2500 or more; or,
   b. A population of fewer than 2500 but having a population density of 1000 persons per square mile, a closely settled area containing a minimum of 50 percent of the population, or a cluster of at least 100 housing units.
2. Contiguous unincorporated area which is connected by road and has a population density of at least 1000 persons per square mile.
3. Other contiguous unincorporated area with a density of less than 1000 persons per square mile, provided that it:
   a. Eliminates an enclave of less than 5 square miles which is surrounded by built-up area.
   b. Closes an indentation in the boundary of the densely settled area that is no more than 1 mile across the open end and encompasses no more than 5 square miles.
   c. Links an outlying area of qualifying density, provided that the outlying area is:
      (1) Connected by road to, and is not more than 1½ miles from, the main body of the urbanized area.
      (2) Separated from the main body of the urbanized area by water or other undeveloped area, is connected by road to the main body of the urbanized area, and is not more than 5 miles from the main body of the urbanized area.
4. Large concentrations of nonresidential urban area (such as industrial parks, office areas, and major airports), which have at least one-quarter of their boundary contiguous to an urbanized area.

Urbanized Area Central Cities

The central cities of urbanized areas are those named in the titles except where regional titles are used. In such cases, the central cities are those that have qualified under items 1 or 2 of the titling criteria.

Counts and data for central cities of urbanized areas refer to the urban portion of these cities, thus excluding the rural portions of extended cities, as discussed above.

Standard Metropolitan Statistical Areas

Definition

The general concept of a metropolitan area is one of a large population nucleus, together with adjacent communities which have a high degree of economic and social integration with that nucleus. The standard metropolitan statistical area (SMSA) classification is a statistical standard, developed for use by Federal agencies in the production, analysis, and publication of data on metropolitan areas. The SMSAs are designated and defined by the Office of Management and Budget, following a set of official published standards developed by the interagency Federal Committee on Standard Metropolitan Statistical Areas.

Each SMSA has one or more central counties containing the area's main population concentration: an urbanized area with at least 50,000 inhabitants. An SMSA may also include outlying counties which have close economic and social relationships with the central counties. The outlying counties must have a specified level of commuting to the central counties and must also meet certain standards regarding metropolitan character, such as population density, urban population, and population growth. In New England, SMSAs are composed of cities and towns rather than whole counties.

The population living in SMSAs may also be referred to as the metropolitan population. The population is subdivided into "inside central city (or cities)" and "outside central city (or cities)." The population living outside SMSAs constitutes the nonmetropolitan population.

New SMSA Standards

New standard for designating and defining metropolitan statistical areas were published in the Federal Register on January 3, 1980. The SMSAs recognized for the 1980 census comprise (1) all areas as defined on January 1, 1980, except for one area which was defined provisionally during the 1970s on the basis of population estimates but whose qualification was not confirmed by 1980 census counts; and (2) a group of 36 new areas defined on the basis of 1980 census counts and the new standards that were published on January 3, 1980.

The new standards will not be applied to the areas existing on January 1, 1980, until after data on commuting flows become available from 1980 census tabulations. At that time, the boundaries, definitions, and titles for all SMSAs will be reviewed.*

Standard Consolidated Statistical Areas

In some parts of the country, metropolitan development has progressed to the point that adjoining SMSAs are themselves socially and economically interrelated. These areas are designated standard consolidated statistical areas (SCSAs) by the Office of Management and Budget, and are defined using standards included as part of the new SMSA standards described above.

Relationship between Urbanized Areas and Metropolitan Areas

Although the urbanized area and the metropolitan area are closely related in concept, there are important differences. The urbanized area has a more limited territorial extent. The urbanized area consists of the physically continously built-up territory around each larger city and thus corresponds generally to the core of high and medium population density at the heart of the metropolitan area. In concept, a metropolitan area is always larger than its core urbanized area, even if the metropolitan area is defined in terms of small building blocks, because it includes discontinuous urban and suburban development beyond the periphery

*After the 1980 census the relationship between central urban (core) and adjacent counties was undertaken, and as a result a revised set of criteria and redefinitions were applied, and the word "standard" was dropped. On June 30, 1983, SMSAs and SCSAs were redesignated as Metropolitan Statistical Areas (MSAs), Consolidated MSAs (CMSAs), and Primary MSAs (PMSAs). These and other redefinations will appear when the 1990 census is published.

of the continously built-up area. The metropolitan area may also include some rural territory whose residents commute to work in the city or its immediate environs, while the urbanized area does not include such territory. In practice, because the SMSA definitions use counties as building blocks, considerable amounts of rural territory with few commuters are often included. However, even in New England, where cities and towns are used as building blocks, SMSAs are generally much larger in extent than their core urbanized areas.

The new standards provide that each SMSA be associated with an urbanized area. However, the reverse is not true—there are some urbanized areas that are not in any SMSA. This situation occurs when an urbanized area does not qualify as an SMSA of at least 100,000 population (75,000 in New England), and the urbanized area has no city with at least 50,000 population.

In addition, some SMSAs contain more than one urbanized area. This occurs when:

1. Two or more urban concentrations not far apart and of generally similar size have separate urbanized areas but qualify as a single SMSA (for example Greensboro, High Point, and Winston-Salem, North Carolina). Often the SMSA title includes the name of the largest city of each of the component urbanized areas.
2. A very large SMSA includes one or more smaller separate urbanized areas within its boundaries. Examples are the separate urbanized areas around Joliet, Aurora, and Elgin within the Chicago SMSA.

Area Measurements

Area measurement figures for counties and county equivalent areas in the 1980 census were prepared using a process called digitizing. This process involved first verifying and highlighting the county boundaries recognized for the 1980 census on copies of the topographic quadrangle maps produced by the U.S. Geological Survey and relocating those boundaries where necessary. An electronically assisted digitizing device was then used to trace over each county line and to calculate the latitude/longitude values associated with each line. From the latitude/longitude information associated with each county, the total area of the county in square miles was computed. The total area figure derived for each county was subsequently reviewed against similar information from the 1960 and 1970 censuses and other sources, with significant variations in area being rechecked and adjudicated.

Following this review, the total area of the county was apportioned between land and water. No direct measurements were made to determine these values separately; instead, information from which the final figures were compiled was gathered from several other Federal and State agencies. The boundary between inland and other water was part of the original digitizing process and was treated as though it were a county boundary line. After all operations, a mathematical conversion was performed to convert all values from square miles to square kilometers.

Differences between 1980 area figures and those reported in previous censuses are attributable to changes in base map scale and detail, methodology for measurement, and occasionally to county boundary change or relocation.

Census Tracts

Census tracts are subdivisions of an SMSA and contain an average population of approximately 4000 but may range in population from 2500 to 8000. The actual delineation of tracts is carried out by local Census Statistical Areas Committees made up of local data users, with the Bureau providing general guidelines, detailed review, and approval of the plans to maintain an overall uniform standard. Census tracts have been established for all counties within the current 288 SMSAs and in a number of other highly populated counties that have expressed an interest in the program. There are over 40,000 census tracts for the 1980 census. While the basic tenet has been to keep census tract boundaries as stable as possible, some modifications are necessary from time to time if the census tracts are to continue to be useful and usable. Each proposed modification is considered in terms of its impact on the usefulness to the tract program, its effect on the reliability of data, whether other means are available to meet the particular need without having to alter the tract boundaries.

Census Blocks

Each census block is a well-defined piece of land bounded by streets, roads, railroad tracks, streams, or other features on the ground. It is the smallest area for which census data will be tabulated. Only selected statistics based on the complete count part of the census are published. No sample data are available at the block level.

Block statistics will be tabulated for all urbanized areas, all incorporated places of 10,000 or more population, and any other areas that have contracted with the Bureau to provide block-level data. In fact, five States have contracted for block-level data for the entire State. These are Rhode Island, New York, Virginia, Georgia, and Mississippi. An estimated 2.5 million blocks will be defined for 1980. This is an increase of nearly 50 percent over the number of blocks for which data were published in 1970. Each block is identified by a three-digit number which is unique within a census tract. In blocked areas where there are no census tracts (e.g., for most cities outside SMSAs) "block numbering areas" are defined as substitutes for census tracts for use in identifying blocks.

Some blocks defined for 1970 will have new boundaries in 1980, primarily those on the edges of urbanized areas and other areas of new development where the street patterns have physically changed. Wherever a block has been redefined, by splitting or other adjustment, the 1970 block number will generally not be reused so as to help the user notice the change. In most areas, however, block boundaries and numbers will be the same in 1980 as in 1970, except for a few areas where blocks were renumbered by local GBF/DIME coordinating agencies to define more desirable block groups.

Appendix C

Map Sources

Table C.1. Federal Map Products and Sources

Type	Publishing Agency	Available from
Aeronautical Charts	Defense Mapping Agency National Ocean Service	National Ocean Service National Ocean Service
Boundary Information:		
United States and Canada	International Boundary Commission	International Boundary Commission
United States and Mexico	International Boundary and Water Commission	Geological Survey (Denver)
Census Geographic Area (1980)	Bureau of the Census	Bureau of the Census
Census Tract Outline Maps (1980)	Bureau of the Census	Bureau of the Census
Climatic Maps	National Oceanic and Atmospheric Administration	National Climatic Center
Coal Investigations Maps	Geological Survey	Geological Survey (Arlington or Denver)
Congressional Districts	Bureau of the Census	Superintendent of Documents
Electric Transmission and Generation Facilities	Federal Power Commission	Superintendent of Documents
Geologic Quadrangle Maps:		
(Maps east of Mississippi River)	Geological Survey	Geological Survey (Arlington)
(Maps west of Mississippi River)	Geological Survey	Geological Survey (Denver)
Geologic Investigations Maps:		
(Maps east of Mississippi River)	Geological Survey	Geological Survey (Arlington)
(Maps west of Mississippi River)	Geological Survey	Geological Survey (Denver)
Geologic Map of North America	Geological Survey	Geological Survey (Arlington or Denver)
Geologic Map of the United States	Geological Survey	Geological Survey (Arlington or Denver)
Geophysical Investigations Maps:		
(Maps east of Mississippi River)	Geological Survey	Geological Survey (Arlington)
(Maps west of Mississippi River)	Geological Survey	Geological Survey (Denver)
Ground Conductivity	Federal Communications Commission	Superintendent of Documents

(continued)

Table C.1. (Continued)

Type	Publishing Agency	Available from
Highways: State and County	State Highway Departments	State Highway Departments
Historical:		
Reproductions from Historical and Military Map Collections	Library of Congress National Archives	Library of Congress National Archives and Records Service
Selected Civil War Maps (reproduced from originals)	National Ocean Service	Superintendent of Documents
Treasure Maps and Charts (bibliography)	Library of Congress	Superintendent of Documents
Hydrographic information: Bathymetric Maps of United States Adjacent and Continental Shelf	National Ocean Service	National Ocean Service
Nautical Charts of U.S. Coastal Waters	National Ocean Service	National Ocean Service
Great Lakes and Connecting Waters	National Ocean Service	National Ocean Service
River Charts:		
Cumberland River	Corps of Engineers	Corps of Engineers, Nashville
Illinois Waterway to Lake Michigan	Corps of Engineers	Corps of Engineers, Chicago
Mississippi River (Lower)	Corps of Engineers	Corps of Engineers, Vicksburg
Mississippi River (Upper)	Corps of Engineers	Corps of Engineers, Chicago
Missouri River	Corps of Engineers	Corps of Engineers, Omaha
Ohio River	Corps of Engineers	Corps of Engineers, Louisville
Tennessee River	Tennessee Valley Authority	Tennessee Valley Authority
Foreign Waters	Defense Mapping Agency	Defense Mapping Agency Topographic Center
Hydrologic Investigations Atlases:		
(Maps east of Mississippi River)	Geological Survey	Geological Survey (Arlington)
(Maps west of Mississippi River)	Geological Survey	Geological Survey (Denver)
Hydrologic Unit Maps (by State):		
(Maps east of Mississippi River)	Geological Survey	Geological Survey (Arlington)
(Maps west of Mississippi River)	Geological Survey	Geological Survey (Denver)

(continued)

Table C.1. (Continued)

Type	Publishing Agency	Available from
Indian Reservations	Bureau of Indian Affairs	Superintendent of Documents
Landsat Imagery	Earth Observation Satellite Company (EOSAT)	EOSAT (4300 Forbes Boulevard, Lanham, MD 20706, or at EROS Data Center, Sioux Falls, SD 57198)
Land Use and Land Cover Maps	Geological Survey	Geological Survey (NCIC)
Map Projections	National Ocean Service	National Ocean Service
Mineral Investigations Field Studies Maps:		
(Maps east of Mississippi River)	Geological Survey	Geological Survey (Arlington)
(Maps west of Mississippi River)	Geological Survey	Geological Survey (Denver)
Mineral Investigations Resource Maps:		
(Maps east of Mississippi River)	Geological Survey	Geological Survey (Arlington)
(Maps west of Mississippi River)	Geological Survey	Geological Survey (Denver)
Minor Civil Divisions	Bureau of the Census	Superintendent of Documents
Moon/Planetary Maps	Geological Survey	Geological Survey (Arlington or Denver)
National Atlas separate sales editions	Geological Survey	Geological Survey (Arlington or Denver)
National Forest Regions	Forest Service	Forest Service
National Parks: Topographic Maps	Geological Survey	Geological Survey (Arlington or Denver)
National Parks System	National Park Service	Superintendent of Documents
Natural Gas Pipelines	Federal Power Commission	Superintendent of Documents
Oil and Gas Investigations Maps and Charts:		
(Maps east of Mississippi River)	Geological Survey	Geological Survey (Arlington)
(Maps west of Mississippi River)	Geological Survey	Geological Survey (Denver)

(continued)

Table C.1. (Continued)

Type	Publishing Agency	Available from
Orthophotoquads:		
(Maps east of Mississippi River)	Geological Survey	Geological Survey (Arlington)
(Maps west of Mississippi River)	Geological Survey	Geological Survey (Denver)
Polar Maps:		
Antarctic	Geological Survey Defense Mapping Agency	Geological Survey (Arlington) Defense Mapping Agency Topographic Center
Arctic	National Ocean Service Defense Mapping Agency	National Ocean Service Defense Mapping Agency Topographic Center
Population Distribution of the United States (1970)	Bureau of the Census	Superintendent of Documents
Soil Survey Maps	Soil Conservation Service	Soil Conservation Service
Space Imagery Maps:		
(Maps east of Mississippi River)	Geological Survey	Geological Survey (Arlington)
(Maps west of Mississippi River)	Geological Survey	Geological Survey (Denver)
State Maps (Base, Shaded, and Topographic)	Geological Survey	Geological Survey (Arlington or Denver)
State Maps (Geologic)	Geological Survey and various State Geological Surveys	Geological Survey (Arlington or Denver) and various State Geological Surveys
Status Maps:		
Standard Topographic Mapping	Geological Survey	Geological Survey (Arlington or Denver)
Intermediate-scale Topographic Mapping (county and quadrangle)	Geological Survey	Geological Survey (NCIC)
Orthophotoquad Mapping	Geological Survey	Geological Survey (NCIC)
Storm Evacuation Maps	National Ocean Service	National Ocean Service
Time Zones of the World	Defense Mapping Agency	Defense Mapping Agency Topographic Center

(continued)

Table C.1. (Continued)

Type	Publishing Agency	Available from
Topographic Map Indexes (by State):		
(Maps east of Mississippi River, Puerto Rico, and Virgin Islands)	Geological Survey	Geological Survey (Arlington)
(Maps west of Mississippi River, American Samoa, and Guam)	Geological Survey	Geological Survey (Denver)
Township Plates (reproductions): Illinois, Indiana, Iowa, Kansas, Missouri, and Ohio	National Archives	National Archives and Records Service
All other Public Land States	Bureau of Land Management	Bureau of Land Management
Transportation Maps (State)	Federal Railroad Administration	Superintendent of Documents
United States Base Maps	Geological Survey	Geological Survey (Arlington or Denver)
	National Ocean Service and other government agencies	National Ocean Service Superintendent of Documents and/or publishing agency
Urban Atlas (selected Standard Metropolitan Statistical Areas)	Bureau of the Census	Superintendent of Documents
Water Resource Development Maps	Geological Survey	Geological Survey (Arlington or Denver)
Weather Maps	National Weather Service	Superintendent of Documents
World Maps	Defense Mapping Agency	Defense Mapping Agency Topographic Center
	National Ocean Service	National Ocean Service

Source: United States Department of the Interior, Geological Survey, *Types of Maps Published by Government Agencies* (Washington, D.C.: USGPO, 1977); see also *Landsat Data Users Notes* 3 (1988).

Table C.2. Addresses of Government Agencies Producing or Distributing Reference Maps

Defense Mapping Agency
Topographic Center
Attn: code 55500
Washington, DC 20315

Federal Communications Commission
Office of Public Information
1919 M Street NW.
Washington, DC 20554

Federal Power Commission
Office of Public Information
825 North Capitol St.
Washington, DC 20426

Federal Railroad Administration
Office of Public Affairs, RPD-1
400 Seventh Street NW.
Washington, DC 20590

International Boundary Commission
United States and Canada
425 Eye Street NW., Room 150
Washington, DC 20536

International Boundary and Water Commission
United States and Mexico, United States Section
Post Office Box 20003
El Paso, TX 79998

Interstate Commerce Commission
Office of Public Information
Constitution Ave. & 12th St. NW.
Washington, DC 20423

Library of Congress
Geography and Map Division
845 South Pickett St.
Alexandria, VA 22304

State Highway Departments
State Capitals

Superintendent of Documents
U.S. Government Printing Office
North Capitol and H Sts. NW.
Washington, DC 20402

Tennessee Valley Authority
Mapping Services Branch
111 Haney Building
Chattanooga, TN 37401

U.S. Army Engineer District
Corps of Engineers, Chicago
219 South Dearborn St.
Chicago, IL 60604

U.S. Army Engineer District
Corps of Engineers, Louisville
Post Office Box 59
Louisville, KY 40201

U.S. Army Engineer District
Corps Of Engineers, Nashville
Post Office Box 1070
Nashville, TN 37202

U.S. Army Engineer District
Corps of Engineers, Omaha
6014 U.S. Post Office and Courthouse Bldg.
Omaha, NE 68102

U.S. Army Engineer District
Corps of Engineers, Vicksburg
Post Office Box 60
Vicksburg, MS 39180

U.S. Bureau of the Census
Subscriber Service Section (Pubs)
Administrative Service Division
Washington, DC 20233

U.S. Bureau of Indian Affairs
Office of Public Information
1951 Constitution Ave. NW.
Washington, DC 20245

U.S. Bureau of Land Management
Office of Public Affairs
Washington, DC 20240

U.S. Forest Service
Information Office, Rm. 3238
Post Office Box 2417
Washington, DC 20013

U.S. Geological Survey
Branch of Distribution
Box 25286, Federal Center
Denver, CO 80225

U.S. Geological Survey
Branch of Distribution
120 South Eads St.
Arlington, VA 22202

U.S. National Archives and Records Service
Cartographic Archives Division (NNS)
Pennsylvania Ave. at 8th St. NW.
Washington, DC 20408

U.S. National Climatic Center
Federal Building
Asheville, NC 28801

U.S. National Ocean Service
Distribution Division (C-44)
Riverdale, MD 20840

U.S. National Park Service
Office of Public Inquiries, Room 1013
Washington, DC 20240

U.S. National Weather Service
Gramax Building
8060 13th Street
Silver Spring, MD 20910

U.S. Soil Conservation Service
Information Division
Post Office Box 2890
Washington, DC 20013

Source: United States Department of the Interior, Geological Survey, *Types of Maps Published by Government Agencies* (Washington, D.C.: USGPO, 1977).

Index

A

B

C

D

E

F

G

H

I

J

K

L

M

N

O

P

Q

R

S

T

U

V

W